YACOV Y. HAIMES
Risk Modeling, Assessment, and Manag

DENNIS M. BUEDE
The Engineering Design of Systems: Models and Methods

ANDREW P. SAGE and JAMES E. ARMSTRONG, Jr.
Introduction to Systems Engineering

INTRODUCTION TO
SYSTEMS ENGINEERING

INTRODUCTION TO SYSTEMS ENGINEERING

ANDREW P. SAGE
George Mason University

JAMES E. ARMSTRONG, JR.
United States Military Academy

A Wiley-Interscience Publication
JOHN WILEY & SONS, INC.
New York / Chichester / Weinheim / Brisbane / Toronto / Singapore

For ordering and customer service, call 1-800-CALL-WILEY.

Library of Congress Cataloging-in-Publication Data:

Sage, Andrew P.
    Introduction to systems engineering / Andrew P. Sage, James E.
Armstrong, Jr.
        p.   cm. -- (Wiley series in systems engineering)
    "A Wiley-Interscience publication."
    Includes bibliographical references (p.).
    ISBN 0-471-02766-9 (alk. paper)
    1. Systems engineering.   2. Large scale systems.   I. Armstrong,
James E.   II. Title.   III. Series.
TA168.S15   2000
620′.001′1--dc21   99-33715
                                                            CIP

Printed in the United States of America

10 9 8 7 6 5 4 3 2

Contents

Preface

This book discusses some fundamental and introductory considerations associated with the engineering of large-scale systems, or systems engineering. We begin our effort by first discussing the need for *systems engineering* and then providing several definitions of systems engineering. We next present a structure describing the systems-engineering process. The result of this is a *life-cycle model* for systems engineering processes. This is used to motivate discussion of the functional levels, or considerations, involved in a systemic process:

- Systems methods and tools
- Systems methodology
- Systems management

While there will be a number of discussions throughout the book on systems engineering life-cycle processes and systems management—especially in the first, second, and last chapters—our major focus is on (a) methods for systems engineering and (b) problem solving using a systems engineering approach. Problems for student solution will be presented at the end of each chapter.

The major content in the book is that in Chapters 3, 4, and 5. In these chapters we present a number of methods appropriate for

- Issue *formulation*
- Issue *analysis*
- Issue *interpretation*

We will apply these to a variety of situations that should enable us to develop an appreciation for the engineering of large systems, as well as for problem solving in general.

This text is written primarily for upper-division undergraduate students in systems engineering and in engineering management. It is also, we believe, useful as an introductory graduate-level textbook. It should also have value for other engineering areas that offer courses in systems engineering problem solving, systems engineering design, and systems engineering methods. Prerequisites for the text are moderate. It will generally be assumed that the reader has a fundamental background common to beginning upper-division undergraduates in engineering in the United States. This will include differential and integral calculus as well as differential equations. Some introductory knowledge of probability theory is also assumed as well as an understanding of some physical engineering systems.

The book should also be attractive to the many professionals in industry concerned with systems engineering and technical direction-related efforts. These include professionals in such diversified areas as project management, software engineering, information systems engineering, manufacturing, command and control, and defense systems acquisition and procurement.

The following are among the most important objectives for systems engineering, expressed in terms of systems engineering processes:

1. Systems engineering processes should encompass all phases of the system life cycle, or life cycles as the case may be, including transitioning between phases.

2. Systems engineering processes should support problem understanding, as well as communication among all interested parties at all phases in the process.

3. Systems engineering processes should enable capture of design and implementation needs for the systems engineering product early in the life cycle, generally as part of the requirements specifications and conceptual design phases.

4. Systems engineering processes and associated methods should support both bottom-up and top-down approaches to systems design and development.

5. Systems engineering processes should enable an appropriate mix of design, development, and systems management approaches.

6. Systems engineering processes should support quality assurance of both the product and the process that leads to the product.

7. Systems engineering processes should support system product evolution over time.

8. Systems engineering processes should be supportive of appropriate standards and management approaches that result in trustworthy systems.

9. Systems engineering processes should support the use of automated aids for the engineering of systems, such as to result in production of high-quality trustworthy systems.

10. Systems engineering processes should be based upon methodologies that are teachable and transferable and that make the process visible and controllable at all life-cycle phases.

11. Systems engineering processes should be associated with appropriate procedures to enable definition and documentation of all relevant factors at each phase in the system life cycle.

12. Systems engineering processes should be associated with appropriate metrics and management controls.

13. Systems engineering processes should support operational product functionality, revisability, and transitioning, both at the initial time of operational implementation and later at the time that a system is phased out of service or retired, or reengineered for continued productivity and use.

14. Systems engineering processes must support both system product development and system user organizations; they must also be compatible with the environments associated with systems development and operation.

15. Systems engineering processes should support quality, total quality management, system design for human interaction, and other attributes associated with trustworthiness and integrity.

When all of these are accomplished, it will be possible to produce operational systems that are economical, reliable, verifiable, interoperable, integratable, portable, adaptable, evolvable, comprehensible, maintainable, manageable, and cost-effective and that lead to a very high degree of user satisfaction. These would seem to represent attributes for metrics, or to be translatable into attributes for metrics, that can measure the quality of an operational systems engineering product. They can be translated into standards with which to measure system performance and systems engineering process effectiveness. Together with cost information, this will allow us to obtain cost and operational effectiveness of systems engineering products.

Needless to say, we believe that systems engineering is one of the fundamental engineering subject areas. Its role in engineering, as well as in engineering education, is stressed in Chapter 1. Chapter 2 describes systems engineering processes. While the focus in this text is upon systems engineering methods, selection of appropriate methods is necessarily contingent upon the process or product line used to engineer the product. Chapters 3, 4, and 5 each focus on one of the major steps in systems engineering:

- Issue *formulation*
- Issue *analysis*
- Issue *interpretation*

They each describe a plethora of methods for these steps. Technical direction and systems management guide the choice of an appropriate process and methods to be used within this process. The concluding chapter of this text describes some facets of systems management. This is not, however, a principal objective of this text. The major objective, as noted, is an exposition of systems engineering methods.

Fairfax, Virginia   ANDREW P. SAGE
West Point, New York   JAMES E. ARMSTRONG, JR.

INTRODUCTION TO SYSTEMS ENGINEERING

CHAPTER 1

Introduction to Systems Engineering

This chapter and Chapter 6 (the last chapter of this text), each attempt to provide a perspective on all of systems engineering. This is a major challenge. We believe that some introductory comments, followed by another look at the big picture after we have discussed some of the methods based details in our intervening chapters, is an appropriate way to meet this challenge.

1.1 INTRODUCTION

Here, as throughout the book, we discuss some fundamental and introductory considerations associated with the engineering of large-scale systems, or *systems engineering*. We begin our effort by first discussing the need for systems engineering and then providing several definitions of systems engineering. We next present a discussion of systems engineering processes or systems engineering life cycles. A life cycle is the product line, or process, that is used to create a product or service, or perhaps even another process. We will also discuss the three functional levels, or considerations, that are associated with systems engineering:

- Systems methods and tools
- Systems engineering life-cycle processes, or methodology
- Systems management

While there will be some discussions throughout this chapter, as well as in the next and final chapters, on systems engineering methodology and systems management, our major focus in most of the book is on methods for systems engineering and on methods for problem solving through use of a systems engineering approach. Here we wish to provide an overview of where we wish to go. We will also attempt to indicate what it is that systems engineers do in professional practice. We will provide a brief indication of the history of systems engineering and will also discuss some of the challenges and pitfalls associated with systems engineering efforts.

In the next chapter, we will discuss a framework, or methodology for systems engineering. We will indicate that this framework is generally comprised of three fundamental steps:

- Issue *formulation*, such as to identify the needs to be fulfilled and the requirements associated with these in terms of objectives to be satisfied; constraints and alterables that affect issue resolution and generation of potential alternative courses of action
- Issue *analysis*, such as to enable us to determine the impacts of alternative courses of action including possible refinement of these alternatives
- Issue *interpretation*, such as to enable us to rank order the alternatives in terms of need satisfaction and to select one for implementation or additional study

We will present a number of methods appropriate for these three steps in Chapters 3, 4, and 5. This comprises the majority of the content of this effort, which is basically concerned with systems engineering methods for the formulation, analysis, and interpretation of issues. To put these methods-based discussions in perspective, we illustrate how they fit into the overall systems engineering picture. These are the objectives of Chapter 2, which is devoted to systems engineering life-cycle processes, and Chapter 6, which is devoted to systems management.

We will apply these systems engineering steps — Formulation, Analysis, and Interpretation — to a variety of situations that should enable us to develop an appreciation for systems engineering and problem-solving efforts. The primary purpose of the text is to describe a variety of systems methods and illustrate their use in formulation, analysis, and interpretation situations that are associated with problem solution and systems definition, development, and deployment. The text concludes in Chapter 6 with a brief discussion of the role of each of these functional levels in the engineering and management of large-scale systems.

1.2   SYSTEMS ENGINEERS

What do systems engineers do? The answer to this question is ostensibly straightforward. They *define*, *develop*, and *deploy* systems. The systems (products or services) they engineer usually involve many considerations associated with both the user (i.e., client or enterprise desiring the system) and the implementation specialists who ultimately produce or manufacture the system. Thus, systems engineering is a human, organizational, and technology-based effort that is inherently multidisciplinary in nature. Often, it is said that systems engineering deals with systems that are large in scale and large in scope. Sometimes it is said that these systems are "complex," meaning that there are many parts to the system and that the parts are related to each other in sometimes complicated ways*.

Often a systems engineer leads, in a technical direction sense, a team of experts in various areas of disciplinary specialization who work together in efforts that are large in scope, in that they involve human, economic, technical, environmental, and other considerations. But what is a large-scale system? Many examples can be found by looking at big projects that are typically coordinated by the systems engineering department of many modern companies. Airport planning and operations, command-and-control systems, management information systems, software development projects, urban planning, plant layout, and manufacturing operations are a few examples.

Often system engineers build relatively inexpensive models of proposed projects to refine and test new ideas. Such prototypes, or models, can save lots of time and money in that they allow experimentation on a synthetic system, rather than on the "real" system where mistakes can be very expensive and may even be dangerous. System failures may be expensive to diagnose and correct, even though they are often not difficult to detect. Because systems engineers know how to analyze and understand complicated situations, they are often called on to organize knowledge for executive decision makers. In this role, they perform systems analysis to develop and evaluate policies and programs. In this capacity, systems engineers typically function as consultants or as technical direction and staff support to management. Thus, in developing and implementing large-scale systems, systems engineers must also understand and appreciate human, organizational, and behavioral concerns, as well as concerns involving technology.

Systems engineering may also be described as a management technology. Technology is the organization, application, and delivery of scientific knowledge for the betterment of a client group. This is a functional definition of technology as a fundamentally human activity. A technology inherently involves a purposeful human extension of one or more natural processes. For example, the stored program digital computer is a technology in that it enhances the ability of a human to perform computations and, in more advanced forms, to process information. Information is data of potential value in decision making. When information is associated with the context that

---

*There is an evolving field of study, called complex systems or complex adaptive systems, which uses a somewhat different meaning of complexity. In this area, complexity is intended to describe a system that cannot be fully studied by the classical approaches of systems analysis. Both are correct uses of the term.

occurs with experiential familiarity with a task and the environment into which it is embedded, it becomes actionable knowledge.

Management involves the interaction of the organization with the environment. A purpose of management is to enable organizations to better cope with their environments such as to achieve purposeful goals and objectives. Consequently, a management technology involves the interaction of science, the organization, and its environment. Figure 1.1 illustrates these conceptual interactions. Information is the glue that enables the interactions shown in this figure. Information is a very important quantity which is assumed to be present in the management technology that is systems engineering. This strongly couples notions of systems engineering with those of technical direction or systems management of technological development, rather than exclusively with one or more of the methods of systems engineering, important as they may be for the ultimate success of a systems engineering effort. It suggests that *systems engineering is the management technology that controls a total system life-cycle process, which involves the definition, development, and deployment of a system that is of high quality, trustworthy, and cost-effective in meeting user needs.* This process-oriented notion of systems engineering will be emphasized here. Our focus will be on the description of (a) an appropriate methods basis for problem solving and (b) support of this process-related view of systems engineering.

We can think of a physical (or natural) science basis for technology, a human and organizational science basis, and an information science basis. The physical or natural science basis involves primarily matter and energy processing. The management science basis involves human and organizational concerns, both of the technology development organization and of the technology user organization. In many ways, the information science basis is more difficult to cope with than the physical science or management science basis, because information is not a truly fundamental quantity but one which derives from the structure and organization inherent in the natural sciences and organizational sciences and the purposeful uses to which they are put. *This leads us to identify the primary concerns of systems engineers: physical systems*

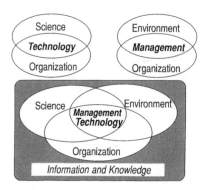

Figure 1.1   Systems engineering as a management technology.

planning, development, and deployment; human and organizational, or enterprise, systems planning, development, and deployment; and information systems planning, development, and deployment.

1.3   THE SYSTEMS POINT OF VIEW

Now that we understand what systems engineers do, we naturally want to know some of the basic ideas that systems engineers use in their work. Because systems engineering is often described as a "new way of thinking," we need to describe what we mean by the systems point of view. This will lead us to see that it is not really new. The systems perspective takes a "big picture" or holistic, or gestalt, view of large-scale problems and their proposed technological solutions. This means that systems engineers not only examine the specifics of the problem under consideration but also investigate relevant factors in the surrounding environment. They realize that problems are embedded in a situation or environment that can have significant impacts on the problem and its proposed alternative solutions. This is not to say that systems engineers do not get very detailed or specific. Far from it. There is also much effort devoted to inscoping, high-fidelity modeling, and specification of system requirements and architecture. The systems viewpoint stresses that there usually is not a single correct answer or solution to a large-scale problem or design issue. Instead, there are many different alternatives that can be developed and implemented depending on the objectives the system is to serve and the values of the people and organizations with a stake in the solution. Let's add more detail to what we mean by the systems point of view.

First, *a system is defined as a group of components that work together for a specified purpose.* This is a very simple but correct definition. Purposeful action is a basic characteristic of a system. A number of functions must be implemented in order to achieve these purposes. This means that systems have functions. They are designed to do specific tasks. Systems are often classified by their ultimate purpose: service-oriented systems, product-oriented systems, or process-oriented systems. An airport can be viewed as an example of a service system. The planes, pilots, mechanics, ticket agents, runways, and concourses are all components that work together to provide transport service to passengers and freight. An automobile assembly plant is an example of a product-oriented system. The raw materials, people, and machines all work together to produce a finished car. A refinery that changes crude oil into gasoline is an example of a process-oriented system. We note here that the systems considered by systems engineers may be service systems, or they may be product systems. The systems may be systems designed for use by an individual or by groups of individuals. These systems may be private sector systems, or they may be government or public sector systems.

This suggests that what is a "system" is very perspective-dependent. An overhead projector may be viewed as a "system." So may the combination of an overhead projector, a screen on which it projects, and a set of overheads. The instructor using the overhead may also be included in the notion of "system." From another perspective, the combination of the overhead, screen, overheads, instructor, and students may be regarded as a "system." Thus, when we use a term such as "engineer a system," we must be very careful to define the nature of the "system" that we wish to engineer and what is included in, and exempted from, the notion of system. We must also be very concerned with the interfaces to the system that we are engineering.

The systems point of view also recognizes that a problem and its solution have many elements or components, and there are many different relations among them. The important aspects of a problem are often a function of how the components interact. Simple aggregation of individual aspects of a problem is intuitively appealing but often wrong. The whole is often not simply the sum of its parts. Often, much more is involved. This does not suggest at all that scientific analysis, in which an issue is disaggregated into a number of component issues and understanding sought of the individual issues, is in any way improper. The following steps are essential in finding solutions to large and complicated poblems:

- Disaggregation or decomposition of a large issue into smaller, more easily understandable parts
- Analysis of the resulting large number of individual issues
- Aggregation of the results to attempt to find a solution to the major issue

This is the essence of the formal *scientific method*. However, interpretation must follow analysis, and meaningful issue formulation must precede it. Also, these formal efforts need to be conducted across a variety of life-cycle phases. And there must be provision for experientially based skills and understanding just as there must be provision for formal rational thought. The systems approach attempts to incorporate all of these.

System components are often of very different types; and it is helpful, from a systems perspective, to distinguish among them. Consider a university as a system for producing educated graduates. Some of the parts of the university system are structural or static components, such as university buildings. As the system is operating, these structural components usually do not change much. Operating components are dynamic and perform processing such as the professors in a university who teach students. Flow components are often material, energy, or information; but in this example, students are the parts that flow or matriculate through the university system. Again, how the components interact is an important aspect of any system, its problems or design issues, and their alternative solutions. For example, grades are one mechanism for inter-action between professors and students. Grades serve a purpose, intended or not, and it is important to understand what purpose, intended and unintended, they serve.

A very important fundamental concept of systems engineering is that *all systems are associated with life cycles.* Similar to natural systems that exhibit a birth–growth–aging and death life cycle, human-made systems also have a life cycle. Most generally, this life cycle consists of definition of the requirements for a system, development of the system itself, and deployment of the system in an operating environment. These three essential life-cycle phases are always needed. Each of them may be described in terms of a larger number of more fine-grained phases, as we describe in Chapter 2. In all types of system evolution, and as we will discuss, there will be a minimum of three phases:

- Definition
- Development
- Deployment

These comprise the essential systems engineering process activities.

This life-cycle perspective should also be associated with a long-term view toward planning for system evolution, research to bring about any new and emerging technologies needed for this evolution, and a number of activities associated with actual system evolution, or acquisition. Thus, we see that the efforts involved in the life-cycle phases for definition, development, and deployment need to be implemented across three life cycles that comprise:

- Systems planning and marketing,
- Research, development, test, and evaluation (RDT&E)
- Systems acquisition or procurement

We will briefly examine these life-cycle phases, especially those for systems acquisition, procurement, or manufacturing in this chapter and in Chapter 2. Even though our primary purpose in this text is to discuss the methods for systems engineering, such discussions would be incomplete if they are not associated with some discussion of systems engineering life cycles, processes, or methodology and the systems management efforts that lead to selection of appropriate processes.

Because large-scale systems are inherently complex in the sense of being comprised of many subsystems*, systems often can be better understood by organizing their parts into groups based on function or some other organizing principle. Often systems are organized into hierarchies. For example, a surface transportation system may be organized into roads, streets, intersections, freeways, and interchanges. The subsystem, interchanges, may be organized into ramp, bridges, and pavement components. The pavement system may be comprised of lanes, shoulders, signs, lights, surface, drainage ditch, and em-

---

*Sometimes the term "system of systems" is used. What is a component, unit, subsystem, or system is again dependent upon perspective.

bankments. We are familiar, if only because of extensive news coverage, with the way in which many of the space shuttle components are organized: the launch vehicle, booster rockets, and flight orbiter. While these are the technological components of these systems, there are obviously human components as well. We see that there are structural, functional, and purposeful points of view from which we can describe each of these several systems. We generally need a structural, functional, and purposeful definition of an entity in order to define that entity in a relatively complete manner.

Because large systems are difficult to describe completely and their behavior changes frequently over time, systems engineers use the concept of the state of a system. Think of the state of a system as a snapshot at an instant of time of a particular system. It is simply a collection of variables that conveniently describe a system at any particular point in time. For example, the systems view of the state of an airport might include such variables as the average delay for incoming and outbound flights, the number of planes waiting to land or takeoff, the number of available parking spaces, and the weather with any flight restrictions. Obviously, the variables we choose to include in our description depends on our purpose for observing or modeling this system over time. If we are modeling an airport to improve passenger convenience, then parking spaces are important. But, if we are concerned about improving air traffic safety, then it is unlikely that we will model automobile parking spaces.

State changes at a point in time are often used to describe particular events of a system such as the shutdown at an airport when the weather has caused all of the runways to close. State and event changes over time are used to describe and explain system behavior. Models of systems generally attempt to imitate or mimic system behavior. Systems engineers classify systems by state behavior for modeling and analysis purposes. For example, systems that have predetermined outputs for specific inputs are deterministic. That is, we know with certainty the behavior of the system given a set of inputs. For stochastic systems we cannot know for certainty which output will occur, but we can assign some probabilities of occurrence to specific outputs. The basic approach of the systems engineer is that the properties of the system or problem at hand determine our analytical approach, not any particular biases towards a favored approach that we happen to hold.

Now that we understand some of the basic ideas underpinning systems engineering, we are ready for some more rigorous definition.

1.4 DEFINITIONS OF SYSTEMS ENGINEERING

When studying a new area, it is important to define it. We provide several definitions. First, we suggest that systems engineering is a management technology to assist and support policy making, planning, decision making, and associated resource allocation or action deployment. Systems engineers accomplish this by quantitative and qualitative *formulation, analysis,* and *interpretation* of the impacts of action alternatives upon the needs perspectives, the institutional perspectives, and the value perspectives of their clients or customers.

The key words in this definition are formulation, analysis, and interpretation. In fact, virtually all of systems engineering can be thought of as consisting of formulation, analysis, and interpretation efforts. We may exercise these in a formal sense, or in an *as if* or experientially based intuitive sense. These are the components comprising a structural framework for systems methodology and design. We need a guide to formulation, analysis, and interpretation efforts, and systems engineering provides this by embedding these three steps into life cycles, or processes, for system evolution.

System management and integration issues are of major importance in determining the effectiveness, efficiency, and overall *functionality* of system designs. To achieve a high measure of functionality, it must be possible for a system design to be *efficiently* and *effectively* produced, used, maintained, retrofitted, and modified throughout all phases of a life cycle. This life cycle begins with need conceptualization and identification, through specification of system requirements and architectures, to ultimate system development, installation or operational implementation, evaluation, and maintenance throughout a productive lifetime.

For our purposes, we may also define *systems engineering* as the *definition, design, development, production,* and *maintenance of functional, reliable, and trustworthy systems within cost and time constraints.* There are, of course, other definitions. Two closely related and appropriate definitions are provided by MIL-STD-499A and MIL-STD-499B [1]:

Systems Engineering is the application of scientific and engineering efforts to.

a. transform an operational need into a description of system performance parameters and a system configuration through the use of an iterative process of definition, synthesis, analysis, design, test, and evaluation;

b. integrate related technical parameters and ensure compatibility of all physical, functional, and program interfaces in a manner that optimizes the total system definition and design;

c. integrate reliability, maintainability, safety, survivability, human engineering, and other factors into the total engineering effort to meet cost, schedule, supportability, and technical performance objectives.

Systems Engineering is an interdisciplinary approach to evolve and verify an integrated and life-cycle balanced set of system product and process solutions that satisfy the customers needs. Systems engineering:.

a. encompasses the scientific and engineering efforts related to the development, manufacturing, verification, deployment, operations, support, and disposal of system products and processes.

b. develops needed user training equipment, procedures, and data.

c. establishes and maintains configuration management of the system.

d. develops work breakdown structures and statements of work, and it provides information for management decision making.

In keeping with the tendency in the U.S. Department of Defense to rely on commercial standards, these standards are no longer operational and have been replaced by comparable commercial standards. Nevertheless, they do provide generally appropriate definitions. These two definitions attempt to combine structural, functional, and purposeful views of systems engineering. It is generally accepted that we may define things according to *structure, function*, or *purpose*.

Often, definitions are incomplete if they do not address structure, function, and purpose. Our continued discussion of systems engineering will be assisted by the provision of a structural, purposeful, and functional definition of systems engineering. Table 1.1 presents these three definitions.

Each of these definitions is important and all three are generally needed, as we have noted. In our three level hierarchy of systems engineering there is generally a non-mutually exclusive correspondence between function and tools, structure and methodology, and purpose and management, as we note in Figure 1.2. A systems engineering process results from efforts at the level of systems management to identify an appropriate methodology, an appropriate set of procedures, or a process for engineering a system. A systems engineering product, or service, results from the use of this process, or product line, together with an appropriate set of methods and metrics. These are illustrated in Figure 1.2.

TABLE 1.1   Definitions of Systems Engineering

Structure	Systems engineering is management technology to assist clients through the formulation, analysis, and interpretation of the impacts of proposed policies, controls, or complete systems upon the need perspectives, institutional perspectives, and value perspectives of stakeholders to issues under consideration.
Function	Systems engineering is an appropriate combination of the methods and tools of systems engineering, made possible through use of a suitable methodology and systems management procedures, in a useful process-oriented setting that is appropriate for the resolution of real-world problems, often of large scale and scope.
Purpose	The purpose of systems engineering is information and knowledge organization that will assist clients who desire to define, develop, and deploy total systems to achieve a high standard of overall quality, integrity, and integration as related to performance, trustworthiness, reliability, availability, and maintainability of the resulting system.

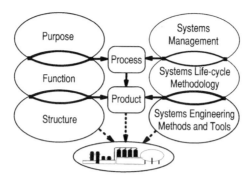

Figure 1.2 The evolution of process and production from purpose, function, and structure and the three levels for systems engineering: Systems management, methodology, and methods and tools.

We have illustrated three hierarchical levels for systems engineering in Figure 1.2. These are associated with structure, function, and purpose as also indicated in Figure 1.2. The evolution of a systems engineering product, or service, from the chosen systems engineering process is illustrated in Figure 1.3. The systems engineering process is driven by systems management and there are a number of drivers for systems management, such as the competitive strategy of the organization. We now expand on this to indicate some of the ingredients at each of these levels. The functional definition of systems engineering says that we will be concerned with the various tools and techniques that enable us to engineer systems. Often, these will be systems science and operations research tools that enable the formal analysis of

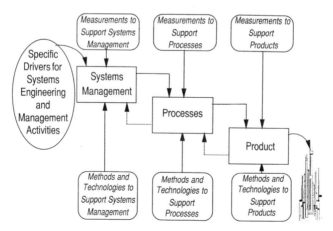

Figure 1.3 A three-level systems engineering perspective on the engineering of systems.

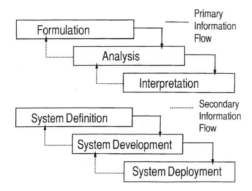

Figure 1.4   The three basic steps and phases of systems engineering.

systems. The basic activities of systems engineers are usually concentrated on the evolution of an appropriate process to enable the definition, development, and deployment of a system or on the formulation, analysis, and interpretation of issues associated with one of these phases. Figure 1.4 illustrates the basic systems engineering process phases and steps. Generally, these are combined to form a nine element matrix structure, as represented in Figure 1.5.

The functional definition of systems engineering says that we will be concerned with an appropriate combination of methods and tools. We will denote the result of the effort to obtain this combination as a systems methodology. Systems engineering methodology is concerned with the life cycle or process used for system evolution. This brings about such important notions as appropriate development life cycles, operational quality assurance issues, and configuration management procedures that are very important [2–4] but

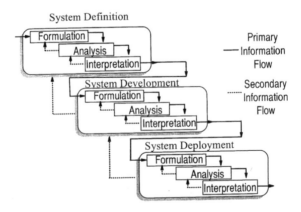

Figure 1.5   A systems engineering framework comprised of three phases and three steps per phase.

which we do not emphasize here. The functional definition of systems engineering also says that we will accomplish this in a useful and appropriate setting. This useful setting is provided by an appropriate systems management process. We will use the term *systems management* to refer to the cognitive and organizational tasks necessary to produce a useful process, methodology, or product line for system evolution and to manage the process-related activities that result in a trustworthy system. More specifically, the result of systems management is an appropriate combination of the methods and tools of systems engineering, including their use in a methodological setting, with appropriate leadership in managing system process and product development, to ultimately field a system that can be used by clients to satisfy the needs that led to its development. There are many interesting issues associated with systems management, but this is not a focus of this introductory text. Figure 1.6 illustrates the notion of systems engineering as a broker of information across enterprises having a need for a system and the implementation specialists who will implement detailed design considerations to bring this about.

The structural definition of systems engineering tells us that we are concerned with a framework for problem resolution that, from a formal perspective at least, consists of three fundamental steps:

* Issue *formulation*
* Issue *analysis*
* Issue *interpretation*

These are each conducted at each of the life-cycle phases that have been chosen in order to implement the basic phased efforts of definition, development, and deployment. There are three major life-cycle phases associated with systems engineering — definition, development, and deployment — as shown in Figures

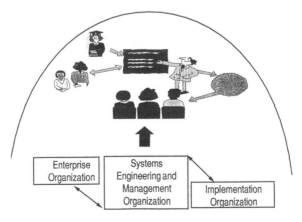

Figure 1.6 Systems engineers as brokers of information and knowledge.

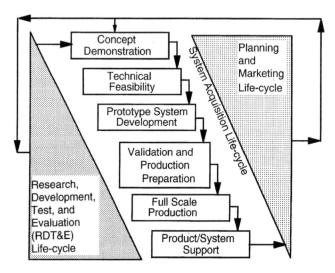

Figure 1.7   Interactions across the three primary systems engineering life cycles.

1.4 and 1.5. These phases may be described in any number of ways, as we describe in greater detail in Chapter 2. There are three general systems life cycles, as suggested by Figure 1.7:

- *Research, development, test, and evaluation (RDT&E)*
- *Acquisition* (or production, or manufacturing, or fielding)
- *Planning and marketing*

  Systems engineers are involved in efforts associated with each of these life cycles, often in a technical direction capacity. We will return to describe the systems acquisition life cycle in much more detail in Chapter 2.

1.5   HISTORY OF TECHNOLOGICAL DEVELOPMENT

Throughout history, the development of more sophisticated tools has invariably been associated with a decrease in our dependence on human physical energy as a source of effort. Generally, this is accomplished by control of nonhuman sources of energy in an automated fashion. Often, these involve intellectual and cognitive effort. The industrial revolution of many years ago represented a major initial thrust in this direction. Two excellent works present discussions of technology evolution in the United States over the past century [5,6] and indicate that it was during that time when the essence of the contemporary systems and information age began in America.

In most cases, a new tool or machine makes it possible to perform a familiar task in a somewhat new and different way, typically with enhanced efficiency and effectiveness. In a smaller number of cases, a new tool has made it possible to do something entirely new and different that could not be done before.

Profound societal changes have often been associated with changes brought about by new tools. In the 1850s, for example, about 70% of the labor force in the United States was employed in agriculture. A little more than a century later, less than 3% is so employed. This 3% is able to produce sufficient food to feed an entire country and to generally produce large surpluses as well. Occasionally, these tools have produced undesired side effects. Two examples of this are (a) pollution due to chemical plants and (b) potential depletion of fossil fuel sources. Occasionally, new tools have the potential for producing significantly harmful side effects, such as those that can occur due to human-originated and technology-abetted operational errors in nuclear power plant operation.

Concerns associated with the *definition, development,* and *deployment* of tools such that they can be used efficiently and effectively have always been addressed, but often on an implicit and "trial and error" basis. When tool designers were also tool users, which was more often than not the case for the simple tools and machines of the past, the resulting designs were often good initially, or soon evolved into good designs. These phased efforts of definition, development, and deployment represent the macrostructure of a systems engineering framework. They each need to be employed for each of the life cycles illustrated in Figure 1.7.

When physical tools, machines, and systems become so complex that it is no longer possible to design them by a single individual, and a design team is necessary, then a host of new problems emerge. This is the condition today. To cope with this, a number of methodologies associated with systems design engineering have evolved. Through these, it has been possible to decompose large design issues into smaller component subsystem design issues, design the subsystems, and then build the complete system as a collection of these subsystems.

Even so, problems remain. There are many instances of failures due to this sort of incomplete approach. Just simply connecting together individual subsystems often does not result in a system that performs acceptably, either from a technological efficiency perspective or from an effectiveness perspective. This has led to the realization that *systems integration engineering* and *systems management* throughout an entire system life cycle will be necessary. Thus, contemporary efforts in *systems engineering* contain a focus on tools and methods, on the systems methodology for definition, development, and deployment that enables appropriate use of these tools, and on the systems management approaches that enable the embedding of systems engineering product and process development approaches within organizations and environments, such as to support the application of the principles of the physical and material sciences for the betterment of humankind.

Figures 1.2 and 1.3 illustrate this conceptual relationship among the three levels of systems engineering, and they also show each as necessary facets that the systems engineering team must apply to evolving a trustworthy system. Each of these three levels—systems *methods*, systems *methodology*, and systems *management*—are necessarily associated with appropriate environments in order to ensure an appropriate systems engineering process, including the very necessary client interaction during system definition, development, and deployment. The use of appropriate systems methods and tools, as well as systems methodology and systems management constructs, enables *system design for more efficient and effective human interaction* [7]. More complete exploration of these facets of systems engineering [2–4] would take us into a more advanced treatment of systems engineering, and to many discussions of systems methodology and systems management, than the introductory level methods focused version of systems engineering that is intended here.

Now let's return to history again to understand how a new type of machine propelled the emergence of systems engineering.

Sometime around the middle of the twentieth century, the use of a new type of machine became widespread. This new machine, the stored program digital computer, was fundamentally different, in many ways, from the usual combination of motors, gears, pulleys, and other physical components that assisted humans, perhaps in an automated fashion, in performing physical tasks such as pulling, or even flying. While this machine could assist in performing functions associated with physical tasks, such as computing the optimal trajectory for an aircraft to move between two locations with minimum energy consumption and cost, it could also assist humans in a number of primarily cognitive tasks such as planning, resource allocation, and decision making.

It is doubtlessly correct to say that the modern stored program digital computer is a product of the physical and material sciences. The internal components of this new machine surely are products of the physical and material sciences. But it is also important to note here that the digital computer is intended, from a purposeful viewpoint, to be used to provide assistance to efforts that can be more efficiently and effectively achieved through more appropriate use of information and knowledge technologies. Systems engineering enables this in an organized framework, and thus provides a very necessary basis for the development of complex systems of humans and machines.

Thus, the purpose of the digital computer is much more cognitive and intellectual support than it is physical support. The computer is an information machine, a knowledge support machine, and a cognitive support machine. It has led to the growth of a new engineering area of endeavor, which involves information and knowledge technology. This new professional area is broadly concerned with efforts whose structure, function, and purpose are associated with the acquisition, representation, storage, transmission, and use of data that is of value for typically cognitive support, but which often results in some sort of ultimate physical effort and human supervisory control of this. There are many contemporary illustrations of this support. They encompass such applications as military command and control systems, flight control and guidance systems for aircraft, and very sophisticated financial accounting support systems.

Associated with this are a plethora of *information technology* products and services that have the potential to profoundly affect the lives of each of us. Clearly this is happening now, and the rate at which these changes will occur is surely going to accelerate. Information technology products and services based on computers and communications, such as telecommunications, command and control, automated manufacturing, electronic mail, and office automation, are common words today. *Computer and communications aided everything*, especially with the merger and integration of computer and communications technologies, will perhaps be a common term tomorrow. The results could be truly exciting: electronic access to libraries and shopping services; individualized and personalized systems of interactive instruction; individualized design of aids to the disabled and handicapped; prediction and planning in business, agriculture, health, and education; and support to knowledge workers in a number of new and classic enterprises. Most of these involve the integration of the information technologies of computers and communications through use of systems engineering approaches for the development of processes and products. This has led to a fundamental change in the way in which systems engineering is accomplished. Some now call the field *computer-aided systems engineering* [8].

Thus, we see that the physical and material science basis for engineering and technology is now augmented by an information science basis made possible primarily through the development of the modern microprocessor and associated computers.

1.6 SYSTEMS ENGINEERING KNOWLEDGE

Figure 1.8 illustrates that systems engineering knowledge is comprised of [2, 9] the following:

1. Knowledge *principles*, which generally represent formal problem solving approaches to knowledge, generally employed in new situations and/or unstructured environments

2. Knowledge *practices*, which represent the accumulated wisdom and experiences that have led to the development of standard operating policies for well-structured problems

3. Knowledge *perspectives*, which represent the view that is held relative to future directions and realities in the technological area under consideration

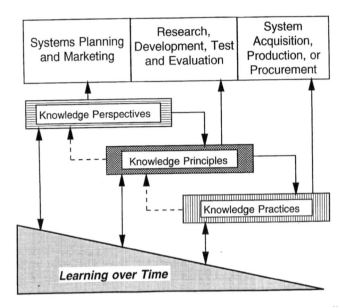

Figure 1.8   Knowledge types and support for systems engineering efforts.

We have encountered each of these three types of knowledge in our daily lives. Knowledge principles include a host of scientific theories. In a sense, these represent the why associated with the functioning of systems. For example, one knowledge principle is that associated with Newton's law. It suggests that force is equal to mass times acceleration and that because acceleration is the derivative of velocity and velocity is the derivative of position, we have

$$f = Ma = M\left(\frac{dy}{dt}\right) = M\left(\frac{d^2x}{dt^2}\right)$$

as a simple differential equation representing the movement of an object. There are a number of assumptions needed in order for this equation to remain valid. First, we must be dealing with movement in one dimension only. Also, the only object to which the force is subjected is a mass. There are no friction elements or springs involved, mass is constant over time, and motion takes place in one-dimensional space. If we relax these latter two assumptions, knowledge principles associated with Newton's law suggest that we may rewrite the differential equation as

$$f = M\left(\frac{d^2x}{dt^2}\right) + B\left(\frac{dx}{dt}\right) + K(x)$$

Given an appropriate force and initial conditions, we can solve this differential equation. This is not an important effort for us here and so we will not do this now. What we have here is a simple *model* of one-dimensional motion. System modeling is a very important activity in systems engineering, and we will devote much of Chapter 4 to this subject.

We could continue to extrapolate on this model of motion, based on Newton's law of mechanics, until we actually come up with a differential equation, doubtlessly a very complicated one, that could be used to predict the motion of an automobile when subjected to various forcing functions due to different time histories of accelerator pedal movement and braking controls. Then we could use this differential equation to project the time that would be required to stop a fancy sports car traveling at 60 miles per hour under a certain type of braking action. We would be using knowledge principles to predict the braking effectiveness of this particular car.

Alternately, we could develop a set of knowledge practices that are based on actual experimental observations of different drivers breaking different cars. Then we could publish such a table. The table might be adopted as a standard, and any particular car that could not stop in the distance specified by the standard might well be subjected to an appropriate repair effort. While the table might have its basis in the physical differential equations for an automobile, there would not necessarily be any reference to these knowledge principles in obtaining the table. The knowledge principles associated with vehicle motion dynamics would be very useful, however, in the design of various subsystems for the automobile.

Knowledge perspectives are needed when we attempt to project various futures for the automobile. For example, we might envision a significant increase in gasoline prices due to an oil embargo. Or we might envision renewed concern for environmental preservation. Each of these could lead to significant interest in smaller size engines, engines that would result in greater fuel use efficiency at the expense of lower power. This could increase the incentives for electric battery-powered automobiles. For these to be cost-effective, there would have to be a technological revolution in battery storage capacities. There would have to be other changes, such as in societal willingness to accept low-power-capacity automobiles.

Clearly, one form of knowledge leads to another. Knowledge perspectives may create the incentive for research that leads to the discovery of new knowledge principles. As knowledge principles emerge and are refined, they generally become embedded in the form of knowledge practices. Knowledge practices are generally the major influences of the systems that can be acquired or fielded. These knowledge types interact together, as suggested in Figure 1.8, and the feedback loops among them are associated with learning to enable continual improvement in performance over time.

It is on the basis of the appropriate use of these knowledge types that we are able to accomplish technological system planning and development and management system planning and development, which lead to a new innovative product or service. All three types of knowledge are needed to cope with the opportunities, challenges, and pitfalls of large-scale systems.

1.7   CHALLENGES AND PITFALLS IN SYSTEMS ENGINEERING

In order to resolve large-scale and complex problems, or manage large systems of humans and machines (technology), we must be able to deal with important contemporary issues that involve and require:

1. Many considerations and interrelations
2. Many different and perhaps controversial value judgments
3. Knowledge from several disciplines
4. Knowledge at the levels of principles, practices, and perspectives
5. Considerations involving product definition, development, and deployment
6. Considerations that cut across the three different life cycles associated with systems planning and marketing, RDT&E, and system acquisition or production
7. Risks and uncertainties involving future events which are difficult to predict
8. A fragmented decision making structure
9. Human and organizational need and value perspectives, as well as technology perspectives
10. Resolution of issues at the level of institutions and values, as well as at the level of symptoms

There are many failures in bringing about a large system due to failure to consider one, or more than one, of these realities.

The professional practice of systems engineering must use of a variety of formulation, analysis, and interpretation aids for evolution of technological systems and management systems. Clients and system developers alike need this support to enable them to avoid several potential pitfalls. These include:

1. Overreliance upon a specific analytical method or a specific technology that is advocated by a particular group
2. Consideration of perceived problems and issues only at the level of symptoms, and the development and deployment of "solutions" that only address symptoms and that fail to address institutional and value issues
3. Failure to develop and apply appropriate methodologies for issue resolution that will allow identification of major pertinent issue formulation elements, a fully robust analysis of the variety of impacts on stakeholders and the associated interactions among steps of the problem solution procedure, and an interpretation of these impacts in terms of institutional and value considerations

4. Failure to involve the client, to the extent necessary, in the development of problem resolution alternatives and systemic aids to problem resolution

5. Failure to consider the effects of cognitive biases that result from poor information processing heuristics

6. Failure to identify a sufficiently robust set of options, or alternative courses of action

7. Failure to identify risks and to manage risks that are associated with the costs and benefits, or effectiveness, of the system to be acquired, produced, or otherwise fielded

8. Failure to properly relate the system that is designed and implemented with the cognitive style and behavioral constraints that affect the user of the system, such as to not properly design the system for effective user interaction

9. Failure to address quality issues in a comprehensive manner throughout all phases of the life cycle, especially in terms of reliability, availability, and maintainability

10. Failure to engineer an appropriate process for evolution of the system

11. Failure to properly integrate a new system together with heritage or legacy systems that already exist and that the new system should support

12. Failure to consider sustainability issues as they affect the system itelf, the consumption of natural resources, and the preservation of the natural environment such as to ensure both intragenerational equity and intergenerational equity

Often, one result of these failures is that the purpose, function, and structure of a new system are not identified sufficiently before the system is defined, developed, and deployed. These failures generally cause costly mistakes that could truly have been avoided. Invariably this occurs because there are one or more failures associated with defining, developing, or deploying a system. These failures result because either the formulation, the analysis, or the interpretation efforts (or all of them perhaps) associated with one or more of the phases of the systems life cycle are deficient. A major objective of systems engineering is to take proactive measures to avoid these difficulties.

In reality, there are many difficulties associated with the production of functional, reliable, and trustworthy systems of large scale and scope. There are many studies which indicate that:

1. It is very difficult to identify the user requirements for a large system.
2. Large systems are expensive.
3. System capability is often less than promised and expected.
4. System deliveries are often quite late.

5. Large system cost overruns often occur.
6. Large system maintenance is complex and error prone.
7. Large system documentation is inappropriate and inadequate.
8. Large systems are often cumbersome to use, and system design for human interaction is often lacking.
9. Individual new subsystems often cannot be integrated with legacy or heritage systems.
10. Large systems often cannot be transitioned to a new environment or modified to meet the evolving needs of clients.
11. Unanticipated risks and hazards often materialize.
12. Large systems often suffer in terms of their reliability, availability, and maintainability.
13. Large-system performance is often of low quality.
14. Large systems often do not perform according to specifications.
15. It is difficult to identify suitable performance metrics for large systems that enable determination of system cost and effectiveness.
16. System requirements often do not adequately capture user needs.

These potential difficulties, when they are allowed to develop, can create many problems that are difficult to resolve. Among these are:

• Inconsistent, incomplete, and otherwise imperfect system requirements and specifications
• System requirements that do not provide for change as user needs evolve over time
• Poorly defined management structures for product design and delivery

These lead to delivered products that are difficult to use, that do not solve the intended problem, that operate in an unreliable fashion, that are unmaintainable, and that, as a result, are not used.

These same studies generally show that the major problems associated with the production of trustworthy systems have more to do with the *organization and management of complexity* than with direct technological concerns that affect individual subsystems and specific physical science areas. Often the major concern should be more associated with the definition, development, and use of an appropriate process for production of a product than it is with the actual product itself, in the sense that direct attention to the product or service without appropriate attention to the process leads to the fielding of a low quality and expensive product or service.

We envision a three performance level hierarchy for systems engineering phased efforts, as we have described. This three-level structured hierarchy comprises a systems engineering life cycle and is one of the ingredients of systems engineering methodology. It involves, as we have already noted:

- System *definition*
- System *development*
- System *deployment*

As we will discuss later, this three-phase life cycle may be expanded into a more detailed life cycle that is comprised of a larger number of phases. Generally the number of phases is in the vicinity of seven, but it may be much larger.

There are many goals that could be stated for a system engineering effort. The following are among the most important of these.

1. The systems engineering process should encompass all phases of the system life cycle, or life cycles as the case may be, including transitioning between phases.

2. The systems engineering process should support problem understanding, as well as communication among all interested parties at all phases in the process.

3. The systems engineering process should enable capture of design and implementation needs for the systems engineering product early in the life cycle, generally as part of the requirements, specifications, and conceptual design phases.

4. The systems engineering process and associated methodology should support both bottom-up and top-down approaches to systems design and development.

5. The systems engineering process should enable an appropriate mix of design, development, and systems management approaches.

6. The systems engineering process should support quality assurance of both the product and the process that leads to the process.

7. The systems engineering process should support system product evolution over time.

8. The systems engineering process should be supportive of appropriate standards and management approaches for configuration management.

9. The systems engineering process should support the use of automated aids that assist in the production of high-quality trustworthy systems.

10. The systems engineering process should be based upon a methodology that is teachable and transferable, and one that makes the process visible and controllable at all life-cycle phases of development.

11. The systems engineering process should have appropriate procedures to enable definition and documentation of all relevant factors at each phase in the system life cycle.

12. The systems engineering process should be associated with appropriate metrics and management controls across each phase and step of the process.
13. The systems engineering process should support operational product functionality, revisability, and transitioning, both at the initial time of operational implementation and later at the time that a system is phased out of service, retired, or reengineered.
14. The systems engineering process must support both system product development and the system user organizations; it must also be compatible with the environments associated with systems development and operation.
15. The systems engineering process should support quality, total quality management, system design for human interaction, and other attributes associated with trustworthiness and integrity.

When all of these goals are accomplished, it will be possible to produce operational systems and products that are economical, reliable, verifiable, interoperable, integratable, portable, adaptable, evolvable, comprehensible, maintainable, manageable, and cost-effective and that provide a very high degree of user satisfaction. These goals represent attributes for metrics, or are translatable into metrics, that can measure the *trustworthiness* of an operational systems engineering product or service.

These objectives relate to quality, effectiveness, and cost. They can be translated into standards with which to measure system performance and systems engineering process effectiveness. The quality and effectiveness information, together with cost information, will allow us to obtain cost and operational effectiveness of systems engineering products and services. This is very important. It is also important that we establish cost-effectiveness metrics for the systems engineering process itself. Thus, we see that we have metrics at three levels:

- System management
- Systems engineering methodological process
- Systems engineering products

Our efforts in this text are, in part, directed toward these ends. An overall major goal for systems engineering is to interrelate people, organizations, and technology, in an appropriate setting or environment, for problem solving. This is done in order to ensure that the resulting systemic problem solution, generally in the form of a product or service, is *trustworthy*. Figure 1.9 indicates some of the attributes of a trustworthy system. A high degree of success across these attributes is generally only achieved by an appropriate interrelationship among people, technologies, organizations, and the environmental setting that surrounds them.

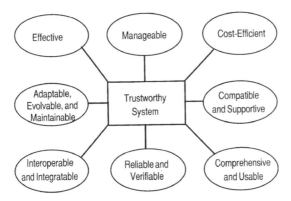

Figure 1.9 Attributes of a trustworthy systems engineering product or service.

In general, we may approach issues from an inactive, reactive, interactive, or proactive perspective [2,3].

- *Inactive.* This denotes an organization that does not worry about issues and that does not take efforts to resolve them. It is a very hopeful perspective, but generally one that will lead to issues becoming serious problems.

- *Reactive.* This denotes an organization that will examine a potential issue only after it has developed into a real problem. It will perform an outcomes assessment and after it has detected a problem, or failure, will diagnose the cause of the problem and, often, will get rid of the symptoms that produce the problem.

- *Interactive.* This denotes an organization that will attempt to examine issues while they are in the process of evolution such as to detect them at the earliest possible time. Issues that may cause difficulties will not only be detected, but efforts at diagnosis and correction will be implemented as soon as they have been detected. This will involve detection of problems as soon as they occur, diagnosis of their causes, and correction of difficulty through recycling, feedback, and retrofit to and through that portion of the lifecycle process in which the problem occurred. Thus, the term interactive is, indeed, very appropriate.

- *Proactive.* This denotes an organization that predicts the potential for debilitating issues and that will synthesize an appropriate lifecycle process that is sufficiently mature such that the potential for issues developing is as small as possible.

It should be noted that there is much to be gained by a focus on process improvements in efforts from any of the last three perspectives. While proactive and interactive efforts are associated with greater capability and process maturity than are reactive efforts, reactive efforts are still needed [2]. Inactivity is associated with failure, in most cases.

1.8 SYSTEMS ENGINEERING EDUCATION

Systems engineering education is concerned with preparing people for professional practice and research in systems engineering. Needless to say, we believe that systems engineering is one of the fundamental engineering subject areas. Systems engineering education is, we believe, solid general engineering education that transcends the boundaries of engineering to encompass a great many other areas as well. While most engineering specializations are *internally focused* on the specific methods needed to bring about a useful product or service, systems engineering is *externally focused* on the interactions and interplay between the system that is being evolved and the broader environment of which it is a part.

Notions of what comprises an appropriate education are not at all new. Many have written, and for a very long time, about educational requirements and the characteristics of educated people. In 1852, for example, John Henry Cardinal Newman [10] identified 10 distinctive traits of university education. An interpretation of these is as follows:

1. It encourages one to identify and place values upon personally held views and judgments.
2. It encourages search for the truth.
3. It encourages eloquence in expressing truths.
4. It encourages clarity and integrity in observing and valuing judgments.
5. It encourages coherency of expression and communication.
6. It encourages one to solve problems through critical thinking skills and by analyzing courses of action, such as to be able to retain meaningful alternates and to discard spurious ones.
7. It encourages cultural and cross-cultural understanding.
8. It encourages compassion, forbearance and comprehension of the views of others.
9. It encourages basic feelings for and understandings of events in both a social and an historical context.
10. It encourages preparation for work and a lifetime of continued learning.

There are many other similar listings. In a recent Alverno College catalog, for example, eight attributes of a general education are considered essential:

1. Effective communication ability
2. Analytical capability
3. Problem-solving ability

4. Valuing in a decision-making context
5. Effective social interaction
6. Taking responsibility for the global environment
7. Effective citizenship
8. Aesthetic responsiveness

Such listings are pervasive in the corporate world as well. DuPont Engineering has an effort, for example, called the *Discussion of Contribution*. This is part of the Individual Career Management Process at DuPont in which engineers and their leadership review the engineer's work performance for the past year. There are 11 attributes, which can be instrumented as attribute measures:

1. *Leadership.* The ability to provide direction under uncertain conditions and taking of initiative at a very high energy level.
2. *Strategic Ability.* The ability to readily learn and understands concepts outside ones immediate functional area.
3. *Creativity and Innovation.* The ability to use innovative approaches to solve problems and identify the implications of a problem that one encounters.
4. *Risk Taking and Bias for Action.* The ability and willingness to take risks in order to advance new programs.
5. *Decision-Making Ability.* The degree to which one looks for win–win solutions in difficult situations, is open to influence and change, and demonstrates confidence and sound judgment.
6. *Knowledge of Professional Field.* The extent to which one has fundamental understanding of ideas, techniques, and technologies and also seeks out and understands new developments.
7. *Managerial Proficiency.* The extent to which one understands complex issues quickly and takes appropriate action, and the extent to which one has a set of continuous improvement principles to help guide goals and decisions.
8. *Resourcefulness.* The ability to adapt to rapidly changing conditions and to learn from successes and failures; and the extent to which one has a high level of initiative, persistence, and involvement.
9. *Maturity and Stability.* The extent to which one is willing to learn and improve and to serve as a role model for others.
10. *Communication Ability.* The extent to which one expresses ideas and concerns clearly and persuasively, listens effectively, and has flexible and effective communication skills.
11. *Interpersonal Competence.* The extent to which one is sensitive to the needs of people, develops rapport and trust, and solicits interpersonal feedback.

To be most useful, attributes such as these need to be associated with attribute measures. This, along with the associated determination of weights to associate with the attributes, is not necessarily a simple task. We will actually address determination of attribute measures and weights in Chapter 5.

The knowledge and skills relevant to a sound education in engineering in general and systems engineering education in particular come, as we have noted, from the world of knowledge practices, principles, and perspectives. The knowledge principles required for a high-quality and relevant education in engineering come primarily from the humanities and sciences study areas. These include the humanities areas of natural languages, history, and philosophy. They also include the sciences. Science must be broadly interpreted here to include the natural science areas of physics, chemistry, and biology; it additionally includes the behavioral and social science areas of psychology and economics. We will also include mathematical sciences within the category of science. In addition, science includes the applied engineering sciences, especially those of the computer and systems engineering sciences that are so relevant to general problem solving.

Five study areas — humanities, social and behavioral sciences, natural sciences, mathematical sciences, and engineering sciences — are absolutely essential for provision of the general education background and perspective needed for professional practice in engineering, including systems engineering, and for development of abilities to use knowledge principles. It is our strong belief that each of these subject areas is absolutely essential for general education for all as well, even though they are often not included in most taxonomies of general education. It is important to note that this broadened definition of general education is being recognized and encouraged, even for liberal arts education [11], and that there is indeed some progress in fulfilling this need.

Programs that have engineering in their title are potentially subject to accreditation by the *Accreditation Board for Engineering and Technology* (ABET). Accreditation planning for, and implementation of, computer science accreditation was accomplished through ABET, and through the Institute of Electrical and Electronics Engineers within ABET. This is now a separate, but related, accreditation effort within the *Computer Science Accreditation Board* (CSAB). ABET is recognized by the U.S. Department of Education (USDoE) and the Council on Postsecondary Accreditation (COPA) as the sole agency responsible for accreditation of *all* educational programs leading to engineering degrees in the United States.

A number of issues relative to engineering education in general and electrical engineering in particular are discussed in reference 12 and the references therein. Discussions of systems engineering education are contained in references 13 and 14. One of the major new developments in engineering education in the United States is *Engineering Criteria 2000* [15], which is comprised of criteria intended to emphasize quality and preparation for professional practice. The criteria retain the traditional core of engineering, math, and science requirements. However, they also place importance on formal efforts that stress teamwork, communications, and collaboration as well as global, economic, social, and environmental awareness. They are based on the following premises:

- Technology has been a driver of many of the changes occurring in society over the last several years.
- It will take on an even larger role in the future.
- The engineering education accreditation process must promote innovation and continuous improvement to enable institutions to prepare professional engineers for exciting future opportunities.

These criteria are focused on ensuring competence, commitment, communications, collaboration, and the courage needed for individual responsibility. These, augmenting the usual listing of competence and assumption of individual responsibility as the two traditionally accepted key characteristics of a professional, might be accepted as the new augmented attributes of a mature professional. They should truly support the definition, development, and deployment of relevant, attractive, and connected (quality) engineering education that will:

- Include the necessary foundations for knowledge principles, practices, and perspectives.
- Integrate these fundamentals well through meaningful design, problem-solving, and decision-making efforts.
- Be sufficiently practice-oriented to prepare students for entry into professional practice.
- Emphasize teamwork and communications, as well as individual efforts.
- Incorporate social, cultural, ethical, and equity issues, along with a sense of economic and organizational realities and a sense of globalization of engineering efforts.
- Instill an appreciation of the values of personal responsibility for individual and group stewardship of the natural, technoeconomic, and cultural environment.
- Instill a knowledge of how to learn, as well as a desire to learn, and how to adapt to changing societal needs over a successful professional career.

The objective of accreditation is to determine that an educational program meets *minimum* standards. It is not, in any sense, a warrant or guarantee of high quality, regardless of the multiattributed approach we might take to quality definition. A recent ABET definition of engineering is as follows [16].

Engineering is that profession in which knowledge of the mathematical and natural sciences gained by study, experience, and practice is applied with judgment to develop ways to utilize, economically, the materials and forces of nature for the benefit of mankind. A significant measure of an engineering education is the degree to which it has prepared the graduate to pursue a productive engineering career that is characterized by continued professional growth.

While it may appear at first glance that there is not much of a breadth focus in this definition, the ABET document goes on to delineate five outcomes of an engineering education in a very meaningful manner that strongly include the sort of breadth advocated for systems engineering education. These five outcomes, clearly valuable as metrics for outcome assessment purposes, are as follows:

1. An engineer should have the ability to formulate and solve, in a practical way, those problems of society that are amenable to engineering solution.
2. An engineer should have a sensitivity to those socially relevant technical problems that confront the engineering profession.
3. An engineer should have an understanding of the ethical characteristics of the engineering profession and professional practice.
4. An engineer should have an understanding of the engineers responsibility to protect both occupational and public health and safety.
5. An engineer should have the ability to maintain professional competency through a lifetime of learning.

These five purposes of a professional lead to several desirable functional characteristics of professionals that we will soon discuss. Engineering is a professional activity, and standards and practices in engineering education must conform to these notions. They lead to such important activities as curriculum design. Additionally, they imply a subset of criteria for evaluation and outcomes assessment of programs and they become critical success factors for students in these programs. Importantly also, they bring into focus the need for a balance among studies involving knowledge practices, principles, and perspectives.

A number of engineering schools have recently reconsidered their educational program offerings. These range from single programs, such as the electrical engineering undergraduate program at Carnegie Mellon University, to undergraduate engineering programs for an entire school [17], and Figure 1.10 presents a list of critical success factors which is very appropriate for systems engineering education as well as other engineering education specialties [18]. It clearly is an amalgam of many past studies [19], such as the ones cited here. These appear appropriate for all of engineering education, although the weights associated with various factors should necessarily change across degree levels and majors.

Student Critical Success Factors

1. Ability to formulate, analyze, and interpret issues - for issue formulation, analysis, and interpretation, and problem solving and design.
2. Ability to use the information technologies, especially computers and telecommunications, to better process and understand information.
3. Understand knowledge practices, principles and perspectives to enable breadth and depth in a meaningful professional area.
4. Fundamental understanding of all of the foundation areas of general education.
5. Effectiveness in communicating ideas, verbally and in writing.
6. High professional and ethical standards.
7. Mature, responsible, creative, quality conscious mind.
8. Motivation for and capability of achieving a lifetime of learning.
9. Knowledge of management strategies and human behavioral motivations.
10. Appreciation of national and global issues - from a variety of perspectives.

Figure 1.10   Critical success factors for engineering education, including systems engineering education.

There a number of other statements of important facets of an engineering education. The Corporate Roundtable of the Engineering Deans Council (EDC) has identified a set of need items for engineering education which, when implemented, are intended to ensure that engineering programs remain relevant, attractive, and connected. Twelve top-level areas for reshaping engineering curricula are identified in this Engineering Deans' Council Report. These are as follows:

1. Team skills and collaborative, active learning
2. Communication skills
3. A systems perspective
4. An understanding and appreciation of diversity
5. Appreciation of different cultures and business practices, along with an understanding that engineering practice is now global
6. Integration of knowledge throughout the curriculum
7. A multidisciplinary perspective
8. Commitment to quality, timeliness, continuous improvement
9. Undergraduate research and engineering work experience
10. An understanding of social, economic, and environmental impact of engineering decisions
11. Ethics

While all of engineering education should be very focused on developing these desirable characteristics, they are particularly important in systems engineering.

This major relevance to systems engineering occurs because the systems that systems engineers work on are typically large-scale, complex, and multi-disciplinary in nature. In most companies the design, development, and production of such large systems or big projects is coordinated by the systems engineering department. This department will typically include engineers with a variety of specialties, which depends generally on the product lines of the company. The systems engineering department must oversee the development and technical direction of the system fielding from the beginning, and it must ensure that the entire system life cycle of effort is considered.

The problems on which traditional engineers work come from the needs of society. As we stated earlier, the role of an engineer is to evolve appropriate technology through the organization and delivery of science for the presumed betterment of humankind. This is also true of systems engineers. However, the problems will generally be larger and more complex and systems engineers will be very concerned with external interfaces. This is the case because systems engineers are very much concerned with the external behavior of systems and the interactions and interplays associated with the use of systems by people in organizational settings.

In some ways, systems engineering might be regarded as a new profession. However, the basic systems engineering approaches have been with us for centuries, although perhaps not as specified for group problem-solving efforts as they are today. The large-scale logistic and operational problems of World War II gave birth to the fields of operations research and systems engineering. The holistic, or "big picture," view of systems engineers and the quantitative analysis techniques of operations research can be combined to tackle many of the messy problem facing society today. This holistic thought, or the breaking down or disaggregation of a problem into smaller and smaller components such that analysis of the separate parts that comprise the problem is possible, is a fundamental characteristic of operations research and systems analysis. Systems engineering also involves a holistic view of issues in which the big picture is envisioned. Systems engineering espouses both a formal analytic and a holistic (or wholistic) view.

In most, if not all, of the academic life of an engineer, much of the learning effort is concerned with solving problems by applying the "correct" formula. Often, it is necessary to examine only a small portion of a large issue in order to obtain a representation that is capable of quantitative analysis. The typical problem is presented in a half page or less, and the corresponding answer can be given neatly on the other half of the page. In systems engineering, we have tools, and sometimes use formulas, but the problems are usually large and incompletely defined. Technology has advanced so rapidly since the beginning of this century that a new method of approaching problems solution, or system evolution, by large teams of people was, and is, required. This leads to the need for systems methodology and systems management.

A *systems engineering process* (SEP) is an organized approach to creativity. It is not a pointless and unstructured free-for-all, nor is it a strict regimen for formulation, analysis, and interpretation of large issues associated with the definition, development, and deployment of systems. Often, one of the hardest points for many systems engineering students to understand is that, for most systems engineering problems, there is no single solution, and often no single "best" solution. There are alternatives, some of which are better than others from some perspectives. The student of systems engineering should not look forward to problems that are well-defined and that can be solved simply by finding the right "tool."

So what makes a "good" systems engineer? We suggest that you start with an open, inquisitive mind, achieve as high a performance level on as many of the critical success factors delineated in Figure 1.10 as possible, and possess the following characteristics.

1. Good systems engineers have a demonstrated affinity for the systems point of view. They pay much attention to the interaction between the system ingredients, including their interactions and interplay, and between the system and the associated internal and external environment. They emphasize the function, purpose, and structural interrelationships among these elements.

2. The systems engineer must have a facility in human relations. Systems engineers are responsible for the fielding (including technical direction) of products, systems, and services to meet the needs of some sector of society. Humans play a role in all aspects of a system—from establishing a need for a system, to implementing and operating the system, or forming an integral element of the functional system.

3. The systems engineer must be an effective broker of information and knowledge. As you will see, information plays a critical role throughout the entire systems engineering process. Without an ability to gather, organize, assimilate, understand, and apply information, any problem or solution to the problem would be incomplete. This brokerage extends across the enterprise with a need for a system to implement contractors responsible for bringing about detailed design and realization of system elements.

4. A systems engineer is imaginative and creative. Solutions to a systems problem require new perspectives and a propensity for looking beyond the present, to fully assess the scope and impact of proposed solutions.

5. Above all, the systems engineer must exhibit objective judgment and sound appraisal capabilities. Systems engineering itself is a decision-making and decision-supporting process. A major purpose of systems engineering is to aid in the decision-making process. The systems engineer can ill afford to infuse personal values into a problem or solution. An engineer applies science and technology to meet some need of society. Rarely are these needs defined by the values of a single individual.

In summary, systems engineers do not wear intellectual straightjackets. They are at least as concerned with "solving the right problem" as with "solving the problem right."

Engineering is, of course, a profession. Six functional characteristics of professionals have been identified [20]:

- Expertise
- Autonomy
- Commitment
- Identification
- Ethics
- Standards

Expertise refers to the wisdom obtained from lengthy education and experience associated with application of a body of abstract or theoretical knowledge. Autonomy refers to the ability to choose the method of approach used to resolve situations. Commitment refers to dedicated interest in pursuit of knowledge practices associated with the profession. Identification refers to formal professional structures or associations that recognize the profession. Ethics refers to the delivery of services without becoming emotionally involved with the client receiving the services or, otherwise, attempting to create a conflict of interest situation. Standards refers to organizing such as to be able to regulate and safeguard the profession and its members from within the profession itself. Professionals generally adhere to these guidelines for professional conduct. Usually, they expect that others will deal with them in a similar way, especially in matters concerning their own professional employment.

Ethical issues are of major importance in systems engineering, as in other professions. The National Society of Professional Engineers (NSPE) has developed a code of ethics for engineering [21]. It is comprised of three principle parts. There are five fundamental canons of engineering practice. These and the several associated rules of professional practice are illustrated in Figures 1.11 and 1.12. Figure 1.13 illustrates the 11 professional obligations of engineers, as stated by the NSPE. Figure 1.14 illustrates a similar set approved and published by the Institute of Electrical and Electronics Engineers (IEEE) in August 1990. Clearly, these ethics prescriptions are all supported by the ingredients and interactions described in this section.

The systems engineer, like other engineers, uses a set of methods and tools as part and parcel of their profession. Words, graphics, and mathematics are the basic methods of human communications. Together with the information technology tools, primarily computers and communications, these comprise the systems engineer's "toolkit." While most conventional engineering specialties are internally focussed on the particular product being evolved, systems engineers are also externally focussed on the many interactions and interplays that ensure technology, people, and organizational integration. Systems engineers use knowledge perspectives, knowledge principles, and knowledge practi-

- Hold paramount the safety, health, and welfare of the public.
  a. Recognize primary obligation for protection of public safety, health, and welfare.
  b. Approve only those engineering documents which meet these standards.
  c. Do not reveal confidential information with prior consent of client, except as authorized by ethical or legal considerations.
  d. Do not permit associations with those believed engaged in fraudulent or dishonest practices.
  e. Cooperate with proper authorities in furnishing assistance concerning ethics Code violations.
- Perform services only in areas of professional competence.
  a. Undertake assignments only when qualified by education or professional experience.
  b. Do not affix signatures to documents dealing with subject matter in which competence is lacking, nor to any document not prepared under their direction and control.
  c. When signing documents for an entire project, insure that individual segments are signed and approved only by those qualified engineers who prepared the statement.
- Issue only objective and truthful public statements.
  a. Include all relevant information in such statements.
  b. Express public opinions only when founded upon adequate knowledge and competence
  c. Issue no statements for which compensation has been received without explicitly identification of parties on whose behalf they speak, and by revealing the extent of any personal interest.

Figure 1.11 Three of five fundamental canons and eleven rules of professional practice.

ces. Information is the stock to which these tools are applied. As we have noted several times, our principal goal is to develop an understanding of the principles, methods, and tools of systems engineering that provide support for systems engineering at the level of product, process, and systems methodology. Chapter 2 will describe the solution steps for systems engineering problems, in terms of systems engineering life cycles or systems engineering processes, in somewhat more detail than we have done here. The subsequent three Chapters, 3 through 5, will discuss a number of tools especially useful for the issue formulation, analysis, and interpretation steps that are needed for systems definition, development, and deployment.

- Act professionally for each client as a faithful agent.
  a. Disclose all known or potential conflict of interests to employers and clients by informing them of any circumstance which could have the appearance of influencing judgment or service quality.
  b. Do not accept compensation from more than one party for services on the same project unless the circumstances are fully disclosed and agreed to by all interested parties.
  c. Do not solicit or accept considerations, directly or indirectly, from others for work for which you are already responsible.
  d. Those in public service should not participate in decisions relative to services solicited or provided by them in private or public engineering practice activities.
  e. Do not accept contracts from government bodies on which an officer of your organization serves as a member.
- Avoid deceptive employment solicitation acts.
  a. Do not misrepresent your professional qualifications.
  b. Do not offer, give, solicit, or receive, either directly or indirectly, political contributions intended to influence the award of contracts. Do not offer gifts or other considerations in order to receive work. Do not pay fees to secure work except to established commercial or marketing agencies retained for these purposes.

Figure 1.12 Two of five fundamental canons and seven rules of professional practice in the NSPE code of ethics.

1. Be guided by the highest standards of integrity in all professional relations.
2. Strive to serve the public interest at all times.
3. Avoid all conduct or practice which is likely to discredit the profession or deceive the public.
4. Do not disclose confidential information regarding the business affairs or technical processes of present or former clients without their consent.
5. Do not be influenced in professional duties by conflicts of interests.
6. Uphold the principle of appropriate and adequate compensation for those engaged in engineering work.
7. Do not attempt to gain employment, or advancement or professional engagements by untruthfully criticizing others, or by other improper methods.
8. Do not attempt to injure maliciously or falsely, directly or indirectly, the professional reputation of others.
9. Accept responsibility for professional activities.
10. Give credit to those to whom credit is due, and recognize the proprietary interests of others.
11. Cooperate in extending the effectiveness of the profession by exchanging information and experiences with others, and provide opportunities for professional development.

Figure 1.13 Eleven professional obligations of engineers, from the NSPE code of ethics.

1.9   OTHER SYSTEMS ENGINEERING AND RELATED SYSTEMS THEORY STUDIES

There have certainly been other works in, and related to, systems engineering. It is neither possible nor desirable to cite all of these here. A partial listing and description of the textbooks available provides an indication of the breadth of the field as well as suggestions for reference and additional study. Specific reference to various tool-and-methods based discussions will be provided in our efforts to come.

1. Accept decision-making responsibility in a way that is consistent with public safety, health, and welfare, and properly disclose any known factors that endangers this.
2. Avoid real or perceived conflict of interest situations whenever possible, and disclose them to potentially affected parties when they exist.
3. Be honest and realistic when making claims and estimates.
4. Reject bribery in all of its forms.
5. Improve my understanding of technology, its appropriate application, and the potential consequences of its use.
6. Maintain and improve my professional competence and undertake assignments for others only if qualified by education or experience, and after full disclosure of any limitations in this regard.
7. Seek, accept, and offer honest criticism; acknowledge and correct errors; and credit the contributions of others properly.
8. Treat all persons in a fair manner regardless of such factors as race, religion, gender, disability, age, or national origin.
9. Avoid injuring others (property, reputation, or employment) by false or malicious action.
10. Assist colleagues and coworkers in professional development, and in following the code of ethics.

Figure 1.14 Ten professional obligations of engineers, from the IEEE code of ethics.

There has been much work devoted to systems engineering over the past 25 years. The first available text describing systems engineering from a perspective similar to that taken here was probably that of Hall, who has published a large amount of early seminal work on systems engineering [22,23], as well as a more recent text [24] that takes on a rather philosophical and reflective flavor. There are a number of older books published 20 and more years ago [25–29] that are written primarily from the perspective of operations research, control theory, or a mathematical theory of systems and do provide a basis for formal systems engineering methods. In 1976, Warfield [30] published a definitive work relating systems engineering to societal issues and which also described some of his seminal contributions to structural issues in systems engineering. In 1979, Churchman [31] described characteristics of the systems approach to problem solving.

There are a number of other efforts [32,33]. Blanchard and Fabrycky [34] have published several editions of an excellent work on systems engineering which contains a first-rate treatment of systems engineering analysis methods. Another excellent analysis-based text is that of de Neufville [35]. Some of the efforts at the International Institute for Applied Systems Analysis [36,37] have been concerned with analysis and with the several pitfalls potentially associated with analysis [38].

Two texts by Sage [2,3] present systems engineering from systems methodology and systems management perspectives. An earlier text discusses a variety of methods and tools useful for implementing a methodology for large systems [39]. To a considerable extent, the present work is an update of this earlier effort. Applications of systems engineering to software intensive systems are discussed in reference 40, and decision support applications are discussed in reference 41. A recent edited *Handbook of Systems Engineering and Management* [4] contains a variety of discussions of methods as well as process and systems management perspectives.

The English school has produced several texts [42,43] that emphasize what is called *soft systems methodology*. While conceptually similar to the methodology portion of this book, the soft systems methodology works do not seem to have gained great favor in the United States. Possibly, this is because there seems to have been little effort at present to extend these approaches into metrics and systems management areas. In reference 44, Keys presents a discussion of potential relations between operations research and systems thinking. Jackson presents a relative comprehensive overview of the systems movement, with special emphasis on UK efforts, and includes a comparative study of soft systems methodology and what is denoted as the *hard systems methodology* of systems engineering, systems analysis, and operations research in reference 45. Nadler has been very active in industrial engineering efforts that emphasize systems engineering, and he has produced a definitive text in this area [46], as well as related papers. Eisner [8] has written a text called *Computer Aided Systems Engineering*. A very respectable work in every way, it is very concerned with provision of the quantitative background for the study of systems engineering methods.

A text by Walter Beam [47] is concerned with pragmatic systems engineering approaches that are directly relevant to detailed design considerations in communication and computer systems. Another recent work on systems architecture and systems design [48] is concerned with the system level details of telecommunication system design. Two recent books on systems architecture by Rechtin [49,50] provides a number of excellent discussions on systems level architecture for large communication and computer systems, including technical direction aspects.

Blanchard [51] focuses on engineering management practices in his recent effort in this area. Design-to-cost is the primary subject in the engineering management text of Michaels and Wood [52], which includes an excellent compendium of recent approaches in this specialized area of systems management. Rouse [53] addresses systems management and systems design issues in his excellent work, but more from the perspective of a systems engineer with interest in human–system interaction issues. Pragmatic aspects of systems engineering and management are emphasized in two recent works [54,55]. Analysis-based efforts are emphasized in reference 56.

Workers in the field of software engineering have discovered systems engineering, and a not inconsiderable number of very useful systems engineering developments are occurring there. This was a major motivating force behind a recent book in this area [40]. There are other related books in the software area which take a systems approach. Some have such systems engineering-like names as *Introduction to Systems Analysis and Design* [57], *Managing the Systems Life Cycle* [58], and *A Unified Methodology for Developing Systems* [59], even though the dominant concern in all of these is software.

There are a number of books in the field that are collections of papers. Typical of these are two recent works on systems design, one edited by Rouse and Boff [60] and the other by Newsome et al. [61]. There are other works that are concerned with such issues as developing new knowledge and skills through information technology [62] such as to support generally distributed decision making and cooperative work [63], strategies for innovation [64], and the catalysts for change [65] that enable the creation of successful products, systems, processes, and organizations. Again, these excellent works are more concerned with systems management than is this book. The continuing growth of interests in systems-engineering-related studies has led to the formation of the International Council on Systems Engineering, or INCOSE (Suite 107, 333 Cobalt Way, Sunnyvale, CA 94086). The journal of the Council, *Systems Engineering*, began publication by John Wiley and Sons in 1998 and provides a very useful compendium of systems engineering efforts as does the proceedings of their annual conferences.

1.10 SUMMARY

In this chapter, we have presented a relatively broad overview of systems engineering. We are aware that some of the discussion may seem a bit overwhelming on first reading. It is hoped that we have been able, however, to impart enough of a broad scope flavor of systems engineering efforts that the much more narrowly constrained discussions of systems engineering methods and tools to follow are more meaningful.

One of the most important concerns affecting the professional practice of systems engineering is the essential notion that systems engineering does not deal exclusively with physical products. While the ultimate aim of many systems engineering efforts may well be to produce a physical product or service, there are humans and organizations involved, as well as technologies. This is suggested in Figure 1.15. This representation also indicates the levels at which problem or issue resolution may be sought:

- Symptoms
- Institutions and infrastructures
- Values

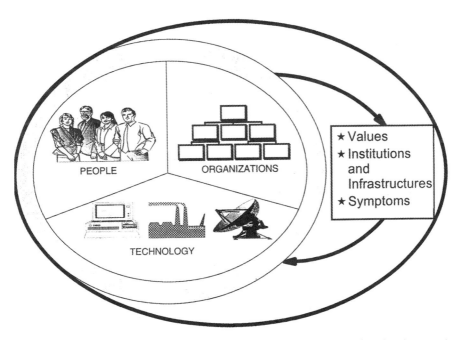

Figure 1.15 Systems of engineering as people, organizations, and technology and levels for issue resolution.

We strongly believe that technology is necessary to ameliorate the problems just delineated. However, there is much evidence that successful application of technology to real-world problem areas must consider the levels of symptoms, institutions and infrastructures, and values or we will generally be confronted with technological solutions that are looking for problems [66]. Too often, problems are approached only at the level of symptoms: bad housing, inadequate health-care delivery to the poor, pollution, hunger, and so on. "Technological fixes" are developed, and the resulting hardware creates the illusion that solution of real-world problems requires merely the outpouring of huge quantities of funds. Attacking problems at the level of institutions and infrastructures, such as to better enable humans and organizations, would allow the adaptation of existing organizations, as well as the design of new organizations and institutions, to make full and effective use of new technologies and technological innovations and to be fully responsive to potentially reengineered objectives and values. Thus, human and organizational issues are all important in seeking technological solutions. We will provide additional discussion concerning these important aspects of systems engineering in Chapter 6.

We encourage you to return to these discussions, after you have finished Chapters 2 through 5, and perhaps at the same time that you are reading the continuation material in Chapter 6.

PROBLEMS

1.1   Recall a contemporary societal or technological problem that you are reasonably familiar with. Analyze this problem in terms of the framework for systems engineering presented in this chapter. What steps do you believe would be necessary to resolve the problem. What relevant elements, associated with your selected problem, immediately suggest themselves? What alternatives exist? What are the critical decisions associated with your problem? What disciplines appear relevant to solution of the problem? What methods and tools do you think might be appropriate for resolution of this problem.

1.2   Why would a phased life cycle be needed for resolution of he problem you have defined in Problem 1.1? Why would systems management be needed? What phases would you recommend for the problem you have identified? What specific systems management roles would exist?.

1.3   From a recent newspaper or magazine, select what appears to be a well-written article of moderate length concerning a controversial societal or technological issue. "Dissect" the article to the point that the issue is formulated according to the systems engineering framework presented in this chapter. Outline the various problem elements. Does your formulation aid in a better understanding of the controversial issue? How much of the controversy arises from or at the level of symptoms? Institutions? Values? What are the technological perspectives, needs perspectives, and value perspectives associated with the controversy and any proposed solution to it?

1.4 Consider a current policy question being faced by a contemporary political body. Describe the policy question in terms of the framework for systems engineering presented here. Pay particular attention to the three principal steps of systems engineering discussed here. What needs, constraints, and alterables can you identify as being part of the issue formulation? What are the appropriate objectives and objective measures that continue the issue formulation step? What are the appropriate activity and activity measures that complete issue formulation? How do these elements fit together such that decisions and policies might be determined?

1.5 Consider a large systems engineering effort which is of interest to you. Describe it in terms of definition, development and deployment efforts. How are the three stages of formulation, analysis, and interpretation conducted at each of these three phases?

1.6 Review the pitfalls in systems engineering described on pages 20 and 21. Discuss the extent to which these could have been or were difficulties for the issue you described in the previous problem. How could these difficulties have been avoided?

1.7 The word system is used in a great variety of contexts. These include systems of equations, control systems, communication systems, banking systems, command and control systems, and systems engineering. Please provide a discussion of the meaning of systems in each of these contexts. How can it be that each of these are called systems?

1.8 The word "system" is used in the previous problem to mean "product system" or "service system." The word "system" can also be used to refer to processes. Describe some processes that are often called systems. What about integrated processes and products? Are these not also appropriately called systems?

1.9 Please write a brief case study of situations in which ethics situations have arisen, and how they were resolved. Please describe one situation that was resolved positively using ethical considerations, and another in which ethical codes were violated.

1.10 The U.S. transportation network is a large-scale, complex system. The major elements of the system are the roads and highways, railroads, airways, and shipping routes. Each component of the system has various strengths and weaknesses. Air traffic, for example, is fast but is expensive and is limited with regard to how much weight can be carried. Ships are slow but are relatively inexpensive and are virtually unlimited as to cargo. One aspect of the system which has significant weaknesses is passenger travel for distances between 200 and 400 miles. Travel by auto is convenient and economical for distances below 200 miles. Travel by air is reasonably convenient and priced for distances over 400 miles. You are employed as a systems engineer working on a design to address the problem of transporting passengers in the 200- to 400-mile range. You are now at the beginning of the design process.

a. Why is the transportation system described in the scenario a system?.

b. Why is this a good systems problem? High-speed railways or a short-range plane service seem to be problems for a civil engineer or a mechanical engineer. What can a systems engineer offer to this problem that the others traditionally do not?

c. Identify some activities that may occur throughout the life cycle of the transportation system described in the scenario. Match each activity with the basic three phases of the systems engineering life cycle in which the activity occurs: design, production, distribution, operation, retirement.

1.11 Reexamine your solution to Problem 1.10 from the standpoint of symptoms, institutions, and values. Indicate how solutions at a symptomatic level may well not be at all appropriate in relieving transportation issues at the level of institutions and values.

REFERENCES

[1]   Mil-STD-499A and MIL-STD 499B, US Department of Defense Systems Engineering Standards, May 1974 and May 1991.

[2]   Sage, A. P., *Systems Engineering*, Wiley, New York, 1992.

[3]   Sage, A. P., *Systems Management for Information Technology and Software Engineering*, Wiley, New York, 1995.

[4]   Sage, A. P., and Rouse, W. B. (Eds.), *Handbook of Systems Engineering and Management*, Wiley, New York, 1999.

[5]   Beniger, J. R., *The Control Revolution: Technological and Economic Origins of the Information Society*, Harvard University Press, Cambridge, MA, 1986.

[6]   Hughes, T. P., *American Genesis: A Century of Invention and Technological Enthusiasm*, Viking Press, New York, 1989.

[7]   Sage, A. P. (Ed.), *Systems Design for Human Interaction*, IEEE Press, New York, 1987.

[8]   Eisner, H., *Computer Aided Systems Engineering*, Prentice-Hall, Englewood Cliffs-NJ, 1987.

[9]   Sage, A. P., Knowledge Transfer: An Innovative Role for Information Engineer-

ing Education, *IEEE Transactions on Systems, Man and Cybernetics*, Vol. 17, No. 5, Sept. 1987, pp. 725–728.

[10] Newman, J. H. Cardinal, *The Idea of a University*, Christian Classics Inc., Westminster, MD, 1973, pp. 177–178.

[11] Goldberg, S., *The New Liberal Arts Program: A 1990 Report,* Alfred P. Sloan Foundation, New York, 1990 (report available from NLA Center, State University of New York, Stony Brook, NY 11794–2250).

[12] Sage, A. P. Electrical Engineering Education, in *Encyclopedia of Electrical and Electronics Engineering*, Webster, J.G. (Ed.), Wiley, New York, 1999.

[13] Sage, A. P., Systems Engineering Education, *IEEE Transactions on Systems, Man, and Cybernetics*, Vol. 29C, No. 1, February 2000. (Also see this special issue for a number of papers on systems engineering education.)

[14] International Council on Systems Engineering, SE Education and Research: A Global View, *INSIGHT*, Vol. 2, No. 1, Spring 1999.

[15] *Engineering Criteria 2000*, Accreditation Board for Engineering and Technology, January 1998.

[16] *ABET 1990 Annual Report*, ABET, New York, 1991.

[17] Christiansen, D., New Curricula, *IEEE Spectrum*, Vol. 29, No. 7, July 1992, p. 29.

[18] Sage, A. P., Systems Engineering and Information Technology: Catalysts for Total Quality in Industry and Education, *IEEE Transactions on Systems, Man, and Cybernetics*, Vol. 23, No. 5, September 1992, pp. 833–864.

[19] Engineering Curriculum Task Force, *Engineering Education: Preparing for the Next Decade*, Arizona State University College of Engineering and Applied Sciences, Tempe, AZ, December 1991.

[20] Kerr, S., Von Glinow, M. A., and Schriesheim, J., Issues in the Study of Professionals in Organizations: The Case of Scientists and Engineers, *Organizational Behavior and Human Performance*, Vol. 18, 1977, pp. 329–345.

[21] National Society for Professional Engineers, *Code of Ethics*, NSPE Publication 2306, NSPE, Washington, DC, September 1991.

[22] Hall, A. D., *A Methodology for Systems Engineering*, Van Nostrand, 1962.

[23] Hall, A. D., A Three Dimensional Morphology of Systems Engineering, *IEEE Transactions on System Science and Cybernetics*, Vol. 5, No. 2, April 1969, pp. 156–160.

[24] Hall, A. D., *Metasystems Methodology: A New Synthesis and Unification*, Pergamon Press, Elmsford, NY, 1989.

[25] Good, H., and Machol, R. E., *Systems Engineering*, McGraw-Hill, New York, 1957.

[26] Chestnut, H., *Systems Engineering Tools*, Wiley, New York, 1965.

[27] Chestnut, H., *Systems Engineering Methods*, Wiley, New York, 1967.

[28] Wymore, A. W., *A Mathematical Theory of Systems Engineering*, Wiley, New York, 1967.

[29] Wymore, A. W., *Systems Engineering Methodology for Interdisciplinary Teams*, Wiley, New York, 1976.

[30] Warfield, J. N., *Societal Systems: Planning, Policy, and Complexity*, Wiley, New York, 1976.

[31]   Churchman, C. W., *The Systems Approach*, Dell Publishing Co., New York, 1979.

[32]   Aslaksen, E., and Belcher, R., *Systems Engineering*, Prentice-Hall, Englewood Cliffs, NJ, 1992.

[33]   Chapman, W., Bahill, A. T., and Wymore, A. W., *Engineering Modeling and Design*, CRC Press, Boca Raton, FL, 1992.

[34]   Blanchard, B. S., and Fabrycky, W. J., *Systems Engineering and Analysis*, 3rd ed., Prentice-Hall, Englewood Cliffs, NJ, 1998.

[35]   de Neufville, R., *Applied Systems Analysis; Engineering, Planning, and Technology Management*, McGraw-Hill, New York, 1990.

[36]   Miser, H. J., and Quade, E. S. (Eds.), *Handbook of Systems Analysis: Overview of Uses, Procedures, Applications, and Practices*, Elsevier Science Publishers, New York, 1985.

[37]   Miser, H. J., and Quade, E. S. (Eds.), *Handbook of Systems Analysis: Craft Issues and Procedural Choices*, Wiley, New York, 1988.

[38]   Majone, G., and Quade, E. S., *Pitfalls of Analysis*, Wiley, New York, 1980.

[39]   Sage, A. P., *Methodology for Large Scale Systems*, McGraw-Hill, New York, 1977.

[40]   Sage, A. P., and Palmer, J. D., *Software Systems Engineering*, Wiley, New York, 1990.

[41]   Sage, A. P., *Decision Support Systems Engineering*, Wiley, New York, 1991.

[42]   Checkland, P. B., *Systems Thinking, Systems Practice*, Wiley, Chichester, 1981.

[43]   Wilson, B., *Systems: Concepts, Methodologies, and Applications*, Wiley, Chichester, 1984.

[44]   Keys, P., *Operational Research and Systems: The Systemic Nature of Operational Research*, Plenum Press, New York, 1991.

[45]   Jackson, M. C., *Systems Methodology for the Management Sciences*, Plenum Press, New York, 1991.

[46]   Nadler, G., *The Planning and Design Approach*, Wiley, New York, 1981.

[47]   Beam, W. R., *Systems Engineering: Architecture and Design*, McGraw-Hill, New York, 1990.

[48]   Chorafas, D. N., *Systems Architecture and Systems Design*, McGraw-Hill, New York, 1989.

[49]   Rechtin, E. R., *Systems Architecting*, Prentice-Hall, Englewood Cliffs, NJ, 1991.

[50]   Rechtin, E. R., and Maier, M. W., *The Art of Systems Architecting*, CRC Press, Boca Raton, FL, 1997.

[51]   Blanchard, B. S., *Systems Engineering Management*, Wiley, New York, 1998.

[52]   Michaels, J. V., and Wood, W. P., *Design to Cost*, Wiley, New York, 1989.

[53]   Rouse, W. B., *Design for Success: A Human Centered Approach to Designing Successful Products and Systems*, Wiley, New York, 1991.

[54]   Martin, J. N., *Systems Engineering Guidebook: A Process for Developing Systems and Products*, CRC Press, Boca Raton, FL, 1997.

[55]   Stevens, R., Brock, P., Jackson, K., and Arnold, S., *Systems Engineering: Coping with Complexity*, Prentice-Hall Europe, Trowbridge, Wiltshire, UK, 1998.

[56] Hazelrigg, G. A., *Systems Engineering: An Approach to Information-Based Design*, Prentice-Hall, Upper Saddle River, NJ, 1996.

[57] Hawryszkiewycz, I. T., *Introduction to Systems Analysis and Design*, Prentice-Hall, Upper Saddle River, NJ, 1988.

[58] Yourdan, E., *Managing the Systems Life Cycle*, Yourdan Press, New York, 1988.

[59] Wallace, R. H., Stockenberg, J. E., and Charette, R. N., *A Unified Methodology for Developing Systems*, McGraw-Hill, New York, 1987.

[60] Rouse, W. B., and Boff, K. R. (Eds.), *System Design: Behavioral Perspectives on Designers, Tools, and Organizations*, North-Holland, New York, 1987.

[61] Newsome, S. L., Spillers, W. R., and Finger, S. (Eds.), *Design Theory '88*, Springer-Verlag, New York, 1989.

[62] Bainbridge, L., and Quintanilla, S. A. R. (Eds.), *Developing Skills with Information Technology*, Wiley, Chichester, UK, 1989.

[63] Rasmussen, J., Brehmer, B., and Leplat, J. (Eds.), *Distributed Decision Making: Cognitive Models for Cooperative Work*, Wiley, Chichester, UK, 1991.

[64] Rouse, W. B., *Strategies for Innovation: Creating Successful Products, Systems, and Organizations*, Wiley, New York, 1992.

[65] Rouse, W. B., *Catalysts for Change: Concepts and Principles for Enabling Innovation*, Wiley, New York, 1992.

[66] Chen, K., Ghausi, M., and Sage, A. P., Social Systems Engineering: An Introduction, *Proceedings of the IEEE*, Vol. 63, No. 3, March 1975, pp. 340–343.

CHAPTER 2

Methodological Frameworks and Systems Engineering Processes

There are a number of frameworks that we might use to characterize systems engineering efforts. We will describe some of them in this chapter. Without a sound and well-understood product line, or process, for the acquisition or production of large systems, it is very likely that there will be a number of flaws in the resulting system itself. Thus, the definition, development, and deployment of an appropriate process, or set of processes, for systems production is very important. Much more is said about this subject in books that deal primarily with systems management and systems engineering processes. [1–3]. To undertake a study of systems engineering methods only and their potential use to support the production of trustworthy systems, without some understanding of systems engineering processes, is likely to lead to very unsatisfactory results. We attempt such an introduction here. In this chapter we will focus our attention on the acquisition, production, or procurement life cycle. We will not discuss the research, development, test, and evaluation (RDT&E) life cycle or the systems planning and marketing life cycle. These are discussed in references 2 and 3 and the references contained therein.

2.1 INTRODUCTION

In this chapter we first discuss the nature of systems acquisition, production, procurement, or manufacturing. We will generally use the word *acquisition* to describe each of these efforts. As systems engineers, we are primarily concerned with the functional and physical architectures that lead to implementation architectures and the resulting detailed design of systems of all types. Design and development is very important, even though the acquisition or production process also depends upon success in the definition and deployment phases of the life cycle. Design and development is a creative process through which system products, presumed to be responsive to client needs and requirements, are conceptualized or specified, manufactured or otherwise produced, and fielded or implemented in such a way that they can be maintained over time. It is clear that this definition of design and development makes it inseparable from the overall life-cycle process of systems engineering, because even though specific design and development activities may not include such early systems engineering efforts as requirements identification or such late efforts as maintenance, an appropriate system development effort must be responsive to and supportive of these other phases in the systems engineering life cycle. There are four primary ingredients in our definition of design and development, and they apply to software design as well as to the design of hardware and physical systems.

1. Development results from specifications or architecture for a product or system.
2. Development is a creative process.
3. Development activity includes design effort, which is conceptual in nature.
4. A successful design and development must be broadly responsive to client needs and requirements.

Good systems engineering practice requires that the systems engineer be responsive to each of these four ingredients for quality design and development efforts. The final ingredient requires of the client a set of needs and requirements for the desired product, process, or system. This information requirement serves as the input to the systems engineering process that leads to design. Systems engineering is creative, and it is a process that is conceptual and pragmatic in nature. The initial result of this creative, conceptual, and pragmatic process is information concerning the specifications or architecture for the product or service that will ultimately be manufactured, implemented, installed, or brought to fruition in some other way. A later result is the system itself and plans for its evolution over time.

   In the detailed efforts that follow, we first discuss systems engineering methodology and systems engineering life cycles as necessary elements in the systems engineering process. This leads to a number of conceptual frameworks for systems engineering processes. Systems engineers provide a needed interface between (a) the client or stakeholder group, or enterprise, to whom an operational system will ultimately be delivered and (b) a detailed design and

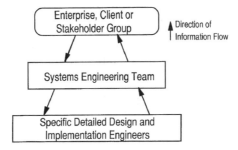

Figure 2.1 The systems engineering team as brokers of information.

implementation group who are responsible for specific system production and implementation. Figure 2.1 illustrates this view of a systems engineering team as an interface group that provides conceptual design and technical direction to enable the products of an internally focused detailed design group to be responsive to client needs. Thus, systems engineers act, in part, as brokers of information between a client group and those responsible for detailed design and system production. There are, of course, other views of systems engineers and systems engineering functions. We will elaborate on many of these in this chapter, and throughout the text as well.

In this chapter, we will discuss the systems acquisition, or production, process in terms of a number of steps and phases that may comprise this process. We will provide a very brief discussion of some modification of the life cycles that will allow for concurrent and evolutionary development. In Chapter 6, we will discuss some systems management factors in systems acquisition or production. This leads naturally into a discussion of a multiple perspective view from which we can visualize systems engineering efforts. As we will then see, it is very important that we address systems engineering issues at the level of values and institutions, as well as at the level of symptoms.

2.2 METHODOLOGICAL FRAMEWORKS FOR SYSTEMS ACQUISITION OR PRODUCTION

In this section we present and explain the complete systems engineering process with emphasis on frameworks for systems methodology and design. The framework consists of three dimensions:

- A *logic* dimension that consists of three fundamental steps
- A *time* dimension that consists of three basic life cycle phases
- A *perspectives* dimension that consists of three stages or life cycles

We discussed this very briefly in Chapter 1. Now, we add some additional depth and breadth to our discussion.

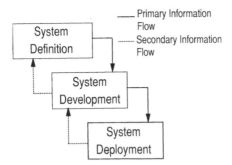

Figure 2.2 Conceptual illustration of the three primary systems engineering life-cycle phases.

As in Chapter 1, we envision a three-performance level hierarchy for systems engineering phased efforts, such as shown in Figure 2.2. This three-level structured hierarchy comprises a systems engineering life cycle and is one of the ingredients of systems engineering methodology. It involves

- System *definition*
- System *development*
- System *deployment*

In Figure 2.2, we illustrate forward flow of information, and feedback from one phase to the other. Usually, this is needed as a situation in which information only flows in one direction from one phase to a subsequent phase will not allow for iterative improvement. It is generally also very desirable to allow for learning as the life-cycle process is iteratively repeated. Figure 2.3 illustrates

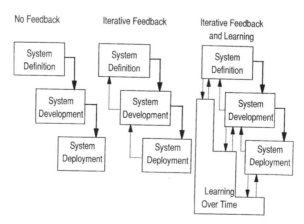

Figure 2.3 Systems engineering life cycles with different patterns of iteration and learning.

three different information flow patterns. Generally, the one with learning and feedback is to be preferred. To expand on this in detail would take us a bit away from our central purpose here. A much more complete discussion is available in references 1–3. We will, however, discuss some extensions of the basic "waterfall" life-cycle model, illustrated on the left side of Figure 2.3, in the concluding portions of this chapter. As we shall soon discuss, this three-phase life cycle may be expanded into a life cycle comprised of a larger number of phases. Generally the number of phases is in the vicinity of seven, although it may contain quite a few more than seven phases.

The structural definition of systems engineering we posed in Chapter 1 tells us that we are concerned with a framework for problem resolution that, from a formal perspective at least, consists of three fundamental steps for a systems engineering activity:

- Issue *formulation*
- Issue *analysis*
- Issue *interpretation*

These are each conducted at each of the life-cycle phases that have been chosen for the definition, development, and deployment efforts that lead to the engineering of a system. Regardless of the way in which the systems engineering life-cycle process is characterized, and regardless of the type of product or system or service that is being designed, all characterizations of the phases of the systems engineering life cycles will necessarily involve:

1. *Formulation of the Problem* — in which the needs and objectives of a client group are identified, and potentially acceptable design alternatives, or options, are identified or generated.
2. *Analysis of the Alternatives* — in which the impacts of the identified design options are identified and evaluated.
3. *Interpretation and Selection* — in which the options, or alternative courses of action, are compared by means of an evaluation of the impacts of the alternatives and how these are valued by the client group. The needs and objectives of the client group are necessarily used as a basis for evaluation. The most acceptable alternative is selected for implementation or further study in a subsequent phase of systems engineering.

Our model of the steps of the logic structure of the systems process, shown in Figure 2.4, is based upon this conceptualization. As we shall also indicate in much more detail later, these three steps can be disaggregated into a number of others. Each of these steps of systems engineering is accomplished for each of the life cycle phases.

As we have noted, there are generally three different systems engineering life cycles. These relate to the three different stages of effort that are needed to result in a competitive product or service in the marketplace:

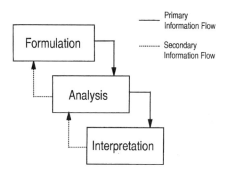

Figure 2.4 Conceptual illustration of formulation, analysis, and interpretation as the primary systems engineering steps.

- Research, development, test, and evaluation (RDT&E)
- System acquisition or production
- Systems planning and marketing

Thus we may imagine a three-dimensional model of systems engineering that is comprised of steps associated with each phase of a life cycle, the phases in the life cycle, and the life cycles that comprise the coarse structure or stages of systems engineering. Figure 2.5 illustrates this across three distinct but interrelated life cycles, for the three steps and three phases that we have described

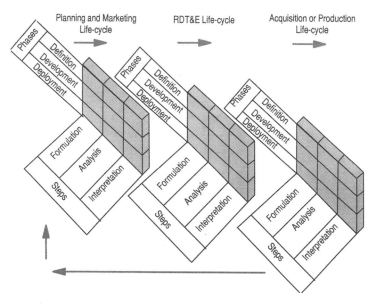

Figure 2.5 Three systems engineering life cycles and phases and steps within each life cycle.

here. This is one morphological framework for systems engineering. As we have noted, it will generally be necessary to expand the three steps and three phases we indicate here into a larger number of steps and phases. Often, also, there will be a number of concurrent RDT&E and systems acquisition efforts that will be needed in order to ultimately bring about a large-scale system.

It is necessary that efforts across the three life cycles of

- systems planning and marketing;
- research, development, test, and evaluation; and
- systems acquisition, production, procurement, and manufacturing

be well-integrated and coordinated or difficulties can ensue. The systems planning and marketing life cycle is intended to yield answers to the question; What is in demand? The research, development, test, and evaluation life cycle is intended to yield answers to the question; What is (technologically) possible (within reasonable economic and other considerations)? The acquisition life cycle is intended to yield answers to the question; What can be developed? It is only in the region where there is overlap, actually in an *n*-dimensional space, that responsible actions should be implemented to bring about programs for all three life cycles. This suggests that the needs of one life cycle should not be considered independently of the other two. Figure 2.6 attempts to represent this conceptually for two-dimensional representations of the possibility space for each life cycle. Only within the ellipse represented by the thicker exterior lines should effort be undertaken.

Each of the logical steps of systems engineering is accomplished for each of the life-cycle phases. There are generally three different systems engineering life

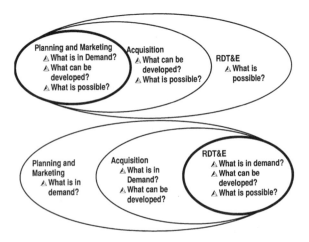

Figure 2.6   Illustrations of the need for coordination and integration across life cycles.

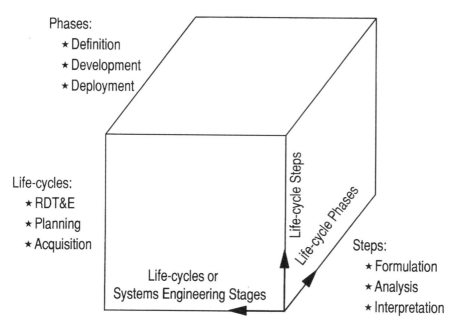

Figure 2.7   Three-dimensional framework for systems engineering.

cycles or stages for a complete systems engineering effort, as we have indicated. Thus we may imagine a three-dimensional model of systems engineering that is comprised of steps associated with each phase of a life cycle, the phases in the life cycle, and the life cycles or stages of a complete systems engineering effort. Figure 2.7 illustrates this framework of steps, phases, and stages as a three-dimensional cube. This is one three-dimensional framework, in the form of a morphological box, for systems engineering.

The words morphology and methodology may be unfamiliar to you. The word *morphology* is adapted from biology and means a study of form. As we use it, a *methodology* is an open set of procedures for problem solving. Consequently, a methodology involves a set of methods, a set of activities, and a set of relations between the methods and the activities. To use a methodology we must have an appropriate set of methods. Generally, these include a variety of qualitative and quantitative approaches from a number of disciplines that enable formulation, analysis, and interpretation of the phased efforts that are associated with the definition, development, and deployment of both an appropriate process and the product that results from use of this process. Associated with a methodology is a structured framework into which particular methods are associated for resolution of a specific issue.

Systems engineering is comprised of much more than just a methodological framework, or frameworks, of course. In an earlier three-level view of systems engineering, we indicated that we can consider systems engineering efforts at the levels of

- systems engineering methods and tools, and associated metrics,
- systems methodology, or life-cycle processes, and
- systems management,

and we suggested Figures 1.2 and 1.3 as illustrative of this representation of systems engineering. These is also an important dimension to a systems engineering framework, as is the situation assessment that occurs for issue recognition, and the individual and organizational learning that should occur as systems engineering efforts evolve over time. An adaptation of Figure 1.3 that would replace the process box with the three-dimensional framework of Figure 2.7 would show some of these enhanced features.

Let us now develop the structured framework of steps, phases, and stages in systems engineering in more detail.

2.2.1   Logical Steps of Systems Engineering

As we have noted, all characterizations of systems engineering will necessarily involve three logical steps [1–4]:

- Formulation of the systems engineering problem under consideration
- Analysis to determine the impacts of the alternatives
- Interpretation of these impacts in accordance with the value system of the decision maker(s), and selection of an appropriate plan of action to continue the effort.

These three steps, or an expansion thereto to more explicitly indicate the actual activities associated with each phase, are conducted at each and every phase of the systems engineering life cycle.

We can expand the three fundamental steps of systems engineering in many ways. However, probably the most useful expansion is the seven steps identified by Hall [5]. Our construal of these seven steps of systems engineering and their relation to the three basic steps of formulation, analysis, and interpretation follows.

Formulation

1. *Problem Definition.* This step involves isolating, quantifying, and clarifying the need that creates the problem and describing the set of environmental factors that constrains alterables for the system to be developed. It involves identifying a set of needs, alterables, and constraints associated with the issue formulation.
2. *Value System Design.* This step involves selection of the set of objectives or goals that guides the search for alternatives. Value system design

enables determination of the multidimensional attributes or decision criteria for selecting the most appropriate system. It involves the identification and validation of a set of objectives and objectives measures.

3. *System Synthesis.* This step involves searching for, or hypothesizing, a set of alternative courses of action or options. Each alternative must be described in sufficient detail to permit analysis of the impacts of implementation and subsequent evaluation and interpretation with respect to the objectives. As part of this step, we identify a number of potential alternatives and associated alternatives measures.

Analysis

4. *System Analysis and Modeling.* As a part of this step, we determine specific impacts or consequences of the alternatives that were specified as relevant to the issue under consideration by the value system. These impacts may relate to such important concerns as product quality, market, reliability, cost, and effectiveness or benefits. There are a variety of simulation and modeling methods, and a great variety of operations research approaches that are of potential value here.

5. *Refinement of the Alternatives.* As part of this step, we attempt to adjust, and hopefully optimize, the system variables in order to best meet system objectives and satisfy system constraints.

Interpretation

6. *Decision Making.* This step involves evaluating the impacts or consequences of the alternatives and, thereby, interpreting the alternatives in terms of the extent to which they achieve objectives. The alternatives are identified in system synthesis and subsequently modeled and refined in the analysis steps of effort, in accordance with satisfaction of objectives associated with the value system. This enables interpretation of these evaluations such that all alternatives can be compared relative to human values. One or more alternatives, or courses of action, can then be selected for advancing to the next step.

7. *Planning for Action.* This step involves communicating the results of the effort to this point and looking ahead to the next phase of the systems engineering life cycle under consideration. This includes such pragmatic efforts as scheduling subsequent efforts, allocating resources to accomplish them, and setting up system management controls. If we are conducting a single phase effort, this step would be the final one. More generally, it leads to a new phase of effort in continuation of a systems engineering life cycle.

Figure 2.8 indicates the waterfall nature of these seven steps of systems engineering and their relation to the three fundamental logical steps. The seven

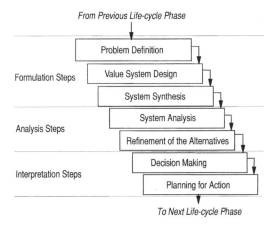

Figure 2.8   Seven steps of the logic dimension of systems engineering.

steps of systems engineering are generally carried out in an iterative fashion for each phase of the systems engineering life cycle. Because the efforts and purposes of the various phases of the life cycle are quite different, the specific tools and methods that are appropriate to accomplish each of the seven steps can be expected to be quite different from one systems engineering process phase to another. It is quite possible to go back to refine and improve the results of any earlier step as a consequence of the results of any later step. We do not show these iterative feedback loops in Figure 2.8. As you will see, and as we have also indicated, it is necessary to conduct each of the seven steps in every phase of the systems engineering life cycle.

2.2.2   Life-Cycle Phases of Systems Engineering

The primary goal of systems engineering is the creation of a set of high-quality and trustworthy operational products and services that will enable the accomplishment of desired tasks that fulfill identified needs of a client group, or user group, or enterprise. Systems engineering is a creative process through which products, services, or systems presumed to be responsive to client needs and requirements are conceptualized or specified or defined, and ultimately developed and deployed. A systems engineering life cycle prescribes a number of phases that should be followed, often in an interactive and iterative manner in order to successfully produce and field a large-scale system that meets user requirements.

There have been a number of process models proposed and used for multiphase life-cycle models in systems engineering. Each of them starts, or should start, by capturing, identifying, or defining user requirements. These user requirements are then converted to technological system requirements and systems management requirements that define the system and will presumably, when satisfied, produce the product or service. These are often called specifications. Following the definition phase, there is a conceptual, or architectural or architecting, design phase and then a detailed design phase, the result of which is an initial working version of a system. This is evaluated and modified to enable ultimate operational deployment of a functionally useful system, a system that fulfills user requirements. Operation includes maintenance and modification, and potentially such other efforts as reengineering or replacing the system as it becomes less useful over time. Together, these phased efforts describe an extended systems life cycle. Hence, a typical systems engineering life cycle has three basic phases — *definition, development*, and *deployment* — as we have often indicated.

For large systems, these three phases need expansion into a number of more finely defined phases. This enables the various phases to be better understood, communicated, and controlled in order to support trustworthy systems engineering efforts. Before describing specific life-cycle methodologies later in this chapter, we present and explain a general and very useful seven-phase life-cycle expansion and its relation to the basic three-phase systems engineering life cycle that we presented in Figures 2.2 and 2.3. Figure 2.9 illustrates the seven-phase life cycle, actually representing acquisition or production, that we now consider.

The seven phases in this life cycle may be described as follows:

System Definition

1. *Requirements and Specifications.* The first part of a systems engineering effort results in the identification of user requirements and the translation of these into technological specifications for a product, process, or system.

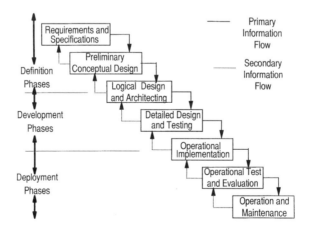

Figure 2.9 One of several possible life-cycle models phase models for systems engineering (acquisition).

The goal of this phase is the identification of client and stakeholder needs, activities, and objectives for the functionally operational system. This means that information is a necessary ingredient and results in the mandate to obtain, from the client for a systems engineering effort, a set of needs and requirements for the product, process, or system that is to result from the effort. This information requirement serves as the input to the rest of the systems engineering process. This phase results in the identification and description of preliminary conceptual design considerations for the next phase. It is necessary to translate operational deployment needs into requirements specifications so that these needs may be addressed by the system design and development efforts. Thus, information requirements specifications are affected by, as well as affect, each of the other design and development phases of the systems engineering life cycle.

2. *Preliminary Conceptual Design and System Architecting.* The primary goal of this phase is to develop several concepts that might work that are responsive to the specifications identified in the previous phase of the life cycle. A preliminary conceptual design, one that is responsive to user requirements for the system and associated technical specifications, should be obtained. Rapid prototyping of the conceptual design is clearly desirable for many applications as one way of achieving an appropriate conceptual design. Several potential options are identified and then subjected to at least a preliminary evaluation in order to eliminate clearly unacceptable alternatives. The surviving alternatives are next subjected to more detailed design efforts, and more complete architectures or specifications are obtained. It is at this phase that the enterprise architecture, and perhaps a preliminary functional architecture, for the system are initially identified. Functional analysis approaches [3, 6] are particularly useful in this phase of effort.

System Design and Development

3. *Logical Design and Product Architectural Specifications.* This phase results in an effort to specify the content and to provide more detail to the associated high level architectures for the system product in question. Specifications are translated into detailed representations in logical form such that system development may proceed. This logical design product is sometimes called a physical architecture, and the product architectural specifications are realized in terms of this physical architecture, sometimes called engineering architecture, of the system that will ultimately be implemented.

4. *Detailed Design, Production, and Testing.* The goal of this phase is a set of detailed design specifications that should result in a useful system product. There should exist a high degree of user confidence that a useful product will result from detailed design, or the entire design effort should

be redone or abandoned. Another product of this phase is a refined set of specifications for the operational deployment and evaluation phases of the life cycle. Again, design alternatives are evaluated, and a final choice is made, which can be developed with detailed design testing, and at least preliminary operational implementation. This results in the implementation architecture for the system. Utilization of this implementation (or detailed design) architecture results in the actual system. Preparations for actual production and manufacturing are made in this phase. The system is produced here, often in an outsourced manner, by an implementation contractor.

System Operation and Maintenance

5. *Operational Implementation.* A product, process, or system is implemented or fielded for operational evaluation. Preliminary evaluation criteria are obtained and modified during the following two phases.

6. *Operational Test and Evaluation* (and Associated Modification). Once implementation has occurred, operational evaluation and test of the system can occur. The system design may be modified as a result of this evaluation, leading (hopefully) to an improved system and, ultimately, operational deployment. Generally, the critical issues for evaluation are adaptations of the elements present in the requirements specifications phase of the systems engineering life-cycle process. A set of specific evaluation test requirements and tests are evolved from the objectives and needs determined in the requirements specifications. These should be such that each objective and critical evaluation component can be measured by at least one evaluation test instrument. If it is determined, perhaps through an operational test and evaluation, that the resulting systems product cannot meet user needs, the life-cycle process reverts iteratively to an earlier phase, and the effort continues. An important byproduct of system evaluation is determination of ultimate performance limitations for an operationally realizable system. Often, operational evaluation is the only realistic way to establish meaningful information concerning functional effectiveness of the result of a systems engineering effort. Successful evaluation is dependent upon having predetermined explicit evaluation standards.

7. *Operational Functioning and Maintenance.* The last phase includes final acceptance and operational deployment. Maintenance and retrofit can be defined either as part of this phase or as additional phases in the life cycle. Either is an acceptable way to define the system life cycle for system acquisition or production.

We may summarize these phased activities as follows. The requirements and specification phase of this systems engineering life cycle has as its goal the identification of client or stakeholder needs, activities, and objectives for the functionally operational system. This results in identification and description of preliminary conceptual design considerations for the next phase. It is necessary to translate operational deployment needs into requirements specifications so that these needs may be addressed by the system design efforts. Thus, user requirements, which are extraordinarily important, and technological system specifications are affected by, as well as affect, each of the other phases of the systems engineering life cycle.

As a result of the requirements and specifications phase, there should exist a clear definition of system design and development issues such that it becomes possible to make a decision concerning whether to undertake preliminary conceptual design. If the requirements specifications effort indicates that client needs can be satisfied in a functionally satisfactorily manner, then documentation is typically prepared concerning specifications for the preliminary conceptual design phase. Initial specifications for the following three phases of effort are typically also prepared, and a concept design team is selected to implement the next phase of the life-cycle effort. This effort is sometimes called functional architecting [7] and sometimes system level architecting [8, 9].

Preliminary conceptual system design typically includes, or results in, an effort to specify the content and associated functional architecture and general algorithms for the system product in question. The primary goal of this phase is generally to develop some sort of functional prototype that is responsive to the specifications previously identified in an earlier phase of the life cycle. An appropriate preliminary conceptual design, is responsive to user requirements for the system and associated technical system specifications. Rapid prototyping of the conceptual design is often useful as one way of achieving an appropriate conceptual design.

The ultimate desired product of this phase of activity is a set of detailed design and architectural specifications that should result in a useful system product. There should exist a high degree of user confidence that a useful product will result from detailed design, or the entire design and development effort should be redone or possibly abandoned. Another product of this phase is a refined set of specifications for the evaluation and operational deployment phases of the life cycle. In the third phase, these are translated into detailed representations in logical form, or physical architecture form, such that system development may occur. A product, process, or system is produced in the fourth phase of the life cycle. This is not the final system design, but rather the result of implementation of the design that resulted from the conceptual design effort of the last phase. User guides for the product should be produced such that realistic operational test and evaluation can be conducted in the next phase.

Evaluation of the detailed design and the resulting product, process, or system is achieved in the sixth phase of this seven-phase systems engineering life cycle. Depending upon the specific application being considered, an entire systems engineering life-cycle process could be called "design," or "manufacturing," or by some other appropriate designator, as we have already noted.

Systems acquisition is an often-used word to describe the entire systems engineering process that results in an operational systems engineering product.

Generally, a system acquisition life cycle (or procurement, or production life cycle) involves primarily knowledge practices, or standard procedures to produce or manufacture a product based on established practices. An RDT&E life cycle is generally associated with an emerging technology and involves knowledge principles. A planning and marketing life cycle is concerned with product planning and other efforts to determine market potential for a product or service, and it generally involves knowledge perspectives. Generally, this life cycle is needed to identify emerging technologies chosen to enter an RDT&E life cycle. It is also needed, particularly in private sector commercial activities, to shape and focus specific configurations for products that are produced or acquired. For overall product trustworthiness and integrity, there needs to be feedback and iteration across these three life cycles. Integration is also needed, such as to ensure that there is symbiosis across the activities at each of these lifecycles, as suggested in Figure 2.6 and the accompanying discussion.

Evaluation of systems engineering efforts are very important. Evaluation criteria are obtained as a part of requirements specifications and are modified during the following two phases of the design effort. The evaluation effort must be adapted to other phases of the systems engineering effort such that it becomes an integral and functional part of the overall process. Generally, the critical issues for evaluation are adaptations of the elements present in the requirements and specifications phase of the process life cycle. A set of specific evaluation test requirements and tests are evolved from the objectives and needs determined in the requirements and specifications phase. These should be such that each objective measure and critical evaluation issue component can be measured by at least one evaluation test instrument.

An important byproduct of system evaluation is determination of ultimate performance limitations for an operationally realizable system. Often, operational evaluation is the only realistic way to establish meaningful information concerning functional effectiveness of the result of a systems engineering effort. Successful evaluation is dependent upon having predetermined explicit evaluation standards and metrics. Reference 3 contains a number of useful discussions concerning system evaluation and the role of measurement and metrics in systems engineering evaluation.

The last phase of the systems life-cycle effort includes final acceptance and operational deployment. Maintenance and retrofit can be defined either as additional phases in life cycles or as part of the operational deployment phase. Either is an acceptable way to define the system life cycle and there are many possible systems engineering life cycles, some of which we will discussed in subsequent chapters.

Later in this chapter, in Section 2.2.4, we show other system life cycles very similar to this one but tailored to specific uses. Before illustrating other phased life cycles and expanding the third dimension of our framework, let's see what the first two dimensions look like.

2.2.3 A Two-Dimensional Framework for Systems Engineering

When we combine the life cycle phases of systems engineering with the steps, we obtain a two dimensional framework for systems engineering. Figure 2.10 illustrates this for the generic three phases and three steps we have considered here. We note that there are nine activity cells in this matrix. Thus, a minimum of nine different activities are needed to complete the life-cycle process illustrated here. We need to associate this methodological framework with systems management and systems methods and tools. Also, we can expand upon the number of steps and phases. Figure 2.11 illustrates this for the three-step and seven-phase model. There are a total of 21 different activity cells in this matrix.

We now combine the seven-step model with the seven-phase model that we have discussed thus far. As Figure 2.12 depicts, this life-cycle model of systems engineering methodology and design consists of seven phases, along with a sequence of steps accomplished within each of the phases. This framework defines a matrix of activities for systems engineering. Each box in Figure 2.12 represents an activity. The phases are the time sequence of activities, and the steps are the logical sequence of activities. This is sometimes called a Hall activity matrix, after its originator, although the particular phases utilized here are not those in the initial Hall effort [4]. There are 49 activities represented in Figure 2.12. Each box requires the systems engineer to select and use appropriate methods and tools to support accomplishment of that particular activity.

Although 49 activities seems like a very large number, the number triples to 147 when we expand our framework by considering the three stages that make up a complete systems engineering effort. For the sake of greater completeness, let us look briefly at these three life cycles.

2.2.4 Life Cycles, or Stages of Systems Engineering

There are three major stages or life cycles in a complete systems engineering effort for technology evolution:

- Research, development, test, and evaluation (RDT&E)
- System acquisition, procurement, or production
- Systems planning and marketing

	Formulation	Analysis	Interpretation
Definition	Activity 1	Activity 2	Activity 3
Development			
Deployment			Activity 9

Figure 2.10 Illustration of nine activity cells for one two-dimensional framework.

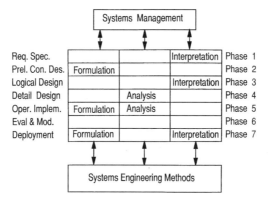

Figure 2.11   The steps, phases, and activity levels in one representative systems engineering framework.

These stages are interrelated, and each is needed to accomplish technological system planning and development and management system planning and development that lead to a new innovative product or service. When we apply the systems engineering framework of life-cycle phases and logical steps to each of the three major stages, we see in Figure 2.13 that we now have as many as 147 activities to concern the systems engineer. This figure represents the two-dimensional framework of Figure 2.12 replicated over the three major systems engineering life cycles. The particular terms we have used to describe the phases of each life cycle are most appropriate for the systems acquisition or production life cycle. We could easily modify these such that they are more appropriate for the RDT&E and planning life cycles. A more complete

Steps / Phases	Problem Definition	Value System Design	System Synthesis	System Analysis	Alternat. Refine.	Decision Making	Planning for Action
Requirements Specifications	*Activity 1*						
Conceptual Design			*Activity 10*				
Log. Design Architecting							
Detail Design and Testing							
Operational Implement.							
Evaluation & Modification							
Operational Functioning							*Activity 49*

Figure 2.12   Seven-phase and seven-step systems framework that defines a matrix of systems engineering activities.

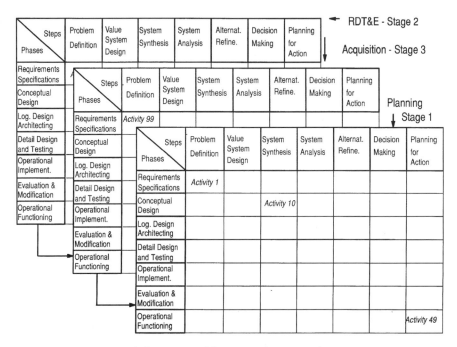

Figure 2.13 Conceptual illustration of three two-dimensional activity matrices as a set of three systems engineering life cycles.

discussion of these other stages, and their life cycles, may be found in references 1–3.

System acquisition, system production, or system procurement are often-used terms to describe the entire systems engineering process that results in an operational systems engineering product. As we have indicated, an acquisition life cycle involves primarily knowledge practices, or standard procedures to produce or manufacture a product based on established practices. An RDT&E life cycle is generally associated with an emerging technology and involves knowledge principles. A marketing life cycle is concerned with product planning and other efforts to determine market potential for a product or service, and it generally involves knowledge perspectives. Generally this life cycle is needed to identify emerging technologies chosen to enter an RDT&E life cycle. It is also needed, particularly in private sector commercial activities, to shape and focus specific configurations for products that are produced or acquired. As you might expect, there needs to be feedback and iteration across these three life cycles. One of the major reasons for this is that a product, or service, will generally fail if it cannot successfully pass three gateways associated with these life cycles. Figure 2.14 presents a three-gateway model that has been used to describe the necessary conceptual relation among these three life cycles [10–12]. This conceptual model involves technology push factors that are

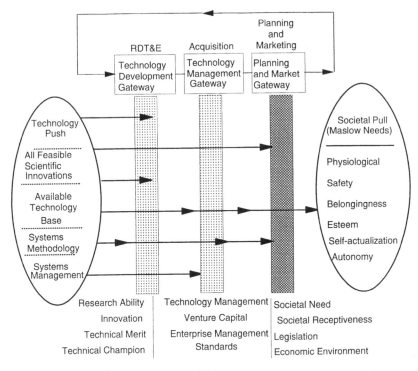

Figure 2.14 A three-gateway model for systems engineering productivity.

often catalysts for RDT&E efforts. It includes societal or market pull factors. These are based on a hierarchy of human needs identified by the psychologist Abraham Maslow [13] a number of years ago. Virtually all studies show that market pull is a much more dominant influencer of commercial technology success than is technology push.

2.2.5 Systems Engineering Processes

Without question, the three-dimensional framework that we present here is a formal rational model of the way in which systems engineers accomplish the three fundamental steps or functions of systems engineering: formulation, analysis, and interpretation. Even within this formal framework, there is the need for much iteration from one step back to an earlier step when it is discovered that improvements in the results of an earlier step are needed in order to obtain a quality result at a later step, or phase, of the systems engineering effort. Also, this description does not emphasize the key role of information and information requirements determination and the embedding of this information within context to result in knowledge. This is concentrated in the formulation step but exists throughout all steps of a systems engineering process.

Even when the realities of information and iteration are associated with the morphological framework, it still represents an incomplete view of the way that people do, could, or should accomplish planning, design, development, or other problem solving activities. The most that can be argued is that this framework is correct in an "as if" manner. We represent one such framework as a three-dimensional morphological box in Figure 2.7. It consists of a number of life-cycle phases and logical steps and also a third dimension, the stages or life cycles associated with the three major efforts involved in technical direction and systems management to ensure apposite systems engineering.

A number of appropriate methods, tools, and techniques are associated with the three functional components or steps within a systems engineering process: formulation, analysis, and interpretation. Systems engineering efforts are very concerned with technical direction and management of the life-cycle efforts of systems definition, development, and deployment. We refer to this effort as *systems management*. By adopting the management technology of systems engineering and applying it, we become very concerned with making sure that correct systems are engineered, and not just that engineered system designs are correct according to some potentially ill-conceived notions of what the system should do. Appropriate tools to enable efficient and effective error prevention and detection in a systems engineering process will enhance the production of system products that are correct. To ensure that correct systems are designed requires that considerable emphasis be placed on the front end of the systems engineering life cycle.

In particular, there needs to be considerable emphasis on the accurate definition of a system, what it should do, and how people should interact with it before one is produced and implemented. In turn, this requires emphasis upon conformance to system requirements specifications, along with the development of standards to ensure compatibility and integratibility of system products. Such areas as documentation and communication are important in all of this. Thus, we see the need for the technical direction and management technology efforts that comprise systems engineering, as well as the strong role for process-related concerns.

Information is the key word in this model of the systems engineering process. Information is an essential feature in the input to the systems engineering process, the process itself, and the product, service, or system that is the output of the process. The development and implementation of sound systems engineering processes is a fundamental goal of systems engineering. These are the perspectives that we take in this text, specifically as they relate to selection of appropriate problem solving and analysis approaches. This information needs to be associated with context such that relevant information can be interpreted and acted upon as knowledge.

There are many descriptions of systems engineering life cycles and associated methodologies and frameworks for systems acquisition — including the phased efforts of definition, development, and deployment — [1–14] — and we have outlined only two of them in any detail thus far in this chapter. These are the generic three-phase model, represented by Figure 2.4, and the seven-phase model represented by Figure 2.9. In general, a simple conceptual model of the overall process, which incorporates all three life cycles or stages of systems engineering, may be structured as the three-dimensional framework of Figure 2.4. In a methodological framework, a very large number of systems engineering methods may be needed to fill in the activity cells associated with this three-dimensional matrix. Choice of an appropriate mix of methods is an important challenge in systems engineering. This choice will depend on the particular life cycle and on the environmental issues surrounding the system being engineered. We present a number of methods in our next three chapters.

2.3   OTHER SPECIFIC LIFE-CYCLE METHODOLOGIES FOR SYSTEMS ACQUISITION, PRODUCTION, OR PROCUREMENT

In this section, we will describe several life cycles for systems acquisition, production, or procurement. In each case, we are concerned with the description of steps and phases in a typical life cycle for acquisition of a major system. Each of these efforts to follow expands upon our earlier discussions of the three-phase life cycle, specifically for system acquisition.

2.3.1   A Seven-Phase Life Cycle for System Acquisition or Product Development

In this subsection, we assume that it has somehow been decided to develop and deploy a product. We assume that the decision has been made relative to who is to do this, and that the requirements for the product to be developed have been identified. The description of the various phases are easily modified to describe more general situations, such as might be encountered in service development, RDT&E, and marketing. Reference 2 and Chapter 2 of reference 3 provide considerable detail on systems engineering life cycles.

Arthur D. Hall [5, 15] reported, many years ago, a seven-phase life cycle of this sort. A more recent work by Hall also comments further on this life cycle [16]. The seven phases of this systems engineering life cycle, which are highly tuned to systems production or acquisition, are illustrated in Figure 2.15. The phases in this life cycle, along with a brief descriptive account of the activities at each phase, are as follows.

System Definition

a. *Program planning* is a conscious activity that should result in formulation of the activities and projects supportive of the overall system requirements into more detailed levels of planning. It is the first phase in this life cycle. The user requirements are presumed known in this life cycle. The

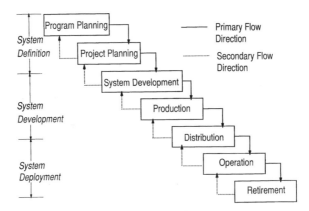

Figure 2.15 The seven phases in the Hall systems engineering life cycle with feedback and interaction between adjacent phases.

program planning phase must, if it is to be successful, also include (i) identification of system level requirements to the extent that they have not been identified from user requirements and (ii) translation of these into the technological system specifications that are planned for later development.

b. *Project planning* is distinguished from program planning by increased interest on the individual specific projects of an overall systems development program. This is the second phase. The purpose of this phase is to configure a number of specific projects, which together comprise the program such that system development can begin.

System Development

c. *System design* results in implementation of the project plans through design of the overall system in detail as a number of subsystems, each of which is described by a project plan. This is the first phase of effort that will ultimately translate the system definition into a product. This phase ends with the preparation of detailed architectures, detailed specifications, drawings, and bills of materials for the system manufacturer or builder.

d. *Production*, for manufactured products, or *construction*, for one-of-a-kind systems, which includes all of the many activities that are needed to give physical reality to the desired system. This could involve, for example, using detailed plans and specifications to construct a new building, manufacture a new product, or produce source code for an emerging software product. A number of related efforts would be needed, such as determining the sequence, materials flow, required shop floor layouts, and the establishment of quality control practices. After completion of this phase of effort, we have a systems engineering product that is capable of being fielded or implemented in an operational setting.

System Deployment

    e. *Distribution*, or *phase-in*, which results in delivering systems engineering products or services to users or consumers. This may involve all kinds of distribution facilities, marketing, and sales organizations.

    f. *Operations*, which is the ultimate goal of system deployment. This phase includes such activities as maintenance

    g. *Retirement*, or *phase-out*, of the system over a period of time and replacement by some new system.

From the perspective of this life cycle, the first phased activity is program planning for production based on a given set of requirements. Clearly, a program plan cannot be established in the absence of knowledge of user requirements for the system. Thus, as we have noted, program planning must necessarily involve information requirements determination and translation of the user requirements for a system into technological specifications. Even though requirements may be specified initially, it is highly desired that these be validated in program planning. In a later chapter, we will discuss some graph-theory-oriented techniques for program planning associated with strategic planning and marketing. These approaches are also useful for information display of planning efforts in other life cycles, as well as for other phases in all three life cycles.

Clearly, we could use somewhat different words to describe these phases of the system acquisition life cycle. For example, the retirement phase includes maintenance, prior to possible system phase-out. The particular representation we have used for Figure 2.15 is called a waterfall life cycle. We have added some feedback to allow interaction among adjacent phases.

2.3.2   A Twenty-Two-Phase Life Cycle

We now describe a life cycle that is comprised of 22 phases. As we will soon see, these phases correspond to the phases in the much-simpler-appearing three-phase model. Here, we identify a set of phases, discussed in reference 17 and also described in reference 1, that might exist when one client or stakeholder group, such as a government agency or private company for example, seeks development by another vendor (to be determined) of a large system. The *system definition* effort, or phase, can be expanded into seven phases. The actors in these phases include system developers, clients for the systems study, and champions or funders of system acquisition efforts.

System Definition

1. *Perception of need* by the customer, or client, or user group.
2. *Requirements definition* of the user group requirements by the user group.
3. *Draft request for proposal* (RFP) by the user group.
4. *Comments on the RFP* by system development industry. The final RFP is typically conditioned by inputs from potential bidders.
5. *Final RFP and Statement of Work* developed by the user group.
6. *Proposal development*, generally in a competitive manner, by the system development industry. Most proposals will include statements regarding

   a. system architecture

   b. selection of approach

   c. work breakdown structure (WBS).

   The work breakdown structure, both in terms of the scheduling of system development effort and in terms of the nature and timing of the work to be done, is a principle vehicle in which the contractor may display knowledge of system development practices, principles, and perspectives. A systems engineering management plan (SEMP) is often included with a contractor proposal. Sometimes, a tentative SEMP is identified by the contracting agency. Associated with the SEMP is a systems engineering master schedule (SEMS) and at least an initial version of a systems engineering detailed schedule (SEDS).

7. *Source selection*, by the customer, client, or user group.

At the end of this phase of effort, a system development contractor has been selected. Formally, this completes the system definition phase of the three-phase system development life cycle. It enables the start of a number of system development phases. We will describe the system development phase in terms of seven more finely grained phases. It is important to remember that some works consider what we call systems development as systems engineering itself. Our definition and interpretation is somewhat broader, although it certainly does encompass the system development phase of systems acquisition.

System Development

8. *Development of refined conceptual architectures* by the selected system development contractor, generally at a very high level such that this effort effectively amounts to program planning.
9. *Partitioning of the system into subsystems*, by the system development contractor. An orderly systems engineering process should include systems management efforts that result in configuration management of the evolving systems engineering product. This systems management

effort includes partitioning of the entire system so integration and interfacing needs are identified, and such that the need for interfacing and integration is minimized to the extent possible.

10. *Subsystem level specifications and test requirements development*, by the systems engineering contractor. A complete systems engineering approach must include means to express and verify subsystem level specifications. These metrics are very important.

11. *Design and development of components*, by the systems engineering contractor or potentially by subcontractors. In general, this will include a variety of hardware and software items:

   a. major hardware,
   b. interfacing hardware,
   c. systems software, and
   d. application software.

12. *Integration of subsystems*, by the systems engineering contractor or possibly one or more subcontractors. This will include integration of

   a. hardware,
   b. software, and
   c. testing practices.

   An appropriate systems engineering life cycle must specifically address the systems integration phase of the life-cycle effort. This is generally a far more exhaustive requirement than that of simply linking together a collection of existing subsystems, such as "reused" program modules. The individual technologies that comprise a system should be addressed specifically to ensure ability to facilitate end-to-end tests and performance.

13. *Integration of the overall system*, by the system engineering contractor. This will include integration of hardware and software, along with appropriate integration testing. The considerations here are similar to those in phase 12, with the exception that not all subsystems will be developed at the same technology level or in the same amount of initial detail. Thus, it will also be necessary to provide for future systems integration efforts.

14. *Development of user training and aiding supports*, by the systems engineering contractor. A key concern in this is that the resulting system be flexible, such that new application requirements can be easily and effectively supported.

Following system development, the system is implemented in an operational environment. The final phase of this systems engineering life cycle begins. Normal operation and maintenance of the system occurs. This will normally lead to system evolution over time. This can lead to the need for major proactive modification of existing systems, including retirement of a system and replacement by a new system. At least eight phases of the life cycle that specifically relate to system deployment can be identified.

System Deployment, Including Operation and Maintenance

15. *Operational implementation, or fielding of the system,* by the system engineering contractor with support from the user or customer group.

16. *Final acceptance Testing,* of the implemented system by the user. Acceptance testing will often result in minor adjustments to system operation. Ideally, minor changes can be made and documented in the acceptance environment.

17. *Operational test and evaluation,* by the system user or an independent contractor that has been selected for this purpose by the user group.

18. *Final system acceptance,* by the client or user group.

19. *Identification of system change requirements,* by the user group. As experience with using the system grows, there will be a need for maintenance in order to better adapt the system to its intended use. This maintenance is quite different from "bug fixes" and repair. It includes the need to evolve the system over time in an adaptive and proactive manner to make it responsive to changing needs, and needs that were present initially but unrecognized in the initial set of requirements specifications. Architectural concerns are of major importance here. Better defined system structures make it easier for the user to identify those portions of the system which would need to be changed to effect a desired alteration in system operation. This could tend to minimize requests requiring detailed changes in a large number of subsystem elements. These needs should be addressed as a part of the configuration management process.

20. *Bid on system changes, or prenegotiated maintenance support,* by a systems maintenance contractor. Generally, changes will be carried out using the same general approach used to develop the initial system. This will often predispose the user group toward returning to the same contracting source that developed the initial system, or to another one that has the same development tools available.

21. *System maintenance change development,* by the maintenance support contractor. Changes will be easier if the "history" of the original system development has been well-captured and documented, and if the system is well-structured and contains an open architecture.

22. *Maintenance testing by support contractor.* The systems engineering process evolves as a linear sequenced process. Errors in an earlier step of the process may cause errors in a later step. This indicates the need for iteration and feedback from later stages in the process to earlier stages. When potential errors are detected, there should be feedback to

earlier activities, and appropriate corrective actions should be taken. Often the only iteration possible is the one of proceeding back to phase 19—system change requirements. If significant defects in the existing system are observed, perhaps brought about by new functional requirements, then iteration back through the entire system acquisition process *may* occur. It appears to be the case in practice, however, that this does not happen often or only happens when very significant performance deficiencies are noted. It is this need for iteration back to any of a variety of earlier phases of a system life cycle that leads to the need for a look at the overall objectives of the system life cycle.

This 22-phase description of the systems engineering life cycle is doubtlessly exhaustive. We have indicated roles for the client or user group, for the systems engineering group or contractor, and potentially for various subcontractors that possess detailed expertise relative to specific technologies. The many phases developed here for the systems engineering, or systems acquisition, life cycle are illustrated in Figure 2.16. We should note that the life cycles of Figures 2.15 begins at phase 8 in the life cycle illustrated in Figure 2.16. For an excellent discussion of system architecture and design principles that, for the most part, follow from the extensive life cycle illustrated in Figure 2.16, the reader is referred to reference 18. This life cycle is very much patterned after

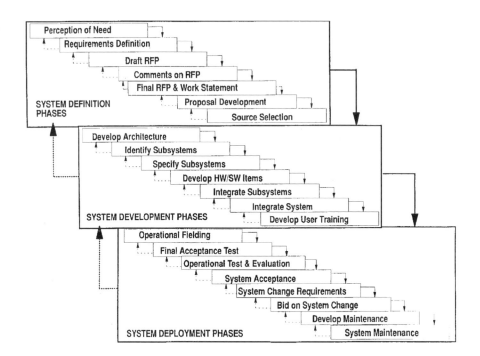

Figure 2.16 A 22-phase systems engineering life cycle.

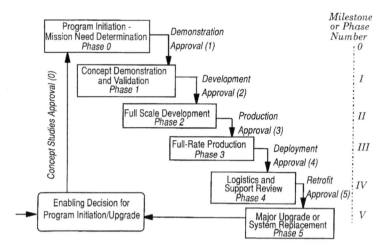

Figure 2.17 The U.S. Department of Defense systems acquisition life cycle.

the U.S. Department of Defense (DoD) systems acquisition life cycle. We now turn our attention to a description of one version of this DoD life cycle.

2.3.3 Defense Systems Acquisition Life Cycles

The life-cycle approach to systems acquisition of the U.S. Department of Defense [19] is illustrated in Figure 2.17. This system acquisition life cycle is comprised of five primary phases.

0. *Milestone 0 — Concept Exploration and Definition/Program Initiation/ Mission-Need Decision.* Mission need is determined and program initiation is approved, including authority to budget the program. A concept-definition analysis is performed. Primary consideration during this initial acquisition phase is given to the following four steps:

0.1. Mission area analysis

0.2. Affordability and life-cycle costs

0.3. Feasibility of a modification to an existing system to provide the needed capability

0.4. Operational utility assessment

1. *Milestone I — Concept Demonstration and Validation Decision.* A concept demonstration/validation effort is performed. If successful, the development continues. Primary attention is paid to seven areas of effort:

1.1. Program alternative tradeoffs

1.2. Performance/cost and schedule tradeoffs, including the need for a new development program versus buying or adapting existing military or commercial systems

1.3. Appropriateness of the acquisition strategy

1.4. Prototyping of the system or selected system components

1.5. Affordability and life-cycle costs

1.6. Potential common-use solutions

1.7. Cooperative development opportunities

The efforts in this phase establish broad program cost, schedule, and operational effectiveness and suitability goals and thresholds, allowing the program manager maximum flexibility to develop innovative and cost-effective solutions.

   2. Milestone II — Full-Scale Engineering and Manufacturing Development Decision. Approval for *full-scale development* (FSD) is made at this life-cycle phase. As appropriate, low-rate initial production of selected components and quantities may be approved to verify production capability and to provide test resources needed to conduct interoperability, live fire, or operational testing. Decisions in this phase will precede the release of the final *request for proposals* (RFP) for the FSD contract. Primary considerations in this phase of activity are described by 13 steps. The goal of the steps are to identify the following:

2.1. Affordability in terms of program cost versus the military value of the new or improved system and its operational suitability and effectiveness

2.2. Program risk versus benefit of added military capability

2.3. Planning for the transition from development to production, which will include independent producibility assessments of hardware, software, and databases

2.4. Realistic industry surge and mobilization capacity

2.5. Factors that affect program stability

2.6. Potential common-use solutions

2.7. Results from prototyping and demonstration/validation

2.8. Milestone authorization

2.9. Personnel, training, and safety needs assessments

2.10. Procurement strategy appropriate to program cost and risk assessments

2.11. Plans for integrated logistics support

2.12. Affordability and life-cycle costs

2.13. Associated command, control, communications, and intelligence (C^3I) requirements

Decisions at this phase result in the establishment of more specific cost, schedule, and operational effectiveness and suitability goals and thresholds than were possible in earlier phases. Particular emphasis is placed on the requirements for transitioning from development to production.

3. *Milestone III — Full-Rate Production Decision.* The decision made at this phase regards activities associated with the full-rate production/deployment phase. Primary considerations at this phase involve 12 steps, many of which are similar to steps in Milestone II:

3.1. Results of completed operational test and evaluation
3.2. Threat validation
3.3. Production cost verification
3.4. Affordability and life-cycle costs
3.5. Production and deployment schedule
3.6. Reliability, maintainability, and plans for integrated logistics support
3.7. Producibility as verified by an independent assessment
3.8. Realistic industry surge and mobilization capacity
3.9. Procurement or milestone authorization
3.10. Identification of personnel, training, and safety requirements
3.11. Cost-effectiveness or plans for competition or dual sourcing
3.12. Associated command, control, communications, and intelligence (C^3I) requirements

4. *Milestone IV — Logistics and Support Review Decision.* The decision process at this phase identifies actions and resources needed to ensure that operations readiness and support objectives are achieved and maintained for the first several years of the operational support phase of the life cycle. Primary considerations at this phase of the life cycle are as follows:

4.1. Logistics readiness and sustainability
4.2. Weapon support objectives
4.3. Implementation of integrated logistics support plans
4.4. Capability of logistics activities, facilities, and training and manpower to provide support efficiently and cost-effectively
4.5. Disposition of displaced equipment
4.6. Affordability and life-cycle costs

5. *Milestone V — Major Upgrade or System Replacement Decision.* The decision process at this phase encompasses a review of a system's current state or operational effectiveness, suitability, and readiness to determine whether major upgrades are necessary or whether deficiencies warrant consideration of replacement. Considerations at this phase of the life cycle are as follows:

5.1. Capability of the system to continue to meet its original or evolved mission requirements relative to the current situation
5.2. Potential necessity of modifications to ensure mission support

5.3. Changes in technology that present the opportunity for a significant breakthrough in system worth

This phase V actually addresses a system many years (normally five to ten) after deployment if the acquisition is *successful*. If there are major flaws, the initially deployed system may not be satisfactory and maintenance may be required from the day of operational implementation. As is now well known, this has been a particular problem with software acquisition.

There has been much present experience, including some criticism as well, of the defense systems acquisition life cycle. This life cycle encompasses the systems engineering life cycle described in MIL-STD 499A and 499B [20], which generally relate most strongly to phases or milestones II, III, and IV of system acquisition. Our definition of systems engineering is, therefore, much broader than that defined by MIL-STD 499A and 499B. It encompasses the definition and deployment phase of system acquisition, and it also includes RDT&E and systems planning and marketing.

This systems acquisition cycle is related to, but different from, the various categories of research and development funded by the DoD. There are six categories identified; each is prefixed by the number 6 in the DoD breakdown structure.

6.1.   *Research* includes all efforts of study and experimentation that are directed toward increasing knowledge in the physical, engineering, environmental, and life sciences that support long-term national security needs.

6.2.   *Exploratory development* includes efforts that are directed toward evaluating feasibility of solutions that are proposed for specific military issues that need applied research and prototype hardware and software.

6.3A. *Advanced technology development* is concerned with programs that explore alternatives and concepts prior to the development of specific weapons systems.

6.3B. *Advanced development* is involved with proof-of-design concepts, as contrasted with development of operational hardware and software. The Milestone I decision, just discussed, occurs during advanced development.

6.4.   *Engineering development* results in hardware and software for actual use that is developed according to contract specifications. A program moves from advanced development to engineering development co-incident with a Milestone II decision by the Defense Acquisition Board.

6.5.   *Management and support* involves the support of specific installations, or those operations and facilities that are required for general research and development use.

2.4   LIFE CYCLES PATTERNED AFTER THE WATERFALL MODEL

As you might suspect, there are a very large number of other life cycles for systems engineering, and a number of them are described in references 1–3. The interested reader should turn to these, and the references contained therein, for a more complete description of systems engineering life cycles. It is always possible to construct a new life cycle from a given one. Figure 2.18, for example, shows an expansion of the basic three-phase life cycle to allow for several concurrent development paths, and then integration of these concurrent developments as the system is deployed.

Since the introduction of the waterfall life-cycle processes into the systems and software engineering community, there have been a number of modifications. It is possible to show [2], however, that these modified life cycles are equivalent to an extended waterfall life cycle. In this section, we will review and interpret some of these efforts and conclude with a discussion of a process taxonomy.

In an excellent work on software quality and management, Ould [21] identifies four process models.

1. The *V process model* is essentially the "grand design" model we discussed earlier. The term "V process model" is used because the waterfall life cycle can be redrawn in a V-like configuration, as we have shown in Figure 2.19. This representation allows us to indicate that this seven-phase life cycle may be viewed from the perspectives of user, systems architect, and programmer. It is also of much interest and relevance to note that these perspectives are dominantly functional, structural, and purposeful. Figure 2.19 illustrates this V process view of the grand design approach.

2. Ould also identifies a *V process model with prototyping*, denoted the VP process model, in which a prototype is constructed of the customer

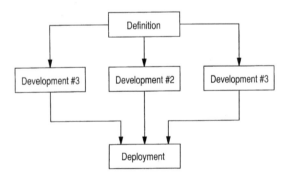

Figure 2.18   Expansion of the basic three-phase life cycle to allow for concurrent development.

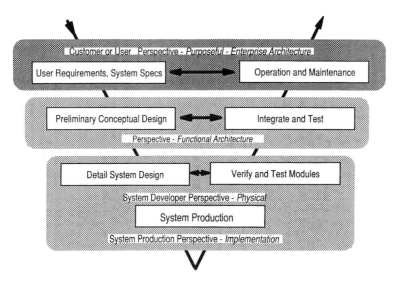

Figure 2.19 Interpretation of the grand design or V process model.

perspective, the systems architecture perspective, and the programmer perspective. The VP process model enables more complete treatment of risks, as well as identification of more appropriate risk management strategies, at each of these first three phases. For the most part, this seems equivalent to the initial spiral life-cycle process model. While an illustration of this need not appear necessarily different from that shown in Figure 2.19, it is also possible to show these first three phases as being comprised of a set of phases, or subphases, that together comprise each of these initial three phases. These define, develop, and deploy prototypes for software requirements and specifications, software system level architectures, and detailed design architectures. Our interpretation of this is shown in Figure 2.20.

3. The *evolutionary process model* is one in which a complete, or virtually complete, software system is developed at each of several repetitions of the life cycle. It is often the case that a mature specification of user requirements cannot be initially obtained. The major purpose of evolutionary development is to allow for the recognition of this, along with the resulting development of a refined software system at each successive repetition, or iteration, of the life cycle. There is no requirement that each life-cycle process be implemented in the same way in evolutionary development. Figure 2.21 indicates the most salient concepts in evolutionary development.

4. In *incremental development*, a kernel and one portion or increment of the system, *a*, is produced on the first iteration of development. Successive

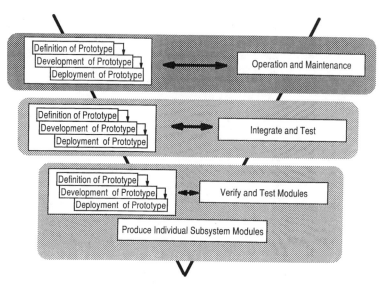

Figure 2.20 Interpretation of the grand design or V process model with prototyping.

portions are added through iteration until the final incrementally built software system is completed. While the evolutionary life cycle is intended to produce a, potentially incomplete, version of the whole system at each phase of development; the incremental life cycle is one in which an intentionally incomplete version is produced initially and successive

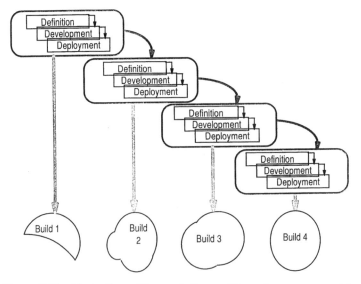

Figure 2.21 Illustration of the evolutionary life-cycle process model.

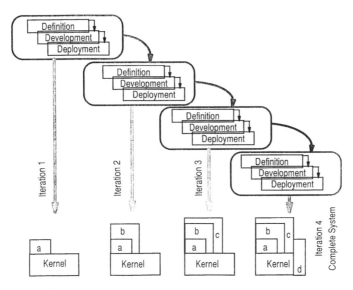

Figure 2.22 Illustration of incremental systems engineering process life cycle.

additional life cycles of effort augment this until the whole system results. Figure 2.22 illustrates a conceptual view of the incremental development life cycle.

Each of these life-cycle models may be thought of as comprised of a number of waterfall life cycles. The actual development strategies may vary from one life cycle to another. For example, reusable software may well be a major part of the prototyping strategy for requirements and specifications identifications in one life cycle, but not in another. There are many activities associated with these life cycles that are not shown in the illustrative figures. This is especially the case with respect to the numerous configuration management and control efforts that are needed. One of the major purposes in use of any of these life cycles is risk management, as discussed in references 1–3 and elsewhere. There have been a number of other items used to describe essentially the same life cycles we discuss in this section. The Software Productivity Consortium used the term *evolutionary spiral process* (ESP) to describe their modifications to the spiral life cycle, which was initially based mostly on a VP development perspective primarily, to enable it to also accommodate evolutionary development. The MITRE corporation has used the term *managed evolutionary development* (MED) [22], and there are doubtlessly other terms used as well for variations of the basic evolutionary approach.

The evolutionary software development life-cycle model is based on the approach of building successively more functional prototypes. The evolutionary, incremental, spiral, and other such approaches can clearly be modeled by an interactive sequence of waterfall models. The concern should not, we feel, be so much directed at the level of process in this regard, when it is found that risk management and trustworthiness issues have not been fully considered, but rather at the level of systems management. Thus, in no way would we encourage stopping or getting off the life cycle [23, 24], but would very strongly encourage appropriate use of systems management to implement high-quality processes that are most appropriate for the engineering of specific systems. This is beyond the scope of our present coverage.

2.5 SUMMARY

In this chapter we have presented a number of life cycles of systems engineering processes. It is normally very useful to think of a systems engineering process as being described by a number of phases that comprise the time dimension of systems engineering. We concluded our discussions with a presentation of some interesting contemporary developments in engineering processes that have been found especially useful for the evolution of software systems.

Some necessary ingredients that must exist in order to develop large systems, solve large and complex problems, or manage large systems are as follows:

1. A way to deal successfully with problems involving many considerations and interrelations, including change over time
2. A way to deal successfully with areas in which there are far-reaching and controversial value judgments
3. A way to deal successfully with problems, the solutions to which require knowledge principles, practices, and perspectives from several disciplines
4. A way to deal successfully with problems in which future events are difficult to predict
5. A way to deal successfully with problems in which structural and human institutional and organizational elements are given full consideration

We believe, strongly, that the three-dimensional systems engineering framework presented in this chapter possesses these necessary characteristics.

Thus, systems engineering is potentially capable of exposing not only technological perspectives associated with large-scale systems but also human needs, and institutional and value perspectives. Furthermore, it can relate these to knowledge principles and practices such that the result is the successful development of high-quality trustworthy systems. We have concentrated on systems engineering frameworks to enable this in the present chapter.

The result of a systems engineering effort is generally a product or service. To focus on this exclusively at the expense of concern for the systems engineering process is, we feel, a mistake. The process must necessarily be

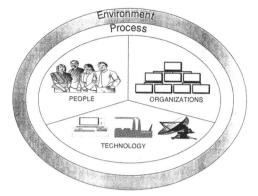

Figure 2.23 Systems engineering as people, organizations, and technology—with a systems engineering process to ensure environmental compatibility.

embedded into the environment that surrounds the systems engineering organization. It must necessarily be concerned with people, organizations, and technology. Figure 2.23 attempts to illustrate this large-scale and large-scope view of systems engineering, along with the systems engineering life cycle.

PROBLEMS

2.1 Please associate a systems engineering method and a systems engineering metric with each of the nine boxes for the systems acquisition life cycle shown as a portion of Figure 2.5.

2.2 Identify a systems acquisition issue of your choice. Write a brief paper outlining the efforts needed at each of the steps and phases of a three-phase, three-step systems acquisition life cycle.

2.3 From references 1 or 2, or other sources, please prepare a discussion of the 2167A software development life cycle. Illustrate how it is a particular case of the waterfall life cycle illustrated here.

2.4 From references 1 or 2, or other sources, please prepare a discussion of the Boehm [25] spiral development life cycle. Illustrate how it is a particular case of the waterfall life cycle illustrated here.

2.5 From references 1 or 2, or other sources, please prepare a discussion of the DOD-STD-498 software development life cycle. Illustrate how it is a particular case of the waterfall, evolutionary, and iterative life cycles illustrated here.

2.6 From references 1–3, or other sources, prepare a discussion of configuration management and configuration control and relate it to the typical phases in a systems engineering life cycle.

2.7 Obtain discussions of such recent commercial systems engineering process standards, such as the IEEE P1220 and EAI 632 standards. Contrast and compare these with the DOD-STD-499 systems engineering life cycle process standard.

REFERENCES

[1] Sage, A. P., *Systems Engineering*, Wiley, New York, 1992.

[2] Sage, A. P., *Systems Management: With Application to Information Technology and Software Engineering*, Wiley, New York, 1995.

[3] Sage, A. P. and Rouse, W. B. (Eds.), *Handbook of Systems Engineering and Management*, Wiley, New York, 1999.

[4] Sage, A. P., Methodological Considerations in the Design of Large Scale Systems Engineering Processes, in Haimes, Y. Y. (Ed.), *Large Scale Systems*, North-Holland, Amsterdam 1982, pp. 99–141.

[5] Hall, A. D., Three Dimensional Morphology of Systems Engineering, *IEEE Transactions on Systems Science and Cybernetics*, Vol. SSC 5, No. 2, 1969, pp. 156–160.

[6] Buede, D. M., *The Engineering Design of Systems: Models and Methods*, Wiley, New York, 2000.

[7] Sage, A. P., and Lynch, C. L., Systems Integration and Architecting: An Overview of Principles, Practices and Perspectives, *Systems Engineering*, Vol. 1, No. 3, 1998, pp. 176–227.

[8] Rechtin, E. R., *Systems Architecting*, Prentice-Hall, Englewood Cliffs, NJ, 1991.

[9] Rechtin, E. R. and Maier, M. W., *The Art of Systems Architecting*, CRC Press, Boca Raton, FL, 1997.

[10] Sage, A. P., Systems Management of Emerging Technologies, *Information and Decision Technologies*, Vol. 15, No. 4, 1989, pp. 307–326.

[11] Sage, A. P., Systems Management for Information Technology Development, in *Expanding Access to Science and Technology: The Role of Information Technology*, Wesley-Tanaskovic, I., Tocatlian, J., and Roberts, K. H. (Eds.), United Nations University Press, Tokyo, 1994, pp. 361–405.

[12] Benson, B., and Sage, A. P., Case Studies of Systems Management for Emerging Technology Development, in W. B. Rouse (Ed.), *Advances in Human Machine Systems Research*, JAI Press, New York, 1994, pp. 261–324.

[13] Maslow, A. H., *Motivation and Personality*, Harper & Row, New York, 1970.

[14] Nadler, G., Systems Methodology and Design, *IEEE Transactions on Systems, Man and Cybernetics*, Vol. 15, No. 6, November 1985, pp. 685–697.

[15] Hall, A. D., *A Methodology for Systems Engineering*, Van Nostrand, New York, 1962.

[16] Hall, A. D., *Metasystems Methodology*, Pergamon Press, Oxford, UK, 1989.

[17] Beam, W. R., Palmer, J. D., and Sage, A. P., Systems Engineering for Software Productivity, *IEEE Transactions on Systems, Man, and Cybernetics*, Vol. 17, No. 2, March 1987, pp. 163–186.

[18] Beam, W. B., *Systems Engineering Architectures and Design*, McGraw-Hill, New York, 1990.

[19] U.S. Department of Defense (DoD) Instruction (DoDI) 5000.2, September 1, 1987.

[20] U.S. Department of Defense, MIL-STD-499A and MIL-STD-499B, *Systems Engineering*, May 1974 and May 1991.

[21] Ould, M. A., *Strategies for Software Engineering: The Management of Risk and Quality*, Wiley, Chichester, UK, 1990.

[22] Reynolds, P. A., Ward, E. S., Gonzalez, P. J., Blyskal, J., Hofkin, R., Garfield, D., and Marple, J., The Managed Evolutionary Guidebook: Process Description and Applications, Report MTR92W0000251, The MITRE Corporation, McLean, VA., March 1993.

[23] McCraken, D. D., and Jackson, M. A., Lifecycle Concept Considered Harmful, *ACM Software Engineering Notes*, Vol. 7, No. 2, 1982, pp. 29–32.

[24] Gladden, G. R., Stop the Lifecycle. I Want to Get Off, *ACM Software Engineering Notes*, Vol. 7, No. 2, 1982, pp. 35–39.

[25] Boehm, B., A Spiral Model of Software Development and Enhancement, *IEEE Computer*, Vol. 21, No. 5, May 1988, pp. 61–72.

CHAPTER **3**

Formulation of Issues

Formulation of issues is a crucially important, as well as one of the most challenging and rewarding, part of any systems engineering effort. As we noted in Chapter 1 and elsewhere [1, 2], the issue formulation step is needed at each of the phases in a systems engineering effort. It is crucial because the main product of a systems engineering issue formulation effort, which is a set of potentially feasible alternatives for the issue associated with the particular phase being formulated, sets the stage for all that is to follow. Eventually, one of the alternatives generated during a responsible issue formulation effort will be selected for implementation. No matter how elegantly the alternatives are analyzed and modeled or how thoroughly you plan for implementation, the value of the entire effort depends on having identified, generated, or developed good alternatives in the first place.

3.1 INTRODUCTION

It is very important, we believe, that texts on systems engineering and problem solving, in general, pay significant attention to issue formulation, as a step in systems engineering that needs to occur in each of the phases of effort. This is so because, in practice, the results of a poor issue formulation effort may be subject to very detailed analysis, and the results of this are carried over into implementation. As a result, we obtain the wrong system! Needless to say, this usually results in a waste of resources and (through engineering) an inferior system, compared to that which might have resulted from a better issue formulation. Issue formulation is critical across the definition, development, and deployment phases of a systems engineering life cycle. This is not a step implemented only as part of the systems definition phase, although it is very important in that phase.

A major intended contribution of this text is a detailed treatment of the all important issue formulation step. This chapter covers issue formulation methods and techniques with examples that should help to unveil much of the mystery surrounding issue formulation. Those who master the art and science of issue formulation are often the most sought-after and highly rewarded systems engineers. A problem resolution effort that is well-begun, in terms of being well-defined, is a problem half-solved. On the other hand, a problem not well-begun is one in which the solution is likely to be improper and unacceptable. The foregoing discussion calls for a definition of the word *problem*, and we provide a simple definition here.

> A problem is the occurrence of an undesired aspect of the current situation that creates a gap between what is occurring and what we would like to have occur.

In general, resolution of a problem requires a decision to implement one of several alternative courses of action that will reduce, in fact hopefully eliminate, the gap between what is occurring and what we would like to have occur. A problem is, or should be, defined in such a way that we are able to obtain a basic understanding of the undesired situation as well as obtain the most appropriate perspective for thinking about it. A *problem*, or an *issue* as a generally synonymous term, can usually be viewed from several perspectives, and the solution that results will often be a strong function of the perspectives from which a problem is viewed and formulated.

Issue formulation begins with a definition of the problem or issue to be resolved. This includes an assessment of the situation surrounding the problem. Situation assessment is important in problem solving and issue resolution. There are many aspects of the environment surrounding an issue that we need to know about in order to be able to generate potentially feasible alternative courses of action that can lead to issue resolution. Many of these factors are dynamic, uncertain, and subject to change with the passage of time. Situation assessment means that we need to make a careful appraisal of the various factors that influence a particular observed state or condition. Generating good alternatives, ones that truly satisfy the needs of our customers in an effective, efficient, equitable, and explicable manner, depends on an in-depth situation assessment. We will examine situation assessment first and then proceed to other aspects of issue formulation. In particular, we will examine a variety of methods associated with various aspects of issue formulation.

3.2   SITUATION ASSESSMENT

A complete situation assessment consists of three basic steps that are intended to answer the three questions highlighted in Figure 3.1. The first part is the identification of the (immediate) past in terms of the question, *What was?* The second part of a situation assessment examines the current status or position

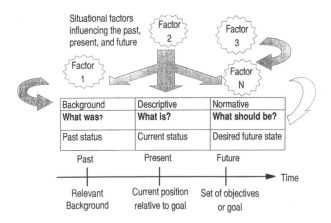

Figure 3.1   The scenarios of a complete situation assessment.

of things relative to the goal. It is a descriptive component and answers the question, *What is?* The third or normative component of a situation assessment answers the question, *What ought to be?* Understanding of the major, ideally all, factors influencing the past, present, and future is needed for a good situation assessment. There is a horizon component of situation assessment and it answers the question, *What factors over time matter and when do they matter?* We have a complete situation assessment once we know the following:

- What do we want to accomplish?
- What is the past and current state of affairs relative to our goal brought about by the changed situation?
- What are the relevant factors that influence the situation and their associated time frame.

There are a large variety of environments in which various systems must operate. A number of different perspectives—such as technical, economic, legal, political, cultural, and social—can be important for understanding a system's environment.* When an adequate appreciation of all the important perspectives is achieved, then we say we have preserved the contextual integrity of the issues under consideration. The problem of issue definition, and the resulting identification of worthwhile alternatives, depends on an accurate and timely situation assessment that preserves this contextual integrity.

Accomplishing a high-quality situation assessment depends on several factors. First, the experiential familiarity with the issues or problem under consideration is very important. For example, it is unlikely that someone with

---

*Often, the term *environment* is used to refer to the *physical environment*, or *natural environment*, which refers to air, land, water, and biota and the state of these. Here, we refer to environment in a more general context as the broad milieu of factors surrounding an issue.

very little background knowledge about nuclear power plants would be able, without considerable study, to develop useful alternatives for ensuring safe operation of such a facility. As we have already noted, knowledge is information in a relevant context.

But you need more than just knowledge about the specific problem or issue at hand. Familiarity with the environment into which the specific issue or problem is embedded is also crucial to an understanding of context. For example, the design of a successful information system for airline reservation agents is made much easier by someone with an in-depth knowledge of the various tasks that such agents perform and the environment that agents must cope with as they perform these tasks. You could learn a lot about designing such a system from someone who is an expert at developing human–computer interfaces. But you can also appreciate how helpful it would be to interview some experienced ticket agents and to find out about the difficulties they experience when talking to customers and booking their flights. It is the job of systems engineers to make sure that the relevant knowledge of experienced people is brought into a project and used in issue formulation. Capturing this knowledge often requires the systems engineer to skillfully conduct both individual interviews and group meetings. A discussion of some useful techniques for this will be presented later in this chapter.

On the other hand, extensive familiarity with an existing system may sometimes blind someone to truly new and innovative approaches. People with extensive experience with an existing system may find it difficult to think about new solutions. It is also the job of the systems engineer to overcome those mental stumbling blocks that can be caused by the sort of thinking that anchors too strongly on existing approaches. Applying a sound systems engineering process, with an appropriate formulation effort associated with each of the phases of the process, will help the systems engineer avoid these pitfalls. Therefore, even if systems engineers lack specific knowledge about the particular domain of a project, they can make important contributions to the effort by using an appropriate process for problem solution or issue resolution and by putting their systems management knowledge to good use in selection of this process.

Besides talking to key people, another way of accomplishing a situation assessment is to observe the current system, if one exists. An objective appraisal of the problems with the current system can be very helpful. A simple listing of these observations in a log book by date and time is often worthwhile. Be sure to note the main ideas of people managing and working with the system. Often, problems will not be evident on the first visit. Subsequent visits will reveal more. People also are more open with you after they get to know you and feel comfortable with you. Information gained by observing the current system and talking to people involved with it is most valuable for writing a descriptive scenario.

A *descriptive scenario* tells why the current system came into being, how it works, and what problems are associated with it. You should include the main actors or stakeholders involved with the current system in your descriptive scenario. These stakeholders will usually be owners, users, customers, clients, managers, maintainers, administrators, and regulators of the current system. But, for a complete situation assessment you will also need a normative scenario.

A *normative scenario* describes how the stakeholders want the system to be in the future. This is not as easy to accomplish as it might appear to be. We will often need to distinguish between realistic needs and frivolous wants. Often also, important stakeholders have conflicting opinions on what the future should be, or on the cause of the present difficulty. Some have unrealistic expectations of what technology can deliver in the time frame under consideration. Others are reluctant to see any changes because they think their job or relationship to a system may be put in jeopardy. A skilled systems engineer must broker these various interests to create enthusiasm and support for a technically feasible, trustworthy project. One way of gaining support for an improvement to an existing system or for a completely new system is to involve stakeholders in the process. Helping people visualize the future in meaningful ways will go a long way toward creating a useful normative scenario. But how can you see into the future?

If there is not a current system to observe, then you need to create a way of seeing how an entirely new system might work. A number of useful ways exist for this. A group of people with relevant knowledge can meet and talk through how a new system might work. Some people can be assigned roles to play. For example, a new software system for managing home security could be thought about by people pretending to use the new system as homeowners about to go on vacation. "Now I am using the system to set lights to go on and off at various times in different rooms so it looks like we are at home" says one participant. As ideas emerge, the system engineer can record them.

Many other ways exist for examining possibilities for new systems. Prototypes [3,4] can be built and potential users can experiment with them through simulation. A prototype is a first-cut approximation of what a new system might be. The purpose of a prototype is to quickly gain information from the user about what seems to work and what is needed. For example, a prototype of a new information system might be a set of screens that a user could view and comment as to how useful the screens are in helping provide needed information.

A model of the situation, and the resulting simulation of operations using the model, can emulate the operation of a new system. To build a simulation requires a model or some representation of the new system. The usual purpose of a simulation is to get an idea of how the new system might actually work if it were put into operation. For example, a city wanting to build an airport at a new location might like to know if the land available can accommodate the number of planes that they anticipate will be landing and taking-off from their new facility. A simulation of an airport that fits within the city's land constraint can be tested to see if any airport layouts can successfully meet anticipated passenger and freight demands. We discuss modeling of systems in our next chapter.

To a systems engineer, accomplishing a situation assessment means writing both a descriptive and a normative scenario. It also means making a careful appraisal of the factors that matter as we think about how to get from where we are now to where we want to be tomorrow. Once you understand where you are now, where you want to go, and what factors are important to consider, you are well on your way to one of the main tasks of systems engineering — identification of a set of potentially feasible alternatives that can get you from today to a better future.

3.3 APPROACHES TO SITUATION ASSESSMENT

In this section we will discuss several contemporary approaches to situation assessment. Noble [5] says that situation assessment includes the following:

1. An estimate of the purpose of activities in the observed situation
2. An understanding of the roles of the participants in these activities
3. Inferences about completed or ongoing activities that cannot be directly observed
4. Inferences about future activities

One does not always have to react to a situation. One can simply observe it, assess it, or ignore it, as well as react to it. Also, we may attempt to respond to a situation in an interactive manner by attempting to be responsive to change needs at an early date and develop alternatives to deal with undesired change as these changes occur rather than wait until a crisis situation has resulted and then react to the crisis. Also, we may attempt to be *proactive* and predict the possibility for difficulties and identify a suitable process that attempts to prevent these difficulties from ever occurring.

The notion of a threshold is often very important in situation assessment. We might have a threshold for a situation, and observing this threshold might be all that is required. When the situation descriptor values exceed this threshold, we might then decide to assess it in greater detail. If it is benign, we may choose to ignore it; if it is not benign, we may decide to react to it or continue to observe it. A reasonably good illustration of this is the situation assessment of temperature of the human body. One of the potential difficulties relative to observing a situation is that we wish to be able to interactively and proactively deal with issues and not simply to react to difficulties after they have eventuated.

The general situation assessment process [6,7], illustrated in Figure 3.2, may be described by the following eight steps:

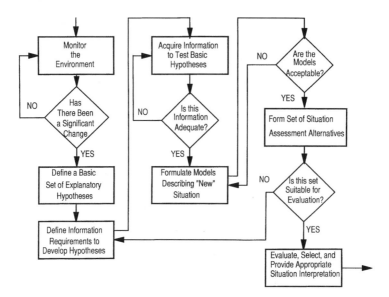

Figure 3.2   Flow chart for situation assessment.

1. We make general observations of as many pertinent aspects of the situation around us as is possible in order to identify or confirm the descriptive scenario or status quo.

2. We see if this information indicates a change in the current situation. If there is no change, we go back to step 1. If there is a change, we proceed to step 3.

3. We identify or formulate a number of alternative hypotheses that may describe the new situation, or descriptive scenario.

4. We attempt to determine if any more information inputs are required for deciding which alternative hypothesis is the most likely one.

5. We obtain the required information from such input sources as a priori predefined stored information, human observations, human knowledge, or sensor measurements.

6. We process this information and knowledge to confirm or disconfirm each of the alternative hypotheses.

7. We identify the hypotheses* with the highest probability(ies) of being true and identify the associated situation assessment with some appropriate quantification of our confidence in the assessment.

8. We determine if the assessment confidence for describing the situation is sufficiently high. If it is not, we go back to step 3 or 4 and repeat the

---

*There are a variety of approaches for hypothesis test and assessment, and they are generally discussed in most elementary statistics texts. We will also comment briefly on hypothesis testing in our next chapter.

assessment process until the confidence threshold is achieved. When it is, we have assessed the situation and we go on with the rest of the problem-solving or issue formulation process.

We will now examine two approaches that have been suggested for situation assessment. In the first of these, DuBois et al. [8] developed MEDICIS, a knowledge-based system that tries to provide an efficient and flexible tool for assisting physicians in performing a medical diagnosis. It has two subsystems:

- A knowledge acquisition and control module
- A consultation module dedicated to each application

A general diagnosis is accomplished in the following manner:

1. A physician examines the patient and detects signs and symptoms that suggest some sort of patient health situation.
2. The observed signs and symptoms evoke one or more possible diagnoses, or hypotheses.
3. The physician compares the patient's state with each hypothesis generated in step 2.
4. If one or more hypotheses are sufficiently confirmed, the physician states a diagnosis.

From this follows efforts that lead to a *correction* to the diagnosed illness. Because the hypotheses are for separate pathologies, more than one hypothesis can be true. MEDICIS consists of an overall architecture, a knowledge representation system, a knowledge acquisition system, and a consultation system. Some of the functional characteristics of the prototype system are as follows:

- An expected diagnosis is always proposed by the system and almost always is selected by a multicriteria decision analysis, a subject we discuss in Chapter 5, using a system called PROMETHEE.
- In cases where one or more characteristic pathological signs are observed, the expected diagnosis is always selected by the multicriteria decision analysis.
- On several occasions, the system has drawn the user's attention to a pathology that had not been considered by the physician.

In the second situation assessment system, Dutton and Duncan [9] presented a model of how decision makers interpret strategic issues. The diagnosis is based on three critical events:

- Activation
- Assessment of urgency
- Assessment of feasibility

A strategic issue is an event that can have the potential to influence an organization's current or future strategy. The triggering and situation assessment of strategic issues is called a *strategic issue diagnosis* (SID), which is, in our terminology, equivalent to a strategic situation assessment. When using the SID, changes in the decision environment are sensed and assessed; on the basis of this situation assessment, courses of action are chosen which initiate or impede strategic change.

In general, issue assessment is based on two assertions:

1. The need for choosing a new course of action is apparent.
2. It is feasible to deal with the issue.

The resulting issue assessment creates the basis for a momentum for change, and the forces for organizational responses are actuated. Urgency and feasibility assessments build momentum for change and determine if the course of action chosen by the decision makers will be incremental or radical, as we illustrate in Figure 3.3.

The critical elements of this model are activation, assessments of urgency, time pressures, visibility, responsibility, and assessments of feasibility. *Activation* is caused by a perceived inconsistency or imbalance that indicates that the costs of inaction are too high to delay consideration of the issue. Critical sources of strategic issues are those individual actors, parties, and organized groups or institutions that have bearing on the policies and actions of the organization. We will refer to these people as *organizational stakeholders*. Another type of trigger or activation occurs as a result of a gap analysis which establishes a need to face an impending deficiency in the organization.

Assessments of urgency result from estimation of the perceived importance of taking action on an issue. Urgency arises from the following critical dimension:

- Time pressures such as deadlines
- Visibility to important internal and external constituencies such as unfavorable publicity
- Attributions of responsibility by certain managers for the issue's occurrence

Assessments of feasibility involve identification of the likelihood of successfully resolving the strategic issue. This is based upon the notion that diagnosis or situation assessment influences organizational action. The more severe the assessments of urgency and level of feasibility, the greater will be the commitment or momentum by a decision maker for change. Where the momentum for change is low, less radical changes are more likely; and as the momentum for change increases, more risky and costly changes are likely to occur, as shown in Figure 3.3.

The situation assessment process can be used to determine the information requirements for a large-scale system operation, such as a military or corporate command and control system. The process can also be used for a number of other applications, such as to develop information, alternative hypotheses, and alternative options for selecting emerging technologies that have the best benefit, cost, and risk tradeoffs for developing into commercial products for national and international markets. In the analysis of various emerging technologies for potential for being developed into commercial products, services, or systems; situation assessment is an important part of the initial effort [10].

One of the most difficult problems in complex decision situations is gathering the appropriate information and properly assessing the situation. Mason and Mitroff [11,12] are among the many who have commented on this reality. These authors identify three types of errors that can occur in assessing a situation:

1. Type I errors are errors that result from incorrectly assessing that there is a problem when there is no problem.
2. Type II errors are errors that result from incorrectly assessing that there is no problem when there is a problem.
3. Type III errors are errors that result from correctly assessing that there is a problem, but incorrectly identifying the nature of the problem.

Type I errors are generally called false alarms, and Type II errors are called misses or missed problem errors. Type III errors cannot generally be overcome by even the best methods of problem solving because one is solving the incorrect problem. Only by some rare fortuitous circumstance would one ever

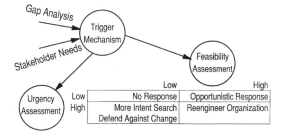

Figure 3.3 Assessments in strategic issue diagnosis.

end up with an appropriate solution for the real problem at hand after misdiagnosis and implementing a course of action for the wrong problem. A major reason for situation assessment is to avoid Type III errors.

Type III errors occur most frequently when problems are ill-structured. In this context, it is likely that the most difficult step in the decision process would be the problem definition or situation assessment. Complex problems are those problems that require multidisciplinary perspectives if Type III errors are to be minimized. A problem is *ill-structured* when one or more of the decision alternatives, decision outcomes, values or weights of importance of the outcomes, or the probabilities of the various outcomes that result from implementing various alternative courses of action are either unknown, imprecisely known, or uncertain. Briefly, an ill-structured problem is one in which it is difficult, if not impossible, to reach a consensus regarding the basic definition of the problem and its possible solutions. Virtually all real-world problems are ill-structured, and it is these problems for which structured situation assessment aids are most needed. We have outlined some of the considerations that need to be given to establishing situation assessment aids for these problems. Our discussion is taking us a bit beyond the systems engineering methods we have established thus far. The presentation we provide here is, nevertheless, important because it indicates the need for the methods we discuss in this and the next two chapters, and how they might be brought to bear on important problems. Additional discussions concerning situation assessment may be found in the references cited thus far in this chapter.

A situation assessment aid is necessarily a prescriptive model. As such, it also has normative and descriptive components. The study of the decision process for descriptive, normative, and prescriptive approaches is highlighted in efforts by Bell et al. [13] and other recent efforts [14–16]. The discussions in these works come from decision specialists, theoretical and behavioral, who are interested in proposing rational procedures for decision making (how people should make decisions if they wish to obey certain fundamental laws of behavior) and who wish to describe behavior. Psychologists are primarily interested in how a person makes decisions in the real world, whether or not rational, and in determining the extent to which the person's behavior is compatible with any rational decision model. They are also interested in learning the cognitive capacities and limitations of ordinary people to process the information required of them if they do not naturally behave rationally, but wish to. Thus, psychological studies of decision making are basically descriptive studies. Systems engineers are concerned with improving the decision-making process through use of innovations in information technology. These are fundamentally prescriptive studies, although a good prescriptive approach must be aware of descriptive reality. A prescriptive theory should be designed to operate on the set of available data so that good decisions can be prescribed even if information on alternative courses of action or situation hypotheses are incomplete. We will expand on these thoughts somewhat in Chapter 5 in our discussion of issue interpretation and decision analysis.

3.4   PROBLEM OR ISSUE IDENTIFICATION

The toughest part of most systems engineering efforts is identifying the problem correctly. Many engineering efforts, although they may represent great feats of engineering skill, fail because they were designed to solve the wrong problem or because the implemented solution created more problems. For example, consider the Aswan Dam in Egypt built in the 1960s. Hundreds of then Soviet engineers set about and successfully built one of the biggest dams and hydroelectrical facilities in the world. However, as the Egyptians soon found out, the effort was, in many respects, a grand failure because it destroyed much of the precious, arable land in the Nile River Valley. The Nile River's natural flooding process had continually replenished the soil downstream and made it possible for many crops to grow in fertile river-bottom land. Once the dam was in place, the natural floods no longer replenished the land and vast amounts of formerly rich agricultural land was lost forever. Egypt, which had been able to feed itself, soon became dependent on food exports. Furthermore, the dam flooded forever many priceless ancient ruins under deep water that were upstream from the massive dam.

Why didn't the engineers consider the upstream and downstream impacts of the project? We can guess that the narrow perspective taken by the engineers led them to conclude that the problem was to block tremendous amounts of water behind a dam for generating electricity and preventing floods. This led to the "solution" of one problem and the creation of more significant problems. A broader perspective of the problem would have accounted for the upstream ancient ruins and the downstream arable lands and, potentially, averted this travesty. A systems engineer must always consider the potential new problems or ramifications caused by any proposed solution to the original problem.

As the Aswan Dam case shows, identifying a problem or issue for a large-scale project is not easy because there is not just one simple question to solve or difficulty to resolve. The large-scale projects that systems engineers work on require the identification of multiple questions or difficulties that must be answered or resolved. And there is not a single correct answer. Instead, there are usually a number of alternatives that can be devised depending on which difficulties the stakeholders believe to be most important. Usually, some success on one important aspect of a large-scale project will come at the expense of success on another part of the project. These tough decisions often require very difficult tradeoffs. As you will learn in Chapter 5, one of the important skills of a systems engineer is to understand how to analyze, assess, and communicate these tradeoffs.

3.4.1   Scoping and Bounding the Problem

How can you successfully identify the relevant questions and difficulties in a large-scale project? Properly scoping and bounding a systems engineering effort is the answer. To scope a project means to understand why the project is necessary, what the stakeholders intend to accomplish with the project, and how to measure project success. In systems terms, this means identifying the needs, objectives, and criteria for the project. The needs tell why the project is necessary. Related to the needs, the objectives describe in detail everything that the project is intended to accomplish. Criteria measure success in achieving the objectives.

To bound a project means to understand the limitations associated with the project, the changes that can be made to achieve desired objectives, and the important quantities that are likely to change as a result of the project. In systems terms, this means identifying the constraints, parameters, and variables for the project. Constraints are the limits that must be observed for the project. Constraints include realistic considerations related to things such as money, time, people, organizations, and society. For example, most projects have budget, time deadlines, and environmental impact constraints.

Parameters are factors that define an alternative and determine its behavior. The value to which parameters are set restrict what results are possible to achieve with an alternative. For example, in the design of a mass transit system for a city, some parameters might be the number of buses and trains. How many people can be moved by any specific mass transit alternative depends on the value to which these parameters are set — how many buses and trains there are for an alternative. Once a system is in operation, parameters do not change much. Every day that the city operates the mass transit system, the same number of buses and trains are part of the system. Parameters are factors that the systems engineer manipulates to create alternatives and alter their performance. To distinguish parameters from the quantities associated with mathematical models, you can refer to them as alterables.

In contrast to parameters, response variables are likely to change once the system is in operation. Variables are measurable quantities that you want to monitor as the system operates. For example, the number of passengers that use the buses or trains each day is a response variable that changes as the system operates. For large-scale systems the number of such variables can be overwhelming. That is why we use the concept of state variables.

State variables are a collection of response variables that we choose to monitor to inform us about the status of a system. The specific variables chosen depend on why we want the information. If we are system engineers investigating the adequacy of a currently existing mass transit system, then we probably want to know a number of variables at different times of the day and week. For example, two state variables might be the number of passengers per route and the number of operational buses and trains.

3.4.2 Systems Definition Matrix

A convenient way for the systems engineer to document the scoping and bounding of a problem is to use a systems definition matrix like the one shown in Figure 3.4. The matrix has two main sections. One section defines the scope

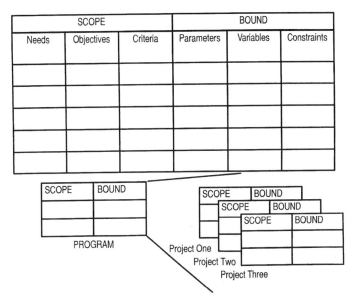

Figure 3.4 Systems definition matrix.

of the project by listing the needs, objectives, and criteria. The other section lists the parameters, variables, and constraints and defines the bounds of the project. The simple framework of the systems definition matrix is useful as a checklist for information gathering, documentation, and communication.

The simple matrix suggested here can be expanded to include more features and elements of a proposed system or project. For example, the purposes and basic functions of a system, its inputs, outputs, major components, human agents, environment, super and lateral systems, and interfaces and controls could all be dimensions of the system that are described in the matrix.

If you want to put more details in the matrix, you can create a systems definition matrix for each subsystem of the larger system. For a multiproject program, you might want to complete a system definition matrix for each major project in the program. And for a multistage project, you may want to complete a matrix for each phase of the project. Additionally, you can use these matrices to help you identify and specify the interactions among the various components.

The system definition matrix, when complete, should give you and those that you need to communicate with a thorough picture of the basic ingredients of an existing or proposed system. It can help you to discover insights into the feasibility and internal consistency of a proposed design. Also it can help you identify different ways that you can make improvements to an existing system. Before we explore more about how to create different alternatives, let's focus in on two very important elements of the problem formulation effort, namely, needs and constraints.

3.4.3   Needs and Constraints: Identification and Analysis

Because the formulation of a successful system depends on meeting the effective needs of the customer, it is no wonder that often there is great importance placed on identifying and analyzing the needs for a project. Such efforts have become so important that they have been given their own category: *needs analysis.*

What are needs and how are they analyzed? First, a need is a lack of something desired or required. Maslow [17] defined a hierarchy of needs for people that shows how basic lower-level physical needs such as hunger or thirst must be satisfied first before higher-level emotional and mental needs such as love and self-esteem become important. These needs are as follows:

- *Physiological needs,* such as food, clothing, and health
- *Safety needs,* such as stability, security, and order
- *Belongingness needs,* such as affection and affiliation
- *Esteem needs,* such as self-respect, prestige, and success
- *Self-actualization needs,* enabling realization of development potentials

The needs hierarchy provides a useful framework for evaluating societal and marketplace demands for the goods and services that can be provided by systems engineering based approaches. There are other human needs that could be identified, and added to this hierarchy, such as the need generally expressed by professionals for autonomy.

So as Maslow has shown, needs stem from being human. But humans also desire to engage in many different activities. Many needs come from the desire to improve or enhance human activities. People are always wanting to do things faster, easier, cheaper, and better, in some way. Technology itself is often defined as the application of scientific knowledge to enhance performance of some human activity. As we know from the increasing pace of advances in technology, needs are limitless. But notice that we put the word "effective" in front of needs in our definition of successful design.

What is an effective need? This is best answered by what one is not. An effective need is not a primitive need, or "want." Primitive needs are the unsupported opinions of people involved with a system. Primitive needs are usually narrowly focused assertions or beliefs that people have. These opinions normally reflect their position or role within a system. For example, in an assembly facility that is experiencing problems, primitive needs from assembly line workers may focus on management expertise as the problem that must be addressed. Management may assert that the workers are the problem.

Where does the truth lie? Like any good detective, the system engineer cannot begin to say until the investigation has been broadened and evidence gathered and objectively analyzed. This means the systems engineer should look beyond just the immediate, obvious clues at hand. Perhaps there are factors in the environment of the assembly plant itself, beyond the immediate control of either the assembly workers or line management, that are the culprits. A needs analysis will broaden the search beyond primitive need statements and look for evidence to support claims of effective needs. For instance, one strong piece of evidence that an effective need exists to bring a new system into being or to improve an existing system is that there is someone willing to pay for it! However, the presence of such a willingness to pay is neither necessary nor sufficient to assure effectiveness of a need. Other evidence can be gathered from historical accounts or logs of system performance.

If a preponderance of the major stakeholders of a system are saying that a need exists to improve some aspect of a system, then it is more likely that an effective need exists. For example, if postal patrons, mail delivery people, and postmasters are all talking and writing about an alarming increase in lost and stolen mail, then evidence is beginning to indicate an effective need. Still, you want to gather better evidence. This means that collecting and analyzing data are often necessary. "Legal authorities report a 25% increase of reported lost and stolen mail incidents this year" is a compelling piece of evidence. The bottom line is that effective needs must be supported by organized, compelling evidence. A good systems engineer cites evidence linked with sound rationale when presenting the results of a needs analysis.

Effective needs are needs that meet three basic conditions. First, they are needs that can be supported by organized evidence. Analyzing past, present, and future trends related to the tangible need in question can help provide evidence. For example, market research and business trends, if properly focused, can reveal important insights about the need for new products and services. This organized evidence must in turn be tied to convincing rationale. For example, if market research shows that people are willing to spend some of their disposable income on a new entertainment service, there should be some compelling rationale that explains why people would want to spend their money for this new service instead of some other leisure pursuit or existing entertainment system. This leads to the next condition, which stipulates that someone must be willing to pay for it! There is not much reason to consider great plans for designing a new system or starting a large improvement project if no one can raise the funds to pay for them.

The last condition simply says that effective needs must be needs that can actually be carried out or satisfied. No matter how much evidence we compile or how much money we raise, there are some needs that are beyond the human grasp or, at least, are beyond the realm of what is currently possible. Many schemes that play on the hopes and fears of people and attract much attention and money do not meet this last condition. The systems engineer must have the integrity and objective sound judgment to identify futile projects that will only waste scarce resources.

3.4.4   Input–Output Matrix

A very useful device for thinking about the needs and constraints for a proposed system is a basic system input–output model. The familiar system

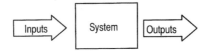

Figure 3.5 Basic input–output model of a system.

input–output model appears as shown in Figure 3.5. The square box in the input–output model represents the system. The main purpose of the system is to take inputs, shown as arrows flowing into the box from the left, and transform them into outputs. Outputs are shown as arrows exiting the box to the right.

Ostrofsky [18] shows how to create an input–output matrix that examines different types of inputs and outputs over the system's total life cycle. Constructing one of these input–output matrices is fairly simple as shown in Figure 3.6. You divide the inputs into two thought-provoking categories, intended and environmental. Intended inputs are ones that the system's designers and operators can completely determine or control. This includes things such as facilities, resources, procedures, organizational structure, and raw materials. Environmental inputs are inputs that may occur but often cannot be controlled. However, it may be possible to partially control, influence, or mitigate the effects of environmental inputs. Some examples are weather, customer demand, labor relations, and government regulations.

To stimulate your thinking about the system's outputs, you might divide the outputs also into two categories, desired and undesired. Desired outputs are the results or main products of the system. They are the outputs that justify the existence of the system. As you specify the desired outputs, you should ask yourself if these outputs are truly needed. Also, you should consider if there are other desired outputs that are also needed. Usually you desire to, somehow,

Life Cycle Phases	Inputs		Outputs	
	Intended	Unintended	Desired	Undesired
Requirements				
Design				
Production				
Distribution				
Operations				
Maintenance				

Figure 3.6 Input–output matrix.

"maximize" the desired outputs. Undesired outputs are the byproducts of the system's transformation process and often they suggest constraints that the new system must meet. Usually you need to minimize undesired outputs. Examples of undesired outputs are waste products, pollution, accidents, and excess inventory. For example, when you realize that the operation of a system produces pollutant byproducts, then that suggests that you should investigate possible constraints related to these byproducts that government regulations or industry standards may stipulate.

To make a complete input–output matrix, you should consider the four input–output categories for each phase of the systems life cycle as shown in Figure 3.6. To begin such a matrix, you will sometimes find it easiest to start with the operations phase of the life cycle. Beginning with the operations phase is recommended because defining what a new system should accomplish while it is operating is usually more straightforward than thinking about the different inputs and outputs for the other life-cycle phases. Once you define the inputs and outputs for the operations phase, you can use that definition as a reference to help stimulate thinking about the other phases.

While searching for evidence to support effective needs, a systems engineer will come across constraints or limits related to the effort. For example, when investigating the question of whether or not someone is willing to pay for a project, information about how much someone is willing to pay will logically follow. Therefore, a budget constraint has been tentatively identified. In similar manner, other important constraints will emerge during a needs analysis effort. These constraints deserve careful scrutiny of their own and should be thoroughly explored before too much work is expended to meet what turns out to be an artificial constraint.

Arthur Hall, who has accomplished much early seminal work in systems engineering [19,20], recommends examining a wide range of boundary conditions for discovering the constraints needed for evolution of a functioning system. Here is a list and short explanation of his recommended questions that systems engineers should strive to answer concerning needs and constraints.

1. *Situation.* What kind of systems design situation are you considering? Are you designing a new system, or is this a project to improve an existing system? If it is an improvement, what aspect of the life cycle of the existing system needs to be improved? Is it the operating, maintaining, or manufacturing of the system that must be improved?

2. *Expertise.* What field of design is most critically related to the need implied by the situation? Is it increasing the functionality, performance, and attractiveness of the system, or is it a matter of reducing the cost or some combination of all of these factors?

3. *Risk.* Do the stakeholders for the proposed system want to be bold and search for new possibilities, or are they more comfortable with less risk and content to align closely with the existing environment?

4. *Spillover Effects.* What are the expected impacts of the resulting system on other systems, on other areas in the environment, and on the rest of the business sponsoring the venture?

5. *Knowledge.* What is the current state of knowledge about the proposed system and its environment especially relevant available technologies.

6. *Viewpoints.* What are the views that different classes of consumers or users have about the proposed system's features and costs?

7. *Experience.* How much experience does the systems team have at developing similar systems or projects?

8. *Kind of Need.* What sort of need does this system satisfy? Is it an isolated need, or does it interact with other needs? Will this need still exist if the related needs disappear or are satisfied in some other way?

9. *Frequency.* What is the frequency of this need? Is it something the consumer wants to have over and over again, or is it a once-and-for-all need?

10. *Urgency.* How urgent is the need for the consumer? What are the time limits that the consumer has for making a decision about how to satisfy this need?

11. *Limits.* What are the physical limits related to this need such as size or weight constraints?

12. *Tolerances.* Are there any tolerances that must be observed when satisfying this need such as speed or capacity constraints?

One of the concerns in attempting to identify needs, as part of an issue formulation effort, is that of distinguishing needs to achieve some purposeful objective as contrasted with wants, some of which may be frivolous. Examining potential needs from the perspective of responses to these 12 questions should help us make this distinction.

3.5   VALUE SYSTEM DESIGN

The second logical step of the formulation portion of a systems engineering effort is called value system design. It is probably the most controversial and crucial step of the entire process because several very important products result from this step.

First, the system engineer defines objectives and structures them in an objectives tree such as we illustrate conceptually in Figure 3.7. These objectives are an organized picture of what the stakeholders intend to accomplish with a given project or program. Selecting objectives is crucial because if we do not choose the correct objectives for the system, it is highly unlikely that we will be able to engineer a "correct" system.

Because stakeholders have different viewpoints about what objectives to

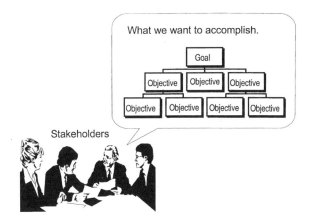

Figure 3.7 One reason for an objectives tree — to enhance communications.

pursue and which objectives are most important, developing objectives may create some controversy. Selecting objectives to pursue represents a claim by the group devising the objectives that it is possible to attain value from achieving them. To value something is to appraise the worth of an object, event, or condition. Value is a relative term and is used here to refer to various outcomes that result from alternatives in a decision problem.

In large-scale public and private systems projects, decision makers must frequently allocate resources among objectives that compete for the same resources. Hence, a value judgment or decision to adopt a particular alternative is an expression of preference for a particular set of outcomes. Value judgments are often hard to discuss because they can involve both objective and subjective points of view. However, there is also good news here. The systems engineer finds many benefits from resolving these disagreements early in the project's life. If you have a clear, agreed-upon idea of what you are trying to accomplish from the beginning, the chances of accomplishing it are much greater. Because systems engineers are concerned with the relative value of competing alternatives, the term value system refers to the set of objectives used for decision-making purposes relative to choice among alternative options.

The second important product of value system design is the definition of a set of objectives measures that may be translated into a set of design criteria. These objectives measures are crucial because you use them to evaluate the effectiveness of alternative courses of action. In other words, these criteria measure your success or failure in achieving the project's objectives. Developing useful objectives measures requires the systems engineer to first help stakeholders identify and define worthwhile objectives.

How do you develop and structure objectives? First, you get objectives defined by meeting with groups of stakeholders and recording and refining their concerns. It is best to first capture lists of possible objectives without comparing them or judging them in any way. This encourages the free flow of ideas and helps people reveal their agendas relative to the project under consideration. As ideas begin to take shape, you want to structure each objective in this form:

To (action word) + (object) + (qualifying phrase)

Having people state objectives in this form is helpful for several reasons. It is much easier to understand precisely what they mean. Stated in this form, it is clear what action they want to undertake, what objects will be impacted by the action, and what qualifications they consider to be important. It is also easier to compare different objectives when they are stated the same way. This will help you structure the objectives in a hierarchy because objectives stated in the same format make it easier to identify them as higher or lower level objectives.

How can you get started thinking about what objectives to choose? Fortunately, experience has shown that certain kinds of objectives appear frequently in many different types of systems. Here are a number of different types of objectives that Hall [19,20] has found to appear most often.

1. *Profit Objectives.* Profit is typically defined as revenue minus costs and is measured in dollars. Revenue is computed in terms of the number of units of product or service sold multiplied by the unit price of the goods or services. How profit varies over the life of a proposed project is important to consider because stakeholders may expect some new systems to generate a lot of profit very quickly. On other projects, the stakeholders may be content to generate modest profits over a longer time.

2. *Market Objectives.* Even for nonprofit or public enterprises, it is always important to think about objectives that relate to the amount of product or service delivered per unit of time as a function of time. Sometimes it is measured as a proportion of all competitors or market share. This means that market objectives can sometimes be independent of profit motives.

3. *Cost Objectives.* Because budgets are limited, cost is an important factor to consider in almost every system. Minimum cost is often a decision criterion, especially when gross income is fixed. Minimum initial cost may be an appropriate objective if capital budgets are tight. However, be careful not to emphasize cost to the extent that you are "penny wise and pound foolish." Minimizing annual costs, costs of money, depreciation, taxes, and maintenance can be most important if the stakeholders are responsible for the system during its useful life. Also, remember that most of the time you get no more than what you pay for. Lower cost often means lower quality, which can cost more in the long run. To avoid low initial cost yet expensive projects you should minimize the overall life cycle cost of a system, considering both initial costs and

annual costs, while ensuring that your system satisfies all performance objectives.

4. *Quality Objectives.* Hall says that quality has both objective and subjective aspects. For example, he says that the objective quality of a television picture is measurable by such things as the number of scanning lines, the size of the screen, and the scan rate. The subjective quality of things is the human response that people have to objective quality in a given environment. For example, the quality of light emitted by a lantern used in a tent may be very pleasing in a camping environment, but it would be very unsatisfactory from a quality standpoint as lighting for your home. Quality objectives are often subjective, especially in the beginning of a project. Understanding what features in a system are important from the user's perception of quality is key to a project's success. The systems engineer will find it a challenge to give quality attributes a physical interpretation and make them quantifiable as the project matures. One way to overcome this challenge is to get end users to experiment with prototypes of the proposed system and to provide feedback on the different features of the design.

5. *Performance Objectives.* These objectives depend strongly on the type of system under consideration because performance relates to the main purpose of a system. Hall uses the example of a communication channel that must include objectives about attenuation and phase characteristics and about signal-to-noise ratios. Normally you want to maximize performance objectives. Such terms as figures of merit, measures of productivity, measures of effectiveness, measures of performance, measures of efficiency, and efficiency factors all relate to a system's performance. Speed, accuracy, response time, throughput, and other concepts are frequently appropriate.

6. *Reliability Objectives.* These objectives concern the probability that the system or its components will operate properly for some specified time under normal operating circumstances. Care must be taken to define the intended operating environment correctly; otherwise the system may fail to operate in the users' environment.

7. *Competition Objectives.* Often there are two sides that can determine the success of a new or improved system. When this is the case, as in many military or business systems, objectives must be stated in terms relative to the competition. For example, market share is an objective that considers capturing a certain amount of the total market available.

8. *Compatibility Objectives.* Making sure the new system can work well with existing systems in the anticipated environment is an important goal in many situations. Sometimes backward compatibility, making sure the new system can work well even with older "legacy systems," is important.

9. *Adaptability Objectives.* These objectives speak to how well a system can adjust to a changing environment. For example, building into new systems the capability to handle growth in demand over time is very important especially for customer service type systems such as telephone and cable communications systems.

10. *Flexibility Objectives.* While adaptability means changing to accommodate growth patterns that place more demands on a system's productivity, flexibility objectives are about converting a system to new or multiple uses. These are very important in assuring extended system life.

11. *Permanence Objectives.* Systems can quickly become old or outdated due to the fast pace of advances in technology. This is especially true today of information technology products and services. For example, software that cannot be modified to incorporate the latest features desired by users will quickly become unwanted. Avoiding technical obsolescence is the goal of permanence objectives.

12. *Simplicity Objectives.* Straightforward designs that are easy to explain and implement often lead to elegant solutions.

13. *Safety Objectives.* These should be a consideration in every system, but how much emphasis to place on these objectives is open to debate. People disagree about safety objectives because they are difficult to measure. For example, how do you measure the cost of an accident, especially if human or environmental damage occurs?

14. *Time objectives.* Objectives related to time are almost always part of any set of objectives. Also, time interacts strongly with many other types of objectives. Many objectives are a function of time or require some qualifying phrase about time to make them clear.

Hall [19] also suggests 10 activities that represent an effective guide to developing and understanding a set of objectives.

1. *List the Objectives.* Create a list of all the possible objectives that are important to the project's stakeholders and others that the systems engineering team believe to be important for the project's success.

2. *Identify Means and Ends.* Place the related chains that link means and ends in a hierarchical structure.

3. *Check Relative Importance.* Test to see that objectives are at the correct level of relative importance.

4. *Test Logic at Each Level.* Test to see that the objectives at each level are logically consistent. Identify the inconsistent objectives because they represent tradeoff relations that will have to be carefully understood and specified.

5. *Define Tradeoffs.* Define the terms of trade for related variables. This may be as simple as finding the derivative of one variable with respect to another. Also state the limits of the variables where the trade is valid.

6. *Use Experts.* Use experience from experts, or the best available representatives of stakeholders, who have worked on similar systems to be sure the set of objectives is complete.

7. *Define Criteria.* Define objectives measures or criteria to measure each objective.

8. *Check Needs and Constraints.* Check objectives to be sure that they link to the needs, or that they will help to satisfy effective needs, and that they are within the project's other constraints.

9. *Account for Risks and Uncertainties.* Remember to account for risks and uncertainties in the way you state and measure objectives.

10. *Settle Value Conflicts.* Isolate logical and factual questions from purely value questions. Have all stakeholders' interests represented? Avoid dogmatism, dictatorial methods, and premature voting.

3.5.1 Objectives Hierarchies or Trees

Structuring objectives into a hierarchy takes considerable thought, but, as you will soon see, it has enormous payoffs. The first step in constructing an objectives tree is to examine how the objectives relate to each other. A simple self-interaction matrix can be useful for involving a group in the effort. A "+" in the matrix means that there is a positive or enhancing interaction between the two objectives: Progress on one objective will help achieve progress on the other. Decide on an appropriate definition for this contextual relationship. Use something simple and clear like "will assist in" or "contributes to." This contextual relationship is portrayed in the tree by the upward-pointing arrowhead lines or branches of the diagram. A "0" in the matrix means that the two objectives do not affect each other. Putting a "−" in the matrix means that there is negative or inhibiting interaction between the two objectives: The two objectives compete or involve tradeoffs with each other. Progress on one objective will hinder progress on the other objective.

As the tree nears completion, the top-level objectives usually describe more lofty goal aspirations that reflect strongly held values and effective need statements. These top objectives are normally difficult to measure directly. Lower-level objectives get progressively more specific and are easier to measure. The lowest-level objectives suggest activities or functions that stakeholders want to pursue. Lower-level objectives are the means for accomplishing the ends of the higher-level objectives. Therefore, a lower-level objective is a means to a higher-level objective if achieving the lower-level objective results in or helps to achieve the higher-level objective. As you examine a completed objectives tree, you should see a causal relation between ends and means. Figures 3.8 through 3.12 represent a generic cross-interaction matrix of objectives, an objectives tree, a set of objectives measures, the form in which an objectives tree generally appears, and a suggested scorecard to use for operationalizing the objectives measure concept.

Objectives	Objective 1	Objective 2	Objective 3	Objective N
Objective 1	NA	+	-	0
Objective 2	+	NA	+	+
Objective 3	-	+	NA	+
Objective N	0	+	+	NA

+ => Objectives tend to reinforce each other
0 => No apparent relationship between the objectives
- => Objectives tend to compete with or inhibit each other

Figure 3.8   Cross-interaction matrix of objectives.

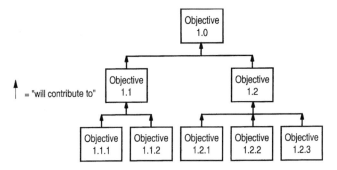

Figure 3.9   Objectives tree.

Objective	Criteria Name	Definition	Units of Measure
1.0 To maximize the capacity of the new airport sufficient to meet future demand.	1.0 Annual Service Volume (ASV)	1.0 The number of operations (landings, take-offs, and touch-and-gos) that the runway configuration can accommodate during one year.	1.0 Operations per Year

Figure 3.10   Matrix of objectives measures.

Figure 3.11   The hierarchical content of an objectives tree.

There are five tests of logic you can use to check your objectives tree. First, when we look up any one branch of the tree, higher-level objectives should tell "Why?" for lower-level objectives. Next, when we look down any branch of the tree, lower-level objectives should explain "How?" for their higher-level objectives. Third, when we look across any one level of any branch, the set of objectives should pass the "Enough?" test. This means that the set of lower-level objectives in question, if successfully achieved, should be sufficient to convince the stakeholders that success on the higher-level objective would have to follow.

The fourth test again looks across any one level of any branch and asks the question, "Extras?"—Are any of the lower-level objectives not needed or extraneous to the higher-level objective? Last, the fifth test of objective tree logic, "Owners?", requires the systems engineer to identify the owners of each objective. If no stakeholders want to claim an objective, then you must seriously consider why it is included in the tree at all. Be careful in doing this because some owners may not have a voice in your meeting. Yet, they may have a very strong stake in the outcome such as many systems where the public

Alternatives	Criteria or Objectives Measures			
	Operational Performance	Life Cycle Cost	Reliability, Availability, and Maintainability	Logistics and Supportability
New and Unique				
Off-the-Shelf				
Modified or Improved				
Baseline or Existing System				

Figure 3.12   "Scorecard" matrix.

has a strongly held interest. In these cases, you may want to have someone represent these interests even if just as a surrogate. Also, it is wise to omit the ownership of objectives during the structuring process itself. Sometimes people will compete to have the objective they own near the top of the tree, despite logic to the contrary, when an ownership tag is attached.

Besides these five tests of objectives tree logic, why?, how?, enough?, extras?, and owners?, you may find a few more dead branches to prune out of your tree. People often want to include in an objectives tree the various tasks that a systems engineering team needs to accomplish in support of a project. An example of this is an objective "To publish a final technical report" or "To prepare an in-progress review briefing by the end of the quarter." Such objectives are project milestones that belong on a Gantt chart of the systems engineering team, but they do not make any contribution at all to showing your stakeholders everything they intend to accomplish with a project. They want you to improve or create a successful plan for a large-scale system, not to generate paper! So leave the byproducts off the objectives tree.

Another problem with objectives is that, after careful inspection, you may find objectives that are duplicates of each other. It is best to define objectives that are significantly different from each other. If you don't eliminate them, they cause you to double count or overemphasize one particular aspect of your design. Not all objectives' structures will be in the form of a tree. More generally, we establish objectives' hierarchies. An objective hierarchy may be converted into a tree by partitioning objectives that would otherwise appear in more than one branch of a tree into two or more objectives that are independent of one another.

3.5.2   Objectives Measures or Design Criteria

Now that you have a logical, complete objectives tree, what can you use it for? First, the objectives tree itself is a coherent statement of everything that the project is intended to accomplish. The tree helps the systems engineer clearly communicate with stakeholders. It is easy to see the relative priority of objectives. More important objectives are near the top of the tree. Also, the tree helps the systems engineer identify tradeoffs. By looking across any one level of the tree, it is normally possible to identify objectives that compete with each other in some way. For example, an objective to minimize cost will normally compete with an objective to maximize performance because lower costs usually result in reduced performance.

The objectives tree also helps to uncover hidden agendas of stakeholders. If you ask stakeholders to review and reach agreement on an objectives tree, then they usually make sure that everything they hope to accomplish with a project is shown on the diagram.

One of the most important uses of the objectives tree is that it helps you to define and develop objectives measures or criteria. These criteria are important because you use them to measure the success of alternatives in achieving the objectives. Even though higher-level objectives may be difficult to measure, lower-level objectives are usually much easier to measure. Hence, the systems engineer can develop a combination of lower-level criteria to measure the more difficult higher-level objectives. You should strive to develop at least one criteria for each objective.

How do you create criteria? Useful criteria are objectives measures that take a quantifiable form with both a clear definition of the measure and the units associated with the measure. Criteria are often associated with counting such things as money, time, people, and products. Useful criteria to consider are: economic concerns; reliability, availability, maintainability, and supportability (RAMS) concerns; and operational and logistics issues. For example, an economic criterion for the objective to minimize the cost of a system is the life-cycle cost (LCC) of the system. The unit of measure for this criterion is the present worth of the alternative in dollars. Operational criteria relate to the main purposes of a system. For an airport, operational criteria would quantify the main purposes of an airport such as the number of flights per unit time and the number of passengers and amount of freight per unit time that the airport can accommodate. Logistical criteria count the number of people and the amount of supplies that are necessary to keep the system in operation. RAMS criteria quantify (a) the durability of the system when it is put into operation and (b) the probability that the system will not fail. The criteria you use depend on the type of system you are designing and the objectives for the system. For information systems, the functionality and usability of the system are important attributes that may require task-oriented criteria measured in user tests or experiments with prototypes. Can the user perform tasks quicker or better with the new system?

One useful technique for developing criteria is to list objectives in the first column of a table and then to brainstorm definitions and associated units of measure in two companion columns. Eventually, the systems engineer puts together a scorecard matrix that shows (a) the feasible alternatives for a project down one side and (b) the design criteria across the top. For every column in the matrix, the systems engineer uses a model or simulation to evaluate how well each alternative scores on each criteria. Because the alternatives are significantly different from each other, the models will vary somewhat for each alternative, although they will be generally the same. Chapter 4 discusses many useful techniques for modeling and analyzing alternatives on various criteria.

3.6 FORMULATION OF ISSUES EXAMPLE

This section presents a detailed example of systems engineering formulation efforts. This example, as presented originally by Sage [27], shows the products of the problem definition, value system design, and system synthesis steps as listed below.

Twelve products of the problem definition step are:

1. A well-conceived problem title,
2. A descriptive scenario that explains the nature of the problem, how it came to be a problem and presents relevant history and data,
3. An understanding of what disciplines or professions are needed to attack the problem,
4. An assessment of scope,
5. A determination of the societal sectors involved,
6. An identification of the stakeholders involved in the problem solving situation,
7. An identification of needs,
8. An identification of alterables (variables and parameters),
9. An identification of major constraints,
10. Some partitioning of the problem into relevant elements,
11. Some isolation of the subjective elements of the problem,
12. A description of interactions among relevant elements of the problem.

The four products of the value system design step are:

1. A list of well-defined objectives that represent everything the stakeholders want to achieve in resolving the problem or by addressing the issues,
2. An ordered hierarchical structure, graph or tree of these objectives.
3. An identification of the interactions relating the objectives to needs, constraints, and alterables,
4. A selection of a relevant set of measures to use to determine the attainment of objectives and related activities.

The three products of the system synthesis step are:

1. A set of alternative approaches for attaining each objective with associated activities and activities measures,
2. A detailed description of each alternative approach,
3. A definition of measures of attainment for each alternative approach,
4. An identification of the linkages among objectives, objectives measures, activities, and activities measures.

Note in the following example, objectives measures, activities, and activity measures are all developed to satisfy the intentions expressed by the system objectives.

Example 3.6.1. Energy Supply and Demand

Descriptive Scenario of the problem: There are three fundamental levels to the energy problem, namely,

1. *Energy reserves*, the deposits of fuel, such as coal, crude oil, natural gas, uranium, and the rates at which they are being discovered and transformed into refined energy.
2. *Refined energy*, the consumption of energy reserves, production of electricity, gasoline, home fuel oil, etc., and the rates at which they are being produced.
3. *Consumed energy*, the use to which refined energy is put and the rate at which it is being consumed.

An energy crisis or fuel shortage or even just large oil and gas price increases can affect everyone in the United States. It is important to ask: When did it all begin, when will it happen again, and when will it end? It probably began when people first began to consume wood by making a fire to cook their food and to protect themselves from a hostile environment.

The first alternative to wood as an energy source was coal, which was first used in the late 1800s. This solved some of the immediate problems with wood but created a few more problems, as well. Coal was subterranean, which made it harder and more costly to obtain than wood. Either the surface of the soil was stripped away to reveal the coal, which left the land scarred and barren, or tunnels were dug to reach deep deposits of coal. Also, when coal was burned, a residue of ash remained, which had to be disposed. Pollutants, such as sulfur oxides, were added to the atmosphere. An alternative to coal had to be found, no so much because of a dwindling supply as for environmental reasons.

The next solution, or energy source, emerging in the early 1900s, was crude oil and, closely related to it, natural gas. When products refined from crude oil were burned, they had two advantages over coal, namely, that very little residue remained and that they added less pollutants to the air. However, oil was usually found at much greater depths than coal, which increased the cost of exploration greatly and made it even more difficult to remove.

Crude oil, however, did have many attractions, which led to increased production and consumption. Whereas coal was usually stored in large quantities before it could be distributed to the consumer, generally by way of coalbarge, railroad cars, and ultimately truck, oil could be transported from its source by pipeline. Transportation by pipeline permits a continuous supply of crude oil to be available to the processor, that would be free from most environmental interferences such as bad weather.

Pipelines, however, are expensive to construct and maintain and must satisfy federal requirements for environmental protection. Exploration for crude oil and transportation of this oil is not an inexpensive undertaking. In order to make it more economically attractive to potential drillers, the U.S.

government had developed certain policies, such as the oil depletion allowance and crude oil import quotas, to help protect the drillers' investments.

According to the oil companies, the lack of sufficient incentives for expanded exploration for crude oil, the extremely high cost of exploration, and the lack of sufficient refineries to process the crude oil, together with an increased population with an increased energy demand per capita, and costly government regulations, have all contributed to reliance on foreign crude oil imports to the extent that U.S. demand far surpasses U.S. supply. In the fall of 1999, U.S. demand for crude oil averaged 19 million barrels per day while U.S. supply averaged only 6 million barrels of crude oil per day.

As more exploration is conducted and oil discovered, the number of barrels of oil per foot drilled has been steadily decreasing. More and deeper wells have had to be drilled to reach the available crude oil. Similarly, natural gas is being consumed at a faster rate than new reserves are being found. Huge Alaskan reserves of natural gas are not used due to pipeline restrictions.

Even if additional discoveries of crude oil and natural gas are made, the solution will only be a temporary one since all these fuels are fossil fuels which, because they are the result of subjecting decayed animals and vegetables to high temperatures and pressures, took many years to form. At the present consumption rate it does not appear that fossil fuels can replenish themselves. Moreover, the pollution byproducts of combustion of fossil fuels makes environmentalists lobby that much harder for government-sponsored discovery and use of new and cleaner fuels. Investigations have been underway for many years into replacement fuels or sources of energy, such as solar energy, nuclear energy, geothermal energy, and hydropower. Each of these sources of energy has its own limitations in that it may be extremely difficult to convert a given energy source to a convenient form or to store it for future use. Possibly its technical feasibility may not yet have been proven. Even if these problems are solvable, these sources may not be economically competitive.

3.6.1 Fuel Reserves: Fossil Fuels

Coal. Current estimates indicate that coal reserves of 1.5 trillion tons will last about 500 years. However, much of the coal has a high sulfur content, which is unacceptable because of its high pollution rate. The state of New York is bringing litigation against Indiana and Ohio for operating several coal-burning power plants that cause polluted air to drift into the skies over New York State. Since 1970, more than 530 million tons of coal was consumed in the United States and the growth rate of coal consumption continues at about 3% a year. The U.S. is the "Saudi Arabia" of coal, but cleaner burning technologies for coal are needed.

Oil. In 1970, approximately 4 billion barrels of oil were consumed in the United States. The U.S. (1999) consumes nearly 20 million barrels per day. U.S.

stockage levels in 1999 were 310 million barrels of crude and 203 million barrels of gas—not nearly enough to withstand another energy crisis. Pollution attributable to oil is of a different nature than that due to coal. Most of the pollution associated with oil, however, comes from the production process at the refineries. In the production of usable petroleum products, liquid, solid, and gaseous wastes are produced which can pollute land, water, and air. Oil spills in the ocean occur due to the large amount of crude oil that is transported by ship, such as the well-known tragedy of the Exxon Valdez. Finally, thermal and air pollution occur when the oil is consumed in either electric generating plants or automobiles. Although progress has been made in creating cleaner burning gasoline engines for cars and trucks, the sheer volume and concentration of modern traffic continues to make "smog" a big-city issue.

Natural Gas. Natural gas has a big advantage over coal and oil in that its consumption contributes to pollution problems only by adding nitrogen oxides to the atmosphere and by increasing thermal pollution. Natural gas has been replacing coal and oil as a fuel because of some advantages in transporting, storing, and using it. Oil and gas experts think that Middle East natural gas may power the next millenium. Per unit volume, however, oil outputs thousands more units of usable energy.

Shale Oil. Shale oil deposits contain anywhere from 10 to 100 gallons of oil per ton of shale. Much of this oil is not considered a practical source of energy since extraction measures, such as super heated steam treatment of a shale oil bed, can be very costly. If current reserves of coal, oil, and natural gas become depleted or the cost of crude oil increases by 150 percent, then shale oil deposits would become profitable. Estimates of shale oil reserves range from 160 to 600 billion barrels, which represent a potential reserve of about 35 to 120 years of supply.

3.6.2 Fuel Reserves: Nonfossil Fuels

Uranium. Nuclear-produced energy in 1970 made up approximately 0.3% of the energy being consumed in the United States. Uranium is used primarily as a fuel for the production of electricity. Older reactors use uranium fuel inefficiently and if fast-breeder reactor technology had not been developed, uranium reserves would have been depleted in little more than a decade. The fast-breeder reactor takes uranium-238 and converts it to plutonium, which undergoes fission with fast neutrons and produces 2.5 neutrons for every neutron consumed. This in essence multiplies our fuel supply more than a hundredfold. A breeder reactor consumes 1.3 tons of uranium per million kilowatt-years of electricity produced, compared with the water-cooled reactor, which consumes 171 tons. Although theoretically uranium could supply our energy needs for thousands of years, there are several serious issues. First, disposal of radioactive wastes from nuclear-produced energy is a primary concern. Radioactive fission products are stored in solution for five years, then the solution is evaporated and the resulting solid fission products are sealed and stored where they must be safeguarded for 1000 years until the radioactivity decreases to a safe level. The more nuclear fuel we use in fission reactors, the more serious the problems will be in disposing of the radioactive wastes. Many consumers are less than enthusiastic about nuclear-produced energy. Huge cost overuns and delays in constructing some nuclear power plants frustrated many people who were willing to take the safety risks associated with living in the vicinity of these facilities in exchange for cheap, reliable electricity. When consumers had to pay as much or more for electricity generated from nuclear power plants, there was a general revulsion against the nuclear industry. The Chernobyl disaster and other widely publicized nuclear power plant safety and radiation exposure incidents made matters worse.

Geothermal Energy. Geothermal energy is that obtained by using the heat stored in steam or, hot water subterranean reservoirs at a depth of 2000–8000 feet, and at temperatures of 400–700°F. Present worldwide geothermal power capacity is estimated to be 700,000 kilowatts. United States capacity is approximately 82,000 kilowatts. These reserves will grow with the discovery and development of new sources.

Hydropower. Hydropower is the cheapest source of electricity now being used. However, most sites at which it would be advantageous for hydroelectric plants to be built have already been developed, so that additional development of hydropower is limited. Further, some dams are being evaluated for removal due to the adverse affect some hydropower dams have had on aquatic life and also new perspectives on flood control.

Solar Energy. There is almost an unlimited supply of solar energy available. Converting it economically into a useful form and storing it for future use, however, are two major hurdles still to be overcome despite much research progress. Solar energy conversion systems that are effective for large-scale use are yet to be developed.

Wind Energy. Wind mills and modern day wind farms can provide substantial amounts of energy but are dependent on location and weather. Further, wind farms take up relatively larger areas of land and mark the land with towers, shafts, and large propellers. There is also a significant maintenance burden associated with all of the mechanical parts associated with a wind farm operation.

3.6.3   Refined Energy

Refined energy is the energy ultimately delivered to the consumer. It can be in the form of electricity, gasoline, fuel oil, or natural gas delivered to the home or factory. Availability of refined energy is subject to many constraints, some by the federal government, some by energy-producing companies, and some by the consumer of energy. The consumer is the one who creates the demand, sometimes ignoring supply problems and the pollution effect of energy use.

Energy-producing companies, like many others, are in the business of supplying a product or a service for economic return. They measure the direction they should take in the discovery and development of energy by the demand of the buying public and policies issued by federal or local government.

3.6.4 Consumed Energy

The United States now has 6% of the population of the world but uses 35% of the energy produced. The United States also provides almost 50% of the world's food and many other goods and services. The consumption of electricity in the United States has been increasing approximately 9% annually, which means that every 8 years electrical capacity must double. To continue this trend for the next 20 years means there will be a need for 300 electric power plants of 3000-megawatt capacity, which will need 7 million additional acres of land for transmission and 5×10^4 gallons of water to cool the plants. The water requirement alone is more than the entire runoff from the continental United States.

Depletion of our oil reserves will make the United States more dependent on imported oil, mainly from the Arab Middle East. The 1970 import quota was set at 25%, and in 1973 increased to 37.5% of total consumption. Today more than 50% of U.S. total consumption comes from imported oil. The Arab states' oil embargos in late 1973 and 1974 created a major shortage of refined crude in the United States. The Gulf War in 1990 was largely fought over strategic control of Middle East oil reserves. The strong dependence and interaction of the entire U.S. economy with energy economics and supply continues to be a major factor that affects prices and wages in the United States today.

There are many problems and many proposed solutions to the very complex problem of energy supply and demand. Although the preceding discussion has only scratched the surface of this issue, it is a good example of the level of detail required to begin to have an adequate descriptive scenario of a large-scale, complex problem. Now we will continue with our formulation of the energy supply and demand example.

The needs, alterables, and constraints as they apply to the energy problem will be identified in the problem definition step. To help bound the problem for this example, we will make the assumption that the fundamental problem to be resolved is to develop a sound energy policy. Specifically, our problem definition, value system design, and system synthesis efforts will focus on the effects of the following policies on energy supply and demand:

Gasoline rationing,

Gasoline taxes,

Alternative energy source development,

Other.

3.6.5 Problem Definition—Identification of Needs

Needs are a condition requiring supply or relief and indicate a lack of something required, desired, or useful. For the problem of energy supply and demand, the needs are those of society or the general public in the United States and that the more basic needs, such as food, clothing, shelter, light, heat, and water, are being satisfied or will be satisfied when these additional needs are met. Needs for the energy supply and demand problem for this example are:

N_1 Adequate gasoline supply for automobiles

N_2 Adequate oil supply for industry

N_3 Adequate home heating fuel

N_4 Development of energy reserves in the continental United States, the outer continental shelf, and Alaska

N_5 A clean environment

N_6 More domestic refining capacity

N_7 Incentives for industrial exploration

N_8 Ability to refine foreign crude

N_9 An equitable crude import policy, including possibilities of taxes, quotas, and importing less crude

N_{10} Suitable refinery sites

N_{11} Efficient energy utilization and conservation, including possibilities of more, higher gas mileage cars, better home insulation

N_{12} Supertanker ports

N_{13} Reasonable fuel prices

N_{14} Reduce import bill

N_{15} Capital and incentives for building new refineries

N_{16} Automobile emission controls; lower octane, lead-free gasoline, cleaner burning engines

N_{17} More competition among foreign energy producers

N_{18} Equitable rationing of oil products during crises

N_{19} Retain freedom and independence from foreign suppliers

N_{20} Adequate supply of refined energy

N_{21} Curtail excess of demand over supply

N_{22} Reduce oil consumption

N_{23} Continue to improve real economic growth

N_{24} Continue to reduce unemployment

N_{25} Increase oil storage facilities

N_{26} Reduce population growth

N_{27} Develop new energy sources and forms

Given this list of 27 needs, the next step is to construct a self-interaction matrix. Entries in the matrix are either a "1" or a "0" to indicate an interaction or relationship exists or little or no interaction exists between the two needs being considered by each cell in the matrix. This exercise is left to the reader but to continue our example, we can determine four top-level needs from analyzing the self-interaction matrix of needs. For simplicity, these four top-level needs will be used as representative of the entire set of needs in the analysis to follow. These reduced top-level needs are:

N_1' Adequate supply of refined energy
N_2' Adequate supply of energy reserves
N_3' Efficient energy utilization
N_4' A clean environment

Definition of Top-Level Needs. *Refined energy* is the form of energy delivered to industry and the customer, such as electricity, gasoline, or house fuel oil. Much of the refined energy is used to provide some of the basic needs, such as light, heat, pure water, processed food, manufactured clothing, and shelter.

Energy reserves are the deposits or sources of fuel, such as coal, crude oil, natural gas, or uranium, which are converted into refined energy.

Efficient energy utilization means improving efficiency in developing energy reserves, processing raw energy, converting energy into useful forms, and minimizing consumption of refined energy, as well as general economic considerations in energy use.

A clean environment implies a reduction in or elimination of the pollution caused in the three stages of energy: development, production, and consumption. It includes reducing the level of pollutants in the air and water below a level determined by scientific studies to be hazardous to the health and well-being of all living creatures and the elimination and improvement of land pollution from such energy production activities as strip mining and storage of nuclear waste byproducts.

It should be pointed out that any systems engineering formulation effort such as depicted in this example requires iteration. For example, the identified needs may not be limited to these four or to the 27 more detailed needs. Other needs may become known as the problem progresses toward a solution, and these recently identified needs may then be added to the needs self-interaction matrix. The amount of effort spent in refining the formulation steps should be reasonable—something that can be handled with existing time, knowledge, and computational resources.

3.6.6 Problem Definition—Identification of Alterables

To define a problem satisfactorily, a systems engineer must determine the alterables, sometimes referred to as parameters and variables. Alterables are those elements of the problem which if changed have a significant affect on the identified needs and objectives. Alterables can be divided into two groups, controllable and uncontrollable. Controllable alterables are those that can be changed or modified to achieve a particular outcome. Uncontrollable alterables may be such things as being beyond the state-of-the-art or being dependent upon the weather. It is the controllable alterables that are of special concern, and these are used for further analysis.

A reasonable set of alterables for the energy supply-and-demand problem are:

A1   Types of refined energy produced and developed in various quantities
     — electricity, gasoline, home fuel oil and others
A2   Development of new energy sources — nuclear solar, geothermal and
     others
A3   Distribution and uses of refined energy — transportation, industry,
     homes, offices and businesses, basic necessities
A4   Types of existing energy reserves selected for development — deposits of
     coal, crude oil, natural gas, and uranium, and sources of solar and
     geothermal power
A5   Amount of foreign energy purchased for domestic consumption
A6   Uses of energy reserves — production of electricity, gasoline, home fuel
     oil and others
A7   Contribution to pollution
A8   Energy converter design — the design of automobile transportation
     engines, insulation for homes, and other new technologies or products
A9   Demand for energy consumption — the usage of electricity, gasoline,
     home fuel oil, solar energy, or geothermal energy
A10 Restrictions on use — the controls imposed on the consumption of the
     various types of refined energy. Taxes and rationing during crisis
     would be included in this alterable

Again, a self-interaction matrix is constructed of the alterables. This exercise is left to the reader. It is important to note that from this self-interaction matrix, there are many interactions among the alterables, which is what makes energy supply and demand a particularly challenging problem.

3.6.7   Problem Definition—Identification of Constraints

It is very important to determine the limitations or constraints under which the needs can or must be satisfied and the range over which the alterables can be varied. A set of constraints for the energy supply and demand problem is

C1   Safety — providing controls so that in obtaining, producing, and
     consuming energy there will be a minimum of danger or harmful effects
C2   Supply and reserve availability — convenient access to refined energy
     and energy reserves

C3   Source locations — the actual location of energy reserves and the distance to processing facilities especially foreign energy sources

C4   Reasonable costs — this applies to the price paid by industry and the consumer for energy, as well as costs of refining energy and obtaining the energy reserves with a fair return on investments

C5   Low pollution — the determination of safe levels of pollution and ensuring that in meeting the needs, pollution is kept within these safe levels

C6   Technological — sources of energy available for use but for which no known method of delivery or conversion is available

C7   Available funding — the funds available for satisfying the needs

C8   Government regulations and policies

C9   Time allowable — for implementation, building, and discovery

C10   Maintenance of a high standard of living

C11   Inflation, recession unemployment considerations

Now the constraints self-interaction matrix would be determined. From constructing this matrix you will find, for example, that many of these constraints do not interact with each other.

3.6.8   Problem Definition — Identification of Societal Sectors

There are four basic societal sectors involved although we would want to identify more subdivisions of these

Underdeveloped nations,

Europe,

Mideast oil-producing nations,

Others.

3.6.9   Value System Design — Defining Objectives

This step of problem definition involves defining objectives, determining their interactions, and ordering them in a hierarchical structure. The objectives are related to needs, alterables, and constraints. Importantly, a set of measures is determined so that the attainment of the objectives can be measured. To satisfy the needs under the constraints imposed and within the range of the alterables, the following objectives for value system design are identified

O1   To have an adequate gasoline supply

O2   To have an adequate oil supply for home and industry

O3   To develop U.S. energy reserves

O4   To have a clean environment

O5   To establish more U.S. refining capacity

O6   To provide incentives for industrial exploration
O7   To be able to refine foreign crude oil
O8   To find an equitable crude import policy
O9   To find suitable refinery sites
O10   To utilize energy efficiently
O11   To build supertanker ports
O12   To minimize the effect of oil prices on inflation
O13   To have a reasonable balance of payments
O14   To provide incentives for building refineries
O15   To control automobile and industrial pollution
O16   To minimize capital flow to hostile foreign interests
O17   To allocate energy equitably to both industry and consumer
O18   To contain rising oil prices
O19   To keep our freedom
O20   To maintain a comfortable standard of living
O21   To prevent an excess of demand oversupply
O22   To have healthy economic growth
O23   To reduce unemployment
O24   To increase oil storage facilities
O25   To reduce population growth
O26   To develop new energy sources
O27   To use different forms of energy
O28   To modify our living, working, and traveling habits
O29   To develop a sound energy policy

3.6.10   Value System Design—Objectives Measures

Most of the objectives previously listed can be evaluated directly. Others, such as O29, can be assessed only in terms of percentage of fulfillment of most of the other objectives. There does not have to be a one-to-one correspondence between objectives and objectives measures. In fact, if there were, then there would be little use for the objectives-objectives measures cross-interaction matrix. Some appropriate objectives measures are:

OM1   Number of known energy reserves, number developed and
      in process of being developed
OM2   Clean air and water indices
OM3   Number and size of refineries
OM4   Capital available for exploration and building
OM5   Number of refineries able to refine foreign crude oil
OM6   Amount of capital leaving country to purchase foreign crude
OM7   Relative amounts of domestic and foreign crude refined in
      the United States
OM8   Amount of taxes on crude and refined products
OM9   Number of available refinery sites

OM10   Available port areas, ports in use or being established
OM11   Balance of payments
OM12   Gross national product
OM13   Automobile and industrial emission measurements
OM14   Average automobile engine performance and efficiency tests
OM15   Percent crude going for gasoline, industry, home
OM16   Percent unemployed
OM17   Percent population increase
OM18   Percent of travelers in work-related or recreational-related travel using personal automobile versus some form of mass transit
OM19   Percent population at different income levels and percent of income spent on energy.

A cross-interaction matrix of objectives and objectives measures can be used to help structure a hierarchy of objectives measures and to identify the interactions among the two sets of elements.

3.6.11   System Synthesis

System synthesis answers two questions: What are the alternatives or courses of action for attaining each objective? How is each alternative described? The answers to these questions include the activities and the objectives, the activities and the constraints, and the activities measures and the objectives.

System Synthesis—Identification of Activities. The first step is to identify the activities or alternatives (set of activities) for attaining each of the objectives. After listing the activities, the activities self-interaction matrix is constructed to show the relationships between the activities. As was done for the objectives, activities measures are used to measure the degree of accomplishment of the activities. For our example problem of energy supply and demand, the activities are:

AC1   Gasoline rationing at the pump
AC2   Controls on amounts of crude going for different uses
AC3   Energy taxes
AC4   Import quotas
AC5   Conserve and increase efficiency of existing energy reserves, such as opening idle oil reserves, switching to coal where feasible, driving less, switching to mass transit, better home insulation with tax incentives
AC6   Energy independence program to include increased exploration and development of reserves, building pipelines and huge refineries and storage facilities
AC7   Crash program for developing new forms of energy and for developing technologies that make better use of existing forms of energy

AC8 Legislative program to reduce energy consumption

AC9 The "do nothing" policy — let things work out without government intervention just ensure equitable treatment for producers and consumers

AC10 Encourage more competition in the energy sector

AC11 Remove restrictions on land exploration, offshore drilling, pipeline construction and other "red tape" safety and environmental regulations that may not be necessary.

Activity measures for some of the activities are:

AM1 Percent fulfillment of demand by fuel needs

AM2 Rationing coupons/dollar or gallon limit

AM3 Amount of taxes

AM4 Price of fuel nationally and internationally

AM5 Amount of foreign crude; amount imported; percent domestic versus percent imported

AM6 Percent crude refined domestically

AM7 Amount of tax incentive

AM8 Annual 'barrels of crude' cost of environmental, safety and other government regulations versus benefit

AM9 Number of competitive firms in energy sector

Descriptions of alternatives can be formed by logically grouping several activities together and then iteratively developing more details to adequately describe each alternative.

Each alternative should contain in the description the agencies and societal sectors involved and agents responsible for carrying out each activity.

It is important to identify all the stakeholders involved: the government, industry, foreign populations, the general U.S. public and specific groups within each of these. The government, for example, includes all its "cross-purposed" agencies and political interests. Industry might include management, labor, and even the scientific community.

In many ways the most important interaction matrices are the activities and objectives self-interaction matrices and the activities-objectives cross-interaction matrix. This is so because it is vitally important to ensure that:

1. Each objective can be potentially achieved by at least one activity.

2. Each need can be satisfied by achieving at least one objective.

3. An activity achieving an objective which satisfies a need must be capable of doing this within the range of alterables and without violating the constraints of problem definition.

In summary, the approach demonstrated by this example shows what type activity is necessary to attain an objective that will satisfy a need under certain constraints and within a certain range. Moreover, it indicates the measures necessary to determine the degree of accomplishment of the activities and the objectives. The formulation method of problem definition, value system design, and system synthesis is extremely useful in helping people and organizations gain insight into the approach to be used to find viable solutions to complex, large-scale problems.

Many elements have been omitted from this example to keep the discussion reasonably simple, but even so the dimensions of the problem have become quite large. The set of self- and cross-interaction matrices have not been shown because of the considerable size of the resulting matrices. These can be conveniently put into a spreadsheet or other computer-aided application to facilitate presentation, documentation, and the complete examination of all linkages among the problem's many elements.

Importantly, the example shown here would be an early iteration of the issue formulation effort. In an actual systems engineering effort, it is essential to iterate these logical steps and refine the problem definition, value system design, and system synthesis steps. It would then be possible to become much more specific with respect to particular special problem features and associated activities and alternatives.

Further, it is ill-advised to select activities or alternatives to pursue without some knowledge of the quantitative influence of the activities on fulfillment of the needs and the measure of satisfaction obtained from this fulfillment. Chapter 4 addresses how to determine these quantitative influences.

3.7 RELATIONSHIPS BETWEEN ISSUE FORMULATION AND DESIGN AND DEVELOPMENT EFFORTS

Going back and repeating the logical steps of systems engineering is very important for achieving results that are successively closer to what is actually desired. As you learn about functional decomposition, business activity models, and quality function deployment, you will see how helpful they are in going back and refining your problem definition and value system design efforts. Also, you will find these formulation techniques helpful for moving forward to the final logical step of formulation, namely, system synthesis. Generating potentially useful alternatives and building descriptions of these alternatives in terms of detailed system requirements and specifications is a very important systems engineering activity.

3.7.1 Functional Decomposition and Functional Analysis

Another benefit from construction of an objectives tree is that the lower-levels of the tree often suggest functions for the new or improved system. A function is something that the system must do or accomplish to achieve its purposes. Because purposeful activity is a basic characteristic of any system, the systems engineer designs a system to accomplish specific tasks or functions. A function is a definite, purposeful action that a system must accomplish to achieve one of the system's objectives. For example, a weapon system normally has a loading function and a firing function. Maintenance actions are generally part of a functional description of a system. For a car, refueling and changing flat tires are two examples of functions related to maintenance activities. The systems engineer can use the lower-levels of the objectives tree to help build a functional description or decomposition.

There are several reasons why a systems engineer needs to understand a project from a functional viewpoint. First, a functional description of a system allows the systems engineer to design or plan a system independent of any specific technical solution. It is often very important not to identify any specific technical solution until later in a project. This ensures that developers do not adopt a specific technical implementation approach until all the different technical possibilities have been adequately evaluated. A functional decomposition provides reasons for the different physical components or equipments selected to implement the system. When someone asks why certain equipment is needed, the system engineer can trace the specification for it to a specific function. For example, this sensor is needed because it performs the detection function, or a certain radar is needed because it performs the tracking function.

Next, thinking of a system in functional terms provides a basis for developing innovative alternatives. Many new ideas have come about from systems engineers experimenting with the reallocation, redistribution, or the duplication and propagation of the functions of a system. One example from naval warfare is the aircraft carrier that married the functions of an airbase (storage, arming, takeoff, landing, and retrofit of warplanes) with the functions of a ship (seaborne transport). The result was an innovation in warfare that permitted vast advances in the ability to project naval power both at sea and onshore.

Another example, this one from the computer industry, is the early Apple laser printer. As Apple engineers and scientists were thinking about how to improve the speed and memory capability of their computer, they came to the realization that a lot of the computer's processing power and memory was being taken up by functions associated with the printer. So someone had the brilliant idea of reallocating those functions to the printer itself. The result of this innovation was the first printer with both memory and processing power. In fact, early Apple laser printers were some of the most powerful "computers" on the market. Soon every printer manufacturer followed Apple's lead. Many innovations in the information technology market today came from this kind of thinking. It is easy to see how different combinations of information technologies are leading to new products and services. The telephone, television, video camera and recorder, compact disc stereo, and personal computer are all being put together today in different ways.

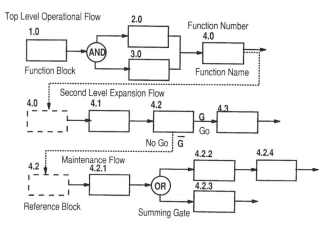

Figure 3.13   Functional flow block diagram.

Before you can begin to rearrange functions, you need a basic functional decomposition of the system. One way to accomplish this is by constructing a functional flow block diagram of the system. A flow diagram consists of multiple levels of function blocks connected by arrows. The arrows that connect the blocks show the flow or sequence of activities. Inputs enter from the left, and outputs exit from the right of function blocks. In cases where the output from a function block carries a "Go, No-Go" connotation, "Go" outputs exit from the right and "No Go" outputs exit from the bottom of a function block. By convention, flow is from left to right and must occur in either a parallel or series sequence. Figures 3.13 through 3.15 represent prototypical functional flow diagrams. In many cases, logic is used in these representations. An **"and"** connector is used, for example, to denote that B and C are parallel functions that are accomplished after function A and also indicates that they both must be accomplished before function D. An **"or"** connector is used to indicate that the flow of activity can proceed by any one of several different paths.

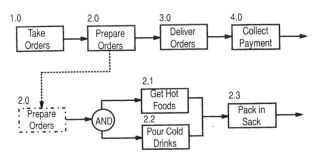

Figure 3.14   Fast-food functional flow diagram.

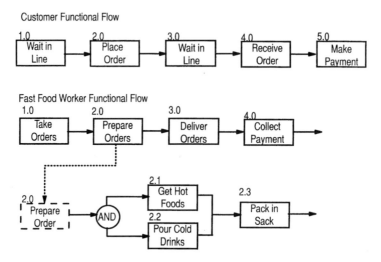

Figure 3.15   Customer and fast-food functional flow diagrams.

You need to carefully label the flow diagram. Each function block has a unique level that marks its level and its sequence within the level. To expand a function into its subfunctions, you create another level and start it with a reference block or interfacing function. This reference block is just the function that is being expanded marked by a broken outline as shown in Figures 3.13 through 3.15. To further tie the expansion to its function of origin, you draw a dotted arrowhead line from the function of origin to the reference block. Note how you number the expansion-level function blocks with consecutive numbers with periods separating each of the various levels.

Consider a straightforward functional flow diagram of a simple fast-food service system, as in Figures 3.14 and 3.15. As shown from the perspective of the servers, the top-level activity flows from taking food orders, then preparing the order, and finally delivering the order and receiving payment. Note how preparing the order is decomposed into several subfunctions at another level. Depending on the order, either servers heat hot foods or pour cold drinks or both. Then the foods are packaged together into one order before delivery to the customer. Look at how including the customer's perspective can change the flow diagram. As you see in these figures, the waiting function is placed in the flow of activity. This waiting function is very important for a fast-food restaurant to consider. The next time you visit a fast-food establishment, compare how fast-food chains have tried to deal with the customer waiting function in different ways. This simple example, although easy to understand, is far from the many levels and hundreds of functions that often are part of a flow diagram in many large-scale systems. But, even from this short introduction to flow diagrams, you will find it relatively easy to read and navigate your way through actual flow diagrams of real systems.

Functional analysis is closely related to functional decomposition. It is also performed as a part of the conceptual design process and addresses the logic structure that the system must achieve in order to achieve its desired outputs. In other words, functional analysis addresses the transformations that are needed in order to turn the available inputs into the desired outputs. There are a number of functional analysis diagramming techniques that have been developed for this purpose, and examining the logical or conceptual architecture of a system prior to development of the physical architecture or detailed architecture is now a very well accepted principle in systems engineering.

There are four elements that should be addressed in any functional analysis approach:

1. The functions are first represented as a hierarchical decomposition. There is a top-level function for the system, and this top-level function is partitioned into a set of subfunctions that use the same inputs and produce the same outputs as the top-level function. Each of these subfunctions can, if desired, be partitioned further until the needed understanding of the systems functions is obtained.

2. Functional analysis diagrams represent flows of data or physical items across the functions in any given portion of the functional decomposition. It is common that one function will produce outputs that are not explicitly useful outside the boundaries of a given system. These outputs are, however, needed in order to produce the needed external outputs.

3. Some functional analysis diagrams will also contain processing instructions that contain information needed for the functions to transform inputs into outputs.

4. The fourth and final element is the control flow that is used to sequence activation and determination of the various functions such that the overall process is both efficient and effective. This control flow determines whether the functions work serially or whether they are processed in a parallel or concurrent fashion. This control flow information indicates whether the various functions are activated only once or many times and also indicates the circumstances that determine when one function is activated rather than another function.

The major functional analysis approaches in use today are based on the following:

- *Structured analysis*, such as represented by the Structured Analysis and Design Technique (SADT), which has been used for integrated computer-aided manufacturing (ICAM) as a definition language for manufacturing systems. More recently, this has been represented by IDEF, which stands for *int*egrated *def*inition language. There are several variants. IDEF0 is an approach that is focused on the functional models of a system. IDEF1 is used to represent the information needed to support the functions of a system, and IDEF1x is a semantic data model that is based on relational data base and entity relationship models. In IDEF0 a numbered function or activity is represented by a box and described by a verb–noun phrase. In accordance with this modeling technique, inputs enter from the left of the box, controls enter from the top, resources enter from the bottom, and outputs leave from the right side of a box. A flow is represented by an arrow or arc.

- Data flow diagrams (DFDs) represent a very early modeling and diagramming technique. The basic ingredients in this approach are functions or activities, data flows, stores, and terminators.

- Control flow diagrams (CFDs) are sometimes used in conjunction with DFDs. They may be superimposed on a DFD, or they may appear concurrently with it. Control flow information is that information which is transmitted between functions or between a given function and an element external to the system.

- Functional flow block diagrams (FFBDs) represent the very first approach used for functional decomposition. There are four types of allowed control structure: series, concurrent, selection, and multiple exit. A set of functions defined in a series control structure must all be executed in the order shown. It is not possible for a second function to begin until the first is finished and control is passed from left to right along arcs from the outside to begin the first function. When a first function has been completed, control passes from the right face of the associated function box and into a subsequent function.

- N2 charts may be completed with FFBDs in order to depict items or data that represent the inputs and outputs of the functions in the functional architecture. The name N2 is used because the chart that represents N functions will be comprised of N2 boxes that illustrate the flow of items within the N functions.

Figures 3.13 through 3.15 represent a set of functional flow block diagrams. Figure 3.16 represents the inputs, controls, outputs, and mechanisms (ICOM) building blocks that comprise the SADT and IDEF approaches. Figure 3.17 represents a typical node tree showing decomposition relationships across activities. Figures 3.18 and 3.19 represent top-level activities in a typical node tree and an $IDEF_0$-like diagram for processing a fast-food order.

A definitive work on functional analysis [21] expands upon many of the discussions presented here. Closely related to functional analysis and decomposition of technologically based systems are techniques for understanding management systems and business operations, including business process reengineering. As noted in reference 2, we may approach reengineering at the level of product, process, or systems management. Let us now turn to a brief discussion of reengineering using functional analysis and decomposition.

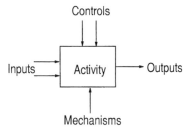

Figure 3.16 ICOM: The basic building block of an activity model.

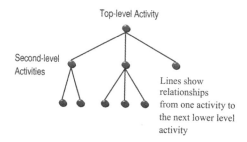

Figure 3.17 Node tree: Depiction of decomposition relationships among activities.

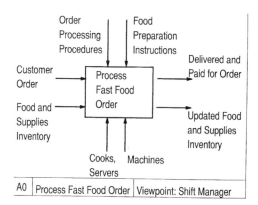

Figure 3.18 Context diagram: The top-level activity in a node tree.

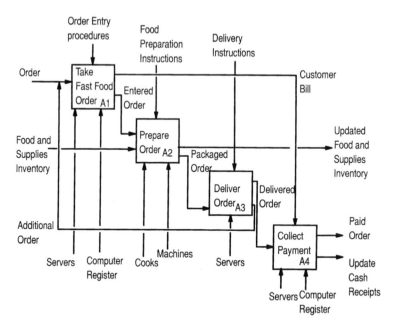

Figure 3.19 Decomposition diagram of process fast-food order.

3.7.2   Functional Analysis and Business Process Reengineering

Business process reengineering systematically examines a set of interrelated business activities and their transactions for several purposes. First, it helps to identify the most important activities so that they receive proper attention and priority effort. The most important activities are ones that add cost and value to the process. Second, it can identify extraneous or redundant activities that can be eliminated. Also, it is possible to cost activities and find new ways of accomplishing activities to save time, effort, and money. Most importantly, it is possible to construct a coherent, effective data model or information architecture to support your business activities using this approach. Building a business activity model also helps to provide a common language that makes communication easier about how a business performs its repetitive processes both within the business enterprise and for external uses. Among the watch words of business process reengineering are prioritize, simplify, save, automate, and communicate.

The basic element of an activity model or diagram is the activity box and its accompanying arrowhead lines input–control–output–mechanisms (ICOMs) identified with arrows. Similar to the functional flow block diagram, each activity box represents an activity or business process that, if appropriate, can be broken down into subordinate activities. Activities can be processes, functions, tasks, or activities but are simply referred to as activities. Use active verbs in the present tense to define activities. When you break down an activity into lower-level component activities, avoid the tendency to model the formal structure of the organization or lines of authority. Instead, you are encouraged to model the organizations processes independent of any one functional area or organizational entity.

ICOMs show the four possible roles of information or materials relative to the activity box as shown in Figure 3.16. Inputs enter the left of the box and are the information or material used to produce the activity's output. The activity acts upon or changes the input to produce the output. Controls enter the box from the top and are information or material that limit or constrain an activity. Controls regulate an activity as it transforms inputs into outputs. Outputs exit from the right of a box and are the information or materials that result from an activity.

Mechanisms enter the bottom of a box and are the resources that perform processing or provide energy to an activity. Examples of mechanisms are people or machines. When trying to decide if something is a mechanism or an input, it may help to realize that a mechanism should never be part of the output. In contrast to mechanisms, inputs are always part of the output. When an activity model is complete, you have a very graphic portrayal of the various activities in the enterprise, their relationship to each other, and how they use information, materials, people, and energy.

A complete business activity model usually has four parts. First, it has a node tree that is a graphic table of contents for the entire model. It shows the hierarchical decomposition of the activity model from the highest activity to all the subordinate, component activities. Usually, you label each activity in the tree using a nested decimal system starting with $A0$ for the top activity, $A1$, $A2, \ldots, An$ for n second-level activities, and then $A1.1$, $A1.2$, $A1.3, \ldots A1.n$ for the next level down. Figure 3.17 represents a generic node tree.

The second part of an activity model is a context diagram and is represented in Figure 3.18. It is a single, large ICOM diagram of the highest activity. Sometimes a business process modeling team may find it difficult to determine what their topmost activity should be for their project. The systems engineer can frequently break this logjam by asking the team to consider the level above their project because identifying activities at that level will put their work in context and make it easier for all to understand what the appropriate highest activity should be for the task at hand.

The third part of an activity model is the decomposition diagrams or ICOMs which show how the various activities are broken down into lower-level component activities. By breaking a complex activity into simpler parts, it is easier to understand the smaller parts, and to examine the details of each part. The model also has text that explains the overall diagram, what occurs in each activity, and how the activities interact. Be sure to include a glossary that defines terms and a legend to explain labels. Figure 3.19 represents a decomposition diagram for the fast-food processing example considered here.

Sometimes you may want to highlight a particular portion of the activity model. For example, you may want to show a "zoom-in" of only two activities to make a point. These diagrams are often part of a complete activity model and are called for-exposition-only (FEO) diagrams. This represents a relatively complete system or organization activity model. Let us see how we might use it.

The Corporate Information Management Process Improvement Methodology for US Department of Defense (DOD) Functional Managers [22,23] cites six uses of activity models:

1. Defining program requirements and scope
2. Discovering and validating business rules
3. Documenting the AS-IS environment
4. Specifying the TO-BE environment
5. Providing a framework for other types of analysis
6. Providing a basis for other modeling techniques and simulation

An activity model helps to define the requirements and scope of a business process organizational effort or a system product evolution effort. The activity model clearly shows which business activities and business transactions are under study. Because an activity model reflects a particular viewpoint from within the organization that owns the process, all members of the business process reengineering team can easily understand the context and purposes of the effort.

Because an activity model shows the relationship between the activities and their associated ICOMs, it is possible to define the business rules of a process and to create data models for the business's knowledge infrastructure or information architecture. Business rules identify, describe, and specify the relationships among the various entities of an organization. Entities are things that an organization worries about. They are part of the organization's knowledge base because the organization must communicate and measure these entities in some way. Examples of entities are people, places, things, ideas, and events. Attributes describe each entity. Business rules are written sentences that express the relationship and cardinality between entities. Taken together, the business rules of a process determine the use and creation of the organization's data related to the activities in a business process. Hence, you can use an activity model to help you build a data model for a business process.

The AS-IS activity model is very important. Because you build an activity model by having a reengineering team examine an existing business process, the activity model is an accurate description of how an organization currently operates. This activity model that describes the current state of the business environment is called an AS-IS activity model. The AS-IS model helps greatly to identify problems. The cost and value of each activity is easy to identify using the AS-IS model. By comparing the value of an activity's inputs to the value of its outputs and totaling the cost of an activity's mechanisms, you can determine the value-added of an activity. The AS-IS activity model is an accurate way for a systems engineer to portray the descriptive scenario or current status of a business process. Then, you can use the AS-IS model to identify problem processes or activities that cost too much or add too little value. This leads to thinking about ways to redesign current business practices.

Following this, we determine the TO-BE activity model. Activity models that show new ways to accomplish a business process are TO-BE models. TO-BE models portray the normative scenario and can help you develop specifications for streamlining processes, eliminating redundancies, automating activities where appropriate, and finding alternatives for making substantial improvements to a business. As different alternatives develop in the creative minds of the business reengineering team, several different TO-BE models are built, one for each alternative.

One important byproduct of this effort is an activity-based costing (ABC) analysis. You can also use both AS-IS and TO-BE activity models as a framework for other types of analysis. This is especially true when you are comparing several TO-BE alternatives to the baseline AS-IS model. One example of the type of analysis that can be based on an activity model is ABC, one of many cost management systems [24]. In ABC you identify the costs of activities and their outputs as a way of measuring the value or need for each activity. These cost assessments help to support design decisions about which activities to automate, eliminate, or improve. One of the most important results of ABC is the determination of value added or non-value added activities. A non-value added activity has input costs and activity costs that, when summed together, are greater than the worth of the output product or service. Such non-value added activities are candidates for elimination or reduction because they cause delays or duplication without benefit. In contrast, the systems engineer targets value-added activities for improvements such as automation that can reduce costs.

There are five steps in an ABC analysis, as discussed in considerably greater detail in references 2 and 24 and the references provided therein:

1. Analyze activities.
2. Gather costs.
3. Trace costs to activities.
4. Establish output measures.
5. Analyze costs.

We describe each of these steps in detail here:

   1. *Step 1: Analyze Activities.* The activity analyst, or reengineering team determines which activities to include in the ABC effort. Then the ABC

team classifies activities as primary, secondary, required, or discretionary. A primary activity is one that directly supports the organization's mission. Secondary activities support primary activities. Activities that the organization must perform are required, while discretionary activities are ones that are truly optional.

2. *Step 2: Gather Costs.* The next step gathers costs for each activity. The team uses the activity model as a basis for conducting interviews and requesting accounting information from key people involved with the activities under study. Much of the cost information gathered is based on the historical expenditures of the organization. When possible, use cost methods that the organization has found useful. Still, it will probably be necessary to estimate some costs.

3. *Step 3: Trace Costs to Activities.* To trace all of the gathered costs to their appropriate activities is the next step. Use this simple rule of thumb for tracing costs: Outputs consume activities that in turn have consumed costs associated with resources. To calculate total costs consumed by an activity, multiply the percent of time expended by an organizational unit on each activity by the total input cost for that organizational unit. Look at who participates in each activity and how much of their time is spent on this participation. Look at the resources used for each activity and determine how much of these resources are used and what they cost. Assign the total input cost of each secondary activity to the primary activity it supports. Check your work by seeing if the sum of all activity costs equals the sum of all organizational costs.

4. *Step 4: Establish Output Measures.* Next, you calculate the unit cost of each activity by establishing output measures. To calculate the activity unit cost, divide the total input costs by the primary activity output volume. Be sure to include all assigned costs from secondary activities in the total input costs. To determine the primary output volume, examine each activity's outputs and choose one of its outputs as the primary activity output. Choose an output that has a volume or quantity that can be easily measured. Now that you have the unit costs for each activity, it is a straightforward task to prepare a table or bill of activities. This lists a set of activities and the amount of each activity consumed based on the activity unit cost with a total cost for the entire bill of activity. You can prepare a bill of activity for any desired cost object such as a process, a product, or a service.

5. *Step 5: Analyze Costs.* The fifth and last step in ABC analysis uses the classification of activities with the activity unit costs and bills of activity to find ways to reengineer the business processes under study. Non-value-added activities are segregated, and their associated non-value-added costs are analyzed so that the needless expenditures of time and money are eliminated or reduced. Value-added activities typically get organized

into a Pareto analysis,* so significant "cost drivers" can be identified and alternative ways can be found to reduce their costs by automation support or other means.

There are a number of approaches that are similar to ABC that are also useful in the issue formulation efforts. These include economic analysis (EA) and functional economic analysis (FEA). While ABC focuses on past and present costs and the AS-IS activity model, you can use similar procedures to conduct an EA of different investment alternatives based on TO-BE activity models. EA efforts usually investigate alternatives for a single initiative or information system. Larger and broader efforts that encompass an entire functional area are called functional economic analyses (FEAs). They consider benefits and risks as well as costs of planned business process reengineering projects. Risks can include such factors as technical risk, which examines the advantages and disadvantages of using proven or new technology, or schedule risk, which looks at the size of the project and the ability of contractors and in-house personnel to meet a project's schedule. One typical measure used in FEAs is the tooth-to-tail ratio. This is the ratio of operational costs over management and support costs. Business operations with low overhead have a high tooth-to-tail ratio.

Simulation is an alternate and useful approach. Besides using activity models for data modeling and ABC-, EA-, and FEA-type efforts, all of which are steady-state or static representations of a business process, you can use an activity model to build a simulation of a business process. A simulation helps to account for the importance of how activities change with time and can show the sequencing of activities. Most of the elements in an activity model translate easily to a simulation model with the straightforward addition of attributes to activities like activity duration and sequence information. Simulations help to identify the dynamics in business processes that cause bottlenecks or other resource allocation or flow problems which may not be evident in static models. We will discuss modeling and simulation further, but primarily from the point of issue analysis, in our next chapter.

3.7.3   Quality Function Deployment

Issue formulation and the associated definition of a system, or organization, that is to be realized is closely related to quality assurance and management because the initial efforts at issue formulation will do much to influence the quality of the resulting product or organization. One of the most widespread

---

*Pareto was an early twentieth-century Italian economist. He postulated the law of maldistribution, which says that 80% of the effort expended in a process is often caused by only 20% of the input. More importantly here is the concept of Pareto optimality, which suggests that in a multidimensional optimization problem a noninferior, or Pareto optimal, solution is one in which it is not possible to make improvement in performance on one of the dimensions without degrading performance on at least one other.

approaches to implementation of TQM is that of quality function deployment (QFD) [25]. QFD was reportedly first used in 1972 by Mitsubishi, at its Kobe shipyard, and then adopted by Toyota, who developed it in a number of ways [26]. We will examine some of the uses of QFD here.

The purpose of QFD is to promote integration of organizational functions and to facilitate responsiveness to customer requirements. QFD is generally said to be comprised of structured relationships and is intended to be used by multifunctional teams. The members of the multifunctional teams attempt to ensure that all needed information regarding customer requirements is identified and used in the best way, that there exists a common understanding of decisions, and that there is a consensual commitment to carry out all decisions.

Interaction matrices, as discussed earlier in this chapter, and in some detail in references 1 and 27 and elsewhere, may be of considerable assistance in this effort. A QFD matrix represents an actual or conceptual collection of interaction matrices that provides the means for transitional and functional planning and communication across groups. Basically, it attempts to encourage an early identification of potential difficulties at an early stage in the life cycle of a system. It encourages those responsible for fielding a large system to focus on customer requirements and to develop a customer orientation and customer-motivated attitude toward everything. Thus, the traditional focus on satisfying technical system specifications is sublimated to, but not replaced by, the notion of total satisfaction of customer requirements.

Two QFD matrices have been suggested. One is a "house of quality" effort matrix, and the second is a "policy assessment" matrix. Generally, the associated interaction matrices might appear as illustrated conceptually in Figure 3.20. This figure presents an illustrative picture of the house of quality and house of policy, which follows illustrative house-of-quality depiction of the various interactions that influence quality.

The generic efforts involved in establishing QFD matrices are as follows:

1. Identify customer requirements and needs. Establish these in the form of weighted requirements expressing their importance.
2. Determine the systems engineering architectural and functional design characteristics that correspond to these requirements.
3. Determine the manner and extent of influence among customer requirements and functional design characteristics and how potential changes in one functional design specification will affect other functional design specifications.

As a related and more strategic matter, it would be necessary to determine (a) the extent to which the specific systems fielding effort under consideration is supportive of the long-term objectives of the organization and (b) the competitive position and advantage to be obtained through undertaking the development. We see that this QFD approach has uses in strategic management to

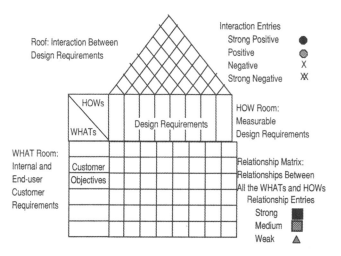

Figure 3.20   QFD chart, which relates customer whats to designer hows.

determine if the management controls or systems management effort associated with the development of the system under consideration is supportive of the overall development strategies of the organization.

The various interaction matrix entries might well be completed with binary ones and zeros to indicate interaction or no interaction. Alternately, there would be major potential advantages in indicating the strength of interaction. Hypothetically, at least, the QFD matrices provide a link between (a) the semantic prose that often represents customer requirements and (b) the functional requirements that would be associated with technological system specifications. In a similar manner, an interaction matrix can be used to portray the transitioning from functional requirements to detailed design requirements. Thus, they provide one mechanism for transitioning between these two phases of the systems life cycle. One goal in this would be to maintain independence of the various functional requirements to the extent possible. A measure of this independence can be obtained from the density and location of the interactions among the functional requirements. In this discussion, we have mentioned the words requirements and specifications numerous times. To understand and appreciate systems engineering to the fullest, you need to learn what these two terms are all about.

3.8   THE SYSTEMS ENGINEERING REQUIREMENTS STATEMENT

Requirements and specifications are the product of all of the initial definition efforts that lead to development configurations decisions relative to a system. This means that the conceptual design or plan for a system is comprised, at this conceptual level, of a set of requirements and specifications for that system.

But what is the difference between requirements and specifications? When system engineers use the term requirements, they are usually referring to the broad, top-level statements that describe the purposes and functioning of a system and its major components in support of client or enterprise needs and objectives. Requirements relate to the effective needs that have been identified in a needs assessment. In contrast, specifications normally refer to the technical details of the system's physical elements that will ultimately be implemented in terms of components. Sometimes systems engineers use the term requirements to encompass the technical specifications, but the two are relatively different, though related, constructs. For this reason, it is useful to speak in terms of user requirements, system specifications, and implementation or component specifications.

Early in the acquisition life cycle of a system, it is very useful to bring together all of the information generated during the definition phase effort into one summarized statement that helps to define the system. This statement is called the *systems engineering requirements statement*. It contains the effective needs, objectives, criteria, parameters, variables, and constraints of the client that will be incorporated into the "system" that has been identified. It describes the logical relationship of how objectives can, if achieved, satisfy the effective needs, and how the objectives can be measured by the objective measures or design criteria. It also discusses the key performance parameters that define the basic operating characteristics of the system, the important state variables that the system must monitor and control, and the various constraints that must be observed for the system design. You should include the major advantages and disadvantages of the different alternatives and technical approaches identified by the feasibility studies in a statement concerning issue formulation.

The purpose of this statement is to inform the clients and important stakeholders of the overall concept of the system in an easy-to-understand way. This gives them a chance to make changes in the direction of the design effort early in the system's life, when changes are least expensive, before major commitment of resources. Information about what major functions the system will perform, how the system will accomplish its objectives, when it will be deployed, where the system will operate, and who will use the system is all open to discussion and approval with the clients for the systems engineering effort [28]. To develop a systems engineering requirements statement we need to accomplish the formulation, analysis, and interpretation steps that comprise the definition phase of system acquisition. In this chapter, we are primarily concerned with the formulation step.

3.8.1   User Requirements

Finding out what users really want or need is difficult for several reasons. Information about user requirements often has a lot of uncertainty associated with it. For example, even though an effective strategy is to ask users about their requirements, there are many perceptual and human information processing biases that may limit the value of information obtained from observing or interviewing users [29]. Sometimes users focus on symptoms related to their true problem and ignore the root causes of their difficulties. This means that the systems engineer frequently has to help users detect the existence of the true problem and properly diagnose its causes.

There are many difficulties faced by systems engineers in problem identification and diagnosis due to different types of human error. Sage [1] mentions four types. First, people can set improper thresholds for recognizing problems. If people set a high threshold, then they may not detect true problems or they may suffer a long time delay and an increase in the problem symptoms before they recognize it. Another difficulty is the failure to generalize. This occurs when people treat related problems as though they were separate. Often people are too reactive and wait too long to address problems instead of looking ahead to anticipate likely difficulties and taking action to avoid them. Fourth, people too frequently ignore evidence that they should consider. This happens especially when they think they know what the problem is and they ignore evidence to the contrary. Despite these difficulties, there are a number of useful strategies for determining user requirements.

Davis [30] describes four approaches that can help overcome some of the human limitations we just mentioned. The first way is to interview users and structure your questions and discussions so that you compensate for the user's biases. In the second strategy, you carefully observe existing systems that are very similar in nature and purpose to the one you are designing. You can review plans and reports about the existing systems to help you develop user requirements for the new system.

The third strategy synthesizes information requirements from a set of characteristics that describe the system to be developed. And the fourth strategy is to use experimentation. In this approach, a prototype or simulation is used in actual or simulated operational settings to provide more information about requirements.

It is very important to have a good set of top-level system requirements to guide decision making throughout the remainder of the efforts, which include development and deployment of the system. We can approach requirements, as with most of the other systems engineering efforts from either a top-down or a bottom-up perspective. Usually, the requirements identification is more efficient and of higher quality if a top-down approach is used. The top-level system requirements should focus on the user. The system engineer derives these requirements from at least five sources:

1. Needs analysis
2. Problem definition
3. Operational concept
4. Maintenance concept
5. Functional analysis

You already know about a needs analysis, problem definition efforts, and the functional analysis. So, let us describe the operational concept and the maintenance concept here.

The operational concept depends on a thorough understanding of the missions that the system must perform and the environment that the system will have to operate in. Hence the systems engineer works closely with users or their representatives to write the operational requirements. Together, they describe the major functions and operational characteristics desired for the system. How the system operates, where in the operating environment the system will be distributed, how long the system must operate, and how effective the system's performance must be are all part of the operational concept. Care is taken not to specify technical solutions but to describe the performance desired of the new or improved system. Usually the operational concept describes typical mission profiles or operational scenarios associated with use of the system.

The maintenance and support concept is put together to establish how we envision keeping the system in operation and restoring it to operational use should it break down. These requirements describe the effectiveness of the support elements of the system. Maintenance levels, repair policies, anticipated major maintenance procedures, and logistical requirements related to people and support equipment are each part of the maintenance and support concept.

3.8.2 System Requirements and Specifications

Generally, we use the term user requirements, or system requirements, to denote the needs of the user(s), or the user enterprise, for the system as expressed in terms of generally purposeful requirements. After the requirements have been identified, there is a need to translate these into system specifications, which are the functional statements that the system will satisfy. The requirements that come from the needs analysis, problem definition, operational concept, maintenance and support concept, and the functional analysis may be translated into a series of specifications that detail the performance, effectiveness, and related characteristics of the different components of the system. There are many different formats and classifications for specifications. One often used in the Department of Defense usually has at least five different types of specifications [23]. These are as follows:

1. System specifications
2. Development specifications
3. Product specifications
4. Process specifications
5. Material specifications

The most important specification, from a systems engineering viewpoint, is the system specification. It is crucial because it serves as the top-level technical guide for all other design specifications. It is normally the job of a systems engineer to prepare it. All significant design characteristics and features that apply to the system level should be included. The system specifications typically include [21] the following:

1. A definition of the system as an entity
2. The major characteristics of the system
3. Related technical performance measures for system evaluation
4. Some general criteria for design and assembly of the system
5. Major data requirements for the system
6. Logistics and producibility considerations
7. Test and evaluation requirements
8. Quality assurance provisions

Thus, we see that the system definition includes a general description of the system, its operational requirements, its maintenance concept, its functional block diagrams, the system interface criteria, and the environmental conditions for system operation.

System characteristics usually include the system's performance characteristics, physical characteristics, effectiveness requirements, reliability, maintainability, human factors, supportability, and transportability requirements.

Design and assembly requirements cover the manufacturing requirements, materials, processes, and parts used in the fabrication of the system and its major components, mounting and labeling requirements, safety restrictions, interoperability, and compatibility standards. Documentation and data requirements define the data collection, distribution, processing, and reporting provisions needed for the system. The logistic requirements detail the system's maintenance requirements, its supply support considerations, test and support equipment stipulations, personnel and training needs, facilities and equipment required for logistical purposes, and packaging, handling, storage, and transportation provisions, and any other resources or services associated with the system's logistical requirements.

Test and evaluation requirements specify the standardized testing, self-test and built-in test provisions, and fault detection and diagnosis requirements and evaluation procedures that should be used throughout the system's development. Quality assurance provisions specify the design reviews, audits, standards, system integration requirements, and configuration management procedures that will be used to ensure a high-quality system.

Once you specify the top-level system requirements, you can allocate them to the various subordinate level specifications: the development, process, product, and material specifications. Development specifications contain details about various components that require research and development efforts. In contrast to these, product specifications contain details about system components that already exist and can be purchased off-the-shelf from various vendors. Examples of the type of standard components that are usually available commercially in current inventory are cables, plugs, receptacles, displays, computer software applications and utilities, standard hardware assemblies, microchips, motors, and various tools and spare parts.

Process specifications detail the technical requirements for any services that must be performed on the various components of the system. Some examples of such processes are machining, welding, painting, and marking. Material specifications detail the technical requirements for the materials and mixtures that are used in making the products or in performing the processes for the system's components. Paints, chemicals, dyes, plastics and composites, asphalt, and wiring are some examples of things that material specifications may cover.

As you can imagine, the requirements and specifications for a system grow considerably as the systems engineering effort continues from very high-level functional issues to physical system issues to implementation issues. Although the system engineer is responsible for the system specifications, which are prepared near the end of the definition phase or beginning of the development phase, there are second-level implementation specifications that will be developed later in the development by experts in appropriate engineering disciplines such as mechanical or electrical engineers. The systems engineer will often be involved in technical direction and integration efforts to ensure that the subordinate implementation specifications comply with the system level specifications.

To help keep track of all of these specifications, the systems engineer often prepares a specification or documentation tree. This tree is based on the system's physical hierarchy that shows how the system is subdivided into subsystems, major components, assemblies, subassemblies, and individual units. The systems engineer builds a specification tree by identifying the different types of specifications that should go with each element shown on the system hierarchy. You number and title each specification according to its name and indenture on the system hierarchy. The tree, when completed, shows the hierarchy of specifications from the system specification on down. The specification tree also helps to resolve conflicts that arise over which specification should take precedence. This specification tree is closely related to a number of other systems engineering management aids, such as a work breakdown structure [1,2].

Blanchard [28] is an excellent source for learning more about specifications, so let's review his major points in a few short lists. Specifications are documents for

1. Acquiring new equipment
2. Procuring off-the-shelf components
3. Contracting for new products
4. Testing of items or products
5. Verifying that items meet standards
6. Building new facilities

7. Calling for production runs
8. Controlling design details, performance, and quality

They can apply to

1. Contracts
2. Major contractors
3. Subcontractors
4. Suppliers and vendors of goods and services

Specifications should not contain

1. Management information
2. Statements-of-work
3. Procedural data
4. Schedules
5. Cost projections

Here are some different types of specifications:

1. General pecifications
2. Program peculiar specifications
3. Military specifications and standards
4. Company standards
5. Industry and trade standards
6. International specifications and standards

Specifications can cover the

1. System
2. Components
3. Products
4. Processes
5. Materials

Specifications should be

1. Oriented on requirements
2. Written in clear and concise language
3. Consistent with the system hierarchy
4. Organized into a documentation tree

3.9   GENERATION OF ALTERNATIVES OR SYSTEM SYNTHESIS

One of the biggest payoffs from making a good objectives tree is the way you can use it to help you generate alternatives. How do you do this? It is a relatively easy down-and-across technique. Working down first, you create a list of activities for each lower-level objective in the tree. This list of activities is a set of different ways for accomplishing each lower level objective. Once you have a good list of activities for each lower-level objective, then you create alternatives by combining these activities together by working across the lists. You can select one activity from each list or you can select several. Each synthesis or combination of activities represents one draft alternative that possibly merits further consideration and refinement.

Consider a simple example of this approach. Suppose that we have an objectives tree that shows what a group of stakeholders hopes to accomplish with a recycling project. Underneath each lower-level objective you should identify a list of activities that represent different ways of achieving the lower-level objectives. Then circle all activities, or combinations, that support attaining all lower-level objectives. These circled activities represent one possible system synthesis or alternative. There are many more possible alternatives that we can generate from these lists of activities. You can appreciate the power of this approach. It helps you quickly generate many different alternatives.

Of course, the quality of alternatives generated in this way depends heavily on the quality of the lists of activities that were used to create the alternatives in the first place. For this reason, most large-scale projects are such that it is highly desirable, if not necessary, that systems engineers responsible for technical direction of effort involve a group of experts to create these lists of activities and to synthesize them into viable alternatives. This aspect of systems engineering is sometimes approached through such group dialogue techniques as brainstorming, brainwriting, and Delphi.

3.9.1   Brainstorming and Brainwriting

Often there is no one person with sufficient knowledge about a complex situation to develop a set of elements which describes it. Usually, available information about a complex situation is incomplete, uncertain, or even wrong. The model that represents a complex situation may also be unverified and perhaps incomplete. Developing useful alternatives requires finding a way to overcome these imperfect information problems and find as much truth as we can.

We can classify the information about a complex situation into three types: speculation, opinion, and knowledge. As speculation becomes knowledge, the probability of truth varies from 0 to 1. We would rather not base decisions on speculations because the probability of truth is so low. However, sometimes the knowledge about a complex issue is such that we have little detailed, reliable information. Therefore, in actual practice, systems engineers must find ways to gather people's opinions and develop information from them such as to inform and enlighten a complex situation. Hence, systems engineers often work with groups to solicit better information.

The appeal of groups is based on the idea that "two heads are better than one." Committees, juries, boards of directors, and panels are all examples of this. They represent some of the many mechanisms for pooling minds to generate ideas. We hope that their ideas are based on wise collective opinions or expert knowledge, each with a high probability of being correct.

However, the systems engineer must realize that there are both advantages and disadvantages to reliance on group opinions. On the favorable side, many times a group will interact to compensate for the bias of individual members of the group. Or knowledge of one group member can compensate for ignorance or speculation on the part of other members. Also, people that participate in a group tend to develop a stronger stake in the group's enterprise. One way of getting more support for a large project is to include many potential stakeholders in the planning group for such an effort.

Unfortunately, there are drawbacks to having a group involved. When too many diverse interests go unchecked in a group, you may end up with the camel outcome: "A camel is a horse designed by a committee." Sometimes the opinion of the group can be strongly influenced by the person who talks the most and the loudest. This can be upsetting to a group because there is little necessary correlation between loudness or talking a lot and knowledge. Another problem is that group activity can lapse into a "bull session" unless the group is well-organized. "Bull sessions" create lots of talk about matters of individual and group interest but avoid discussing the main issues of the group's charter. Often there is strong pressure for group conformity and avoidance of unpopular viewpoints. Martino [31] provides a good list of group disadvantages that the systems engineer should try to control when working with committees to evolve expert opinion:

1. *Misinformation.* There is as much misinformation available to a committee as there is available to any of its members. Hopefully, correct information held by some group members cancels out wrong information held by another group member. There is no guarantee that this will happen.

2. *Social Pressure.* A group often exerts strong social pressure on its members and encourages all members to agree with the majority even when several individuals feel that the majority view is wrong.

3. *Vocal Majority.* Often the number and volume of comments and arguments for and against a position is more influential in determining results than the validity of these. Again, a strong, loud, vocal majority may overwhelm the group.

4. *Agreement Bias.* A group is often more concerned with reaching an

agreement than reaching well-thought-out conclusions. The result is usually a statement of mild philosophical views that can offend no one rather than concrete, specific suggestions that may offend some group participants.

5. *Dominant Individual.* The dominant individual often has an undue impact on the final results of the group unless there is strong and impartial group leadership. This dominance may be due to active and loud participation, a persuasive personality, or extreme persistence.

6. *Hidden Agendas.* Hidden agendas and vested interests on the part of some group members may lead to a game in which the objective is convincing the group of their view, rather than striving for what might be a better group decision.

7. *Premature Solution Focus.* The entire group may possess a common bias for a particular alternative or technology.

So how does a systems engineer cope with the challenges of group dynamics? Some of the drawbacks of groups may be made better by the process of brainstorming. A classical brainstorming exercise involves a small group, a well-defined problem, prior awareness of the problem by the group, a leader, a secretary, and a blackboard. Classical rules for a brainstorming session, which normally lasts for 60 to 90 minutes, are that the leader reminds the group of the problem at hand and the rules for brainstorming. The leader's role is to ensure that all participants join in the discussion. Leaders should suppress their own ideas as long as the group is generating ideas. If the group has a lull in their efforts, then the leader may inject new ideas to spark the group forward. The leader does not allow criticism of any ideas. The group is encouraged to keep the ideas short and to save full details for later. As participants talk, the leader writes short, two-word descriptions of all ideas on the blackboard. The secretary captures more details. To stimulate new ideas, the leader may review aloud the ideas and elements already generated by the group.

Many variations of the classical brainstorming exercise are possible. For example, participants may supply ideas in writing to the leader before the meeting. The group hears these ideas and discusses them without necessarily knowing who thought of the ideas. The classical brainstorming approach, if led properly, can minimize the negative aspects of misinformation, dominant individuals, and hidden agendas. These are disadvantages 1, 5, and 6 in the foregoing list. Anonymous brainstorming, as just described, can minimize undue majority social pressure and overemphasis on consensus as well, or disadvantages 2 and 4.

It is possible to modify the brainstorming exercise in other ways. You can do this by not announcing the problem to the group prior to the meeting. This helps to get people to reveal what they really think because they do not have an opportunity to coordinate their position on issues beforehand. Also, you can hold a series of brainstorming meetings instead of just one session. In the first meeting, the leader presents ideas and elements that the group criticizes in all possible ways. In later meetings, the group finds alternatives to the difficulties generated in the first meeting. Any of these modifications to brainstorming still have advantages and disadvantages.

One very popular way to modify brainstorming is to replace most of the verbal communication with writing. This is often called brainwriting. In this approach, participants write a number of relevant ideas, often limited to three, on a sheet of paper. The paper is then passed to another group participant, and the ideas originally developed by one group member are further developed by another group member. These new ideas and elements are added, and the augmented pages are passed to another individual.

When the brainwriting cycle is complete and each person receives the sheet of paper they first used, the beginning phase of the session is over. The group leader collects the sheets of paper in preparation for the next phase. In this next phase, the written contribution from the first phase is circulated to the entire group. The object of the second phase is to revise the ideas or elements developed in the first phase. This process can be repeated in similar phases until enough information is collected to serve the meeting's purpose.

Again, several variations to brainwriting are possible. The papers may be put into a pool rather than circulated to the participants, and an individual who puts a sheet in the pool takes another sheet from the pool and further expands upon the ideas and elements on that sheet. Deciding what sort of brainstorming or brainwriting exercises to use depends on the particular group you are working with and the issues you want them to consider.

3.9.2 Groupware

Computer and communications technology may be used to increase the speed and productivity of group dialogue and to make it possible to have larger brainstorming or brainwriting groups. In fact, a number of researchers and private companies have made it their business to specialize in providing hardware, software, and expertise for groups. This area of specialization which helps people work together better is called collaborative computing. Collaborative applications for groups, usually called groupware [32], help people work in groups by making sharing ideas easier in three ways: common task sharing, environment sharing, and time and place sharing. Common-task-sharing systems help many people in a workgroup to work on the same task. Environment-sharing systems help people stay abreast of a project and what other participants are doing on the same project. Time- and space-sharing systems help people to work together in several different modes: same time and place (synchronous), same place at different times (asynchronous), same time at different places (distributed synchronous), or different times and places (distributed asynchronous). Johansen [29] has identified no less than 17 approaches for computer support in groups in his discussion of *groupware*. These are important, and it is of value to describe them here.

1. *Face-to-Face Meeting Facilitation Services.* This is little more than office automation support in the preparation of reports, overheads, videos, and the like that will be used in a group meeting. The person making the presentation is called a *facilitator* or *chauffeur*.

2. *Group Decision Support Systems.* By this, Johansen essentially infers a support system with a single video monitor that is under the control of a facilitator or chauffeur.

3. *Computer-Based Extensions of Telephony for Use by Work Groups.* This involves use of commercial telephone services or use of private branch exchanges (**PBX**). These services exist now, and *Northern Telecom Meridian* is an example of a conference calling service.

4. *Presentation Support Software.* This approach is not unlike that of approach one, except that computer software is used to enable the presentation to be contained within a computer. Often, the presentation material is prepared by those who will present it, and this may be done in an interactive manner to the group receiving the presentation.

5. *Project Management Software.* This is software that is receptive to presentation team input over time, and which has capabilities to organize and structure the tasks associated with the group, often in the form of a Gantt chart. This is very specialized software and would be potentially useful for a team interested primarily in obtaining typical project management results in terms of PERT charts and the like.

6. *Calendar Management for Groups.* Often, individuals in a group need to coordinate times with one another. They indicate times that are available, potentially with weights to indicate schedule adjustment flexibility in the event that it is not possible to determine an acceptable meeting time.

7. *Group Authoring Software.* This allows members of a group to suggest changes in a document stored in the system, without changing the original. A lead person can then make document revisions. It is also possible for the group to view alternative revisions to drafts. The overall objective is to encourage group writing and to improve the quality and efficiency of group writing. It seems very clear that there needs to be overall structuring and format guidance that, while possibly group determined, must be agreed upon prior to filling out the structure with report details.

8. *Computer-Supported Face-to-Face Meetings.* Here, individual members of the group work directly with a workstation and monitor, rather than having just a single computer system and monitor. A large screen video may, however, be included. Although there are a number of such systems in existence, the Colab system at Xerox Palo Alto Research Center [33] is probably the most sophisticated of these. Generally, there are both public and private information contained in these systems. The

public information is shared, and the private information, or a portion of it, may be converted to public programs. The private screens normally start with a menu screen from which participants can select activities in which they engage, potentially under the direction of a facilitator.

9. *Screen Sharing Software.* This software enables one member of a group to selectively share screens with other group members. There are clearly advantages and pitfalls in this. The primary advantage to this approach is that of sharing information with those who have a reason to know specific information and not having to bother others who do not need it. The disadvantage is just this also, and it may lead to a feeling of *ganging up* by one subgroup on another subgroup.

10. *Computer Conferencing Systems.* This is the group version of electronic mail. Basically, this is a collection of Decision Support Systems with some means of communications among the individuals that comprise the group. This form of communication might be regarded as a *product hierarchy* in which people communicate. Normally, communication across the lowest levels in a product hierarchy does not occur. It is easy to recognize that systems such as these create channels of communication that may not be in conformity with organizational charts, unless these are considered when accomplishing the system design.

11. *Text Filtering Software.* This allows system users to search normal or semistructured text through the specification of search criteria that are used by the filtering software to select relevant portions of text. The name of the original system to accomplish this was *electronic mail filter* [34], although there is now also emphasis on an *information lens* that will enable the system to obtain information matching rules that are specified by the system user.

12. *Computer-Supported Audio or Video Conferences.* This is simply the standard telephone or video conferencing, as augmented by each participant having access to a computer and appropriate software.

13. *Conversational Structuring.* This involves identification and use of a structure for conversations that is presumably supportive of the task, environment, and experiential familiarity with these of the group participants [35] Although structured conversations might provide for enhanced efficiency and effectiveness, there may be a perception of unwarranted intrusions that may defeat the possible advantages.

14. *Group Memory Management.* This refers to the provision of support between group meetings such that individual members of a group can search a computer memory in personally preferred ways through the use of very flexible indexing structures. The term *hypertext* [36] is generally given to this flexible information storage and retrieval. One potential difficulty with hypertext is the need for a good theory of how to prepare the text and associated index such that it can be indexed and used as

we now use a thesaurus. An extension of hypertext to include other than textual material is known as *hypermedia.*

15. *Computer-Supported Spontaneous Interaction.* The purpose of these systems is to encourage the sort of impromptu and extemporaneous interaction that often occurs at unscheduled meetings between colleagues in informal setting, such as a hallway. The need for this could occur, for example, when it is necessary for two physically separated groups to communicate relative to some detailed design issue [37]

16. *Comprehensive Work Team Support.* This refers to integrated and comprehensive support, such as perhaps might be achieved through use of the comprehensive DSS design philosophy described in reference 38.

17. *Nonhuman Participants in Team Meetings.* This essentially refers to the use of unfacilitated DSS and expert systems that automate some aspects of the process of decision making.

According to Johansen, the order in which these are described above also represents the order of increasing difficulty of implementation and successful use.

Identifying or formulation of an issue or problem is the first step in any problem-solving process. A problem statement should always be identified prior to the application of solution methods. Often, there is considerable merit in identifying the problem in terms of a number of interdependent elements that can be characterized as problem definition elements (needs, constraints, alterables), value system design elements (objectives or objectives measures), or system synthesis elements (activities or controls and activity measures). Too often a notion of a problem is attacked with a large outpouring of energy only to end up back at the starting point, because an immature issue formulation effort has led to solution of the wrong problem. After providing some encouragement for these efforts, we discuss one representative issue formulation method that may be used to aid a decision group in beginning to formulate the problem or issue chosen for study.

One of the difficulties in coping with a complex situation is that often there is no individual with sufficient knowledge concerning the situation to develop a set of elements which describes it. Usually a most difficult problem area concerns the unsatisfactory nature of available information, because data are often incomplete, imprecise, or otherwise faulty.

It has often been assumed that *"two heads are better than one."* Corollary to this would be the statement that *"N heads are better than N − 1 heads."* Committees, juries, boards of advisors, boards of directors, citizens' groups, and the like are representative of the many mechanisms for pooling minds to generate ideas that hopefully represent either expert knowledge or collective opinion, each with a high probability of being correct.

There are both advantages and disadvantages to reliance upon group opinion. Many times a group will interact to compensate for the bias of individual members of the group, and knowledge of one member of the group may well compensate for ignorance or speculation on the part of other members. It is unfortunately true, however, as we and many others have noted, that "*a camel is a horse designed by a committee*." Furthermore, opinion can be highly influenced by the individual who talks the most and the loudest. This influence of one or more dominant individuals can be most upsetting to a group, because there is little necessary correlation between loudness or frequency of speech and knowledge. Furthermore, unless the group activity is well-organized, a *bull session* may well result in which much more discussion deals with matters of individual and group interest than with the problem at hand. This type of difficulty could well be called *noise*. Also, there is often strong pressure for group conformity and avoidance of unpopular or minority viewpoints.

As an illustration of one approach that might potentially be utilized as part of an issue formulation effort and supported in a group decision support system (GDSS), let us consider the *nominal group technique* (NGT). This is a structured group approach for the purpose of generating ideas. In addition to idea generation, NGT is designed to provide for initial screening of ideas. The screening phase consists of a discussion and clarification portion of the meeting, followed by an optional prioritization of ideas.

NGT emphasizes independent or silent idea generation in a group environment. A *trigger question* that concisely and clearly describes the issue is posed to the group, generally by a facilitator, in order to elicit responses in the form of ideas. The process is designed to overcome some of the problems associated with interacting groups, such as unbalanced participation among group members or reduced creativity in element generation due to interference from other group members.

The process usually begins with approximately six to ten participants, selected on the basis of their background with regard to the issue and their motivation, sitting around a table in plain view of each other. Each participant has writing instruments and paper. The trigger question is posed by the group leader, or facilitator. The group leader is a person selected on the basis of familiarity with the nominal group technique and general leadership ability. After hearing the trigger question the group members record a list of ideas on their papers.*

At the end of a specified time, usually 5 to 15 minutes, this phase terminates and the *screening* of generated ideas phase of the NGT process begins. Initially, a structured sharing of ideas takes place in order to clarify, and possibly combine, ideas. This structured sharing of ideas is conducted in *round robin* fashion, with each member presenting a single idea from their *private list*. Ideas are recorded on a flip chart or board by a scribe. This effort continues until all ideas are recorded. After all ideas are recorded, each is discussed and clarified.

---

*An appropriate trigger question might be, What are appropriate objectives for regional mobility that we might support over the next year?

It is possible to record ideas for greater clarity or to combine very similar ideas. In essence, this discussion step is the first group interface with each individual's private list of ideas.

Evaluation and criticism of ideas are not appropriate during the discussion, or subsequent voting, portions of this exercise. Rank ordering the ideas is itself a limited evaluation of ideas with relation to each other. However, the primary purpose of NGT is to generate a list of ideas for evaluation and not primarily to decide among alternative courses of action. The prioritized list, therefore, indicates the order in which the group feels the ideas ought to be addressed. The list is generally not an indicator of the worth of ideas or elements in resolving the issue under consideration unless the process is specifically guided in that direction by the leader. Generally, this is not an appropriate use of the nominal group technique.

A most important step in the NGT is the formulation of the *trigger question*. Unless carefully conceived, problems of interpretation of the question may defeat the intended purpose of the effort. After the trigger question is posed, ideas will be silently generated, and then discussed. Discussion and even preliminary evaluation of these ideas should be postponed until the entire list is compiled. Any discussion or evaluation prior to that stage can lead to intimidation or fear that an idea would be unpopular, as well as to minimization of hitchhiking new ideas onto previously stated ones. As a consequence of this, there would be fewer ideas generated than if discussion and evaluation is postponed.

The final step of nominal group technique is one of clarification, amplification, and prioritization of generated (and hitchhiked) ideas. This phase can very easily enter into an infinite circular loop where discussion of minor points or wording can erode available time. Nominal group technique follows a regulated systemic approach until priority and weights are applied. Members may be inclined to retain "pet" ideas and exclude unpopular ones from further consideration even though further analysis might otherwise prove their worth. Due to the subjectivity of this final stage, it may be appropriate to terminate the nominal group process after generation of ideas.

There are many advantages and disadvantages of group techniques. An optimal group technique would result from a maximization of the advantages and a minimization of the disadvantages of several approaches. This is quantitatively infeasible, but we must nonetheless accept the fact of tradeoffs or compromise to reach a feasible approach for collective inquiry.

The dynamics of a group will ideally produce more ideas, lessen or eliminate biases (in both questions and answers), and provide open and agreeable interaction processes. Some of the advantages of a group are as follows:

- Potentially greater sum total of knowledge and information.
- Greater number of alternative approaches to the problem.

- Participation in decision making increases acceptance of the final choice.
- There is a better understanding and comprehension of the decision.

Conversely, some disadvantages to a group effort may be as follows:

- Social pressure
- Individual domination
- Desire to *win* the decision
- Excessive amiability and esprit de corps

Each of these may result in the replacement of critical independent thought by irrational and dehumanizing actions against outgroups or groupthink.

Consequently, efforts to evaluate the product of group effort are important. Success of a group process is evaluated by comparing the end results with the original desires or purposes that led to conduction of the process. The success of the collective inquiry effort can be evaluated using the following criteria:

1. *Creative.* Was a sufficient number of creative ideas for the issue obtained?
2. *Effective.* Were suggestions and conclusions feasible and understanding of the issue increased?
3. *Efficient.* Were the objectives accomplished within the optimal time span?
4. *Organized.* Did every member do their job? Together the "whole" team completed the task.
5. *Cooperative.* Did members work together constructively toward a group goal or did the meeting become an arena for individual competition?
6. *Participative.* Did every member contribute their knowledge and/or perspective?
7. *All Encompassing.* Did the group overlook relevant items?

After concluding that these criteria were satisfactorily met, the appropriate fundamental components of issue formulation also need to be examined to determine if they were well executed. Figure 3.21 illustrates characteristics of group issue formulation, and it includes individual issue formulation as a special case. The components may be evaluated as follows.

1. *Formulation.* Was a descriptive account of the problem attained? Does it seem to be relatively unbiased? Does it reflect input from the whole group? Does it identify sources of problems and needs rather than a simple listing of the symptoms? Are the ideas clearly identified and stated? Are the ideas responsive to the needs of the issue? Has enough information been gathered and disseminated?
2. *Analysis.* Were forecasts of alternatives or assessment of impacts postu-

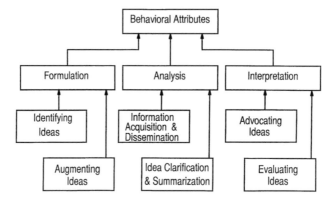

Figure 3.21   Characteristics of group idea (issue) formulation.

lated? Have the ideas been clarified and examined for feasibility or linkages?

3. *Interpretation.* Were alternatives carefully considered and ranked? Is it clear which activities aim at which objectives? Have utility, feasibility, and normative values all been considered? Has a plan been designed for action and implementation of chosen alternatives.

Both the extremes of emphasis and deemphasis on each of the fundamental components would cause ill effects on a group meeting. Therefore, it is appropriate to assign a "weight" to each component of the meeting and each subcomponent such as to develop an attribute tree for performance evaluation of the NGT session. It would be possible to use to evaluate several NGT sessions and, on the basis of this evaluation, to select the characteristics of an NGT session that produced the most appropriate results and use it as a paradigm for future sessions. In this way, we are potentially able to design the components of the meeting to obtain as high a score as possible for meeting effectiveness.

These factors strongly influence the roles of the individual in an NGT effort. NGT is generally very useful for the following activities:

• Situation assessment
• Gathering information or sensing information distribution to others
• Potential plan identification or generation
• Evaluation of potential plans
• Resolution of conflicts with respect to information and activities
• Execution of selected plans or action alternatives

One example of a groupware application is GroupSystems V from Ventana in Tucson, Arizona. This package provides a group decision support environment on desktop or laptop computers arranged in a meeting room configuration and linked by an Ethernet or LAN. Basic tools for group processes in GroupSystems V are a meeting manager, group link, agenda, briefcase, electronic brainstorming, categorizer, vote, topic commenter, group dictionary, alternative evaluation, and policy formulation. Most of these tools are self-explanatory. Meeting manager helps the session leader manage face-to-face workgroups by initiating certain planned activities and by recording and printing session reports. Group Link is a utility that people can use away from a meeting room, such as at their desk, to collaborate with other people on group projects. Briefcase provides several convenient tools that workgroup participants may find handy during a meeting such as an electronic calculator, calendar, and notepad.

The electronic brainstormer in GroupSystems V allows people to work together in parallel by using the computer to capture and broadcast their comments and ideas to others simultaneously and anonymously. Vote allows people in a workgroup to express their evaluation of issues in seven quantifiable ways: rank order, multiple choice, agree or disagree, yes or no, true or false, 10-point scale, and allocation of points. Results from a vote session are electronically recorded, tabulated, and displayed graphically or printed in text on the screen or on paper. Policy formulation helps a group to jointly write and edit statements by an electronic, interactive process of review and revision.

As you might expect, companies like Ventana are improving their groupware by adding more functions and features. For example, advanced tools for GroupSystems V include group outliner, survey, group writer, idea organization, group matrix, stakeholder identification, and questionnaire. Again, most are self-explanatory, but here is a short description of a few of them.

Group matrix helps a group fill in the boxes of an interaction matrix. Participants choose a relationship between the ideas or items in rows and columns by entering a numeric value or predefined word. The system displays the average group response and indicates by colors which cells have consensus or disagreement so the group can focus its work. Stakeholder identification helps a workgroup to carefully identify and consider the impact of stakeholders on a proposed plan. But, Ventana's GroupSystems V is only one example of the many groupware products on the market. While GroupSystems V provides mainly text-based interactions for groups, other groupware applications are using the power of computer graphics.

An example of the kind of groupware graphics power available today is CM/1 developed by Corporate Memory Systems Inc. of Austin, Texas. CM/1 is a Windows-based hypertext tool for creating a graphical discussion tree with a workgroup [39]. The tree helps a group explore complex issues with unknown answers by structuring the discussion using issue nodes, position nodes, and argument nodes. Discussions start with the creation of an issue node that is labeled with an open-ended question. Then group participants offer their solutions by creating a position node and labeling it with their idea. The link between an issue node and a position node is a "responds to" arc or line. Argument nodes "support," "object to," or "specialize" (refine) a position. Participants can raise new issues that expand or challenge previous issues, positions, or arguments. Eventually, the group can enter a decision node to show that an issue has been resolved.

The graphical discussion tree encourages a logical, focused approach to group work and promotes understanding among participants. Reference nodes and note nodes allow participants to put in links to documents and other evidence to support their arguments. CM/1 and GroupSystems V are just two examples of the many computer-based workgroup applications that you might want to consider. Groupware can help the groups you work with generate ideas, develop action plans, refine work, make decisions, and negotiate compromises.

3.9.3 Delphi Methods

In traditional group discussion methods, effective communication is sometimes inhibited by the disadvantages inherent in groups. These include many psychological factors, problems with dominant individuals, the consequent problem that all points of view are not heard, and the lack of documented, concise statements about what actually happened in the group dialogue. Brainstorming and brainwriting exercises try to get the most from group dialogue by a systematic and controlled exploration of the elements and factors of a particular issue. Still, even computer-assisted brainstorming and brainwriting sessions suffer from many group drawbacks.

However, the procedure known as Delphi, developed in large part by the Rand Corporation, eliminates many of the disadvantages of group dialogue [40]. The Delphi approach does this because of three features not found in either classical brainstorming or brainwriting. These are anonymity, iteration with controlled feedback, and statistical group response. Anonymous response is made possible by the use of a formal questionnaire to get responses from group members. Iteration and controlled feedback happens because Delphi is a systematic exercise conducted in several iterations. Between each iteration there is carefully controlled feedback. At the conclusion of the final iteration, group opinion is combined from individual opinions to form a statistical group response based on the relative expertise of members with the issue under consideration. These three essential features of the Delphi technique make it a very successful method of soliciting and refining group opinion.

Although a Delphi exercise is really a modification of both brainstorming and brainwriting, there are two primary variations that make it different. First, there are simultaneous individual contributions from each participant at every step, without these participants' having knowledge of inputs supplied by others for that particular step. Second, the sources of all inputs are anonymous. This anonymity of input is maintained through the entire Delphi dialogue.

The Delphi approach tries to minimize the biasing effect of irrelevant dialogue between individuals and of group pressure toward conformity. Delphi helps to ensure that group interactions about the issue at hand compensate for the biases of individuals, and that the knowledge of several members of the group will compensate for ignorance on the part of others. Now let's describe the classic Delphi approach and then present several well-known variations.

To begin a Delphi sequence, a Delphi group leader interrogates a group using a series of questionnaires. Each iterative summation of a questionnaire to a group is referred to as a round in the original Rand terminology. The questionnaire distributed to the group asks questions of the group; it also provides information to group members concerning the degree of consensus that has resulted from previous rounds. It also gives diverse arguments presented anonymously by various group members. A director leads the group or panel through each round. Each round calls for different activities on the part of either the panel, the director, or both. The subject area for the panel to deliberate is clarified before the first round. The director resolves any questions about the rules of the Delphi exercise with the panel members before the first round.

The questionnaire distributed during the first round is completely unstructured. It requests the group to discuss elements or make a forecast about the specific problem under consideration. If the group is truly expert and very knowledgeable about the subject area under discussion, then this approach to round one is most useful. It allows the group to utilize its expertise to maximum advantage. If the questionnaire used during round one is too structured and restricted, the group may well overlook elements and events discussed in the questionnaire.

After the panel responses have been returned to the Delphi exercise director, they must be consolidated and placed into a single set. Individual members of the panel may present their discussion of events, elements, and forecasts in the form of a narration. The director will then disaggregate the pieces into a set of discrete events. Other individuals may perhaps have given a list of events and elements arranged in chronological form. The main task of the director at the end of round one is to identify events, elements, and forecasts; consolidate those that are similar; eliminate those that are unimportant for the purpose at hand; and prepare a final list of elements, events, and forecasts in the most clear fashion possible. This final list becomes the input or second questionnaire for round two.

In round two, the panel examines the list of consolidated events prepared from round one and estimates dates when these elements or forecasts may occur. Panel members give reasons why they expect these dates to be correct. After these forecasts and estimated dates from round two are complete, the director prepares a consolidated statistical summary of the panel's opinion. A brief discussion of the reasons for the group's is also prepared. From this information, the director prepares the questionnaire for round three.

The third round questionnaire consists of the list of elements, events, forecasts; the group median date and the upper and lower quartile dates for the occurrence of each event, element and forecast; and a summary of the reasons for the choice of dates. This questionnaire starts round three. The panel reviews the arguments and prepares a new estimate of the date on which each event, element, or forecast may occur. If any of the estimated dates falls later than the upper quartile or earlier than the lower quartile, the panel members of the group must give reasons to justify this view as well as to comment on the opposing views of others.

If an individual's estimate is earlier or later than that of three-fourths of the group, the individual must justify this estimated date and indicate why previous group arguments in support of an earlier or later date are wrong. These arguments typically reference outside factors that other individuals in the group neglect and cite facts that others may not have considered. Therefore, individuals in the group are just as free to raise objections and make arguments as they would in any face-to-face confrontation. However, their arguments are now anonymous.

Results from the round three questionnaire now go to the Delphi group leader. Preparation of a questionnaire for round four is similar to that for previous rounds. Group estimates must be summarized; new medians and quartiles must be computed; summary arguments for both sides of any dispute must be prepared. All of these are put together for the round four questionnaire.

After the panel members get the new questionnaire (a new list of events, elements, and forecasts; a statistical description of the estimates of dates on which these will occur; and summary arguments), the group makes a new forecast. In classic Delphi, round four is the final round. Often there is no need for the Delphi group leader to analyze the group's arguments at the end of round four. However, if the panel has not been able to reach a consensus, the leader may want to get final arguments from both sides in a dispute. With these final arguments, the director can prepare an effective statement about this lack of consensus as the final result of the Delphi exercise.

The final Delphi output is a written report from the director with a list of events, elements, and forecasts with the median and quartile dates estimating when they will occur and a summary of relevant arguments. It is not necessary that any particular event, element, or forecast last through all four rounds of the Delphi exercise unchanged. If the group agrees that a particular event, element, or forecast will never take place, then it can be dropped at the round where this agreement occurs. It does not need to appear in later rounds. Sometimes originally stated events, elements, or forecasts must be restated to make them clearer, or perhaps one will be divided into several events. Some events or elements may need to be combined. In general, most Delphi exercises observe an outcome with three characteristics:

1. *Wide Initial Distribution of Responses.* On the first round, there is a wide distribution of individual responses to the questionnaire.

2. *Convergence of Responses.* The distribution of individual responses converges in later rounds as iteration and feedback take hold.

3. *Accuracy*. The group response becomes more accurate from round to round. Accuracy is defined here as the median of the final individuals' responses that can never be reduced to zero because we are usually talking about future events.

Many variations on the original Delphi approach are possible. You can appreciate that a very useful modification is to use a computer-based approach to help make the job of the director easier and faster. Most variations preserve the three fundamental principles of a Delphi exercise: anonymity, iteration, and statistical response. You can also understand that some panelists who are not familiar with statistics may at first find it difficult to make assessments using statistics-based methods. Again, the computer with its graphical display capabilities can make it easier for participants to understand statistical terms like quartile and median. Other variations actually modify one or more of the Delphi principles. Here are some variations you may want to consider:

1. Begin with a blank sheet of paper. The group is now completely unstructured and no precise guidelines tell the participants how or where to start. This approach may allow for more creativity. A drawback of this variation is that the elements, events, or forecasts produced by the group may be totally irrelevant to the director's purpose.

2. Start with a list of events, elements, or forecasts generated by external processes prior to the Delphi exercise. Use this list to begin the Delphi sequence. This is the same as accomplishing round one with another group and then transferring the results from round one with the first group to a second group to start round two. This variation helps to focus the Delphi exercise, but it also may inhibit the creativity of the second group. If you use more than one group in a Delphi effort, there is no reason why some of the same people cannot be members of more than one group.

3. Use ratings of individual expertise. If the panelists vary considerably with respect to their expertise about the issue being considered, then the director may ask each individual to indicate, on a simple scale, their level of expertise on each question they answer. The director combines the individual estimates in a weighted average using the self-rating of expertise as the weights. In this way, answers from panelists with the most knowledge count for more.

4. Provide background briefings to panelists. When highly technical issues influence the results of the Delphi exercise, it may be very helpful to brief all the panelists with necessary background information. For example, when determining futures for oil supply and demand, it might be very helpful to inform the group of likely developments in solar and other alternative energy sources, especially if the group lacked expertise in this area. This provides the entire group with a starting baseline of knowledge and can avoid needless arguments about technical details.

5. Attach names to responses. You may want the influence or position of the originator of an idea to sway the judgment of others. This defeats one of the basic principles of Delphi which holds that the purpose of anonymous responses is to judge responses solely on their merits. However, one panelist may be so truly expert in the area that is under discussion that it may be beneficial to identify that expert's responses.

6. Reduce feedback. If you eliminate all feedback, then individual opinions on the second and subsequent rounds might just be a repetition of the first round's responses without any reexamination of these initial responses. This could cause the reinforcement of wrong responses by repetition. It is the well-known phenomenon that "if you tell a lie often enough, then people will soon begin to believe it to be true." Sometimes feedback may result in overconvergence to a given median. This happens when people want to avoid providing an argument for not shifting to the median. One way to overcome this is to only provide a portion of the feedback, eliminate the previous group median information, and give only the upper and lower quartile information.

Which variation is right for your project depends a lot on the problem under consideration and the people you are bringing together to work on it. Here are some good guidelines that should help you successfully conduct a Delphi exercise [41].

1. Get willing agreement from the individuals that you want to have serve on the Delphi panel. A few willing participants is better than a bigger group with a "bad apple."

2. Explain the Delphi procedure completely to the group. You may find it helpful to run a practice round or two on an easy problem so that everyone understands the rules.

3. State events, elements, and forecasts in simplest form. Avoid using compound events.

4. Avoid ambiguous statements of events, elements, and forecasts.

5. Make the questionnaire as easy as possible.

6. Use no more questions than an individual can adequately consider. Fewer is better than too many.

7. Explain why contradictory events, elements, or forecasts are included, so that panelists will not think that the director is trying to "trap them."

8. The Delphi group leader, or director, should never inject personal opinions into group feedback. If the director knows that the group has overlooked significant factors, then the group output should be discarded and the group should be considered unqualified. Repeat the exercise with a more qualified group. The director must not meddle in the deliberation of the group.

9. Compensation of group members for work load involved in a Delphi exercise should be appropriate to the service rendered as well as to the type of organization (profit or nonprofit, for example) requesting the Delphi exercise.

10. Remember that the knowledge that the panelists have about the issues under discussion and the clarity of the panelists' thinking processes play the most important part in determining the quality of the Delphi results. So strive for the best qualified panel that you can find.

3.9.4 Morphological Box Approach

One very useful technique for stimulating a group's creative thought is called Zwicky's morphological box [42]. Zwicky makes a strong case that people can become very creative if they use techniques that force them out of their traditional modes of thinking. His morphological box approach tries to do this by getting people to think of new ideas at the subsystem level and then putting these ideas and elements together into new and untried combinations. The following steps may be followed.

1. Define all of the functional classes that make up the basic subsystems of the problem under consideration. Again, the lower levels of the objectives tree can help to suggest what some of these functional classes should be. A generally useful approach is based on partitioning the problem into a number of functional classes or major subsectors. Note that these sectors are not technological solutions in and of themselves. They focus individually upon the various activities that must be accomplished in order that the composite of them resolves the issue at hand. In the next step of Zwicky's box, you concentrate on all the different ways including technological activities for accomplishing each sector.

2. In the next step, sector brainstorming, the group focuses on one sector at a time and creates a list of activities for accomplishing that one sector. This is important because you do not want to limit your thinking about one sector because of preconceived constraints about how one sector relates to another. You are looking for new ideas, not the standard conventional solutions in operation today. Brainstorming, brainwriting, and Delphi are all useful techniques that you can use to help a group create these sector lists. Because the ideation of activities for each sector is an outscoping mental effort, there should be no criticism of ideas for the list while it is being created. The group might identify conceivable ways of accomplishing each sector without regard for practical limits or consideration of the other sectors. After we create exhaustive lists for each sector, the group is ready to try to synthesize or forge the ideas from the lists together into a set of workable alternatives.

3. The next step involves combining elements. After we pick at least one element from each sector list, a combination or synthesis of elements is

put together to form an alternative. At this stage, it is again important not to criticize these combinations even though some will seem ridiculous at first glance. All possible combinations should be exhausted so that no new directions or innovations are overlooked. The important point is that creativity often results from exploring the unusual or searching in new directions, and that is why every combination of the morphological box should be explored. At this point, you might ask why we bothered with the morphological boxes or sector lists. Why didn't we just go directly into creating complete alternatives? The reason is that this latter approach may not be as creative. With Zwicky's morphological box, you ideally have many alternatives, some of which were never even imagined as the individual sector lists were being created.

4. The next step involves making combinations work. After you assemble all of the possible combinations from the elements in the different sector lists, the group now focuses its creative energies on making the combinations work. The group should strive to think about how to make even the most unusual combinations work. As work progresses, you can discard combinations that are impractical and build more detail into the descriptions of promising alternatives. These promising alternatives will undergo more scrutiny in feasibility studies.

5. Variations are needed. As you can imagine, there are many successful ways to use brainstorming, nominal group, brainwriting, and Zwicky's morphological box together to work on tough issues. For example, you may want to use brainstorming to create the different functional classes or sectors in Zwicky's box. Then, you may want to involve the group in a brainwriting exercise to develop the lists of ideas for accomplishing each sector. Putting the elements from the lists into different combinations and elaborating how each combination might work could be done by a Delphi exercise.

Here are some important variations on Zwicky's box that you may find useful especially in corporate settings.

1. Match Combinations to Objectives. One way of determining if the combinations from a Zwicky's box approach will work is to simply compare them to the objectives and attributes of the corporation. If the combinations seem to support the corporation's objectives and match up well with the firm's attributes, then it is likely that those combinations deserve serious attention.

2. Order the Selection. Another way to change Zwicky's procedure is to impose some ordering discipline on the selection of elements from the various morphological sectors or categories. A benefit of this approach is that it reduces the number of impractical combinations generated.

The drawback to this approach is that it clearly limits the consideration of many ideas, especially unconventional ones.

Needless to say, we need to formulate alternative courses of action appropriately if we are to solve the right problem.

3.10 FEASIBILITY STUDIES

Many systems engineers use the term feasibility study to encompass the entire formulation effort because the end product of formulation is a set of feasible alternatives. However, since we have discussed most of the formulation effort already, this section uses the term feasibility study to focus on how to screen a set of alternatives to eliminate those that are not viable.

3.10.1 Feasibility Screening

In the last section, you learned how to combine the ideas of experts and stakeholders to generate many alternatives. Now you need a way to scope down all of these alternatives to a sensible number of viable alternatives. We call this type of effort a feasibility screening because it examines alternatives in sufficient detail to determine if they meet the minimum requirements of the project's stakeholders. A feasibility screen is not an effort to fine tune the alternatives and make them the best they can be. Nor is it an effort to select the best alternative. Such efforts come later in the analysis and interpretation of alternatives. In a feasibility study, you want to identify those alternatives that cannot meet even the minimum requirements of the project's stakeholders and society. Another goal of a feasibility study is to make changes, if possible, to alternatives identified as infeasible so that they are viable. However, the overall result of a feasibility study is usually a tremendous reduction in the number of alternatives.

Why do you want to reduce the number of alternatives by conducting a feasibility study? First, analyzing and interpreting alternatives takes considerable time, money, and effort. It would be a waste of valuable resources to analyze and interpret alternatives that could never be placed into operation anyway. Second, because design is an iterative process, developing high-quality alternatives takes lots of working and reworking of each alternative. Scattering this effort across many alternatives reduces the amount of work that can go into developing each alternative to make it the best it can be. Most importantly, you conduct a feasibility study to make certain that flawed alternatives are not eventually selected for implementation. How sad it would be to create a new systems design only to realize, after lots of time and money, that the design will never work in actual practice because of some limitation that could have been discovered much earlier.

Now that you understand why you want to do a feasibility study, how do you go about it? The first challenge that you must overcome is the sheer number of alternatives that exist at the beginning of a feasibility study. Don't worry. This is something to be happy about. It means that you and your group of stakeholders and experts have been very successful and creative! One way to reduce the number of alternatives to consider at the very beginning of the feasibility study is to categorize or group the alternatives into similar sets. For example, a project that is developing alternatives for a computer system for its sales force might have a very large number of alternatives. But a grouping of the alternatives into several categories, such as desktop, laptop, subnotebook, and handheld, might prove very useful. Now your efforts can focus on picking representative alternatives from these four categories for further study.

If major variations within or across the categories of alternatives exist, you may need to consider more than one representative alternative from each category. For example, installing a multimedia capability on a computer device is a significant variation that will impact these four categories in different ways. Therefore you may want to consider two alternatives from each of the four categories, one with multimedia capability and one without it. The main point to remember is that it is much easier to work with a few representative alternatives from a large category than to try and work with every single possibility. If the alternative you have chosen is truly representative of the entire class, then you often find that if it is not feasible, then the entire class is also not feasible. One way of testing whether or not you have divided the alternatives into proper representative categories is to determine if the categories are significantly different from each other. Alternatives should be significantly different from each other, and therefore categories of alternatives should be very significantly different from each other.

3.10.2 Architecture and Standards

Another good way to categorize alternatives and to reduce the number of alternatives quickly at the beginning of a feasibility study is to use considerations that involve architecture and standards. Architecture is the scheme of arrangement of the components of a system, and it describes features that are repeated throughout the design and explains the relationship among the system's parts. Architecture can be very useful for categorizing or eliminating alternatives. In the example of a computer for a sales force, you could categorize alternatives by their hardware or software architectures such as by their central processing units or by their operating systems. Also, you may be able to eliminate a large number of alternatives because, for example, corporate headquarters may have placed a constraint on what is possible by imposing a certain architecture as a company-wide standard. In our earlier discussions, we noted the difference between conceptual or system-level architectures and detailed architectures. Figure 3.22 illustrates both conceptual and detailed architectures and some components of these architecture types. Many authors

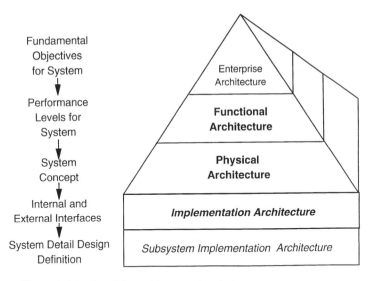

Figure 3.22   Possible architecture levels in systems engineering.

define three levels of architecture: functional, physical, and implementation or operational, or variants of these [43]. These are the major architectural levels although additional ones can be defined as illustrated in this figure. One of the purposes of an architecture is to support ultimate development of a Systems Engineering Management Plan (SEMP), such as illustrated conceptually in Figure 3.23. We will comment further on systems engineering management plans in Chapter 6.

A standard is an accepted or approved way of doing things. Government can require and enforce standards such as the safety belts and air bags in automobiles. Industry can recommend standards such as the RS-232 interface cable for computers. Some standards are *de facto*. They simply evolve into common practice because it makes good business sense to adopt them.

Standards are very useful because they make it possible for many things to work together better and help achieve many economies. For example, you know when you buy an electrical appliance that you can take it to any office or home in the United States and it will easily plug into the electrical system because of the standard plug, receptacle, and power source. You can imagine the frustration of users if they had to worry about whether or not some electrical appliance would work in their house or they had to buy some expensive adapters to make machines work at the office.

Standards can also limit innovation. Perhaps there are easier and safer ways to accomplish hooking up an appliance to a home power source, but the standard may inhibit the use of such new ways. So you have to be careful to balance the benefits and costs of following different standards. Also, standards

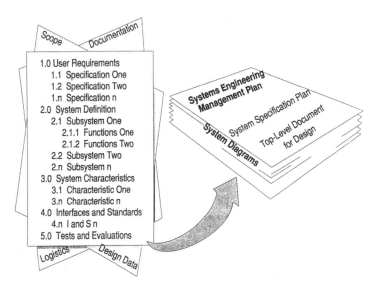

Figure 3.23 Generic systems engineering management plan.

must be evaluated periodically to see if they are really necessary. If simple guidelines or functional standards will do, then they should be used in lieu of rigid technical standards.

But oftentimes, there are many useful standards that ought to be followed as part of a large-scale design project. So another way of reducing the number of alternatives you must consider is to screen out those alternatives that cannot be adapted to standards that you know the project must use. Most relevant systems engineering standards are process standards rather than product standards.

A more detailed discussion of architectures and standards from a systems engineering perspective is contained in references 1, 2, and 44. These subjects are generally outside of the scope of this methods based book.

3.10.3 Feasibility Criteria

You can now see that much of the feasibility study involves identifying criteria that can be used to screen the alternatives, eliminating those that cannot meet the minimum criteria. What are these feasibility criteria? Although it depends on the specifics of the system being designed and the system's environment, there are a number of different categories of feasibility criteria that are useful to consider. For example, the Accreditation Board for Engineering and Technology cite economic, safety, environmental, and social impact as important feasibility considerations in engineering design.

Ostrofsky [18] recommends screening a set of candidate systems by eliminating those that are not physically realizable, economically worthwhile, or financially feasible. He defines physical realizability as whether or not the components of each subsystem are physically compatible. One way of doing this, as shown by Ostrofsky, is to make a two-column matrix: Each component is listed in one column, and the corresponding incompatible components are listed in the second column. Then you can use this matrix to screen each alternative to identify compatibility problems and make decisions to either eliminate the alternative from further consideration or modify the alternative so it is physically realizable. For example, an alternative can possibly be made compatible by buying or making a translator or adapter to overcome the incompatibility problem between two components.

Economic worthiness examines the expected costs and potential revenues and benefits of the alternatives. This efforts requires forecasting the future and understanding how to evaluate cash flows and incorporate risk considerations. You can learn these and other analysis techniques in Chapters 4 and 5. Closely related to economic worthiness is financial feasibility. In short, this really examines the budget required to accomplish each alternative and determines if the stakeholders funding the project are willing to pay for it. This is necessary because even though an alternative may eventually pay off in the long term, the shorter-term outlays may exceed what is financially feasible to do.

Another very important feasibility category is legal acceptability. The systems engineer must be careful to consider the various legal ramifications of large-scale projects. No doubt that there are many local, state, and government agencies and private organizations that can have a significant impact on a project's viability. Again, careful planning in the early stages of an effort can avoid lots of wasted time and money that can be caused by regulatory work stoppages or litigation.

A thorough feasibility screening effort will examine each of these suggested categories and develop from them appropriate specific feasibility criteria. For each criteria, you establish a minimum or maximum level of attainment that you can use to make a "Go" or "No Go" decision. We might devise a feasibility screening matrix that summarizes these kind of results. In this, there would be a recapitulation column that makes a decision about the overall viability of each alternative. A "No Go" in any column for an alternative results in a "No Go" overall for that alternative. Only alternatives with an overall "Go" are carried forward to the next step: modeling and analysis.

3.11   SUMMARY

In this chapter we have presented a relatively wide-ranging discussion of some of the principles of issue formulation. This step, together with the subsequent steps of analysis and interpretation, comprise one of the systems engineering phases. We have emphasized the definition phase in many of our discussions. We suggest working a good number of the problems and exercises that now follow.

PROBLEMS

3.1 Consider the following mythical situation: Thirteen mythical colonies exist on the eastern coast of a large mythical continent. In recent years, certain actions taken by the mythical parent country of these colonies have led to much tension. A number of the inhabitants of these colonies have decided that the situation is becoming intolerable, and armed conflict has already broken out. A Continental Congress has been convened to consider various possible courses of action, including dissolution of the political ties with the parent country. As an exercise, examine this mythical problem from the standpoint of program planning. Perform the problem definition, value system design. and system synthesis steps of systems engineering, identifying mythical needs, alterables, constraints, objectives, objectives measures, activities, activities measures, and societal sectors and determining program planning linkages for them.

3.2 The following is a possible set of objectives associated with Little League baseball. These objectives can be ordered into several different hierarchies of importance, depending on whose viewpoint is being considered. Using the criterion "objective A is more important than objective B," structure these objectives into hierarchies as they might be viewed by a player, a coach, and a dominant parent. Discuss the differences that result.

O1 To provide children with healthful exercise
O2 To improve physical coordination of children
O3 To teach confidence and self-reliance
O4 To let children have fun
O5 To build better future citizens
O6 To get children out of the house
O7 To teach cooperation and teamwork
O8 To win baseball games
O9 To provide parents with vicarious ego gratification
O10 To provide coaches with experience
O11 To bring together communities in common activities
O12 To build a better society
O13 To provide a starting point for future professional baseball players
O14 To build friendships through friendly competition

3.3 Study one or more of the following problems from the point of view of program planning in which you attempt to identify relevant elements that comprise the issue formulation elements (problem definition, value system design, system synthesis) for the following issues:

(a) The world food crisis

(b) The housing crisis

(c) The future of space exploration and activity

(d) The future of intelligent transportation systems in this country

(e) Sustainable development and earth systems

(f) Illegal drugs and drug trafficking

(g) Illegal money laundering

(h) Proliferation of weapons of mass destruction throughout the world

Develop a set of relevant elements for the issue formulation elements that you can use for structuring, modeling, and decision analysis.

3.4 Consider a current policy question being faced by a contemporary political body. Describe the policy question in terms of the framework for systems engineering presented here. Pay particular attention to the three principal steps of systems engineering discussed here. What needs, constraints, and alterables can you identify as being part of the issue formulation? What are the appropriate objectives and objective measures which continue the issue formulation step? What are the appropriate activity and activity measures that complete issue formulation? How do these elements fit together such that decisions and policies might be determined? You may recognize this as problem 1.4 from Chapter 1. Reexamine this problem using the knowledge you have acquired in this chapter and determine appropriate program planning linkages across the many elements that you have identified.

3.5 Sometimes needs are characterized as primitive needs and effective needs. Could a situation arise in which the primitive need and the effective need are the same? Explain how it could happen? There are also wants, and one of the tasks of a systems engineer is to assist a client in expressing a set of requirements that are, in effect, a blend of these. It is quite possible for the primitive need and the effective need to be the same. One often sees it when the problem is very limited in scope (a small problem), or when the client is objective and can take a holistic view of the problem. However, you should keep in mind that most clients seldom have a good initial understanding of their own problem. Generalization of the primitive need is generally necessary. Oftentimes the primitive need reflects the client's opinion, as well as associated biases and/or perceptions on the problem. Consequently, to act on the client's primitive need may not always solve the real problem. It is up to the systems engineer to consider the big picture and determine the "true" underlying problem or client's (effective) need. Discuss the relationship between needs of various types, wants, and system requirements.

3.6 Consider the use of email for a Delphi exercise. Please write a description of an email based Delphi exercise.

3.7 The Commission on the Year 2000, American Academy of Arts and Sciences, has presented the following list of values in its first set of working papers. The numbering begins with the value that has the least number of letters and proceeds to the value that has the most number of letters. There is no other significance associated with the numbering.

(a) Privacy
(b) Freedom
(c) Equality
(d) Education
(e) Rationality
(f) Law and order
(g) Social adjustment
(h) Personal integrity
(i) Ability and talent
(j) Pleasantness of environment
(k) Efficiency and effectiveness of organizations

Using these values as an element set and the relation "is necessary for," generate a reachability matrix and an interpretive structural model of your personal perception of value structuring for this issue.

3.8 Identify an appropriate case study area involving the need for issue formulation. Contrast and compare the brainwriting, nominal group, and Delphi approaches to identifying the issue formulation variables.

3.9 Imagine that this is the year 2025. What will be the structure, function, and purpose of an automobile in year 2025? Suggest brainstorming, brainwriting, nominal group and Delphi approaches to identification of likely futures for the automobile in 2025.

3.10 Consider two activities, such as acquiring housing (house, condominium, rental apartment) and acquiring transportation to work (car, mass transit). Consider the relationship between the two (such as money tradeoff and distance from work—you can assume that, with a shorter distance between home and work, a particular kind of housing increases in price and transportation price falls), and, including elements pertaining to this relationship, make a schematic study of the two separate problems utilizing the functional analysis approaches described here. Then schematize the interaction between programs by building various interaction matrices.

3.11 Discuss the use of the QFD approach for the issue described in the previous problem.

3.12 Discuss the differences between an objectives tree and an objectives hierarchy. Illustrate how you can convert an objectives hierarchy to an objectives tree. Illustrate this by considering a case study.

3.13 Using the descriptive scenario shown in Section 3.6, identify and list the problem elements involved in "Energy Supply and Demand."

3.14 Construct a causal loop diagram of the problem elements from problem 3.13 and write a paragraph describing the relationships illustrated by the diagram. A causal loop diagram, discussed in Chapters 4 and 5, illustrates the influences or causes across elements.

3.15 Construct a self-interaction matrix of the 27 needs of the "Energy Supply and Demand" example in Section 3.6. Explain the significance of the result.

3.16 Construct a self-interaction matrix of the 10 alterables of the "Energy Supply and Demand" example in Section 3.6. Analyze the resulting matrix. Which alterables are highly related, which ones are relatively separate or isolated from others? Determine the range of each alterable.

3.17 Construct a self-interaction matrix of the 11 constraints of the "Energy Supply and Demand" example in Section 3.6.

3.18 Construct and analyze a self-interaction matrix of the 29 objectives of the "Energy Supply and Demand" example in Section 3.6.

3.19 Construct an objectives hierarchy of the 29 objectives of the "Energy Supply and Demand" example in Section 3.6. Identify objectives that compete or are trade-offs for one another.

3.20 Construct an objectives–objectives measures cross-interaction matrix for the 29 objectives and the 19 objectives measures for the "Energy Supply and Demand" example in Section 3.6.

3.21 Using the cross-interaction matrix from problem 3.20, structure a hierarchy of objectives measures. Discuss how higher level objectives can be measured using different sets of lower level objectives measures.

3.22 Construct a self-interaction matrix of the 11 activities shown in the "Energy Supply and Demand" example in Section 3.6. Explain the resulting matrix structure.

3.23 In a spreadsheet, construct a cross-interaction matrix of activities and activity measures for the "Energy Supply and Demand" example in Section 3.6.

3.24 Develop three alternatives for the "Energy Supply and Demand" example in Section 3.6 by logically grouping several of 11 activities together and then iteratively adding more details to each description of an alternative.

3.25 For the alternatives developed in problem 3.24 identify the agencies, societal sectors, other stakeholders, and agents responsible for carrying out each activity within each alternative.

3.26   Construct the very important activities and objectives cross-interaction matrix using the 29 objectives and 11 activities of the "Energy Supply Demand" example shown in Section 3.6. Comment on the adequacy of the activities. Can each objective be potentially achieved by at least one activity?

3.27   Construct a cross-interaction matrix of the 27 needs and 29 objectives of the "Energy Supply and Demand" problem in the example of Section 3.6. Can each need be satisfied by achieving at least one objective? Which needs do you think will be the most difficult to satisfy?. Explain why.

3.28   Select what you think are the three most important activities from the 11 activities shown in the "Energy Supply and Demand" example in Section 3.6. For each activity you selected, write a short discussion that describes whether it can be carried out within the range of the alterables and without violating the constraints in the "Energy Supply and Demand" example.

3.29   Construct self- and cross-interaction matrices for the 12 products of the problem definition step, the 4 products of the value system design step, and the 3 products of the system synthesis step. Explain why the resulting structure highlights the need for iteration in a systems engineering formulation effort.

REFERENCES

[1]   Sage, A. P., *Systems Engineering*, Wiley, New York, 1992.

[2]   Sage, A. P., *Systems Management for Information Technology and Software Engineering*, Wiley, New York, 1995.

[3]   Boar, B. H., *Application Prototyping*, Wiley, New York, 1984.

[4]   Lantz, K. E., *The Prototyping Methodology*, Prentice-Hall, Upper Saddle River, NJ, 1990.

[5]   Noble, D., Schema-Based Knowledge Elicitation for Planning and Situation Assessment Aids, *IEEE Transactions on Systems, Man, and Cybernetics*, Vol. 19, No. 3, May 1989, pp. 612–622.

[6]   Smith, C. L., Jr., and Sage, A. P., Situation Assessment in Command and Control, in Andriole, S. J. and Halpin, S. M. (Eds.), *Information Technology for Command and Control*, IEEE Press, New York, 1991, pp. 449–466.

[7]   Smith, C. L., and Sage, A. P., A Theory of Situation Assessment for Decision Support, *Information and Decision Technologies*, Vol. 17, 1991, pp. 91–124.

[8]   DuBois, P. Brans, J., Cantraine, F., and Mareschal, B., MEDICIS: An Expert System for Computer-Aided Diagnosis Using the PROMETHEE Multicriteria Method, *European Journal of Operational Research*, Vol. 39, 1989, pp. 284–292.

[9]   Dutton, J., and Duncan, R., The Creation of Momentum for Change Through the Process of Strategic Issue Diagnosis, *Strategic Management Journal*, Vol. 8, pp. 279–295, 1987.

[10] Sage, A. P., Systems Management of Emerging Technologies, *Information and Decision Technologies*, Vol. 15, No. 4, 1989, pp. 307–325.

[11] Mason, R., and Mitroff, I., A Program for Research on Management Information Systems, *Management Science*, Vol. 19, 1973, pp. 475–487.

[12] Mason, R., and Mitroff, I., *Challenging Strategic Planning Assumptions*, Wiley, New York, 1981.

[13] Bell, D., Raiffa, H., and Tversky, A. *Decision Making: Descriptive, Normative, and Prescriptive Interactions*, Cambridge University Press, Cambridge, UK, 1988.

[14] Matheson, D., and Matheson, J., *The Smart Organization: Creating Value Through Strategic R&D*, Harvard Business School Press, Boston, MA, 1998.

[15] Hammond, J. S., Kenney, R. L., and Raiffa, H., *Smart Choices: A Practical Guide to Making Better Decisions*, Harvard Business School Press, Boston, MA, 1999.

[16] Goldstein, W. M., and Hogarth, R. M. (Eds.), *Research on Judgment and Decision Making: Currents, Connections, and Controversies*, Cambridge University Press, New York, 1997.

[17] Maslow, A. H., *Motivation and Personality*, Harper & Row, New York, 1970.

[18] Ostrofsky, B., *Design, Planning, and Development Methodology*, Prentice-Hall, Englewood Cliffs, NJ, 1977.

[19] Hall, A. D., *A Methodology for Systems Engineering*, D. Van Nostrand, New York, 1962.

[20] Hall, A. D., *Metasystems Methodology*, Pergamon Press, Oxford, UK, 1989.

[21] Buede, D. M. *The Engineering Design of Systems: Models and Methods*, Wiley, New, York, 2000.

[22] D. Appleton Company Inc., *Corporate Information Management: Process Improvement Methodology for DOD Functional Managers*, 2nd edition, D. Appleton, Fairfax, VA, 1993.

[23] Office of the Secretary of Defense, ASD (C3I), The DoD Enterprise Model Volume I: Strategic Activity and Data Models, Volume II: Using the DoD Enterprise Model—A Strategic View of Change in DoD, U.S. Department of Defense, January 1994.

[24] Cooper, R., and Kaplan, R. S., *The Design of Cost Management Systems*, Prentice-Hall, Upper Saddle River, NJ, 1991.

[25] Clausing, D., Total Quality Development: *A Step by Step Guide to World Class Concurrent Engineering*, ASME Press, New York, 1994.

[26] Hauser, J. R., and Clausing, D., The House of Quality, *Harvard Business Review*, May–June 1988, pp. 63–73.

[27] Sage, A. P., *Methodology for Large Scale Systems*, McGraw-Hill, New York, 1977.

[28] Blanchard, B. S., *Systems Engineering Management*, Wiley, New York, 1998.

[29] Sage, A. P. (Ed.), *Concise Encyclopedia of Information Processing in Systems and Organizations*, Pergamon Press, Oxford, UK, 1990.

[30] Davis, G. B., Strategies for Information Requirements Determination, *IBM Systems Journal*, Vol. 21, No. 1, 1982, pp. 4–30.

[31] Martino, J. An Experiment with the Delphi Procedures for Long Range Forecasting, *IEEE Transactions on Engineering Management*, Vol. EM-15, Sept. 1968, pp. 138–144.

[32] Johansen, R., *Groupware: Computer Support for Business Teams*, Free Press, New York, 1988.

[33] Stefik, M., Foster, G., Bobrow, D. G., Kahn, K., Lanning, S., and Suchman, L., Beyond the Chalkboard: Computer Support for Collaboration and Problem Solving in Meetings, *Communications of the ACM*, Vol. 30, No. 1, January 1987, pp. 32–47.

[34] Malone, T. W., Grant, K. R., Turbak, F. A., Brobst, S. A., and Cohen, M. D., Intelligent Information Sharing Systems, *Communications of the ACM*, Vol. 30, No. 5, May 1987, pp. 390–402.

[35] Winograd, T., and Flores, F., *Understanding Computers and Cognition*, Ablex Press, San Francisco, 1986.

[36] Nielson, J., *Hypertext and Hypermedia*, Academic Press, San Diego, 1989.

[37] Goodman, G. O., and Abel, M. J., Communications and Collaboration: Facilitating Cooperative Work Through Communications, *Office Technology and People*, Vol. 3, No. 2, August 1987, pp. 129–146.

[38] Sage, A. P. *Decision Support Systems Engineering*, Wiley, 1991.

[39] Gottesman, B. Z., Software — Best of a New Breed-Groupware: Are We Ready?, *PC Magazine*, Vol. 12, No. 1, June 15, 1993, pp. 276–284.

[40] Dalkey, N. C. An Experimental Study of Group Opinion — The Delphi Method, *Futures*, September 1969, pp. 408–426.

[41] Martino, J., *Technological Forecasting for Decision Making*, American Elsevier, New York, 1972.

[42] Zwicky, F., *Discovery, Invention, and Research*, New York: Macmillan, 1969.

[43] Sage, A. P., and Lynch, C. L., "Systems Integration and Architecting: An Overview of Principles, Practices, and Perspectives," *Systems Engineering*, Vol. 1, No. 3, 1998, pp. 176–227.

[44] Sage, A. P., and Rouse, W. B. (Eds.), *Handbook of Systems Engineering and Management*, Wiley, New York, 1999.

CHAPTER **4**

Analysis of Alternatives

The analysis step of systems engineering relies on two important products from the results of an issue formulation effort:

- A detailed description of feasible alternatives
- A set of criteria to evaluate the alternatives

Figure 4.1 shows that you want to evaluate each feasible alternative, or alternative course of action, in terms of the various problem definition and value system design elements that have been identified. There will also, as we noted in our earlier chapters, be a number of objective measures and alternatives measures that will be needed to instrument the sort of scoring matrix suggested in Figure 4.1. This figure is a relatively static and simplistic representation of what we need to accomplish in an actual situation. It is entirely possible that the alternatives are so diverse that different modeling techniques might be needed to suggest measures of the various elements associated with the rows of this figure. But, the figure does convey the notion of "scoring" of the alternatives. Usually, you will not be able to obtain a score, or valuation, of each alternative without knowing many details about how each of the alternatives impacts the various functions that a system is to perform. This is a task for modeling and analysis, and this is what we discuss in this chapter.

4.1   INTRODUCTION

After an issue has been successfully formulated, it is generally subject to analysis. Your objective in systems analysis is to utilize knowledge such that

	Alternative 1	Alternative 2	Alternative 3	Alternative 4	Alternative 5
Need 1					
Need 2					
Constraint 1					
Constraint 2					
Alterable 1					
Alterable 2					
Alterable 3					
Objective 1					
Objective 2					
Objective 3					

Figure 4.1 Hypothetical matrix of problem definition and objective elements and scores for each alternative.

you thoroughly understand how each alternative impacts the needs, constraints, alterables, and objectives that were specified as part of the issue formulation effort. Your analysis work will, ideally and eventually, fill in each square of the matrix in Figure 4.1 so you know how each alternative scores on all the criteria. In this way, we will have identified the impacts of each alternative. Chapter 5 shows how to combine these individual criteria scores into an overall system score and how to incorporate the client's preferences into a useful decision-making structure. In this chapter we describe a variety of modeling techniques that are typically very useful in systems engineering analysis efforts. Then the chapter concludes with ways for using the information gained from modeling efforts to refine and improve the alternatives.

The analysis portion of a systems effort usually consists of two steps:

- Impact identification
- Optimization or refinement of individual alternatives

First, the options or alternatives identified in issue formulation are analyzed to assess the expected impacts of their implementation. Basically, you want to know if your alternatives will make a real difference. Will the alternatives, when implemented, satisfy the customer's effective needs and impact your client's objectives? How will the alternatives score on the criteria? This is often called impact assessment. The second part of an analysis effort leverages what you have learned from the impact assessment to improve and refine your viable alternatives. You try to maximize system performance of each alternative by adjustment or refinement of the parameters. This alternative refinement effort is often called *optimization*.

Forecasting is an essential ingredient of impact assessment. There are many complications associated with forecasting the impacts of alternative courses of action. Among these are uncertainty and imprecision concerning important future events, uncertainty concerning institutional changes, and uncertainty concerning values and changes in values. A great many approaches have been developed and used for forecasting. There are two general classes of methods that may be used for forecasting: expert opinion methods and modeling or simulation methods. We have discussed some expert opinion methods in Chapter 3. Models are extraordinarily useful in systems engineering. Virtually all of this book is based on models, especially models of the systems engineering process and models that are useful for measurement and evaluation of alternatives.

Expert opinions are really mental models that are formed on the basis of experience with a particular issue. A difficulty sometimes with mental models is that they are usually difficult to communicate, or explain in understandable terms, to others. Also, there may be difficulties in accomplishing sensitivity analysis of a mental model to changed conditions and assumptions.

Frequently, all the elements of a mental model are difficult to identify, and the interaction between elements of such a model is not clearly indicated. Assumptions are sometimes treated as if they are fixed elements or relations that are not subject to change. As we have noted, it is not easy for us to communicate mental models to others. We cannot manipulate mental models effectively. Each of these defects taken separately creates confusion; but as these defects combine and interact, understanding of complex systems is severely hampered. *We postulate that systems engineering models transform mental models into forms that are better defined, have clearer assumptions, are easier to communicate, and are more effectively manipulated.* This assumes that the mental model is an appropriate one.

A great deal of intellectual activity has been stimulated by attempts to construct models of systems. *A simple definition of a model is that it is a set of assumptions that describe how something works.* More formally, a model is an abstract generalization or representation of a system. Any set of rules and relationships that describes something is a model of that thing. A system model may be viewed as a physical arrangement, a flow diagram, or a set of actions and consequences that can be shown graphically through time as a simplified picture of reality. When we model systems, we enhance our ability to comprehend their behavior and to understand their interrelationships, as well as our relationship to them. A typical result of a systems engineering model is the opportunity to see a system from several viewpoints or to view issues from multiple perspectives. Systems engineering, operations research, and related areas have made many contributions toward the improvement of clarity in modeling.

There are many types of models. An *iconic model* is a scaled physical representation of an object. A *visual model*, is a graphical or pictorial representation of an object. A *symbolic model* may be *syntactic* or *mathematical*. A mathematical model may be completely analytical, or it may be a simulation model. A mathematical model may also be *deterministic*, which means that everything is known and certain relative to the model and its inputs, or

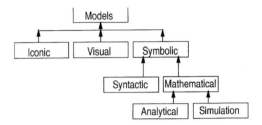

Figure 4.2 Illustration of some of the many different types of models.

stochastic, which means that there are uncertainties associated with some parameters of the model or the inputs to the model, in nature. We may use any combination of words, mathematics, and graphics to establish a model. A model may be based on continuous-time events or discrete-time events. Figure 4.2 illustrates this structure of models. In general, a model contains a *structure* and *parameters* within this structure.

Simulation is the process of conducting experiments with an identified model of a system, which is sometimes called a simulation model, for the purpose of better understanding the system. The major purpose of a simulation is to enable study of the response of a system through study of the response of a model of that system. Often, you will construct a model in order to make relative comparisons of the impacts of various alternative courses of action in satisfying needs and meeting the objectives that were identified as part of the formulation step that exists as the first step for each phase of a systems engineering effort. Figure 4.3 illustrates the generic input–output structure of a model and indicates that there are some elements of the structure yet to be specified.

Improvements in modeling have become more important as systems have become more complex. Usually systems evolve as an aggregate of subsystems that interact with one another to create an interdependent whole. If we understand subsystems and their interactions more clearly, the possibility of their functioning appropriately will be much enhanced. As a prerequisite for such functioning, the complex nature of each subsystem must be understood and the interdependence of subsystems with one another must be explored and comprehended. It would, therefore, seem apparent that a representation of

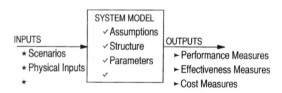

Figure 4.3 Input–output structure of a model.

systems and their interactions will enable increased communication between those concerned with systems behavior and those concerned with system fielding.

Many formal systems engineering approaches are based on simulation and modeling concepts and methods. Simulation and modeling methods are based on the conceptualization and use of an abstraction, or model, which behaves in a way similar to the real system. Impacts of policy alternatives can then be studied through use of the model, something that usually cannot be done through experimentation with the real system. A relatively large number of modeling approaches are described in an excellent but now dated work by Gass and Sisson [1] as well as in an encyclopedia devoted to modeling and simulation, edited by Atherton and Borne [2].

There are three basic purposeful categorizations of models:

- Descriptive models
- Predictive or forecasting models
- Policy or planning models

Representation and replication of important features of a given problem is the object of a descriptive model. Good *descriptive models* are very valuable because they reveal much about the structure of a complex issue and demonstrate how the issue formulation elements impact and interact with each other. An accurate descriptive model must be structurally and parametrically valid, as we will soon discuss. One of the primary purposes behind constructing a descriptive model is to learn about the impacts of various alternative courses of action. A descriptive model is one that is designed primarily to replicate features of a given problem or system. In a descriptive model it is usually sufficient to know, for example, that $y = x^2$. This equation itself says nothing about whether x causes y or y causes x and is thus insufficient for predictive or forecasting purposes. Yet, it may be quite valid in a descriptive sense.

In building a *predictive or forecasting model*, we must be concerned with determination of proper cause-and-effect or input-and-output relationships. We must know causal relationships; for example, a doubling of x will cause y to quadruple. Independent, or exogenous in the sense that they arise external to the system, variables must generally be separated from dependent, or endogenous in the sense that they follow as a consequence of the input to the model as modified by the system, variables. If the future is to be predicted accurately, we must have a method to determine exogenous or independent "given" variables accurately, the model structure must be valid, and parameters within the structure must be accurately identified. Often, it will not be possible to accurately predict all exogenous variables, and, in that case, conditional predictions can be made from scenarios. Consequently, predictive or forecasting models are often used to generate a variety of future scenarios, each a conditional prediction of the future.

Policy or planning models are useful for much more than predictive or forecasting purposes, although any policy or planning model is also a forecasting model. The policy or planning model must be evaluated ultimately in terms of a value system. Policy or planning efforts must not only predict outcomes of implementing alternative policies, but also present these outcomes in terms of a value system useful for ranking, evaluation, and decision making.

Verification of a model is necessary to ensure that the model behaves in a fashion intended by the model builder, as well as accomplishes the intended goal. If we can confirm that the structure of the model corresponds to the structure of the elements it is representing, then the model is verified structurally. But this offers no assurances that specific predictions made from the model are valid. This is the case because the exact parameters of the model have not been confirmed. While verification compares a model against its requirements specifications, validation is concerned with comparing a model with real-world results. The term *accreditation* is sometimes used to describe a model that is suitable for use because it has achieved some identified degree or level of verification and validation.

Normally, three techniques are used to test the *validity* of a model. First is a *reasonableness* test. This test attempts to determine from knowledgeable people that the overall model, as well as model subsystems, respond to inputs in a reasonable way. The model should also be valid according to statistical time series used to determine or estimate parameters and variables within the model. Also, the model should be valid in the sense that the policy interpretations of the various model parameters, structure, and recommendations are consistent with the ethical, professional, and other standards of the group affected by the model. *It is very important to verify and validate a model from structural, functional, and purposeful perspectives.*

Creation of a process to produce information concerning the consequences of proposed alternatives or policies is the principal objective of systems analysis. Descriptive and normative scenarios are available from the problem definition, value system design, and system synthesis steps of systems engineering. In the systems analysis step, we wish to compare and evaluate alternative policies with respect to a value system.

This process is greatly enhanced by having one or more "models" of the issue or problem that is under study. A model is a substitute for reality, but hopefully it is descriptive enough of the real system elements under consideration that we can use the model to answer relevant questions. We want to pose policy questions using the model, and from the results obtained we want to learn how to cope with that subset of the real world being modeled. However, a model must depend on more than the particular problem elements being modeled. It must also be strongly dependent on the value system and the purpose behind utilization of the model. We want to be able to determine correctness of predictions based on usage of a model and thus be able to validate the model. Given the definition of a problem, a value system, and a set of feasible alternatives, we wish to be able to design a model consisting of relevant elements of these three steps and to determine the results of implementing proposed alternatives or policies.

There are *three essential steps in constructing a model*:

1. Determine those problem definition, value system, and system synthesis elements that are most relevant to a particular issue or problem.
2. Determine the structural relationships among these elements.
3. Determine, or identify, the parametric coefficients within this structure.

These steps are necessarily accomplished as part of any model construction effort and are illustrated in Figure 4.4. We could associate a fourth step with this, namely, model verification and validation. In reality, these might be viewed as the development part of an overall effort concerning model acquisition. The first phase would be model definition in which a purposeful definition of the anticipated results of the modeling effort would lead to a set of model specifications. Thus, we may envision a three-phase effort involving definition, development, and deployment for the engineering of a model of a given situation, such as represented in Figure 4.5.

We may *classify models* according to whether they are *quantitative* or *qualitative*. A quantitative model is a mathematical model whose behavior is completely determined by the assumptions used in constructing it. In strategic situations, there are many elements which are difficult to describe adequately in quantitative model form. A qualitative model is one whose behavior is described by deductions based on assumptions and subjective judgment about the complex issue under consideration. Verbal or mental models are examples of qualitative models. Although quantitative models will be emphasized in this chapter, there is a great need for nonquantitative models in many systems analysis efforts, and some theory and applications of causal loop diagrams are presented in this chapter to partially fill this need.

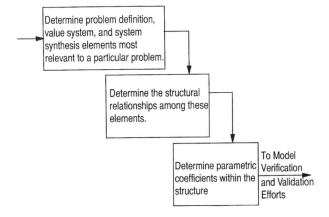

Figure 4.4 Essential steps in construction of a model.

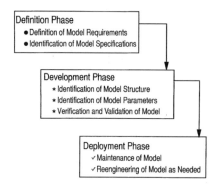

Figure 4.5 Waterfall model of a three-phase modeling effort.

There are not many realistic situations when we have complete certainty about the outcome of alternative courses of action. Often, we must deal with information that is imperfect, from any of several perspectives. Inferential activities based on imprecise, incomplete, inconsistent, or otherwise imperfect knowledge are becoming more important in the design, implementation, and operation of systems that support enhanced human functioning. Inference is concerned with the generation of theories and hypotheses beyond those originally given. In planning and decision-making activities the information that is usually available initially is limited as to allow satisfactory performance of judgment and choice. Hence, inference is an essential activity for humans, as well as for systems intended to aid humans in information processing activities.

In this chapter we first provide a brief overview of analysis approaches that can be used with imperfect information. This will include a discussion of cross-impact analysis and some related issues concerning hypothesis testing, inference, and logical reasoning. Then we examine some approaches for structural modeling using trees, causal loop and influence diagrams. We present a section with many details on system dynamics models and extensions. Next, we will have a look at some models for cost and reliability, availability, and maintainability. Finally, we turn our attention to a discussion of deterministic and stochastic network flow and graph models, queuing models, discrete event simulation models, time series and regression analysis, and the evaluation of models of large-scale systems. Our chapter concludes with a discussion of model verification and validation, sensitivity analysis of models, and the refinement of alternatives.

4.2 ANALYSIS OF SYSTEMS WITH UNCERTAIN AND IMPERFECT INFORMATION

Several approaches for making inference from available information have been developed, ranging from strict probabilistic Bayesian reasoning to less mathematically rigorous approaches. Analysis of systems based on these methods sometimes reveals discrepancies in terms of equivalence of the results obtained due largely to the differences in the underlying assumptions in which the various approaches are based. Among these assumptions are the following:

1. The way in which the uncertain information is represented
2. The assumptions that form the basis for processing information
3. The inferential structure used for this processing
4. The treatment of inconsistent information

A summary of contemporary efforts involving inference mechanisms for information processing in systems, concentrating on the extent to which these mechanisms can be Bayesian, is presented in reference 3. We shall not be able to explore this interesting subject in any detail here and simply assume Bayesian processing of uncertain information.

4.2.1 Cross-Impact Analysis Models

A primary shortcoming of such issue formulation methods as brainstorming, brainwriting, nominal group, and Delphi is the lack of any explicit consideration of the existence of a structure and linkages between events, elements, and forecasts. In generally, identification of the structure of an issue and determination of the parametric interrelationships among coefficients in this structure is not a major purpose of issue formulation. It is a major purpose of issue, or impact, analysis.

In addition to a knowledge of structure, such that we are not restricted by having to project events as if they are independent of one another, it is highly desirable to also be able to deal with uncertainties. There is no convenient way, however, to specifically indicate the randomness or probability of the occurrence of events, elements, and forecasts using most issue formulation approaches. This is needed, and so we turn to issue analysis. *When particular events occur or do not occur, the probability of other structurally related events occurring is typically altered or impacted.* We desire to be able to determine this impact. Quite some time ago now, various authors [4–9] developed a matrix of probabilities connecting events as a method of considering these possible impacts. Much of the original cross-impact matrix analysis methodology will be developed and refined in this section. We will also present several examples. A number of other interesting examples are presented in reference 10.

Before presenting some illustrative examples, let us look at just a little bit of theoretical development. We consider two events e_i and e_j. We denote the probability of occurrence of these two events as $P(i)$ and $P(j)$. Considered together, these two events may be totally uncoupled, coupled, or totally included, as indicated in Figure 4.6. Totally uncoupled events are those events whose occurrence or nonoccurrence has no effect upon occurrence or nonoc-

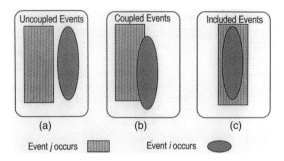

Figure 4.6 Event relationships for uncoupled events, coupled events, and the included event case where the occurrence of j guarantees the occurrence of i.

currence of other events in the event set. Coupled events are those whose occurrence or nonoccurrence will affect the likelihood of occurrence or nonoccurrence of other events in the element set. A totally included event is one that is entirely contained in another event such that if one event occurs, then the totally included event must occur. If an event does not occur, then a totally included event cannot occur. It is necessary to consider conditional probabilities and cross-impact analysis to deal with coupled events. Uncoupled and totally included events are somewhat easier to deal with and may not need to be subjected to a cross-impact analysis. However, events of importance are rarely uncoupled or totally included with all other events. Bayes' rule, discussed in any elementary probability text, states that the probabilities of events e_i and e_j must be related by

$$P(i \mid j) = \frac{P(j \mid i)}{P(j)} P(i) \tag{4.2.1}$$

In cross-impact analysis we consider two types of connecting modes for impacting events: *enhancing* and *inhibiting*. If the probability of event i occurring conditioned upon the knowledge that event j has occurred or will occur is greater than the probability of event i occurring, we say that event e_j is *enhancing* e_i. If the probability of event i occurring conditioned upon the knowledge that event j has occurred or will occur is less than the probability of event i occurring, we say that event e_i is *inhibiting* event e_j. Finally, if the probability of event e_i occurring is independent of the occurrence or nonoccurrence of event e_j, we say that event e_j is independent of event e_j. In symbols we have

If $P(i \mid j) > P(i)$, then e_j enhances e_i.

If $P(i \mid j) < P(i)$, then e_j inhibits e_i.

If $P(i \mid j) = P(i)$, then e_j is independent of e_i (and e_i is independent of e_j).

Figure 4.6(a) represents the case where events i and j are uncoupled and nonindependent, and $P(j|i) = 0$. Thus event i is completely inhibiting to event j. Figure 4.6(c) represents the totally included and nonindependent case where $P(j|i) = 1$, such that event i is completely enhancing to event j. The much more interesting case is represented by Figure 4.6(b). The relationship between events i and j may take any of the forms shown in Figure 4.7. In this figure the probability of each of the various events is presumed equal to the fraction of the total area the event space occupies. The three cases of Figure 4.7 are

$$\frac{P(j|i)}{P(j)} = 1, \quad \text{independent events}$$

$$\frac{P(j|i)}{P(j)} > 1, \quad \text{event } i \text{ enhances event } j$$

$$\frac{P(j|i)}{P(j)} < 1, \quad \text{events } i \text{ inhibits event } j$$

It is of value to determine bounds on enhancement and inhibition of event j upon event i. From the laws of joint probability we have

$$P(i) = P(ij) + P(i\bar{j}) \tag{4.2.2}$$

where $P(ij)$ is the probability of occurrence of events i and j, and $P(i\bar{j})$ is the probability of occurrence of event i and nonoccurrence of event j. Using the conditional probability law allows us to write this as

$$P(i) = P(j)P(i|j) + [1 - P(j)]P(i|\bar{j}) \tag{4.2.3}$$

Now because we must have $0 \leqslant P(j) \leqslant 1$ and $0 \leqslant P(i|\bar{j}) \leqslant 1$, we see that the second term on the right in the above equation must be positive, and we have

$$P(i) \geqslant P(j)P(i|j) \tag{4.2.4}$$

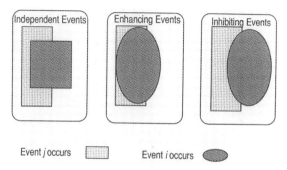

Figure 4.7   Event relationships for independent, enhancing, and inhibiting events.

Thus you can easily see that an upper bound on the conditional or impacted probability $P(i\,|\,j)$ is given by

$$P(i\,|\,j) \leqslant \left[\frac{1}{P(j)}\right][P(i)] = aP(i) \tag{4.2.5}$$

The number a must be positive and greater than 1. Thus you can show that this upper bound must appear as in Figure 4.8.

From the probability law for compound events, we have

$$P(i \cup j) = P(i) + P(j) - P(ij) \tag{4.2.6}$$

where $P(i \cup j)$ is the probability that event i or j or both will occur. We may rewrite this as

$$P(i \cup j) = P(i) + P(j) - P(i\,|\,j)P(j) \tag{4.2.7}$$

We know that this probability must be greater than 0 and less than 1. Thus we have

$$P(i\,|\,j) \geqslant \frac{P(i) + P(j) - 1}{P(j)} = \frac{P(i) - 1}{P(j)} + 1 \tag{4.2.8}$$

and

$$P(i\,|\,j) \leqslant \frac{P(i) + P(j)}{P(j)} = 1 + \frac{1}{P(j)}P(i) \tag{4.2.9}$$

You may easily show that the inequality of Eq. (4.2.5) is stronger than that of Eq. (4.2.9), so that we do not need to consider Eq. (4.2.9) any further. Equation (4.2.8) represents a lower bound on the impacted or conditional probability $P(i\,|\,j)$ in terms of the initial probability of the impacted event $P(i)$. This lower

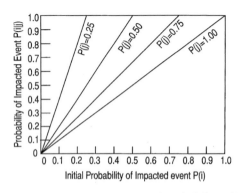

Figure 4.8   Upper bound for $P(i\,|\,j)$ in terms of probability of impacted event.

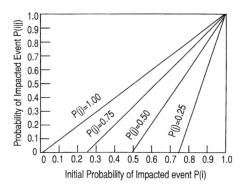

Figure 4.9 Lower bound for $P(i|j)$ in terms of probability of impacted event.

bound is illustrated in Figure 4.9.

Now, we know that the event j will either be enhancing to, inhibiting to, or independent of event i. If event j enhances event i, then you know that $P(i|j)$ is greater than $P(i)$. The limiting case for enhancement occurs when we have independent events and $P(i|j) = P(i)$. The bound for enhancing event j is, therefore, from Eq. (4.2.5),

$$P(i) \leqslant P(i|j) \leqslant P(i)\left[\frac{1}{P(j)}\right], \qquad \text{for } j \text{ enhancing} \qquad (4.2.10)$$

For inhibiting event j we know that $P(i|j)$ is less than $P(i)$, with equality occurring when events i and j are independent. The bounds for inhibiting event j are determined from this inequality and Eq. (4.2.8) as

$$1 + \frac{1}{P(j)}[P(i) - 1] \leqslant P(i|j) \leqslant P(i), \text{ for } j \text{ inhibiting} \qquad (4.2.11)$$

We can also consider probability bounds for nonoccurrence of impacting events e_j. It is straightforward for you to show that these bounds are

$$P(i) \leqslant P(i|\bar{j}) \leqslant \frac{P(i)}{1 - P(j)}, \qquad \text{for } j \text{ inhibiting} \qquad (4.2.12)$$

$$1 - \frac{1 - P(i)}{1 - P(j)} \leqslant P(i|\bar{j}) \leqslant P(i), \qquad \text{for } j \text{ enhancing} \qquad (4.2.13)$$

We must use these bounds to ensure consistency of probabilities. If we specify three of the four probabilities $P(i)$, $P(j)$, $P(i|j)$, and $P(j|i)$, then the other, unspecified probability is immediately determined from Eq. (4.2.1). Even if only two of these probabilities are specified, the third may not be chosen arbitrarily.

We must be sure that the bounds just developed are satisfied.

Example 4.2.1. Let us consider a particular case where $P(j) = 0.25$. For the case where j *enhances* the occurrence of event i, the probability bounds are determined from Eqs. (4.2.10) and (4.2.13) as

$$P(i) \leqslant P(i \mid j) \leqslant 4P(i)$$

$$\tfrac{4}{3}P(i) - \tfrac{1}{3} \leqslant P(i \mid \bar{j}) \leqslant P(i)$$

These bounds are depicted in Figure 4.10(a).

For the case where event j inhibits event i, we have

$$4P(i) - 3 \leqslant P(i \mid j) \leqslant P(i)$$

$$P(i) \leqslant P(i \mid \bar{j}) \leqslant \tfrac{4}{3}P(i)$$

and these are depicted in Figure 4.10(b). In attempting to elicit individual or group response concerning probabilities we must make sure that the response is somehow constrained to be within these bounds or we will violate fundamental laws of probability.

At this point we may well question the use to which these probabilities might be put. We are considering the occurrence and nonoccurrence of four possible events $\overline{ij}$, $\bar{i}j$, $i\bar{j}$, and ij. Generally these will be future events. For convenience let us assume that occurrence or nonoccurrence of event j is at an earlier time than occurrence or nonoccurrence of event i. The *event tree* of Figure 4.11 depicts the possible outcomes and the associated probabilities necessary to determine the probability of the four possible futures. We may compute bounds on the future probabilities, assuming that j is the impacting

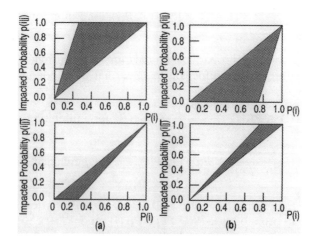

Figure 4.10   Impacted probability bounds for Example 4.2.1.

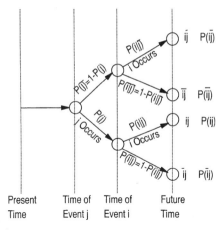

Figure 4.11 Event tree for two-event model.

(enhancing or inhibiting) event. Using Bayes' rule and Eqs. (4.2.10) and (4.2.13) yields

$$P(i)P(j) \leqslant P(ij) \leqslant P(i)$$

$$P(i) - P(j) \leqslant P(i\bar{j}) \leqslant P(i)[1 - P(j)]$$

$$P(j) - P(i) \leqslant P(\bar{i}j) \leqslant [1 - P(i)]P(j)$$

$$[1 - P(i)][1 - P(j)] \leqslant P(\overline{ij}) \leqslant [1 - P(i)]$$

for the case where j is *enhancing*, and it yields a similar set of four equations where j is inhibiting.

If we specify $P(i) = 0.5$ for this example, we then have

$$0.5 \leqslant P(i \,|\, j) \leqslant 1$$

$$0.33 \leqslant P(i \,|\, \bar{j}) \leqslant 0.5$$

for event j enhancing to event i, and we have

$$0 \leqslant P(i \,|\, j) \leqslant 0.5$$

$$0.5 \leqslant P(i \,|\, \bar{j}) \leqslant 0.67$$

for event j inhibiting to event i. We determine the corresponding bounds on the various compound probabilities as

$$0.125 \leqslant P(ij) \leqslant 0.25$$

$$0.25 \leqslant P(i\bar{j}) \leqslant 0.375$$

$$0 \leqslant P(\bar{i}j) \leqslant 0.125$$

$$0.375 \leqslant P(\overline{ij}) \leqslant 0.5$$

for event j enhancing to event i, and

$$0 \leqslant P(ij) \leqslant 0.125$$

$$0.375 \leqslant P(i\bar{j}) \leqslant 0.5$$

$$0.125 \leqslant P(\bar{i}j) \leqslant 0.25$$

$$0.25 \leqslant P(\bar{i}\bar{j}) \leqslant 0.375$$

for event j inhibiting to event i.

For this particular example we can determine the most likely future outcome merely by a specification of $P(i)$, $P(j)$ and a statement, which must be correct if we are to obtain correct bound i, concerning whether event j is enhancing or inhibiting to event i. For the case where j enhances i, the most likely outcome is that event i will not occur and that event j will also not occur. This is entirely reasonable, as you can easily see by examining the foregoing inequalities.

Occurrence of event $\bar{j}$ enhances occurrence of $\bar{i}$, which already has an even chance of occurring. In a similar way the most likely outcome is that event i will occur and event j will not occur when j inhibits i. We should note in concluding this example that specification of the marginal probabilities $P(i)$ and $P(j)$ does not restrict the joint probabilities. Knowledge of whether j enhances or inhibits i, together with the marginal probabilities, does allow us to bound the joint and conditional probabilities. In this specific example, $P(i|j)$ may vary from 0 to 1 because knowledge of the marginal probabilities tells us nothing about the conditional probabilities. Specification of j inhibiting allows us to bound $0 \leqslant P(i|j) \leqslant 0.5$, whereas specification of j enhancing requires $0.5 \leqslant P(i|j) \leqslant 1$.

Let us consider a set of events that may occur at specific times in the future. These events are denoted by e_1, e_2, ..., e_n and have associated (marginal) probabilities of occurrence $P(1)$, $P(2)$, ..., $P(n)$. Suppose that $P(1) = 1$, so that event e_1 occurs with certainty. How does the known occurrence of e_1 affect the probabilities of occurrence of the other events? If there is a cross-impact between events, the impacted probability of individual events will either increase or decrease with the occurrence or nonoccurrence of other elements. In the general case we wish to obtain conditional probability responses to the following question: If event e_j occurs, what is the probability of event e_i, $P(i|j)$, for all i and j? Many will find it difficult to respond to this question as it is posed, and so it is desirable to re-pose the question. Suppose that you ask the following question: If event e_j occurs, is the probability of event e_i occurring enhanced, inhibited, or unaffected? By how much? A cross-impact matrix can be filled in from responses to the above. The responses and entries you might obtain from this query may vary from binary entries $+1$, 0, -1 to integer entries on a scale of 10, 100, or 10^{24}.

For example, a cross-impact matrix might be as follows:

$$M = \begin{bmatrix} - & +3 & -6 & +10 \\ +2 & - & +3 & 0 \\ -7 & +1 & - & 0 \\ +8 & 0 & 0 & - \end{bmatrix}$$

where the ijth element in the matrix represents the relative enhancing $(+)$ or inhibiting impact $(-)$ that the occurrence of event e_i has on event e_j. For this particular illustration, we assume that $+10$ represents maximum enhancement and -10 represents maximum inhibition. Each row in the matrix corresponds to the occurrence of a particular event. Positive numbers in that row indicate that another event, denoted by the column heading, is enhanced by the occurrence of that particular event. Negative entries mean that some event is inhibited by that particular event. An entry of zero means that the event is unaffected by the occurrence of that row event. For instance, the 0 in row 4, column 2 indicates that event e_2 is unaffected by the occurrence of event e_4.

Generally, you would categorize enhancing events as those which are enabling in the sense of the occurrence of one event making it easier for a subsequent, enhanced event to occur. In a similar way, inhibiting events are those that are denigrating in the sense of making other events infeasible, or impractical or antagonistic, in the sense that their occurrence makes it more difficult for the subsequent, inhibited events to occur. For the specific cross-impact matrix under consideration in Figure 4.8 we can state that event e_3 is either antagonistic or denigrating with respect to occurrence of event e_1. Thus event e_3 inhibits e_1.

Some early researchers in cross-impact analysis have assumed that the impacted probability is a quadratic function of the impacting event. To implement this assumption, you might let

$$P(i|j) = P(i) + A_{ij}P(i)[1 - P(i)] \qquad (4.2.14)$$

This is the *shilling fraction*, which is the most general impacted probability $P(i/j)$ such that if $P(i) = 0$, then $P(i/j) = 0$, and if $P(i) = 1$, then $P(i/j) = 1$. For $A_{ij} = 0$, we have independence between events i and j. For $A_{ij} > 0$, event j enhances j event i; and for $A_{ij} < 0$, event j inhibits event i. Individuals participating in a cross-impact exercise would be asked to provide their estimates for the (marginal) probability $P(j)$ of event j as well as their estimate of impact factors that are transformed into a value of A_{ij}.

Other approximations have been suggested. For example, Enzer [11] proposed utilizing an expression of the form

$$P(i|j) = \frac{r_{ij}P(i)}{1 + (r_{ij} - 1)P(i)} \qquad (4.2.15)$$

where r_{ij} is a *likelihood ratio* of impacted odds for event j impacting upon event i. For this expression the impacted probability contours are symmetrical for the required values of r_{ij}. Also, any positive value of r_{ij} may be used in Eq. (4.1.16) to completely span the space of impacted probabilities, whereas we must impose another restriction in Eq. (4.1.15) in order to avoid senseless impacted probabilities that are negative or greater than 1. This restriction makes it impossible to completely scan all admissible $P(i/j)$. The use of the likelihood ratio impacted probability expression of Eq. (4.1.15) does resolve the problem of symmetry; however, neither the likelihood ratio impacted probability expression nor the quadratic impacted probability expression of Eq. (4.1.14) will generally result in probabilities that are correct according to Bayesian probability theory. In order to utilize Eq. (4.2.15), we need to have an appropriate expression for the likelihood ratio. This may be obtained from

$$r_{ij} = \frac{P(i/j)[P(i) - 1]}{[P(i/j) - 1]P(i)} \tag{4.2.16}$$

When you use Eqs. (4.2.15) and (4.2.16) to estimate probabilities, you will not generally obtain the same expressions as you obtain when using Eq. (4.2.14). There is no assurance that either result will be a valid probability, and for this reason we use the shilling fraction (/) to indicate this.

From the statement of appropriate probabilities of occurrence and nonoccurrence of events, a cross-impact analysis will provide a statement of probabilities of future events. Thus cross-impact analysis can be associated closely with a quantitative deterministic systems analysis model in a more complete simulation exercise to evolve time-dependent future predictions. We suggest that you use the following principal steps in conducting a cross-impact analysis.

1. Use collective inquiry methods to generate a set of appropriate events, elements, and relations for a particular problem.

2. Use a structural modeling method to evolve a structure for the events under consideration.

3. Develop a scale so that one may translate expert opinion and knowledge concerning the likelihood of occurrence of events into probabilities. This scale should be discussed, understood, and approved by the group supplying the probability estimates.

4. Obtain an initial set of probabilities of the type in which estimators have most confidence. Every effort should be made to obtain as many of the direct tree-related conditional probabilities as possible. For direct tree-related conditional probabilities not estimated, obtain sufficient marginal and lower-order conditional probabilities, along with impact assessments, to bound the direct tree-related conditional probabilities. These upper and lower bounds are computed.

5. Display the set of direct tree-related conditional probabilities or bounds computed on these probabilities if they are not directly estimated. Obtain final probabilities with which to determine future probabilities.

6. Determine future event probabilities.

7. Incorporate the results of the cross-impact analysis into a system simulation model of realities in which events are assumed to occur or not occur as predicted from the results of the cross-impact analysis. Inputs or policies connected with the events must be determined in order to accomplish this simulation. Iterate these simulation runs for other likely futures.

8. Incorporate the results of the cross-impact analysis and system simulation exercise as the systems analysis phase of systems engineering activities for the particular problem under consideration.

This approach will generally result in suitable cross-impact probabilities.

There are other approaches that might be used. Here are a number of cautions that should be kept in mind when using cross-impact analysis.

1. *Inconsistencies.* There is no guarantee that Bayes' rule will be satisfied if a group picks all four terms $P(i)$, $P(j)$, A_{ij}, and A_{ji} and then uses the quadratic relation of Eq. (4.2.14). In fact, there is no guarantee that the inequalities of Eqs. (4.2.10) through (4.2.13) will be satisfied either. It is for this reason that we use the symbol $P(i/j)$ in Eqs. (4.2.14), (4.2.15), and (4.2.16) rather than $P(i/j)$ and have refrained from calling $P(i/j)$ a conditional probability. There is no guarantee that it will even be a valid probability function unless Eqs. (4.2.1), (4.2.2), and (4.2.7) are satisfied. The inconsistency may or may not be serious [12].

2. *Low- and High-Probability Events.* The relationship between $P(i/j)$ and $P(i)$ determined from Eq. (4.2.14) is not even symmetrical. We can show that, as a consequence, an enhancing event can only make minor changes in a low-probability $P(i)$, and an inhibiting event cannot significantly change a high-probability $P(i)$. A number of other and similar early proposals have similar difficulties, especially with respect to the reality that the probabilities often obtained are not correct according to Bayesian probability theory. Use of Eqs. (4.2.15) and (4.2.16) will lead to a symmetrical relationship. However, there is still no guarantee that Bayes' rule will be satisfied, although it should be "close" to satisfaction.

3. *Nonoccurrence of Events.* Just as the occurrence of an event j influences, in general, the probability of occurrence of event i, so also will the known nonoccurrence of event j influence the probability of occurrence of event i. This must happen unless $P(i/j) = P(i)$, which would indicate that event i is independent of event j but not independent of event $\bar{j}$. This could, of course, occur. But, there is certainly no reason why it must. The classical cross-impact method, represented by Figure 4.11, assumes that nonoccurrence of any event has no impact on any other event.

4. *Time Order of Events.* Time is an important variable, and events must generally be associated with evolution over time. It is quite a different matter to determine $P(i/j)$ when the time of occurrence of i is later than that of j than it is when the time of occurrence of j is later than that of i. It is vitally important to consider the time order of events when they are defined, and failure to do this can lead to crucial errors, as the following example indicates.

Example 4.2.2. Let us consider a cross-impact analysis for two events:

e_1 In 2000, Congress passes a law banning importation of international crude.

e_2 In 2002, there is a dramatic shortage of energy in the United States.

In the classical approach to cross-impact analysis, an expert or a group of experts would be asked to give estimates of $P(1)$, $P(2)$, and the quadratic ratios A_{12} and A_{21}.

Time histories may be associated with these factors if desired. Relations of the form $A_{ij} = -k_{ij}f_{ij}[(t_i - t_j)/t_i]$ have been suggested. Here k_{ij} would be $+1$ if event j enhances event i, and it would be -1 if j inhibits event i. The expression f_{ij} represents an impact factor that has a value between 0 and 1, with the larger number implying greater impact strength. The times t_i and t_j represent the times of occurrence of events i and j. As the time interval between events increases, the strength of the impact is reduced, and it is necessary to restrict $t_i < t_j$.

Difficulties arise with respect to obtaining the correct sign for the impact if we are not careful to note the time of occurrence of the event. For example, if event e_1 occurs and there is a law banning imported crude, then there is a greater likelihood of event e_2, an energy shortage, occurring. The impact of event 1 on event 2 is positive, and event 1 enhances event 2. Now suppose that event e_2 occurs, and there is a dramatic energy shortage. Will Congress pass a law banning imported crude? Certainly the likelihood of this law being passed is much less than it would have been if the energy shortage had not occurred. Thus we might say that event 2 inhibits event 1. A particular cross-impact matrix for this particular example might be obtained from the responses $P(1) = 0.3$, $P(2) = 0.4$, $A_{21} = 0.2$, $A_{12} = -0.8$. The impacted probabilities are, however, such that Bayes' rule is not satisfied. From Eq. (4.2.15), we obtain, $P(1|2) = 0.132$ and $P(2|1) = 0.448$. These are not at all consistent with Bayes' rule. However, they are in accord with our statement that e_1 (banning imported crude) enhances e_2 (energy shortage), and thus we expect $P(2|1) > P(2)$. Also we noted that an energy shortage would inhibit passage of a crude import law. Thus we expect $P(1|2) < P(1)$.

Unfortunately our "experts" have given precisely the wrong sign to the impact direction for the relation of an energy shortage to passage of a law banning imported crude. The correct response should have been, "A shortage of energy will encourage Congress to pass a law banning imported crude!" This seems very improper. But, when we associate the correct dates with the events, the above statement is, "In 2002 a drastic shortage of energy occurred." Was the probability that Congress had a law banning imported crude in the year 2000 enhanced or inhibited by this event? Clearly the probability has been enhanced, and not inhibited as earlier stated. Failure to associate dates with the earlier statement no doubt leads us to believe that Congress is passing the law (event 1) after the occurrence of the energy shortage (event 2), and this is of course incorrect.

There is no real need to even request the "experts" to estimate $P(2|1)$ for this example if we already know $P(1|2)$. Bayes' law could certainly be used to compute $P(1|2)$ from values of $P(1)$, $P(2)$, and $P(2|1)$. Alternatively, we could ask a group for all four responses, and then use Bayes' rule to demonstrate any inconsistency and finally request the group to determine a consensus that satisfies Bayes' rule and the probability inequalities of Eqs. (4.2.10) through (4.2.13). If we accept the responses for $P(1)$, $P(2)$, and A_{21}, then we may compute $P(2|1)$ from Eq. (4.2.15) as $P(2|1) = 0.448$. Bayes' rule then yields $P(1|2) = 0.336$. Thus event 2 is enhancing to event 1, as it indeed should be, considering the dates of occurrence of the events. This turns out to correspond to $a_{12} = 0.1714$, as we see from Eq. (4.2.15).

Use of a causal loop diagram approach to structure these two events using the relationship "attainment or nonattainment of event i is contributory to attainment or nonattainment of event j" leads easily to the structural model of Figure 4.12(a). The associated future tree is indicated in Figure 4.12(b). Event probabilities for this figure are obtained from the calculations of the previous paragraph. It is not possible to use the set of probabilities determined in the first part of this example for the future, because they are inconsistent.

We note that there is nothing to prevent $P(2|\bar{1}) > P(2|1)$ in Figure 4.12(b). Normally we would expect that the probability of an energy shortage in 2002, given that Congress banned imported crude in 2000, should be greater than the

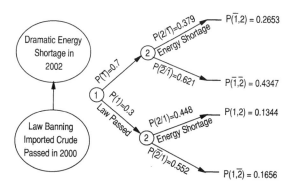

Figure 4.12 Causal loop diagram and event tree for Example 4.2.1.

probability of an energy shortage, given that Congress did not pass the law. The reason for this discrepancy is that we initially requested that the marginal probability $P(2)$ be specified. This is $P(2) = P(12) + P(\bar{1}2)$, or the sum of the probability that Congress passes the law and there is an energy shortage and the probability that Congress does not pass the law and there is an energy shortage. Undoubtedly, this is a difficult probability to estimate, particularly since it is the sum of two future probabilities.

It would seem more appropriate here to obtain estimates for two conditional probabilities $P(2|1)$, $P(2|\bar{1})$ and one marginal probability $P(1)$. In order to do this, we might specify the relations $P(2|1) = 0.448$, $P(2|\bar{1}) = 0.333$, and $P(1) = 0.3$. Then, we obtain $P(2) = 0.3765$. We should be much more pleased with this value of $P(2|\bar{1})$ than with the previous one, and the value of the calculated $P(2)$ is really little different from the previous value. Alternatively, we could start with assumed estimates of $P(2)$ and $P(1|2)$ and then calculate the other probabilities using Bayes' rule.

From our discussions thus far we have seen that it will often be difficult to estimate some of the probabilities that may be used to construct a cross-impact analysis. Certainly, to specify the marginal probability of a single event, when there are many events involved in a given problem, is difficult. For example, to estimate the probability that Congress will pass a law in 2003 banning all forms of home heating not derived from solar energy would be especially difficult. One could only speculate on the probability of this event occurring. However, if one were to condition this event upon many other events such as availability of crude, natural gas and nuclear energy as well as developments in solar energy technology prior to 2003, then a much more reliable estimate of the conditional probability would be possible. Not all possible conditional probabilities are needed to determine future probabilities. In fact, some of these, such as an estimate of the probability of an event conditioned upon an event occurring at a later time, are difficult to obtain, as we have seen in our last example. We suggest a procedure that uses the formulation techniques of the previous chapter to structure events, elements, and forecasts in such a way that a probability tree results. Only the conditional probabilities that are directly relevant to this decision tree are needed to conduct a cross-impact analysis. An example with three events is particularly useful to demonstrate the concepts involved. More general cases associated with the use of hierarchical inference structures to determine relevant probabilities for cross-impact and decision analysis are available and will be suggested later in this section.

Example 4.2.3. We consider the three-event probability or future tree of Figure 4.13. The directly relevant probabilities for evaluation of the possible futures are $P(k)$, $P(j|k)$, $P(j|\bar{k})$, $P(i|jk)$, $P(i|\bar{j}k)$, $P(i|j\bar{k})$, and $P(i|\bar{j}\bar{k})$. All other probabilities needed to specify every future are determined from these probabilities. If expert opinion or knowledge can be used to estimate or determine these probabilities, they should be specified. Specification of other conditional

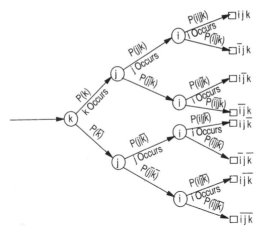

Figure 4.13   Probability tree for Example 4.2.3.

probabilities can lead to these conditional probabilities, which we may call the direct tree-related conditional probabilities. Approximations will generally have to be used to accomplish this, however. Unless we specify all necessary probabilities, then we cannot compute the probability $P(i|jk)$, which is, from Bayes' rule,

$$P(i|jk) = \frac{P(k|ij)}{P(k|j)} P(i|j) = \frac{P(jk|i)}{P(k|j)P(j|i)} P(i|j) \tag{1}$$

Now if we assume that events j and k conditioned upon i are independent, we have

$$P(ij\bar{k}) = 2/5, \quad P(\bar{i}jk) = 1/21, \quad P(i\bar{j}k) = 0, \quad \text{and} \quad P(\overline{ij}k) = 4/15$$

By substitution of the relation $P(jk/i) = P(j/i)P(k/i)$ into Eq. (1), we obtain

$$P(i|jk) = \frac{P(k|i)}{P(k|j)} P(i|j) \tag{2}$$

Thus, we see that if we specify all the single-event conditioned probabilities, we can approximate the needed double-event conditioned probabilities.

Even if a group is able to specify some of the needed probability functions quite accurately, the group will often have little confidence in their ability to specify all needed probabilities. The inequalities of Eqs. (4.2.10) through (4.2.13) have previously been developed to bound conditional densities such that they satisfy valid probability laws. These inequalities are still valid when conditioned on a third event. Thus, where event k occurs before the possible occurrence of event j, we have

$$1 + \frac{1}{P(j|k)}[P(i|k) - 1] \leqslant P(i|jk) \leqslant P(i|k) \qquad j \text{ inhibiting} \qquad (3)$$

$$P(i|k) \leqslant P(i|\bar{j}k) \leqslant \frac{P(i|k)}{1 - P(j|k)} \qquad j \text{ inhibiting} \qquad (4)$$

$$P(i|k) \leqslant P(i|jk) \leqslant \frac{P(i|k)}{P(j|k)} \qquad j \text{ enhancing} \qquad (5)$$

$$1 - \frac{1 - P(i|k)}{1 - P(j|k)} \leqslant P(i|\bar{j}k) \leqslant P(i|k) \qquad j \text{ enhancing} \qquad (6)$$

We may bound conditional probabilities just as we did in Example 4.1.1. Let us consider the case where the direct tree-related conditional probabilities are

$$P(k) = 1/3, \quad P(j|k) = 1/5, \quad P(j|\bar{k}) = 3/5, \quad P(i|jk) = 2/7, \quad P(i|\bar{j}k) = 1/2,$$
$$P(i|j\bar{k}) = 1, \quad P(i|\bar{j}\bar{k}) = 0$$

From these probabilities we can construct the single-event conditioned probabilities. We obtain the results $P(i) = 58/105$, $P(j) = 7/15$, and $P(k) = 1/3$ for the single-event probabilities. For the single-event conditional probabilities, we obtain the results $P(i|j) = 44/49$, $P(i|k) = 16/35$, $P(j|j) = 22/29$, $P(j|k) = 1/5$, $P(k|i) = 8/29$, and $P(k|j) = 1/7$. The true futures obtained from the known direct tree-related conditional probabilities are

$$P(ijk) = \tfrac{2}{105}, \quad P(\bar{i}jk) = \tfrac{2}{15}, \quad P(ij\bar{k}) = \tfrac{2}{15}, \quad P(\bar{i}j\bar{k}) = 0$$

Therefore, the most likely future is occurrence of events i and j and nonoccurrence of event k because it is the future with the greatest probability, or $P(ij\bar{k}) = 2/5$.

Let us now assume that we do not know any probabilities other than the marginal probabilities and the single-event conditional probabilities. From these probabilities we see that event j enhances i, whereas k inhibits i, and k inhibits j. From these probabilities let us determine the approximate two-event conditional probabilities. From Eq. (2) we may easily determine the fact that $P(i|jk) = 2.93$, $P(i|\bar{j}k) = 4/29$, $P(i|j\bar{k}) = 22/29$, and $P(i|\bar{j}\bar{k}) = 21/58$. Unfortunately, these numbers are not very close to the true known values. The true value of $P(i|jk) = 2/7$ actually lies outside of this region. The event j really *inhibits* event i conditioned upon k occurring. Thus we should have used Eq. (5) to obtain $0 \leqslant P(i|jk) \leqslant 16/35$, and the true value of the conditional probability is in the correct range for this inequality.

If we knew that j inhibited the occurrence of i conditioned upon k, then we should have selected a conditional probability $P(i|jk)$ in the range 0 to $P(i|jk)$ in the range 0 to 16/35. For this purely numerical example, there is no truly correct value of $P(i|jk)$ obtainable only from the marginal probabilities and first-order conditional probabilities. Any value of $P(i|jk)$ is acceptable if we assume that j enhances i given k, and any value in the range 16/35 to 1 is acceptable if we assume that j inhibits i. Hence, the approximation of Eq. (2) must be used with considerable caution, because it will often yield a result that violates the inequality bounds of Eqs. (3) through (6). There is little reason to expect that they should be adequate for our use unless the conditions used to derive them, such as independence of j given i and k given i, are satisfied. The inequality bounds of Eqs. (3) through (6) are of value and can be used to pose bounds to a group to help them estimate the probability of an event i conditioned upon n events when we already have estimates concerning the probability of the event i conditioned upon fewer than n events. Thus statements made earlier that probabilities associated with the probability tree should be directly estimated whenever possible appear to be particularly appropriate.

Now that we have examined several situations from the perspective of cross-impact analysis, we summarize the general objectives and approach to be followed. From a statement of appropriate probabilities of occurrence and nonoccurrence of events, a cross-impact analysis will provide a statement of probabilities of future events. Thus cross-impact analysis can be associated closely with a quantitative deterministic systems analysis model in a more complete simulation exercise to evolve time-dependent future predictions. To accomplish cross-impact analysis, you should make every effort to obtain the following:

1. Reasonable estimates for the probability of occurrence of each event considered
2. The degree and nature of the impact that the occurrence and nonoccurrence of each event will have on the likelihood of occurrence of each other event in the complete set of events.

When it is believed that an event has occurred, the probability of occurrence of each of the remaining events is adjusted in accordance with available information. This principle is used repeatedly in order to obtain the probability of occurrence of the various chains of occurring and nonrecurring events. The objectives in using cross-impact analysis are to identify the following:

• The very likely and the very unlikely chains of future events
• A better understanding of how and why some plans and policies will impact outcomes
• Critical events and outcomes

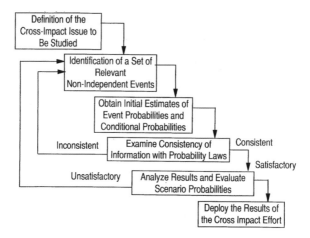

Figure 4.14   Typical effort sequence in cross-impact analysis.

It is necessary to check the information obtained in a cross-impact analysis for consistency with the laws of probability and, especially when the inconsistencies are serious, iterate the process such as to obtain revised estimates of cross-impact probabilities. After information is obtained that does not produce serious violation of the probability laws and when the individual or group responsible for supplying information is satisfied with the veridicality of the information provided, the probabilities of various outcomes, futures, or scenarios are computed. It is especially important to be careful in estimating various conditional probabilities because they may be relatively more important than marginal probabilities in determining event probabilities after a long cascade of conditional events. Figure 4.14 illustrates a suggested process for accomplishing a cross-impact analysis.

Hierarchical inference is a subject closely related to cross-impact analysis. We conclude this section with a presentation of some of the wealth of results available concerning use of both this and related approaches.

4.2.2   Hierarchical Inference and Extensions

Inferential activities based on imprecise, incomplete, inconsistent, or otherwise imperfect knowledge is becoming more important in the design, implementation, and operation of systems that support enhanced human information processing. Inference is concerned with the generation of theories and hypotheses beyond those originally given. In planning and decision-making activities the information that is usually available initially is limited as to allow satisfactory performance of judgment and choice. Hence, inference is an essential activity for humans, as well as for systems intended to aid humans in information processing activities. There are many instances in our lives where we have the need to make inferences. One of the more classic of these is the inference that a patient has a particular disease based upon the evidence of having obtained positive results from several diagnostic tests that "infer" the disease.

Several approaches for making inference from available information have been developed, ranging from strict probabilistic Bayesian reasoning to less mathematically rigorous approaches. Analysis of systems based on these methods reveals discrepancies on the results obtained due largely to the differences in the underlying assumptions in which they are based. Quinlan [13] has contrasted several of these approaches and classifies their dissimilarities in terms of the following:

1. The way in which the uncertain information about propositions is represented
2. The assumptions that form the basis for propagating information
3. The control structure used for this propagation
4. The treatment of inconsistent information

Much research exists on the subject of inference that uses probability theory as the standard for the representation, aggregation, and interpretation of information. However, while such theories have the advantage of modeling the uncertainties and imprecision present in human discourse, their semantic correspondence to natural language expressions is questionable on some occasions. A large number of studies in cognitive psychology indicate that human judgments of probability values are often inconsistent with the simple axioms of probability. A comprehensive review of these efforts can be found in reference 14. Often, these errors are of considerable magnitude and not just small deviations usually expected from intuitive, subjective assessments.

Failure to follow the rules of probability are generally attributed to errors of application and errors of comprehension of such rules. An error of application exists if there is evidence that people know and accept a rule that they did not apply. If people do not recognize the validity of the rule they violated, it is called an error of comprehension. Because both types of errors are described in terms of violations to the rules of probability, we could as well claim that the errors are the result of a misrepresentation of human judgments about uncertainty.

An error of representation refers to the semantic correspondence between the natural language expression and the symbolic representation and rules of aggregation used for inference. Errors of representation may result in a set of inconsistent hypotheses. So, an inferential inconsistency may indicate an error in representation, but the contrary is not true; that is, agreement does not necessarily reflect understanding of semantic principles. Consequently, questions arise concerning how to detect and avoid errors of representation and which framework to use in modeling uncertainty and imprecision.

Inferential activities based on logical interconnection of elements in a hierarchical net or tree are called hierarchical inference. Hierarchical inference usually entails a series of inversion, aggregation, and cascading processes to compute the likelihood of an underlying hypotheses and observable evidences based on their logical relations. Inversion involves reversing the logical relation among elements in the network in order to calculate more easily the desired relation. In a Bayesian model, the process of inversion is represented by Bayes' theorem. When a datum D is perceived to have an impact on the occurrence of an event H, the relation between D and H is given by

$$P(H|D) = \frac{P(H)}{P(D)} P(D|H) \qquad (4.2.17)$$

so the perceived effect of the likelihood of H given D is *expressed in terms of the perceived effect of the likelihood of D* given H. Aggregation is the task of assessing the impact of a set of data on a given hypotheses based on the immediate logical relations between the data $\{D^1, D^2, \ldots\}$ and the hypotheses H. Symbolically we have $P(H|D^1, D^2, \ldots) = R[P(H|D^1), P(H|D^2) \ldots]$, where R is the function that aggregates the local relations $P(H|D^i)$ to form the global relation $P(H|D^1, D^2, \ldots)$. Cascading is the combination of a series of immediate relations on a chain of sequential impacts to assess a global relation. For example, if a datum D is perceived to have an effect on an event E and this in turn effects $H(D \rightarrow E \rightarrow H)$, then the process of cascading consists in calculating $P(H|D)$ based on the local relations $P(H|E)$ and $P(E|D)$.

The general case of hierarchical inference involves a number of processes of inversion, aggregation, and cascading. A node in the hierarchical inference net represents a finite partition of exclusive and exhaustive possible states. It may be a set of hypotheses, a set of observable or unobservable events, or more generally just data.

The impact of a given state D_i on a state A_j is given by Bayes' inversion theorem as

$$P(A_i|D_i) = \frac{P(A_i)}{P(D_i)} P(D_i|A_i) \qquad (4.2.18)$$

Conditioning $P(D_i|A_i)$ on the states of the intermediate node B, in order to calculate $P(D_i|A_j)$ in terms of the intermediate node results, and then inverting and cascading the result gives us the expression

$$P(A_j|D_i) = \frac{P(A_i)}{P(D_i)} \sum_{k=1}^{b} P(D_i|A_j, B_k) P(B_k|A_j) \qquad (4.2.19)$$

Here, it is usually assumed that the relation among the states of adjacent nodes is unaffected by the occurrence of states at other nodes. In this case, we would assume that the likelihood of state D_i, given that B_k occurred, is independent of every state A_j such that we then have $P(D_i|A_j,B_k) = P(D_i|B_k)$, and Eq. (4.2.19) then becomes, for the chain of nodes $A \to C \to E$,

$$P(A_j|E_i) = P(A_j) \sum_{k=1}^{c} \frac{P(C_k|E_i)P(C_k|A_j)}{P(C_k)} \qquad (4.2.20)$$

Equation (4.2.20) is sometimes referred to as the *Modified Bayes' Theorem* [15] and has been used in a class of procedures called *probabilistic information processing* [16] to help people overcome the suboptimum behavior they show when revising probabilities of interrelated events. Use of this equation requires the assessment of large amounts of data that may be very difficult to assess intuitively in complex hierarchical inference structures. For example, the meaning of the likelihood or probability of a new state, given all previous information, is difficult to understand when it comprises the conjunction of a large number of states. In addition, the complexity in the processing, storing, and assessment steps increases rapidly with the number of nodes in the network. This has led to the common criticism of using a formal Bayesian framework for inference. Recent work, especially that by Pearl [17, 18] and Schum [19], indicates that this criticism may not be fully justified. Often, a number of appropriate simplifications are possible.

An interesting, efficient scheme for the propagation of beliefs or evidence in hierarchically organized inference structures has been reported by Pearl [17, 18]. The scheme relies in decomposing an inference task into a series of simpler intuitive inferences linking each stage in the hierarchy to produce a global assessment. The computation of the global assessment is simplified by reformulating the general Bayesian procedure for hierarchically organized inference structures discussed here. Data can be communicated among adjacent nodes and can be used to update the information at every node throughout the network. The decomposed Bayesian processing, characteristic of this scheme, allows updating to be performed by a series of local updating processes between each node and its neighbors, rather than by a central processing as in the general Bayesian framework. The likelihood of the various states of a given node depend on the entire data observed. Hence, the impact of the entire data on a given node can be decomposed in two disjoint sets of data: data obtained from the network rooted at that node, and data obtained from the network above the node. At node A let $D_d(A)$ stand for data obtained from the network rooted at A, which is the data at nodes $B^1,\ldots, B^L$ and nodes in the networks rooted on these. Also, let $D^u(A)$ be data obtained from nodes in the network above A — that is B and nodes above it, as well as $A^1,\ldots, A^M$ and nodes rooted on these. This decomposition prescribes how information obtained from above and below some node should be combined. A series of manipulations leads to

$$P(A_i) = ag(A_i)q(A_i) \qquad (4.2.21)$$

where $g(A_i) = P[D_d(A)|A_i]$ represent the probabilistic support attributed to A_i by the nodes below it, $q(A_i) = P[A_i|D^u(A)]$ represents the probabilistic support received by A_i from the nodes above it, and a is a normalization constant defined as $a = 1/P[D_d(A)|D^u(A)]$. Here, the term $P(A_i)$ represents a conditional probability, it is conditioned on the existing state of knowledge.

Updating the values of g and q at every node in the light of new information allows for the calculation of the probability or likelihood of the state of every node. The calculation of g at a node involves only data obtained from the network rooted at that node. The data obtained from the network rooted at A is equivalent to data obtained from each of the networks rooted at nodes adjacent to A. This says that g can be calculated at a node if the g's of the nodes immediately below it and the conditional probabilities quantifying the relation between these nodes are known. The data above A, $D^u(A)$, required to calculate $q(A_i)$ can be decomposed into two disjoint sets, $D^u(B)$ and D_d, which are called "siblings of A." Following the same reasoning as just used, we obtain a result that enables us to compute $P(A_i)$ and $P(B_j)$ without requiring normalization. These results indicate that information to perform the local processing can be represented at each node by assessed conditional probabilities relating adjacent nodes in the hierarchy and computed values of g and $P(\cdot)$ at each node.

To initialize the inference net for propagation, we need the assessed conditional probabilities at each node. At an observational node, every state is equally likely to occur in the absence of any information; hence $g(\cdot)$ is set to 1 at every observational node. From this, the value of g at every other node can be calculated. From the prior probability at the top node and the computed values of g, the probability of the states of each node can be calculated. Once the net is initialized, the occurrence of a particular state at an observational node will cause g to be updated. This information is then propagated up to update the g's of all other nodes and then down to update the likelihood of the states of each node.

In contrast with strict Bayesian procedures, Pearl's scheme requires only the assessment of a prior probability for the node at the top of the hierarchy — that is, the last stage of the hierarchical inference structure usually representing the hypotheses being studied. The probabilities of all other stages in the structure are uniquely determined by the assessed conditional probabilities at each node, thus reducing somewhat the amount and complexity of prior information required. On the other hand, Pearl's work relies on stricter independence assumptions in order to obtain computationally tractable results, and it also requires prior knowledge about the distribution of the underlying hypothesis being studied.

This sort of "divide and conquer" philosophy pervades the entire Bayesian approach to information processing, including inference and impact analysis. There are three major decision-aiding efforts involved in building up inferences from component parts based upon causal relationships:

- *Cascading*, or chaining a sequence of cause–effect relationships to form a relation
- *Aggregation*, or combining solutions to subsets of a problem to form the overall problem solution
- *Inversion*, or use of causal linkages to arrive at desired diagnostic inferences, as in Bayes' rule

We find these three efforts in both cross-impact analysis and hierarchical inference.

It has often been remarked that Bayesian analysis makes no distinction between *uncertain knowledge* and *imprecise knowledge*, and that the same representational system is used to represent probability and possibility. A related major criticism of Bayesian approaches is the need to identify point values about the probability of events. Usually, a point value assessment of the probability of an event is an overstatement about our actual knowledge of the likelihood of occurrence of that particular event. In response to the need of representing imprecision of Bayesian probability values, Dempster [20] utilized the concept of lower and upper probabilities to deal with the subjective imprecision of uncertainty measures. Shafer [21,22] presents a comprehensive exposition of this novel idea as well as extensions to the theory of inference based on the concept of upper and lower probabilities. The basic idea of this concept is that instead of representing the probability of an event A by a point value $P(A)$, it may be bounded by a subinterval of $[0,1]$. That is, the exact probability $P(A)$ may be unknown but bounded. This kind of representation has solid grounds in the Dempster–Shafer theory of basic probability, and for that reason it has received considerable attention recently. We have but scratched the surface of a new and important area of effort. It underlies much of the current interest in data fusion and situation assessment and other formal efforts to enhance the processing of uncertain and imperfect information. This is a very important subject and is considered in more advanced treatments of this subject.

4.2.3 Logical Reasoning Models and Inference

In many ways, models for logical reasoning have much in common, particularly structurally, with inference analysis models. Of particular interest here is the work of Toulmin [23], which has resulted in an explicit structured model of logical reasoning that is suited for analytical inquiry and computer implementation and serves as a framework for logical thought and evaluation of ideas and systems. The model is sufficiently general that it can be used to represent logical reasoning in a number of application areas.

Starting from the assumption that whenever we make a *claim* there must be some *grounds* on which to base our conclusion, Toulmin states that our thoughts are generally directed from the grounds to the claim. The grounds and the claim are statements that express fact and values. As a means of stating observed patterns of stating a claim, there must be a reason that can be identified to connect the grounds and the claim. This connection is called the *warrant*, and it is the warrant that gives to the grounds–claim connection its logical validity.

We say that the grounds support the claim on the basis of the existence of a warrant that explain the connection between the grounds and the claim. It is easy to relate the structure of these basic elements with the process of inference, whether statistical, deductive, or inductive. The warrants are the set of rules of inference, and the grounds and claim are the set of well-defined propositions or hypotheses. It will be only the sequence and procedures that are used to come up with the three basic elements and their structure in a logical fashion that will determine the type of inference that is used.

Sometimes, in the course of reasoning about an issue, it is not enough that the warrant will be the absolute reason to believe the claim on the basis of the grounds. For that, Toulmin allows for further backing that, in his representation, supports the warrant. It is the backing that provides for the reliability, in terms of truth, associated with the use of the warrant. The relationship here is analogous to the way in which the grounds support the claim. *An argument will be valid and will give the claim solid support only if the warrant is relied upon and is relevant to the particular case under examination.* The concept of logical validity seems to imply that we can only make a claim when both the warrant and the grounds are certain. However, imprecision and uncertainty in the form of exceptions to the rules or low degree of certainty in both the grounds and the warrant does not prevent us on occasions from making a "hedge" or a vague claim. Very commonly, we must arrive at conclusions on the basis of some less-than-perfect evidence; and we put those claims forward not with absolute and irrefutable truth but rather with some doubt or degree of speculation.

To allow for these cases, Toulmin adds modal qualifiers and possible rebuttals to his framework for logical reasoning. Modal qualifiers refer to the strength or weakness with which a claim is made. In essence, every argument has a certain modality. Its place in the structure presented so far must reflect the generality of the warrant in connecting the grounds to the claim, and also with condition of validity of the set of facts as grounds. Possible rebuttals, on the other hand, are exceptions to the rules. Although modal qualifiers serve the purpose of weakening or strengthening the validity of a claim, there may still be conditions that invalidate either the grounds or the warrants, and this will result in deactivating the link between the claim and the grounds. These cases are represented by the possible rebuttals.

The resulting structure of logical reasoning provides a very useful framework for the study of human information processing activities. The order in which the six elements of logical reasoning has been presented serves only the purpose of illustrating their function and interdependence in the structure of an argument about a specific issue. It does not represent any normative pattern of argument formation. In fact, due to the dynamic nature of human reasoning, the concept formation and framing that results in a particular structure may occur in different ways. The six-element model of logical reasoning is shown in Figure 4.15.

It is important that we recognize that there is a difference between probabilistic and logical support. Probabilistic support refers to the increase in likelihood of the occurrence of an event A given that another event B has occurred. That is, A supports B if $P(B|A) > P(B)$. Logical support exhibits the relation of implication between two premises, denoted $A \rightarrow B$, that fails to hold only if the first is true and the second is false. Logical support is transitive. This requires that if $A \rightarrow B$ and $B \rightarrow C$, then it follows that $A \rightarrow C$. When $A \rightarrow B$, it is also true that $\bar{B} \rightarrow \bar{A}$. With these definitions, the distinctions between probabilistic and logical support should be apparent. It is important to note that logical support does not imply conditions that are similar to those that follow from probabilistic support. For example, it turns out that $A \rightarrow B$ is logically equivalent to $\bar{B} \rightarrow \bar{A}$, but does not say anything about the truth of $\bar{A} \rightarrow \bar{B}$ or $B \rightarrow A$. Likewise, probabilistic support is not transitive, and logical support must be transitive. A major point in this distinction that arises here is when to apply these two methods of representation in inferential activities. We are concerned with this issue because, as previous research has shown, the method used to represent human judgments may influence its validity and consistency. Suppes [24] provided an analysis of causality based on probabilistic support relations that finds difficulties in distinguishing spurious from genuine causes as well as direct from indirect causes. After an attempt to overcome these difficulties, by making modifications to Suppes' theory, Ottes [25] concludes by believing that it is impossible to give a detailed account of causation using only probabilistic relations. He points out that the belief that a positive cause raises the probability of that cause necessarily having its effect, fundamental to Suppes' theory, is flawed. He illustrates this point with examples in which an event A is clearly caused by another event B, and yet most people believe that $P(A|B) < P(A)$.

The framework and process for inference support discussed here is applicable to a general class of networks of interrelated propositions. Specifically, it

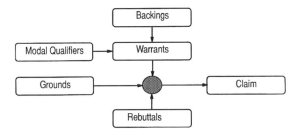

Figure 4.15 The six-element model of logical reasoning.

can be applied to finite connected networks where the number of propositions is finite and where every pair of distinct propositions is joined by at least one chain of relations. The information processing functions associated with the use of the framework for inference described here involves four steps:

- Initial problem framing
- Hypothesis generation
- Parameter value assessment
- Hypothesis evaluation and situation assessment

Your intent in the first step should be to capture those elements and relations that constitute an issue, and to represent them in a form that is suitable for inference. Given the assessed initial problem frame, the task of hypothesis generation involves the generation of reasonable hypotheses that are based on situational perception and information needed for the task at hand. In most cases this involves the specification of alternative hypotheses at each node. Ideally, the set of hypotheses under consideration at each node should be mutually exclusive and exhaustive. This task also involves the selection of the basic premises and possible rebuttals relevant for each inferential link.

The parameter value identification step provides for the continual assessment of the parameters of the inference model. This includes the assessment of probability values of the propositions at each node as well as the probability values representing the uncertain logical relations at each inferential link. These assessments can be related and represented imprecisely in the form of bounded intervals and/or linear inequalities on the set of parameters. Achieving the task of parameter value assessment with minimal imprecision will depend strongly on the quality of the information available and the person's perception and familiarity with the task at hand.

The hypothesis evaluation and situation assessment step involves probability categorization, over a set of alternative hypotheses, of the probable situations as captured by the information that is provided to the inference model. Given the set of consistency relational equations for each link in the inference network, we can calculate the probability values for the propositions at each node. If this turns out to not yield appropriate results, then more precise information is generally needed. There have been a number of applications of Toulmin-based logic in systems engineering [26–29]. It has been proposed as a framework for a decision support system that will aid in resolution of conflict concerning important judgment and choice issues.

4.2.4 Complications Affecting Inference and Cross-Impact Analysis

As we noted before, there are a number of complications that, in practice, may significantly affect our ability to use unmodified Bayesian analysis algorithms. These concern such issues as unreliable observations, potential computational complexity of the Bayesian algorithms, human cognitive biases, imprecise knowledge or lack of knowledge as opposed to uncertain knowledge, and the need to use causal as well as diagnostic inference rules. There are many potential sources of incoherences in individual estimates of probabilities, or in the aggregation of individual estimates to form a group estimate. If we contemplate the use of Bayesian processing schemes, it is necessary that a set of mutually exclusive and exhaustive hypotheses be identified. Precision in formulation or framing of hypotheses is necessary to accomplish the former. We can often use formal approaches to determine whether or not a set of hypotheses is mutually exclusive, but it will often not be possible to determine whether or not we have identified all reasonable hypotheses. Forecasts based upon a presumed set of exhaustive hypotheses have context validity only within that set of hypotheses. It is always necessary to ensure that what has been omitted in issue framing and hypothesis generation is not more important than what has been included.

Even if hypotheses are coherently formulated, there may exist numerous difficulties associated with the estimation of probabilities and likelihood ratios. The needed statistical data, especially in strategic situations, are often unavailable. There are also a number of information processing biases to which humans are prone. Reprints of much of the seminal literature concerning cognitive heuristics and information processing biases are presented in reference 30. One particular difficulty, for example, which greatly affects information analysis, is the tendency of people to be concerned with the probability of obtaining some observed data conditioned upon some hypothesis, $P(D|H)$, but to be not at all concerned with whether or not the probability of obtaining the data under the null hypothesis, $P(D|\bar{H})$, might not be greater. There are many interesting and relevant issues to be explored in human inference analysis and the use of support procedures to aid this. The advanced definitive work by Schum [19] describes much of our presently available knowledge. This subject is very important for such contemporary application areas as situation assessment and data fusion.

4.3 STRUCTURAL MODELING: TREES, CAUSAL LOOPS, AND INFLUENCE DIAGRAMS

In this section we will discuss a number of topics that generally fall under the more generic title of structural modeling, or structured modeling. We have need for various structural representations throughout our efforts, and for this reason we need to be very concerned with approaches that deal with system structures. We will discuss tree structures and causal loop diagrams here. We discussed the related subject of functional analysis in Chapter 3 and will discuss influence diagrams, as a natural extension of causal loop diagrams, here and in Chapter 5.

4.3.1 Tree Structures

The term "tree" is used to indicate (a) a special type of graphical representation that consists of elements represented as vertices or nodes and (b) relations between those elements represented as lines or edges. A tree may represent either a single-source or single-sink structure. It starts from a single source or edge, and branches develop along nodes. Each node can have only one incoming branch, but can have several outgoing branches. There can be only one path between every pair of nodes. Trees can be particularly useful for displaying hierarchies of objectives, organizational structures, or various events that can develop from a single starting state. In general, trees are useful aids for representing the structure of elements and for communicating an understanding of such structures. Several type of trees useful in the study of complicated issues include objectives trees, activity trees, event trees, intent trees, decision trees, and worth or attribute trees.

In constructing a tree, it is first necessary to determine the elements that will be structured and the contextual relation to be used in building, or constructing, the structure. In constructing an organizational tree structure, for example, we would need to know the functional positions associated with the organization. An appropriate contextual relation is "reports to." If we were to use this contextual relation, we would be posing questions of the form "Does the person in position a report to the person in position b?" If the answer is yes, we have a directed line relationship from a to b. If the answer is no, we do not. This question would need to be posed for all a and b and then we might use the mathematics of graph theory, or some related approach, in order to determine the structure.

In general, the elements to be structured are available from an issue formulation effort. The type of relation according to which the structuring is done will, in general, determine the feasibility of representing the structure by some type of tree. The various types of trees are so different that comments concerning their construction and use are perhaps best made in terms of separate discussions of the most common tree types. Of course, not all structural models will look like trees. We will first discuss some specific tree structures and will then examine some more general cases.

1. An objectives tree represents the hierarchical structure associated with objectives or intents. It is sometimes referred to as an intent structure.

   a. In constructing an objectives tree, it is helpful to formulate the objectives to appear in the objectives tree with an appropriate contextual relation in the form to "+(action word) + (object) + (constraint)". A representative example of a formal objective might be "to achieve rapid advancement through competence, commitment, and communications." It is almost mandatory to use the proper semantic form of contextual relation to aid in developing action-oriented statements. The definition of the contextual relation between the objectives should not change during construction of the tree. Typical contextual relations are "will contribute to" in the case of upward pointing associations, and "will be achieved by" for downward pointing associations.

b. Next, elements are arranged in hierarchical order, with the more general goals above the more specific goals. If a specific objective seems to fit into several branches, possibly at different levels, such that the structure is not that of a tree, the objective statement should be reworded, or replaced by several different objectives but related objectives at different hierarchical levels, such that the tree structure is retained. We could simply repeat an objective several times such that it appears as needed in the various branches of the tree. But, we do need to recall in doing this that the resulting structure only looks like a tree.

c. As the tree is constructed, five tests of objective tree logic should be applied. Each test should be made on each statement in the tree as follows:

    i. In *going down* any branch of the tree, each goal or objective statement must answer "how" for each immediately superior objective.

    ii. In *moving up* any branch of the tree, each higher level objective should provide information concerning "why" the objective below it is needed.

    iii. In *reading across* the objectives at any given level and under any one general goal, it should be possible to see that each of these more specific objectives is needed to accomplish the more general objective.

    iv. In *reading across* the objectives at a given level and under any one general objective, we should also ask whether there are other specific objectives at this level that are needed to accomplish the more general objective. If there are, they should be included in the tree.

    v. Often, it will be desirable to *identify the ownership* of all or many objectives in the tree.

Figure 4.16 indicates a typical set of objectives that might be identified in order to accomplish "total quality management (TQM)," a subject discussed in some detail in references 31–33 and in the references contained in these texts. Associated with these objectives is a self-interaction matrix indicating how the objectives interact. Figure 4.17 illustrates how one objective is reached from another and two realizations of an objective tree. Clearly, the minimum edge realization is the simplest objectives tree structure. This is often called a hierarchical tree. The structure can easily be redrawn in a classic tree form by defining additional third-level elements, such that a third-level element is directed

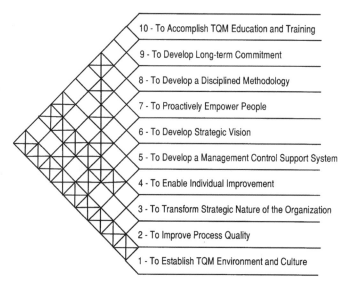

Figure 4.16 Ten top-level TQM objectives and associated self-interaction matrix.

at one and only one second-level element. The resulting minimum edge objective tree structure is illustrated in Figure 4.18. The trees of Figures 4.17 and 4.18 represent a set of top-level objectives for total quality management. There is one top-level objective, three second-level objectives, and six objectives in the third level. The conceptual relationship is "is desirable or necessary in order," and this connects all the objectives in the manner indicated in the figures.

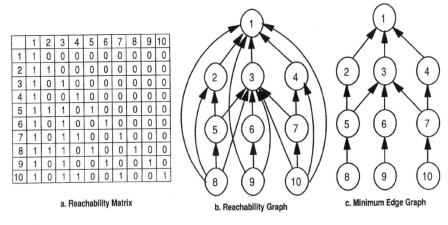

Figure 4.17 Objective tree and directed graph relations for the objectives tree of Figure 4.16.

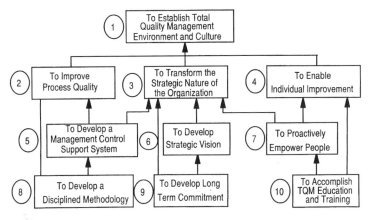

Figure 4.18 Simple objectives tree for total quality management objectives of Figure 4.16.

2. Activity trees can normally be constructed most easily after an objectives tree has been prepared. Frequently, there is a significant correspondence usually found between activities and objectives. Each vertex entry in an activity tree consists of an activity, and the edge relation is that lower-level activities must contribute in some way to the conduct of the higher-level activity. Activity trees are useful in helping to structure a program to determine what is related to what, and to examine how responsibilities can be assigned. Applications include developing responsibility assignment and constructing an activity network.

Objectives can be converted into activities by removing the infinitive preposition "to" and the constraint set. It is interesting to note that project network modeling methods deal extensively with activities. However, the usual representation framework is not directly that of a tree structure, although the structure of activities can often be recast into a tree structure. Often, there are a rather large number of elements in various structural diagrams. Some elements in the structure are sometimes combined in order to simplify the resulting graphical presentation. This must often be done when it would be cumbersome to display a vast number of elements in a crowded visual presentation. Often, there is feedback, especially when the evolution of objective or activity attainment occurs over time. This results in what we will call a causal or influence diagram, such as illustrated in Figure 4.19. For many uses, this would be a more appropriate representation than the tree representation. One of the challenges in many systems engineering efforts is choosing an appropriate framework for representation of salient information. As we will indicate later, this framework has much to do with the manner in which information is thought about and acted upon by the various people involved in the effort.

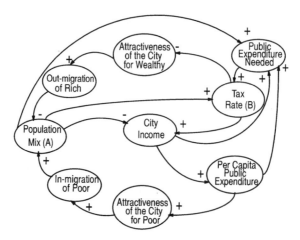

Figure 4.19 Causal loop diagram of some feedback loops in a simple model of urban dynamics.

3. Decision trees and influence diagrams are comprised of decision nodes, event nodes and event outcome nodes, as illustrated in Figure 4.20. The decision node is generally represented by a square. Lines, or edges, in the network emanate from the decision node and represent the possible alternative courses of action.

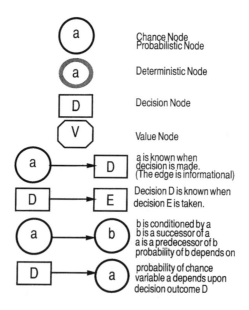

Figure 4.20 Influence diagram conventions and simple structures.

The event node, or chance node, is generally represented by a circle. Edges emanating from the chance node represent the different possible consequences of choosing the alternative leading to that chance node. Probabilities of occurrence are assigned to the consequences; and costs or profits will usually be estimated for all branches of the tree. The probabilities of consequences emanating from a particular chance node must sum to one. Typically, time and activities in a decision tree proceed from left to right.

The final outcome nodes appear at the extreme right-hand end of a decision tree and have incoming edges but no outgoing edges. A decision path represents a possible sequence of decisions and consequences, and it leads, in association with uncertainty considerations, to a particular outcome. Our next chapter will develop notions associated with decisions and consequences, in order to make the tasks of formal decision making more organized and explicit than they otherwise might be.

4. Attribute trees are much like objectives trees. One often-found difference, however, is that attributes tend to be more measurable than objectives. Often, it is desirable to cast attributes in the form of objectives measures.

There are many other types of tree structures, including fault trees, event trees, relevance trees, and preference trees. A hierarchical structure is one in which each level is directed upwards only to higher levels, and downwards only to lower levels. Thus, all tree structures are necessarily hierarchical. But, not all hierarchies are trees.

Trees can be used throughout any step of a systems engineering effort and can be associated with any of the systems engineering phases used as part of the systems engineering process life cycle, but the process of constructing a tree is fundamentally an analysis effort. For example, objectives trees help structure the value elements determined in issue formulation; decision and event trees may be used as aids in the analysis of alternatives; and activity trees may be used as aids to planning for action after a decision has been made.

Various structural modeling approaches are helpful in constructing hierarchies for a large set of elements. An interaction matrix may also be helpful in tree construction. Tree structures may be constructed from the influence diagram approach that we will soon describe, although it is generally a simple matter to manually construct a tree with less than two dozen elements. The more difficult problem is identifying the relevant elements for the tree.

Much additional information concerning hierarchies, tree structures, and their construction is available. The theory of digraphs and structural modeling is authoritatively presented in reference 34, and a number of applications to what is called interpretive structural modeling are described in references 35–38]. Cognitive map structural models are considered in reference 39. A development of structural modeling concepts based on signed digraphs is discussed in reference 40. Geoffrion has been especially concerned with the development of a structured modeling methodology and environment [41–43].

4.3.2 Causal Loop Diagrams and Influence Diagrams

Causal loop diagrams or influence diagrams are a particular form of structural modeling that represents causal interactions between sets of variables. They are particularly helpful in making explicit one's perception of the causes of change in a system, and they can serve very well as communication aids. In this subsection, we will first discuss causal loop diagrams, or causal diagrams, and will then conclude with a very brief presentation of recent efforts involving influence diagrams.

A causal diagram is a graphic representation of those cause–effect mechanisms that are responsible for change, generally change over time. Lines and arrows between elements show the existence and direction of causal influences. An arrow pointing from element A to element B illustrates that some change in a pertinent aspect of A causes a change in a corresponding pertinent aspect of B.

Optionally, we might indicate the nature of the influences by a "+" to indicate that an *increase in A causes a larger value of B*, or a "−" to indicate that *increases in A cause a smaller value of B to occur*. The relevant sign is placed at the arrowhead, just as with signed digraphs. Sometimes, the symbol s is used to represent "same," and the symbol o is used to represent "opposite."

A causal loop diagram is particularly useful for portraying cyclic interactions, or feedback loops, in which influences act to reinforce or counteract change. It is a powerful communication tool, and it aids considerably in structuring discussions or ideas about complex issues. Potential applications for causal loop diagrams include (a) the identification and portrayal of cycles and feedback and (b) the construction of a framework for dynamic modeling. A causal loop diagram is a special sort of influence diagram or a directed graph, or digraph. Causal loop diagrams draw specific attention to interacting loops and to the causal influences these loops portray.

We shall give a descriptive account of the major activities involved in causal diagram construction, along with related hypotheses explaining causal mechanisms of change. A causal structure permits tentative conclusions concerning such major influencers of behavior as critical mechanisms, points of leverage, strong cycles, or reasons for resistance of an organization or system to change. These conclusions may indicate why certain policies are, or could be, highly effective or very harmful, while others are ineffective. Causal diagrams are most appropriate for structuring interactions between relevant issue formulation elements and for preparing for more thorough analysis efforts. Causal diagrams provide "a bridge" between the formulation and the analysis steps of a systems engineering effort.

A causal diagram is a graphic representation of those cause–effect mechanisms that are responsible for change, generally change over time, although it could be change over other variables such as space. Figure 4.19 is a typical causal, or causal loop, diagram. Lines and arrows between elements show the existence and direction of causal influences. An arrow pointing from one element, *A*, to another element, *B*, illustrates that some change in a pertinent aspect of *A* causes a change in a corresponding pertinent aspect of *B*. A pertinent change could be a change in value, magnitude, intensity, size, color, beauty, worth, or any of several other possible descriptors. Optionally, you can indicate the nature of the influences by a "+" to indicate that an increase in *A* causes a larger value of *B*, or a "−" to indicate that increases in *A* cause a smaller value of *B* to occur. You place the relevant sign at the arrowhead end of each line.

Often, causal loop diagrams are useful for analyzing systems with multiple feedback loops. Consider the situation modeled in Figure 4.19 which illustrates the issue of erosion of the tax base in a hypothetical city with a fraction of the population characterized as "poor" and where the remaining fraction is characterized as "wealthy." The tax base erodes through outmigration of the "wealthy," attracted by lower tax rates in the suburbs, and immigration of the "poor" who perceive better public facilities in the city than in the suburbs or rural areas. The city's desired policy alternative is to keep public expenditures at a constant level. With the eroding tax base, due to the outmigration of more wealthy taxpayers and increasing demand for services by immigration of poor people, the city's income is indeed in trouble. You can use this causal loop diagram to investigate some of the mechanisms causing long-term change in such a city. Note that the causal loop model in Figure 4.19 contains a mixture of flows: people, money, and information. You will probably want to make these different types of flows easy to distinguish, and you can do this by using branches of a different color or thickness to represent different flows.

Two major interacting loops can be identified in this figure. The first major loop shows that a continuing outmigration of the "rich" will change the population mix and reduce city income, all other things being constant. As a result, to keep public expenditures constant, the tax rate will be increased. This causes a decrease in attractiveness of the city to the rich, which leads to an increase in outmigration of the rich and a further increase in the fraction of the poor in the city. The second loop shows that immigration of "poor" causes similarly deteriorating conditions. A policy that keeps the public welfare expenditures for each poor person at a constant level results in more and more poor people moving into the city, thereby further draining already dwindling revenues through increased social welfare expenditures. A further complication, even if secondary in importance, is the uncontrolled outmigration of the rich. This could potentially be changed by a policy of diverting part of the city income to cultural pursuits or other projects that the rich generally find attractive, thereby making it a more attractive place for the rich to live, despite high tax rates.

You can use the following steps to aid in the construction of a causal, or cause and effect, diagram.

1. Identify the problem or issue relevant to the system for which a causal diagram is needed.

2. Identify the major groupings of causes and their related major effects.
3. Search for more detailed causes and effects, perhaps through brainstorming or use of the nominal group technique.
4. Eliminate inapplicable causes and their related effects. This needs to be done with care because what may be thought initially to be frivolous causes may well turn out to be quite important.
5. Connect causes and effects together by determining predecessor and successor relationships. An arrow pointing from element *A* to element *B* means that some change in a pertinent aspect of *A* causes a change in a corresponding pertinent aspect of *B*.
6. Determine the nature of the influences of the causes by labeling at the arrowhead a "+" to indicate that an increase in *A* causes a larger value of *B* or a "−" to indicate that increases in *A* cause a smaller value of *B* to occur.
7. If the diagram contains a mixture of flows, such as people, money, and information, then distinguish among the different flows by labels, different colors, or line thickness.
8. Refine the causal diagram as appropriate.

Causal loop diagrams draw specific attention to interacting loops and to the causal influences these loops portray. Typically, applications include the identification and portrayal of cycles and feedback, along with the consideration of a framework for dynamic modeling. The example portrayed in Figure 4.19 is representative of the sort of dynamic situation described by Forrester in his system dynamics model of urban growth and decay [44]. The causal network representation of Figure 4.19 can be converted to a systems dynamics modeling representation, as you will learn in our next section.

As you can appreciate from the "erosion of the city tax base" example, a causal diagram is particularly useful for portraying cyclic interactions, or feedback loops, in which influences act to reinforce or counteract change. It is a powerful communication tool, and it aids considerably in structuring discussions. To make structural modeling much more useful, particularly in quantitative terms, the influence diagram concept was developed and first described by Howard and Matheson [45] and Shachter [46]. It is an approach to structural modeling that was initially developed for decision analysis and allows representation of probabilistic effects in a relatively direct manner. Chapter 5 will provide some illustrations of the use of influence diagrams as part of decision analysis and decision assessment.

An influence diagram is potentially able to represent probabilistic and functional dependencies, and associated information flow patterns, as a directed graph. The basic conventions and notations typically used in influence diagrams are, like those for decision trees, illustrated in Figure 4.20. There are five types of nodes: probabilistic or chance nodes, deterministic nodes, decision nodes, and value nodes. Information flows from one node to the other, as indicated in this figure. There are three types of influence: conditioning influence, information influence, and value influence.

A *conditioning influence* indicates the presence, or absence, of probabilistic dependence. Thus conditioning influence and probabilistic influence are equivalent terms. From elementary probability theory, we know that probabilistic independence, or statistical independence as an equivalent expression, exists among two random variables, x and y if and only if $p(x, y|\&) = p(x|\&)p(y|\&)$. In this expression, $\&$ represents all other existing prior information or conditioning predecessor that is known to, and available at, the explicit nodes in the network under consideration. A major advantage to probabilistic independence is the considerably reduced effort required in measuring, or eliciting, $p(x, y|\&)$ as two single-dimensional probability density functions rather than one two-dimensional density function. Another major advantage to probabilistic independence is that we can obtain the probability density of a single dependent variable in a much simpler manner than would otherwise be possible. This is so because the density of the dependent variable y, which is $p(y|\&) = \int p(x, y|\&)\, dx$ and does not depend upon the x variable as we may easily show from evaluation of the probability expression $p(y|\&) = \int p(x, y|\&)\, dx = p(y|\&) \int p(x,|\&)\, dx = p(y|\&)$. Moreover, probabilistic independence does not exist if it can only be stated that $p(x, y|\&) = p(x|y, \&)p(y|\&) = p(y|x, \&)p(x|\&)$. In other words, it must be true that $p(x|y, \&) = p(x|\&)$ and $p(y|x, \&) = p(y|\&)$ if probabilistic independence exists.

An *information influence*, which is indicated by directed relationships or arrows that lead into decision nodes, indicates a causal ordering of elements. Thus any chance outcome leading into the decision node will necessarily be known before it is needed to make a decision concerning or associated with that decision node.

A *value influence* variable or node represents outcome influencers, or attributes, and their interrelations that enable determination of a single value that reflects the underlying interpretations and value system of the decision maker. We will have a bit more to say about this in Chapter 5 in our discussions of decision analysis and utility.

The concepts of information influence, or decision influence, and value influence, or attribute influence, are relatively straightforward. Some further commentary concerning probabilistic, or chance, influence is needed here because there are a number of points that are often not fully appreciated. For the case of three random variables, x, y, z, it must be true that $p(x, y, z|\&) = p(y, z|x, \&)p(x|\&) = p(x|\&)p(y|x, \&)p(z|x, y, \&)$. But, there are eight possible ways in which $p(x, y, z|\&)$ can be represented as the product of three conditional densities of the form $p(x, y, z|\&) = p(x|(\Gamma_{y,z}, \&)p(y|\Gamma_{x,z}, \&)p(z|\Gamma_{x,y}, \&)$. Six of these are shown in Figure 4.21. These are the six possible conditional density representations of $p(x, y, z|\&)$ that are in the form of an acyclic digraph. The other two representations, $p(x, y, z|\&) = p(x|z, \&)p(y|x, \&)p(z|y, \&)$ and $p(x, y, z|\&) = p(x|y, \&)p(y|z, \&)p(z|y, \&)$, involve cyclic digraphs of probabilities and are, therefore, nonallowable due to the impossibility of eliciting the requisite information, or knowledge, needed for the digraph.

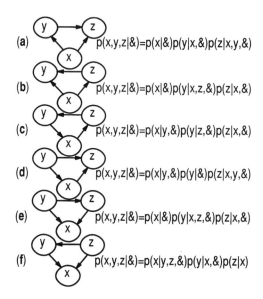

(a)   p(x,y,z|&)=p(x|&)p(y|x,&)p(z|x,y,&)

(b)   p(x,y,z|&)=p(x|&)p(y|x,z,&)p(z|x,&)

(c)   p(x,y,z|&)=p(x|y,&)p(y|z,&)p(z|x,&)

(d)   p(x,y,z|&)=p(x|y,&)p(y|&)p(z|x,y,&)

(e)   p(x,y,z|&)=p(x|&)p(y|x,z,&)p(z|x,&)

(f)   p(x,y,z|&)=p(x|y,z,&)p(y|x,&)p(z|x)

Figure 4.21   Conditioning relations for three-node diagrams.

Figure 4.22 also represents this situation in the form of alternate influence diagrams. If we wished to know the probability density of the random variable z, conditioned on the common information &, we could obtain this from an expression such as

$$p(z|\&) = \iint p(x, y, z|\&)\, dx\, dy = \iint p(x|\&)p(y|x, \&)p(z|x, y, \&)\, dx\, dy.$$

Howard [47] suggests appropriate alternate forms of this expression. He considers an illustration for three related variables, x = age, y = education, z = income. Figure 4.21 has illustrated the relevant acyclic knowledge maps or influence diagrams that result from these observations. Formally, a knowledge map is an influence diagram that has no value nodes and no decision nodes. It is, therefore, equivalent to a probability tree. Some authors call this a *partially formed influence diagram*. The term partial is used because of the lack of an output value node. Inclusion of this would make the influence diagram fully formed. Figures 4.21(a) and 4.22(a) indicate the map most appropriate for the calculation just indicated. All that is desired is the probability density over z as conditioned on the information &. This can be obtained from the expression $p(z|\&) = \int p(y|\&)p(z|y, \&)\, dy$, where we use the fact that $p(y|\&) = \int p(x|\&)p(y|x, \&)\, dx$.

To obtain $p(z|\&)$ through use of this two-stage procedure, we may assess a density function over the variable x and another one over the variable y conditioned on the variable x and the common information. We then need to

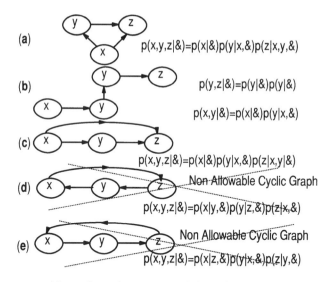

Figure 4.22   Additional conditioning possibilities for a three-chance-node set.

multiply these density functions together and integrate the result to obtain the desired density function $p(y|\&)$ that is needed. The initial density function estimates suggested by Figure 4.22(a) involve a one-dimensional assessment, a two-dimensional assessment, and a three-dimensional assessment. Now, we need a one-dimensional assessment and two two-dimensional assessments. Representation by this latter approach requires the sort of disjoint knowledge map represented by Figure 4.22(b). This is, indeed, a disjoint representation because probability density of the random variable y is called for twice, with different conditioning variables. We need $p(y|\&)$ and $p(y, x, \&)$, and they are not the same. There might be a temptation to use the influence diagram of Figure 4.22(c). This is a well formed and quite proper influence diagram, but it is quite the same as that shown in Figure 4.21a.

In a similar manner, we might attempt a representation in the form of Figure 4.22(d) or Figure 4.22(e). However, these are not correct representations of probabilistic information in the sense that they are cyclic. There is simply no way in which the relevant knowledge needed to assess these probability density functions can be assessed because of this cyclic conditioning. Any of the six representations of Figure 4.21 are acceptable, at least in principle.

So, we see that there are a number of perspectives that we need to take concerning information (i.e., knowledge in the terminology of the initial authors in this area) representations. Doubtlessly, the way information is initially represented and the purposes to which it is to be put influence this consideration strongly. The various approaches to combining probability distributions are quite relevant to this consideration, and an extension of this discussion might well consider some of these [48].

It is very important to note that probabilistic, or conditioning, influence is in no way equivalent to the notion of a causal influence. While two variables that are conditionally influenced by a third variable are necessarily correlated, they are not at all necessarily causally influenced.

Shachter [49, 50] has defined an influence diagram to be a single connected network that is comprised of an acyclic directed graph, together with associated node sets, functional dependencies, and information flows. There are three types of nodes: decision nodes, value nodes, and chance nodes. There are two types of arcs: informational arcs and conditioning arcs. An influence diagram is said to be well-formed, or fully specified, or fully formed, if the following conditions hold.

1. There are no cycles. In other words, an influence diagram is a directed acyclic graph or, in other words, a set of nodes or variables and a set of edges or branches that connect the nodes in a directed sense and where there are no nontrivial paths that begin and end at the same node.

2. There is one, and only one, value node. There is a value function that is defined over the parents of this single value node. A set of nodes P_i are called parents of node n_j if and only if there is an edge e_{ji} for each node n_j that is an element of P_i. In a similar manner, a set of nodes C_i are called children of node n_i if and only if there is an edge e_{ij} for each node n_j that is an element of P_i. A *barren node* is a node with no children. A *border node* is a node with no parents. There is much rather specialized graph theory terminology, and we shall not go this far afield in our discussions here.

3. Each node in the digraph is defined in terms of mutually exclusive and collectively exhaustive states.

4. A joint probability density function is defined over all of the states of the uncertainty nodes.

5. There is at least one path that connects all of the decision nodes both to each other and to the single-value node.

6. There are functions over the parents of each deterministic node that are defined over the parents of those nodes.

Even though there exists a unique joint probability density function for a specific well-formed influence diagram, there may be a number of physical influence diagram realizations for any given joint density function.

There are three steps that will enable identification of an appropriate probability distribution from a well-formed influence diagram.

1. *Barren Node Removal.* All decision nodes and chance nodes may be eliminated from the diagram if they do not have successors. If there is nothing that follows a node, that node can have no effect on the outcome value. Such nodes are irrelevant and superfluous and can be eliminated. Formally, this says that a node *a*, which is barren with respect to node *b*

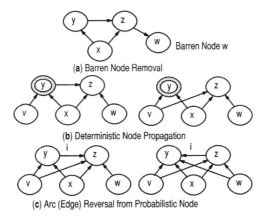

(a) Barren Node Removal

(b) Deterministic Node Propagation

(c) Arc (Edge) Reversal from Probabilistic Node

Figure 4.23   Additional node manipulations.

and c, can be eliminated from an influence diagram without changing the values these nodes take on, such as $p(b|c, \&)$. Figure 4.23(a) illustrates this concept. It is important to note that a barren node is not frivolous, but only irrelevant with respect to a particular set of nodes for which it is barren. This suggests that $p(x, y, z|\&)$ does not depend upon any conditioning upon w in Figure 4.22(a).

2. *Deterministic Node Propagation.* If a well-formed influence diagram contains an arc from a deterministic node a to node b, which may be a chance node or a deterministic node, it is possible to rearrange or transform the influence diagram to one in which there is no edge from node a to b. The new influence diagram will be one such that node b will inherit all conditional predecessors of node a. Furthermore, if node b was deterministic before the transformation, it will remain a deterministic node. Figure 4.23(b) illustrates this concept.

3. *Arc Reversal.* In a well-specified influence diagram in which there is a single directed path or arc from probabilistic node a to node b, the diagram may be transformed to one in which there is an arc from node b to node a and where the new nodes a and b inherit the conditional predecessors of each node. If node b was initially deterministic, it becomes probabilistic. If it was initially probabilistic, it remains probabilistic. Figure 4.23(c) illustrates this concept.

These simple reductions can be grouped together into a series of transformations that potentially resolve influence problems. This leads to three additional steps.

4. *Deterministic Node, with a Value Node as the Only Possible Successor, Removal.* A given deterministic node may be removed from the network.

The given deterministic node is propagated into each of its successors until it has none, and is then barren and can be eliminated from the diagram. Figure 4.23(b) has actually illustrated this. Node y, a deterministic node, is barren after the manipulation leading to Figure 4.23(b).

5. *Decision Node, with a Value Node as the Only Possible Successor, Removal.* A given decision node may be removed from the network. When any conditional predecessors of the value node that are not observable at the time of decision are removed first and when the decision node is a conditional predecessor of the value node, it may be removed. These conditional predecessors are typically the successors of the decision node in question. No new conditional predecessors are inherited by the value node as a result of this operation, and the operation ends when all predecessors to value nodes have been removed. Decision nodes are removed through the maximization of expected value, or subjective expected utility, a subject considered in our next chapter.

6. *Probabilistic Node, with a Value Node as the Only Possible Successor, Removal.* A probabilistic or chance node that has only a value node as a successor can be removed. In some cases, it will be necessary to reverse a conditioning arc between the node and other successors such that the value node inherits conditional predecessors of the node that is removed. Figure 4.23(c) illustrates this concept. Node y can be removed as it is a barren node, after the manipulations leading to Figure 4.23(c).

These three steps follow from the first three stated. A relatively complete discussion of these steps is contained in Shachter [49, 50] and in Call and Miller [51]. Included in these efforts is a discussion of transformations needed to solve inference and decision problems, sufficient information to perform conditional or unconditional inference, and the associated information requirements for decision making, including calculations of the value of (perfect) information. The Call and Miller paper also discusses influence diagram software.

One of the questions that naturally arises is whether an influence-diagram-type representation or decision-tree-type representation is "better." The answer is, of course, quite problem and perspective dependent. Figures 4.24 and 4.25 provide some comparisons of alternative representations of decision situations in terms of influence diagrams and decision trees. It is relatively easy to construct a situation for which one representational framework is the best. In any case, both decision trees and influence diagrams are equivalent to spreadsheet-like matrix representations.

One particularly impressive demonstration of potential superiority of the influence diagram representation is in situations where probabilistic independence exists. Figure 4.26 illustrates this sort of situation in decision tree format and influence diagram format. What is displayed here is a case where $p(c_2|c_1, \&) = p(c_2|\&)$. This independence is clearly illustrated in the influence

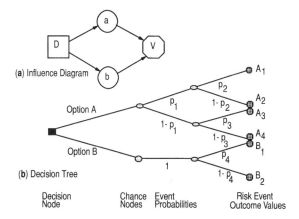

Figure 4.24   Simple influence diagram and associated decision tree.

diagram but not at all evident in the decision tree structure unless we actually examine the probabilities shown in the tree. This effective representation of probabilistic independence also makes it easier to enforce the distinction between probabilistic influences and values influences.

A relatively good illustration of this is provided by the influence diagram of Figure 4.27. There are three decision nodes in this figure: detection (DE), diagnosis (DI), and correction (CO). The function of some directly unobservable functioning of some systems is influenced by some chance mechanism, C, which is also a border node. The detection decision outcome is influenced by the chance mechanism and some additional random mechanism CDE. Depending on the detection result, we enter a diagnostic decision phase, DI. Again, there are chance mechanisms involved that influence the actual diagnostic outcome. Finally, the correction decision phase (CO) produces an

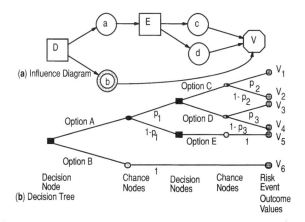

Figure 4.25   Three-decision influence diagram and associated decision tree.

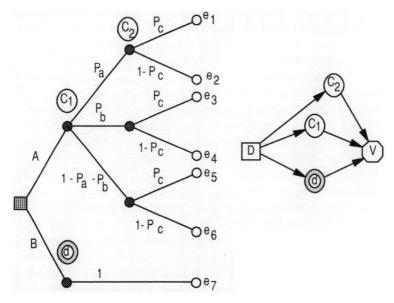

Figure 4.26   Decision tree and equivalent influence diagram illustrating conditional independence of two chance nodes.

outcome that depends on the failure state of the system and the chance result of the corrective effort. This influence diagram is a relatively illuminating and straightforward representation of the decision situation. The decision tree may not be comparably illuminating and straightforward. This difficulty usually increases for more complex decisions and provides some potential and real advantages to the influence diagram approach.

Other efforts [52] have introduced the concept of super value nodes that enable the representation of separable value functions in influence diagrams. Additional extensions to influence diagram efforts [53] remove restrictions 2, 3, 4, and 6 from the definition of a well-formed influence diagram. This enables the expansion of the influence diagram concept to decision processes in which value aspects are critical. In particular, it enables the determination of value-driven clusters and decision-driven clusters of elements such that a relatively

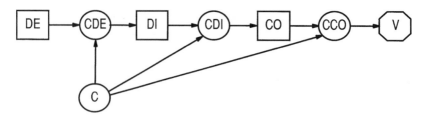

Figure 4.27   Influence diagram for fault detection, diagnosis, and correction.

complex influence diagram may be viewed in an abstracted form that simplifies its representation and presentation. The relationship between dynamic programming and influence diagrams is discussed in reference 54.

4.4 SYSTEM DYNAMICS MODELS AND EXTENSIONS

You probably already know that many large-scale systems in the public and private sectors involve human concerns and a variety of belief and value considerations. Population pressures, urban growth and decay, pollution propagation and mitigation, energy generation and conservation, health care, and many more issues are amenable to analysis using system dynamics models. A major task in these issues is getting the "appropriate numbers" for policy-making purposes, and this requires reasonably accurate predictions and forecasts. It is easy to use the words "reasonably accurate" but in many contemporary areas of concern it is very difficult to make forecasts that are in fact "reasonably accurate."

4.4.1 Population Models

Predictions and forecasts depend to a considerable extent upon population growth and decay. All public and industrial services depend, in a basic way, upon population levels. Population models are therefore of interest in themselves and useful as portions of larger models. Population models are based upon the birth–death equations. These may take either a deterministic or a probabilistic form. We are much more interested in the deterministic equations here and will obtain the deterministic equations from the stochastic equations. Then a discussion of the system dynamics methodology [55–57] and some example modeling efforts will follow.

We want to predict population $n(t)$, which may represent, for example, the number of people, the number of companies in a given industry, or the number of housing units in a subdivision or perhaps an entire city or region. It would appear hopeless to attempt to determine precisely the number of people in a city at a given time because births, deaths, inmigrations, and outmigrations occur at random times. Thus we will attempt to predict instead the expected or average number. To do this we will use the standard definition of expected value:

$$\mu_n(t) = \xi[n(t)] = \sum_{n=0}^{\infty} nP_n(t) \qquad (4.4.1)$$

where $P_n(t)$ is the probability that there are exactly n people (or companies or houses) in the population at time t. We will soon use this relation to develop the deterministic birth–death differential equations. Before doing this let us consider a heuristic derivation of the birth–death equations which will provide insight as we proceed with a more rigorous derivation.

We will let $\beta(t)$ represent the average birth rate per unit person in the population at time t. We let $\Delta(t)$ be the average death rate per unit person in the population at time t. We shall assume a completely closed system, so that there is no inmigration or outmigration. The products $\beta(t)\mu_n(t)$ and $\Delta(t)\mu_n(t)$ will then be the total average birthrate and deathrate. The average rate of population growth is the difference between the total average birthrate and deathrate. Thus we have

$$\frac{d\mu_n(t)}{dt} = [\beta(t) - \Delta(t)]\mu_n(t) \qquad (4.4.2)$$

This is the basic birth–death equation, and it is very useful as you will see later in this section.

While the basic birth–death equation seems intuitively reasonable, there are some questions:

- Should the birthrate and deathrate depend only upon the population at that time?
- Should these rates be independent of the actual sequence and timing of births and deaths that have occurred in the past?

The answers to both of these questions are fundamentally yes, if the probability is associated with a Markov process. A process is Markov if the probability of an event occurring at time t conditioned upon events occurring earlier at time $t - 1$, $t - 2$,... is equal to the probability of the event occurring at time t conditioned only upon the events occurring at the most recent time $t - 1$. By using the Markov assumption, it is possible to rigorously derive the basic birth–death equation.

This basic birth–death equation is the simplest type of dynamic equation we might postulate for the modeling of processes in a dynamic system—in other words, any system that evolves over time. This first-order differential equation, the common mathematical physics flow equation, lacks a spatial element. For symbolic convenience we will now drop the average or mean symbol and simply use x as the population state variable. We wish to consider different classes or subsets of population such as housing, people, or industry. We will use the symbol x_i to represent the ith state variable used to denote a specific category such as housing. We will use a superscript, as in x^k, to represent the region in which the state variable is located. Thus, $x_i^k(t)$ represents the number of units of category i in zone k at time t. The linear birth–death equation representing the way in which a certain category of population in a certain zone changes in time is

$$\frac{dx_i^k(t)}{dt} = \beta_i^k(t) - \Delta_i^k(t) + \Sigma_{j=1, j \neq i}^n [m_{ji}^k(t) - m_{ij}^k(t)] + [m_{Ei}^k(t) - m_{iE}^k(t)] \qquad (4.4.3)$$

In this equation the terms $\beta_i^k(t)$ and $\Delta_i^k(t)$ represent the birthrate and deathrate for category k in zone i at time t. $m_{ji}^k(t)$ represents the migration rate from zone j to zone i, and $m_{ij}^k(t)$ represents the migration rate from zone i to zone j. The terms $m_{Ei}^k(t)$ and $m_{iE}^k(t)$ represent migration to and from an external zone that is outside of the basic n zones under consideration. The various terms on the right-hand side of Eq. (4.4.3) may be very complex linear or nonlinear functions of the various state variables.

Example 4.4.1. We consider development of a simple population model for the United States for the period 1920 through 2000. We propose to use Eq. (4.4.2), in which inmigration and outmigration factors are negligible compared with birthrates and deathrates. Equation (4.4.3) may be rewritten as

$$\int_{x(t)}^{x(t+dt)} \frac{dx(t)}{x(t)} = \int_{t}^{t+dt} [\beta(t) - \Delta(t)]\, dt$$

If we assume that the birthrates and deathrates are constant over a given time interval dt, we may remove the $\beta(t) - \Delta(t)$ term from the integral sign and then integrate both sides of the foregoing to obtain

$$x(t + dt) = x(t)\exp\{[\beta(t) - \Delta(t)]\}\, t \tag{1}$$

which is an approximate difference equation corresponding to the birth–death differential equation that we may propagate to determine the population for given birthrates and deathrates. The data in Table 4.1 represent factual data

TABLE 4.1   Forecast from Simple First-Order Model of U.S. Population (Population times 10^8)

Year	Birthrate	Deathrate	True Population	Forecast Population
1920	0.0237	0.0130	1.057	—
1925	0.0213	0.0117	1.158	1.115
1930	0.0189	0.0113	1.231	1.203
1935	0.0169	0.0109	1.273	1.278
1940	0.0179	0.0108	1.317	1.312
1945	0.0195	0.0106	1.405	1.365
1950	0.0241	0.0096	1.513	1.469
1955	0.0250	0.0093	1.659	1.627
1960	0.0237	0.0095	1.793	1.794
1965	0.0194	0.0094	1.938	1.925
1970	0.0178	0.0097	2.032	2.023
1975				2.116

taken from U.S. census information as well as information computed from our model Eq. (1). In the computations leading to Table 4.1, 5-year predictions of population are made using Eq. (1), and the actual forecast population is replaced by the corrected value when making the next 5-year projection. The model projection is quite accurate, but only 5-year projections are being made.

Can we project the population at year 2000 from the true population in 1970? Alternatively, could we use the data from year 1920 to project the population in 1970? If we use Eq. (1) with $dt = 50$ years and the data for 1920, we obtain a population of 1.805×10^8, which is in error by 22,700,000 people. Urban services based upon such an erroneous projection would undoubtedly be pretty poor. We would be assuming that the birthrate remained constant at 0.0237 and the deathrate constant at 0.0130 from 1920 to 1970. Actually the deathrate has been reduced considerably, and the birthrate for many of the intervening years was greater than 0.0237. If we use data from 1940, we obtain a predicted population of 1.630×10^8 30 years later in 1970. The error is even worse than it was in the previous calculation. In 1940 we had just come out of a great depression, and birthrates were very low, only slightly higher than they are at present.

Because we have little confidence in the ability of this model to project the present known population using data far in the past, we must naturally have considerable reservations concerning the ability of this simple model to project future populations, say 30 years from now, based upon known and available present data. For the year 2000, for example, we predict a population of 3.164×10^8 using 1960 data and 2.591×10^8 using 1970 data. For this simple model we could also postulate various future birthrates for the next 30 years and then determine a population based upon the various assumed birthrates and deathrates. Alternatively, we could seek a more complex model that would determine the birthrates and deathrates as a function of other important variables such as pollution, natural resources available for use, and other considerations that affect sustainability of a region or perhaps the entire planet.

A principal difficulty with Eq. (4.4.2) is that the population modeled by this equation will approach either infinity or zero over time if the birthrates and deathrates are truly constant. A simple way to model the finite environment is to determine a maximum population and then introduce a saturation term such that the difference between birthrate and the deathrate approaches zero as this saturation is reached. Also, the birthrate minus the deathrate must go to zero as the population goes to zero.

Example 4.4.2. In this example we will develop a model for propagation of an epidemic resulting from a contagious disease such as influenza. The population will be aggregated into three categories: x_1, the susceptible population; x_2. the infected population; and x_3, the immune population. Each category of population is a state or level variable influenced by an input or an output. It will be convenient to use a box to represent the level variables. The rate variables control the flow of people from one level variable to another. We assume that

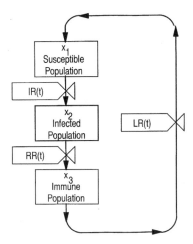

Figure 4.28 Partial systems dynamics description for the epidemic propagation model.

the natural progression is from susceptible to infected to immune to susceptible. Figure 4.28 depicts a convenient block diagram representation that we may use to initiate our construction of a model for the propagation of an epidemic. We assume that there are no exogenous or external influences upon the model, that there are no births and deaths, and that the total population is constant. Thus the birthrate and deathrate terms in Eq. (4.4.3) are assumed to be zero, as are the terms m_{Ei} and m_{iE}. The only task in completing the model is to determine the inmigration and outmigration terms or rate variables m_{ji} and m_{ij}. From Figure 4.28, we see that these are $IR(t)$, the infection rate or number of people infected per day; $RR(t)$, the recovery rate or number of people recovering from the epidemic per day; and $LR(t)$, the loss of immunity rate or number of recovered people losing immunity per day.

It is reasonable to assume that the infection rate increases as the number of infected people increases, and so the number of susceptible people increases. Thus we assume that

$$IR(t) = ax_1(t)x_2(t)$$

where a is, for simplicity, assumed constant. The recovery rate is the ratio of infected people to the average duration, $1/b$, of the disease in days. Thus we have, for a constant b,

$$RR(t) = bx_2(t)$$

The loss of immunity rate is the ratio of the immune population to the period, l/c, of immunity. Thus we see that

$$LR(t) = cx_3(t)$$

for constant c. We could, if we wanted, show flow lines in Figure 4.28 in order to illustrate how the various rate variables are determined in terms of the various level variables. The final differential equations for our model are easily seen to be

$$\frac{dx_1(t)}{dt} = LR(t) - IR(t) = cx_3(t) - ax_1(t)x_2(t)$$

$$\frac{dx_2(t)}{dt} = IR(t) - RR(t) = ax_1(t)x_2(t) - bx_2(t)$$

$$\frac{dx_3(t)}{dt} = RR(t) - LR(t) = bx_2(t) - cx_3(t)$$

These differential equations may be programmed on a computer with appropriate initial conditions $x_1(t_0)$, $x_2(t_0)$, and $x_3(t_0)$ and parameters a, b, and c, and the propagation of the epidemic may be modeled by the differential equations and parameters determined. In this particular model the fact that there are no births, deaths, and migrations allows us to write $x_1(t) + x_2(t) + x_3(t) = $ constant and to use this relation to reduce the order of the differential equation by one. In general there will not be a constant population, and so this reduction in order does not occur. Figure 4.29 illustrates two typical responses from this model. In order for an epidemic to occur, as in Figure 4.29(a), the infected population $x_2(t)$ must initially increase in time such that the derivative dx_2/dt must initially be positive. This will occur when the infection rate is greater than the recovery rate, or when $x_1(t_0) > b/a$. An epidemic can be prevented, in this simple model, by increasing b, which can be done by decreasing the duration of the disease, or by decreasing a, which can be accomplished by decreasing the effect of contact between infected people and susceptible people either by quarantine or by some measure such as an enzyme filter. Figure 4.29(b) shows the results of a simulation in which the epidemic fails to grow.

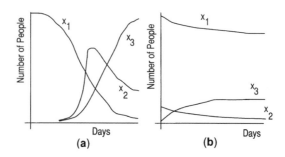

Figure 4.29   Possible simulation results for epidemic model.

4.4.2   System Dynamics

One approach to the development of population models is due to Jay W. Forrester and his coworkers, who use the phrase *system dynamics* to describe their modeling methodology. In system dynamics it is assumed that four hierarchical structure levels can be recognized. These four levels are depicted in Figure 4.30.

In order to develop system concepts sufficiently complete that they can be analyzed by a system dynamics model, we must establish the boundary within which interactions and impacts take place. We choose a derivative variable to control a flow into the state or level variable that integrates or accumulates this level. Information concerning the level is used to control the rate variable. In other words, we define a rate variable as the time derivative of a level or state variable and determine rate variables as functions of level variables.

Example 4.4.3. Let us consider what is doubtlessly the simplest positive feedback problem, that of accumulation of money in a savings account. If there is but a single initial investment of money M_0 in the account, the interest rate is N per year, and the interest is compounded q times per year, then the equation for the growth of money is

$$M(\{k + 1\}T) = M(kT) + TNM(kT) \tag{1}$$

where $T = 1/q$ and $M(0) = M_0$. In Forrester's system dynamics terminology we would identify two variables. The first is a level variable, which would be money. In the particular simulation language known as Dynamo, in which many system dynamics models have been coded, the symbol M.K would be used to represent the difference equation notation $M(kT)$. The other variable

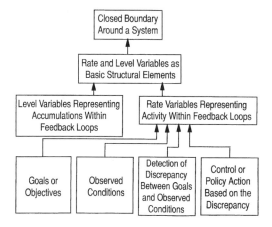

Figure 4.30   Major facets of system dynamics modeling.

would be a rate variable, the money rate or interest rate, which would be written as MR.JK. This symbol is used to indicate that the expression written is for the money rate and that this is assumed to be constant over the time interval from J to K. With the exception of format input and output statements, the complete Dynamo simulation language representation of this problem is the rate variable equation

$$MR.JK = (N)M.J \tag{2}$$

and the level variable equation

$$M.K = M.J + (DT)MR.JK \tag{3}$$

where DT denotes the sampling time or interest compounding time for this example. The Dynamo simulation language will not be developed in this text, but symbols from it will occasionally be used because they are common in the system dynamics literature. Dynamo information, including software, is available from Pugh-Roberts Associates Inc., 41 William Linskey Way, Cambridge, MA 02142, telephone 617-864-8880. Systems dynamics software, known as STELLA, for Windows and Macintosh operating systems is available from High Performance Systems, Inc., 45 Lyme Road, Suite 300, Hanover, NH 03755, telephone 603-643-9636.

If we let the samples become dense—that is, let $T \to 0$, $k \to \infty$, $kT \to t$— then Eq. (1) becomes the continuous-time differential equation

$$\frac{dM(t)}{dt} = NM(t) \tag{4}$$

If the interest is compounded frequently, say daily or weekly, then the solutions obtained from either Eq. (1) or Eqs. (2) and (3) will be essentially the same as that obtained from Eq. (4). All three of these mathematical models may be represented by the symbolic or block diagram representation of Figure 4.31, which depicts a single level or state variable (money) and a single rate variable (money rate). The source of the flow of money exerts no influence on the system in this example. Thus the flow is shown as coming from an infinite source that cannot be exhausted. Any amount needed by the model will come from this infinite source. Lines that indicate information transfer should be distinguished from those lines that represent the flow of the content of a level or state variable. In this example we accomplish this by using a double line to indicate money flow. Any other convenient method of distinguishing information flow from money flow could, of course, be used.

Example 4.4.4 In this example, let us turn our attention to a water pollution problem and an associated system dynamics model. Ordinarily, rivers and lakes can decompose significant amounts of untreated waste products without upsetting the living processes that occur within and around them. However, pollution arises when there is unlimited and uncontrolled dumping of un-

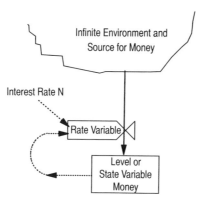

Figure 4.31 System dynamics diagram of interest accumulation.

treated waste in the water. The decomposition processes may, under these conditions, use up considerable oxygen that is dissolved in the water. If the amount of oxygen in the water gets too low, the water is incapable of further decomposition, and the water is unfit for either fish, drinking, or recreational activities. Rivers and lakes replenish their oxygen level by drawing it from the air. The more turbulent the surface of the water, the faster the rivers and lakes take in fresh oxygen.

Part of the solution to this problem of environmental pollution depends upon our ability to develop a predictive model of the effect of proposed solutions to the problem so that decision and policy analysis studies may be conducted. We will bypass the earlier, formulation steps of systems engineering and concentrate upon the systems analysis and modeling portion here. Specifically, we will develop a simplified model to predict how the oxygen content of a river or stream changes as pollutants are dumped into it.

First, we will develop a model for the concentration of waste products in a particular body of water. We will assume that the physical situation is adequately described by the fact that the rate of change of waste in the water is proportional to the rate at which waste is dumped into the water and the product of the amount of waste in the water and the waste chemical composition coefficient of the waste. We define the level variable symbol

$W(t)$ = volume of waste products in water at time t (gallons) and the rate variable symbols

$WAR(t)$ = waste addition rate (gallons per day) at time t. (this is an input or exogenous variable).

$WDR(t)$ = waste decomposition rate (gallons per day) at time t and an auxiliary coefficient.

WCC = waste chemical coefficient (per day)

We may then determine the difference equations corresponding to the verbal statement of the waste product concentration equation as

$$W(t + dt) = W(t) + dt[\text{WAR}(t) - \text{WDR}(t)]$$

$$\text{WDR}(t) = [\text{WCC}]W(t)$$

This is a first-order difference equation that may be solved once we have specified the auxiliary constant WCC, the sample time dt, and the exogenous variable WAR.

From studies of replacement of oxygen in the water we find that the rate at which oxygen is replaced is proportional to the rate of use of oxygen by the wastes and the difference between the amount of oxygen contained in the water and some maximum amount that the water can hold. Although oxygen is constantly being replenished in the water by the air, it is also being used up in the decomposition of waste products process. The rate of use of oxygen by the waste products depends directly on the rate of decomposition of waste. To develop a model for oxygen content in the water we define the level variable

$O(t)$ = oxygen content in the water at time t (cubic feet at a given pressure) and the rate variables

$\text{ORR}(t)$ = oxygen in water replacement rate

$\text{OUR}(t)$ = oxygen usage rate

and the auxiliary constants

OM = maximum oxygen content in water

AT = adjustment time for oxygen replacement or turbulence coefficient

c = oxygen demand coefficient or coefficient representing oxygen requirement for the particular waste products in the water

From the aforementioned verbal description, the difference equations for the oxygen concentration are

$$O(t + dt) = O(t) + dt[\text{ORR}(t) - \text{OUR}(t)]$$

$$\text{ORR}(t) = \text{AT} - 1 \, [\text{OM} - O(t)]$$

$$\text{OUR}(t) = [c]\text{WDR}(t)$$

The source and sink for both oxygen and waste-product concentration are assumed to be infinite in this simple model which has the differential equation or state variable representation

$$\frac{dx_1}{dt} = a_{11}x_1 + u_1$$

$$\frac{dx_2}{dt} = a_{21}x_1 + a_{22}x_2 + u_2$$

where

$x_1 = W(t)$ = waste products
$x_2 = O(t)$ = oxygen content
$a_{11} = -\text{WCC}$
$a_{21} = -[c]\text{WCC}$
$a_{22} = -[\text{AT}]^{-1}$
$u_1 = \text{WAR}$ (input variable)
$u_2 = \text{OM}[\text{AT}]^{-1}$ (input constant)

Figure 4.32 illustrates the system dynamics diagram for this model, and Figure 4.33 represents a control systems-type block diagram, which, while entirely equivalent to the systems dynamics diagram, is not quite as useful for the purpose of displaying the physically different flow quantities, level variables, and information pickoffs as the system dynamics diagram. Much more complex studies of water pollution are currently available. Particularly interesting system dynamics studies are contained in references 58–61.

In each of the examples considered thus far we have developed a solution by determining a closed boundary around the system and then identifying state or level variables and rate variables as the basic components of feedback loops. As the last example has indicated, rate variables are generally made up of information flows based on action taken as a result of a discrepancy between goals (either physical or social) and observed conditions. Thus we see the four levels of hierarchical structure, described in Figure 4.30, inherent in even simple system dynamics models.

As we have seen in our examples thus far, system dynamics models take the form of closed-loop feedback control systems whose dynamic behavior results

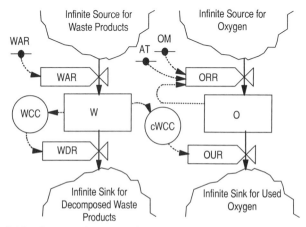

Figure 4.32 Systems dynamics diagram for environmental pollution.

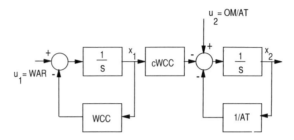

Figure 4.33 Feedback control system block diagram equivalent of system dynamics diagram of Figure 4.24.

from their internal structure. These feedback loops are basic structural elements in systems, and it is the interaction of the many diverse feedback loops that makes the behavior of large-scale systems so complex. All decision processes should be made within feedback loops, because action to implement decisions should depend upon the discrepancy between desired goals and observed conditions.

In system dynamics methodology, level or state variables are changed only by rate variables. The present value of a level variable can, therefore, be computed without knowledge of the present or previous value of any other level variable. In difference equation terminology this says that we agree to write all state or level equations in the form

$$x_i(t + dt) = x_i(t) + dt \sum_{j=1}^{N} a_{ij} x r_j(t) + dt u_i(t) \qquad (4.4.4)$$

such that a level variable is a function only of the previous value of that level variable, the rate variables $x_{rj}(t)$, and the exogenous inputs $u_i(t)$. Thus the present value of a level variable is determined by the past value of the level variable and all rate variables, which are assumed to be constant from t to $t + dt$ in this difference equation formulation. Equation (4.4.4) is just a particular version of Eq. (4.4.2), the basic birth–death equation, and illustrates the two fundamental and distinct types of variables, level variables and rate variables, in a system dynamics model.

Only state or level variables are necessary to completely describe a system. If all state or level variables are known at a given time, all other variables can be computed from them. In particular, we can compute rate variables from level variables. The exogenous inputs $u_i(t)$ are assumed to be known time functions and do not depend upon the state or rate variables. In system dynamics, we assume that rate variables cannot be measured instantaneously but, instead, can be measured only as an average over a period of time. Rate variables cannot directly control other rate variables. There must be an intermediate level between two rate variables. Rate variables can depend only upon level or state variables and auxiliary constants. Rate variables, which are often also policy variables or policy statements in a system, should be of algebraic form. Level and rate variables must alternate in a system dynamics model. Thus rate variables are of the form

$$a_{ij}xr_j(t) = x_j(t)[K_{ij+}A_{ij+} - K_{ij-}A_{ij-}] \tag{4.4.5}$$

such that the fundamental level-variable difference and differential equations become

$$x_i(t + dt) = x_i(t) + dt \sum_{j=1}^{N} x_j(t)[K_{ij+}A_{ij+} - K_{ij-}A_{ij-}] + dtu_i(t) \tag{4.4.6}$$

or

$$\frac{dx_i(t)}{dt} = \sum_{j=1}^{N} x_j(t)[K_{ij+}A_{ij+} - K_{ij-}A_{ij-}] + u_i(t) \tag{4.4.7}$$

Here the x_i, $i = 1,2,\ldots,n$, represent the n level or state variables of the system, such as population, natural resources, or pollution. $u_i(t)$ represents exogenous deterministic or random inputs which could result, in part, from modeling error. The K_{ij+} and K_{ij-} expressions represent cross impacts or the nominal percentage rates of increase and decrease in variable x_i due to level variable x_j. These nominal values are defined in terms of chosen dates at which data are available. The A_{ij+} and A_{ij-} terms are the incentives that modify the normal influx rates according to existing conditions in the system being modeled. These incentives are of value 1 under normal conditions and are expressed as a product of a series of multipliers that quantify the effect of a single-parameter factor upon nominal rates, such as the effect of pollution controls or land-use practices upon the rate of oil field development. Thus these incentives can be written as products of the form

$$A_{ij+} = \Pi M_{ij+}(x_1, x_2,\ldots, x_n, u_1, u_2,\ldots, u_n) \tag{4.4.8}$$

where we note that the multipliers are functions of not only the state or level (vector) variables $x^T = [x_1, x_2,\ldots, x_n]$ but also the exogenous inputs $u^T = [u_1, u_2,\ldots, u_n]$.

One of the major contributions of Forrester is contained in the multiplier functions, which, when combined as in Eq. (4.4.8), form the incentives that are quantifications of the dynamic model hypotheses relating the various effects between states and exogenous variables. If a given multiplier has a value of 1, then conditions, for that multiplier, are precisely the nominal conditions. Values less than or greater than 1 represent a tendency to decrease or increase the associated level variables.

Example 4.4.5. To illustrate further the formulation of system dynamics models involving multiplier nonlinearities of the form of Eq. (4.4.8), let us consider a simplified version of an energy supply and demand example. We choose to investigate some problem elements leading to an impending gasoline shortage.

The short-term dynamic structure of the gasoline supply-and-demand problem appears to involve the following level variables:

Crude oil supply (COS)
Gasoline supply (GS)
Free market price for gasoline (FMP)

Long-term level variables, those which cannot be quickly changed, are as follows:

Number of producing fields (NPF)
Domestic refining capacity (DRC)
Tanker port facilities (TPF)

Because these level variables change very slowly with respect to the first three level variables, it appears desirable to consider these level variables as exogenous inputs for a system dynamics model consisting, in part, of the first three level variables. A separate submodel may then be developed for the long-term level variables. Thus we propose a two-level hierarchical structure for the system dynamics model of the gasoline supply and demand problem, hierarchically structured on the basis of time levels of influence.

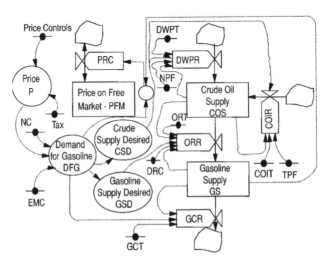

Figure 4.34 Dynamic structure of short-term factors in gasoline supply.

We will develop the short- and long-term demand-supply submodels. The following assumptions lead to the models of Figures 4.33 and 4.34:

1. The demand for gasoline (DFG) is primarily determined by the number of cars in use (NC).
2. The environmental controls (EMC) in place determine the type of gasoline produced.
3. This in turn affects the mileage obtained and consequently influences the demand for gasoline.
4. The demand for gasoline will also depend on the price of gasoline (P).
5. The rate of change in the price of gasoline (PRC) is proportional to the difference between demand for gasoline and supply of gasoline (GS).
6. The price of gasoline (P) is limited by the maximum price (P_{max}) allowed by the governmental price controls.
7. The price is also affected by the tax per gallon rate (TPG) and the free-market price for gasoline (PFG).
8. The oil refining rate (ORR) depends mainly upon the domestic refining capacity (DRC) and the difference between desired supply of gasoline (GSD) and actual supply (GS).
9. However, the supply of crude oil available for refining may also limit the refining rate. This is indicated by the variable RRSM.
10. Similarly, the consumption rate of gasoline (GRC) is determined primarily by the demand for gasoline (DFG).
11. Consumption may, however, be limited by the supply of gasoline available for consumption. This is indicated by the variable CRSM.

In general, variables in the above discussion and the models of Figures 4.34 and 4.35 whose names end in "D" are "desired" quantities. A variable name ending in "R" indicates a rate (an exception is PRC), and a variable ending in "T" indicates an adjustment time for the corresponding rate variable.

A suggested set of difference equations for this model is shown below. To complete the model development we must identify all unknown parameters in these equations and specify all needed functional relations. Expansion to the truly large-scale problem that exists in the real world is a major task, and the simplified model presented here is only a suggestion of the type of approach to be used in systems dynamics model development.

The short-term model of Figure 4.34 is described by

- Auxiliary equations

$$\text{DFG}(k) = f_1(\text{NC})[\text{ECM}]f_2[P(k)]$$
$$\text{CSD}(k) = f_3[\text{DFG}(k)]$$

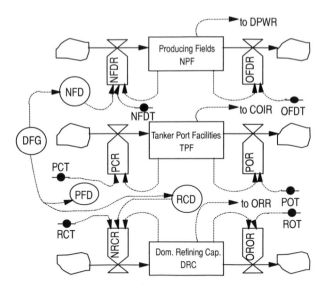

Figure 4.35 Long-term elements in the systems dynamics model for gasoline supply.

$$\text{GSD}(k) = f_4[\text{DFG}(k)]$$

$$\text{PRC}(k) = [\text{PRCT}]^{-1}[\text{DFG}(k) - \text{GS}(k)]$$

$$P(k) = \min\{P_{\max}, PFM(k)\} + \text{TPG}$$

- Rate variables

$$\text{DWPR}(k) = [\text{SWPT}]^{-1}[\text{NPF}(k)][\text{CSD}(k) - \text{COS}(k)]$$
$$\text{COIR}(k) = [\text{COIT}]^{-1}[\text{ITPF}(k)][\text{CSD}(k) - \text{COS}(k)][\text{IQ}]$$
$$\text{ORR}(k) = [\text{ORT}]^{-1}[\text{DRC}(k)][\text{GSD}(k) - \text{GS}(k)][\text{RRSM}(k)]$$
$$\text{GCR}(k) = [\text{GCT}]^{-1}[\text{DRG}(k)][\text{CRSM}(k)]$$

- Level variables

$$\text{GS}(k+1) = \text{GS}(k) + \Delta\text{T}[\text{ORR}(k) - \text{GCR}(k)]$$

$$\text{COS}(k+1) = \text{COS}(k) + \Delta\text{T}[\text{DWPR}(k) + \text{COIR}(k) - \text{ORR}(k)]$$

$$\text{PFM}(k+1) = \text{PFM}(k) + \Delta\text{T}[\text{PRC}(k)]$$

- Variable definitions — auxiliary

NC = number of cars
EMC = environmental control multiplier
CSD = crude oil supply desired

GSD = gasoline supply desired
PRC = price rate of change
P = price
TPG = tax per gallon

- Variable definitions — rate

DDWPR = domestic well production rate
NPF = number of producing fields
IQ = import quota
COIR = crude oil importing rate
TPF = tanker port facilities
ORR = oil refining rate
DRC = domestic refining capacity
RRSM = refining rate-supply multiplier [This variable allows for the fact that refining rate (ORR) may be limited by the supply of available crude oil (COS)]
CRSM = consumption rate-supply multiplier [This variable allows for the fact that consumption rate (GCR) may be limited by the supply of available gasoline (GS)]
GCR = gasoline consumption rate

- Variable definitions — level

COS = crude oil supply
GS = gasoline supply
PFM = price − free market

The long-term model of Figure 4.35 is described by the following:

- Auxiliary equations

$$NFD(k) = f_5[DFG(k)]$$
$$PFD(k) = f_6[DFG(k)]$$
$$RCD(k) = f_7[DFG(k)]$$

- Rate variables

$$NFDR(k) = [NFDT]^{-1}[NFD(k) - NPF(k)]$$
$$OFDR(k) = [OFDT]^{-1}NFP(k)$$
$$PCR(k) = [PCT]^{-1}[PFD(k) - TPF(k)]$$

$$POR(k) = [POT]^{-1}TPF(k)$$

$$NRCR(k) = [RCT]^{-1}[RCD(k) - DRC(k)]$$

$$OROR(k) = [ROT]^{-1}DRC(k)$$

- Level variables

$$NFP(k + 1) = NFP(k) + \Delta T[NFDR(k) - OFDR(k)]$$

$$TFP(k + 1) = TFP(k) + \Delta T[PCR(k) - POR(k)]$$

$$DRC(k + 1) = DRC(k) + \Delta T[NRCR(k) - OROR(k)]$$

- Variable definitions — auxiliary

NFD = number of fields desired
PFD = port facilities desired
RCD = refining capacity desired

- Variable definitions — rate

NFDR = new-field development rate
OFDR = old-field depletion rate
PCR = port construction rate
POR = port obsolescence rate
NRCR = new-refinery construction rate
OROR = old-refinery obsolescence rate

- Variable Definitions — Level

NFP = number of producing fields
TPF = number of tanker port facilities
DRC = domestic refining capacity

This brief study of a system dynamics model of the energy supply-and-demand problem has considered only one aspect of this complex problem. Clearly, computer simulation is needed here, and certainly for any realistic look at this complex large-scale issue. We see here that computer simulation will be an invaluable support to analysis of the impacts of alternative controls or policies.

Nonlinear Functions   The example just presented has introduced the need for nonlinear functions in system dynamics modeling and simulation. One such nonlinearity was previously mentioned, the multiplicative nonlinearity of Eq. (4.4.8). This will allow us to obtain products of state or level variables. Also needed is the ability to obtain a nonlinear function of a single state variable. Of course a program can easily be written on a computer to compute $y = f(x)$ for a specified nonlinear function f. Many individuals and groups will perceive nonlinear functions in the form of a table rather than as some analytical function. Also, when we attempt to validate system models using system parameter identification methods, we must be careful not to overstate the range of the level or state variable over which the identification is accurate. To identify the parameter a in the equation $y = ax^2$ using data over the range 0 to 1 of the x variable and then assume that this equation is valid regardless of x could lead to considerable difficulties. Thus it will often be better to specify the relationship between x and y by means of a table function. Fortunately, most simulation languages and computer routines do include table functions. Nonlinear functions often occur in systems, and these table functions locate, generally by straight-line interpolation, those values that are between points entered in the table. In the Dynamo simulation language, for example, the program statements

$$\text{Y.K} = \text{TABLE(TNAME, XK, N1, N2, N3)}$$
$$\text{TNAME} = \text{Q1|Q2|...|QM}$$

(4.4.9)

would define the auxiliary variable Y(K) or Y.K in terms of the table of Figure 4.36. The name of the table is TNAME, X is the input variable for which the corresponding table entry is to be located, N1 is the value X for the first table entry, N2 is the value X for the last table entry, and N3 is the interval in X between table entries. The Qi are the numerical values of the table function Y at the various values of XK.

Time Delays and Averaging. Often it will be desirable to incorporate averaging or smoothing operations as well as perception time delays in a model. For example, something may happen at time t_1 but we may not be able

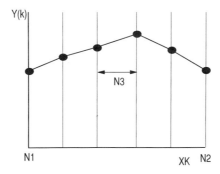

Figure 4.36 Typical table function for system dynamics modeling and simulation.

to determine this through measurement until time $t_1 + T$. Thus, we have need to model a time delay function. It was mentioned earlier that instantaneous values of rate variables could not be measured. Thus we do need a method of smoothing these, and, of course, we may well wish to smooth a level variable as well. Often it is not a level variable that individuals observe but a perceived value of a level variable, which may be the level variable delayed in time. We will now develop expressions for smoothing and perception operations. These will be in the form of difference equations. Thus any smoothing of rate or level variables and any perception delays will necessarily increase the order of the differential or difference equation that represents the system. However, these smoothing and perception operations are not regarded as level variables — at least, not fundamental system level variables.

Many operations may be used to smooth or average a variable $x(t)$. For example, we may use $(1/T) \int_{T-t}^{t} x(\tau)\, dt$ as representative of the average of $x(t)$ over the last T seconds. It turns out that the difference or differential equation necessary to implement this finite time average is difficult because it is of infinite order. It must remove chunks of data as t increases. Also, it is often true that the immediate past is more important, for a variety of reasons, in determining the smoothed value than the distant past. Thus, it may well be preferable for you to use an exponentially weighted time average defined by

$$\mathbf{x}^T(t) = \frac{1}{T} \int_{-\infty}^{t} e^{(\tau - t)/T} x(\tau)\, d\tau \tag{4.4.10}$$

for your smoothing operations. T represents the weighting factor and has the dimension of time. Physically only data from t to $t - T$, which is the data over the most recent T seconds, are given much significance in weighted averages. Thus it is reasonable to call T the smoothing time. If we use Eq. (4.4.10) to obtain the expression for $\mathbf{x}^T(t + dt)$ and make a simple Taylor series expansion, we obtain the approximate difference equation for a weighted time average operation:

$$\mathbf{x}^T(t + dt) = \mathbf{x}^T(t) + \frac{dt}{T} [x(t) - \mathbf{x}^T(t)] \tag{4.4.11}$$

which is equivalent to the differential equation

$$\frac{d\mathbf{x}^T(t)}{dt} = \frac{1}{T} [x(t) - \mathbf{x}^T(t)] \tag{4.4.12}$$

The system dynamics diagram for the smoothing operation is illustrated in Figure 4.37. A dotted line is used for the flow into the level variable (SMOOTH) to indicate that this is a flow of information.

A pure information or material delay is a delay such that if the input is $x(t)$,

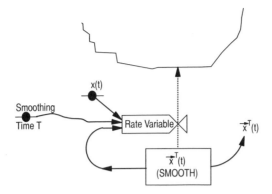

Figure 4.37   Systems dynamics model of smoothing or averaging operation.

the output is

$$x^D(t) = x(t - T) \tag{4.4.13}$$

where T is the amount, in time, of the delay. You may obtain the difference equation corresponding to this by writing the expression for $x^D(t + dt)$ and expanding the result in a Taylor Series and dropping the higher-order terms. After a modest amount of manipulation, we obtain

$$\mathbf{x}^D(t + dt) = \mathbf{x}^D(t) + \frac{dt}{T}[x(t) - \mathbf{x}^D(t)] \tag{4.4.14}$$

This is the same expression as the difference equation for the weighted time averaging operator, Eq. (4.4.13). The system dynamics equation for a single time delay is precisely that of Figure 4.37. As with the smoothing operator, there are valid arguments to suggest that this weighted perception delay is much more typical of human information delay than a pure time delay.

If we have a long information or material delay, it may be advisable to use M delays, each delaying the material or information by an amount T/M. We may, for example, cascade three delays as shown in Figure 4.38. While the system dynamics representation of Figure 4.38 is correct for material delays, it is not physically appealing, because the material does not "flow" from input to output. A somewhat more satisfactory physical picture may be obtained by noticing that the rates associated with each difference equation (4.4.13) are just the difference between the output of the previous time delay segment and the time delay currently being computed. Thus the system dynamics diagram of Figure 4.38 is entirely equivalent to that of Figure 4.39, which is more physically appealing for a physical time delay. For both smoothing and time delay operators, we should note that the initial condition for the difference

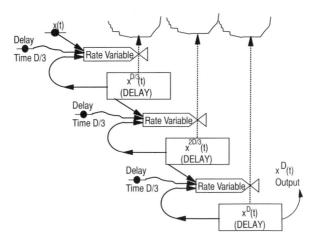

Figure 4.38 Three-stage information perception time delay model.

equations is basically unspecified because no smoothing or time delay has occurred at the time the simulation is started. It is necessary for you to obtain some perceived initial condition, generally the initial value of the smoothed output or the initial perceived delayed output, in order to start the simulation.

Applications. Two of the most widely reported uses of system dynamics modeling methodology have been in the development of models for the city and the world, as well as other models that represent various perceived limits to growth. The initial seminal efforts concerning these models are due to Forrester and his colleagues. In the efforts at modeling an urban city and the world, Forrester organized the structures of an urban city area and the world into system models showing life-cycle dynamics of various forms of growth and decay and conducted various experiments utilizing these models. Perhaps no other single works in system dynamics have generated as much interest, controversy, and follow-up study as have these. Even though these models are about three decades old now, many of the issues they address are still contemporary ones. Doubtless this is due to the value of the efforts themselves,

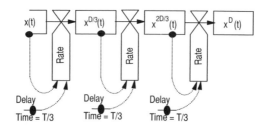

Figure 4.39 Three-stage material flow time delay.

the contemporary urgency of the problems they discuss, and considerable concerns with respect to some of the assumptions used to obtain the models. We will examine salient features of both of these models as we conclude our presentation of the system dynamics approach to systems analysis and modeling.

Example 4.4.6. The urban dynamics model. Forrester's interest in modeling the city is a somewhat abstract one in that he does not fit the data and parameters for his city to any particular city. Effort is primarily directed at discovering the essential features of the city and expressing relationships between these features in mathematical terms as difference equations. The urban dynamics model considers one city only. No spatial behavior is involved, and the city is assumed to be surrounded by a limitless environment that is capable of producing and receiving unlimited quantities of people, jobs, and homes. The model enables studies of the impacting dynamics of population, industry, and housing within this mythical city. Each of these three fundamental sectors is divided into three subcategories. The population subcategory consists of

- Managerial–professional (MP)
- Skilled labor (L)
- Underemployed (U)

Underemployed people are defined as unemployed and unemployable people as well as those in unskilled jobs or marginal activity who might work during periods of great economic activity. Within a given class of workers, each worker is the head of a household with an average family size that demands housing of a specified type.

Each subclass of people has its own housing:

- Premium housing (PH)
- Worker housing (WH)
- Underemployed housing (UH)

Premium housing must be constructed and will eventually decline to worker housing. Worker housing may also be built and will eventually decline to underemployed housing. Underemployed housing may be built, but its principal source is declining worker housing. Underemployed housing declines after a long period and is demolished to make room for new land use. Failure to disaggregate population by age sectors could be cited as one disadvantage inherent in this nonspatial model, because there may be a difference in age distribution between various population categories — unemployed migrants and managers, for example. To fully utilize a city model for projections of residential land use would seem to require disaggregation of both population and housing by other factors in addition to those cited in this work. Aggregation of housing into only three types rigidly attached to a population subcategory and the assumption that each dwelling occupies a fixed land area that is independent of the type of city or the population density of the city or the time at which the housing is constructed would seem not fully in accord with modern practice.

Industry in the urban dynamics model is subdivided into three categories:

- New enterprise (NE)
- Mature business (MB)
- Declining industry (DI)

The three categories of business employ heads of household from the three population categories in differing mixes. New enterprise requires more managers than declining industry, for example. The city is restricted to a land area of specified size. Industry and housing must inevitably compete for this land when the city ages if the land becomes scarce. New enterprises are assumed to be outgrowths of existing industry or attracted from the limitless external environment. As new enterprises age, they become mature businesses and eventually become declining industry. Finally they are destroyed to create vacant land which may be reused.

The model is a twentieth-order dynamic model with nine fundamental level variables and a sampling period of one year, and time constants in the model are such that steady-state behavior is reached in approximately 200 years. There are numerous perception time delays reflecting the time between when conditions actually change and when the change is recognized. Numerous multiplier functions are used in relating various impacting events. Industry and people inmigrate to the city, and new homes are built if the attractiveness of the city relative to its environment is great enough. Outmigration occurs if the city becomes unattractive enough. Each population sector bases a decision to inmigrate or outmigrate on different criteria and values. What may result in attracting inmigration for new enterprises and manager–professionals may well result in outmigration of the unemployed.

Simulation. The Forrester urban dynamics model is a twentieth-order, nonlinear, nonspatial model of an urban system with fixed land area. The model simulates 250 or more years of urban development. Plots are produced for population, housing, and business. Each of these areas is made up of three sectors. Population is broken down into underemployed, labor, and managerial–professional. Housing consists of underemployed, worker, and premium housing. Business is classified as new enterprise, mature business, or declining industry. Equilibrium is reached after about 150 to 200 years using Forrester's original parameter settings and policies, and the variables do not change significantly thereafter. This condition of stagnation is what is believed to be typical of our urban areas today.

The original urban dynamics model allows the user to introduce any of 10 "urban improvement" programs into the system to determine their effect on the urban environment. The basic policies originally available for implementation are as follows:

1. *Underemployed Job Program.* This provides additional jobs for the underemployed such as might occur under a public service job program.
2. *Underemployed Training Program.* This "upward mobility" program allows for the training of a certain percentage of the underemployed to qualify for jobs that would enable them to be classified as labor.
3. *Tax-per-Capita Subsidy Program.* This simulates a flow of tax dollars into the city from sources outside the model area, such as state or federal funds.
4. *Low-Cost Housing Construction Program.* This represents the building of low-cost housing with funds not from the city government, but presumably from some outside sponsor.
5. *Worker Housing Construction Program.* This provides for construction of middle-class housing.
6. *Premium Housing Construction Program.* This provides for construction of expensive housing for manager–professionals.
7. *New Enterprise Construction Program.* This allows the model user to evaluate the effects of new business development.
8. *Declining Industry Demolition Program.* This provides for the blanket removal of a certain amount of declining industry.
9. *Slum Housing Demolition Program.* This represents a "slum clearing" policy that does not explicitly take into account such things as relocation or redevelopment, which are provided by the normal dynamics of the city.
10. *Labor Training Program.* This program is much like the second one, and it allows members of the labor class to move up to managerial–professional status.

The programs that Forrester reports in his 1969 book are only briefly reviewed here. There are a number of reviews of this effort and similar efforts in the literature, including references 62 and 63. It turns out that the most effective of Forrester's programs are as follows:

1. Demolition of slum housing
2. Encouragement of new enterprise
3. Demolition of declining industry

"Effective" is used here to mean that the city is richer in the sense that more tax revenue generators and fewer tax revenue users are present at equilibrium in the city.

There are a number of implicit values incorporated in the urban dynamics model. These are the values of the analyst, for the most part. This is, in this case, not entirely unreasonable because there was no "client" for the study, which might be viewed as more of an effort to demonstrate the power of the systems dynamics modeling approach than as an effort to provide implementable policy decisions. Forrester's implicit values, reflected in his "preferred" urban revival program, include a favorable outlook toward Western concepts of progress and the upward mobility of people. Forrester appears to strongly believe in the precept that "all progress is good." He favors economic self-support, and he incorporates a permanent value structure in the model. Balancing the city budget, decreasing per capita taxes, and increasing the attractiveness of the city for management and labor are all central policies in the Forrester urban model. Thus, in synthesizing programs that correlate well with Forrester's suggested strategies, we must remember that optimizing the wealth of the city is a sort of natural objective if we are to use the model as developed with its given structure and parameters.

The Forrester urban dynamics model is certainly a pioneering venture in the study of urban dynamics. Over the last three decades, it has generated a great deal of new activity in urban modeling and, because of its complexity and the many assumptions made, has generated a great deal of criticism. It appears from Forrester's writing that "expert opinion" rather than real data were used in constructing the many and varied nonlinear relations and multipliers used. He has been criticized for developing a model with built-in bias, for not explicitly stating a value system, and for making untested and possibly invalid assumptions. For example, do increasing taxes always attract the underemployed group and drive away the managerial–professional class? Forrester assumes that taxes are equivalent to public expenditures. Increased public expenditures in the form of welfare, slum housing projects, and the like are attractive to the poor. Some would claim, however, that most city taxes are nonprogressive, hitting lower-income groups relatively more than the managerial–professional group. Do additional public expenditures for police, sanitation, and social services really help the poorer sections of the city more than wealthier sections? Again, some would claim that increased taxes result, in many cases, in better educational facilities and enhanced cultural activities, and that these are more attractive to the managerial–professional class than to the underemployed. Is it the middle-class and skilled workers, Forrester's worker class, that seem to be most violently opposed to new tax increases or is it, as Forrester claims, the wealthier classes? Could a tax that in effect introduces income redistribution cure the ills of the city? Would a tax on natural resource depletion, or pollution, alter the results obtained from use of the model?

Forrester is not particularly concerned with things that are happening beyond his community. Thus, there is little concern for such issues as natural resource use and sustainability. To accommodate such concerns, the dynamics of the city would have to be superimposed on the dynamics of the limitless environment. Do economic crises, wars, and technological and social developments not reflect themselves in such actions and events as inmigration and outmigration, housing construction, and jobs? Should there not be an urban dynamics model for several cities, along with competition among the cities for industries and jobs, such that game-theoretic simulations would be needed? Does the availability of underemployed housing serve as a lure for the poor? Does the lack of availability of jobs in other cities and the hope of finding work in the city play a leading role in migration to the city? Does the model cover the housing situation adequately? In many cases the model predicts an excess of housing for the underemployed. In real life there appears to be a shortage of this type of housing as well as rapidly deteriorating housing. Where are the suburbs? How do racial problems affect the city?

By changing the parameters within Forrester's urban dynamics structure as well as changing the structure itself, quite different model response has been obtained. Various methods to validate urban models have been proposed, and they can be used to validate the structure and parameters within the structure of Forrester's urban model. Forrester's efforts, by stimulating such a vast number of questions concerning the dynamics of social systems, have led to a great deal of further effort toward understanding the dynamics of social systems. Thus we can only regard Forrester's pioneering efforts as invaluable in adding an important new dimension to the solution of many contemporary problems through systems analysis. While the model may well contain improper assumptions, they are there to be viewed and, when they are believed to be improper, changed.

Example 4.4.7. The World Dynamics Model. In 1971, almost three decades ago, Forrester published his text *World Dynamics*, in which the system dynamics methodology was applied to the behavior of the impacting forces of global dynamics. As with the urban dynamics model, statements, observations, and assumptions were the primary inputs determining the structure and parameters of this world model. Forrester indicates that manifestations of stress in the world system are excessive population, rising pollution, and disparity in standards of living. He is concerned with whether these are causes or symptoms of world problems, and one of the major aims of the world dynamics model is to resolve this question.

Three premises appear to be basic to the philosophy used to characterize this world model.

1. Physical attributes of the global system all obey the birth–death population model equation and are characterized by exponential growth.
2. The capacity of the world ecosystem and world resources is such that exponential growth cannot be sustained forever.
3. There is a very long perception delay on the part of society for any problem involving population growth and ecosystem and natural resource limitations.

Any of these three characteristics, taken separately, would be quite serious. When they are combined, we have a complex system of interacting problems, commonly called a "mess." If an accurate model of the global system could be produced and if governments and individuals would believe the model and implement "solutions" demonstrated to be workable by the model, then major advances in policy determination would result, as contrasted with the common approaches of "muddling through" or "try it for awhile and see what happens." To develop a tentative model and draw various conclusions seems to be Forrester's intent. He argues most persuasively that his model is better than any of the existing mental models, with their inherent fuzziness, incompleteness, and possible hidden intents.

The major state and level variables for the world dynamics model are

- Population
- Capital investment
- Natural resources
- Pollution

From these major sectors and their interactions come such other important variables as quality of life, material standard of living, and deathrates from pollution. The four level variables comprised in the world system are shown in Figure 4.40. A rising population creates pressure to increase industrialization to maintain and increase the standard of living. More food must be grown and more land must be occupied. More food, more material goods, better technology, and better health care each encourage a larger population. Population growth, material standard of living increase, pollution increase, and natural resource depletion are all in a cycle, and eventually pollution becomes excessive or natural resources are exhausted, the earth becomes overburdened, the

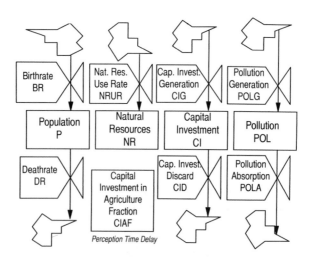

Figure 4.40 Four fundamental level variables in the world model.

standard of living is inevitably reduced, and stagnation occurs. The four fundamental level variables of Figure 4.40 are connected by determining the seven rate variables of this figure in terms of the four level variables. There is one perception time delay, the fraction of capital devoted to agriculture. This perception delay, of 15 years, adds a first-order differential equation to the four first-order level equations. Thus Forrester's world dynamics model is a fifth-order differential equation.

Simulations are available in Forrester [56] for the basic world model with a given set of initial conditions. From these results, it appears that the "world quality of life" has been declining since 1948. Natural resources must monotonically decrease, because there is no physical way, in the model, that natural resources can be "created" through the use of technology. This lack of a technology sector has been one of the major criticisms of the Forrester world model, and several models are now available with a technology sector. Population, capital investment, and pollution all peak between 2030 and 2050 and decline thereafter.

Several possible policies that might pose a cure for the world's ills are proposed and evaluated:

1. In 1970 natural resource usage normal is reduced to 25% of its original value. If everything else remains the same, resource consumption will be reduced 75%. However, natural resource usage is still affected by population and material standard of living. Reducing demand for natural resources allows further growth, because population and industrial growth only consume 25% as much natural resources as before. Population rises, and industrial growth causes pollution to rise more rapidly until there is a substantial crisis around 2050 with pollution 40 times its level in 1970. World population declines, due to death from pollution, to 20% of its peak level.

2. The effects of both natural resources and pollution are ameliorated. Natural resource usage is forced to zero in 1970, and pollution rate normal is reduced to 10% of its value in 1970. Population then rises greatly and stabilizes. Even though pollution is low and natural resources are fixed at their 1970 level, the overall quality of life due to material standard of living rises by a factor of 2.3 from its 1970 value, and quality of life due to pollution and food is essentially unchanged. Quality of life due to crowding has decreased by 60% from its 1970 value. This solution might be acceptable if the policy producing it were not unrealistic. No costs of the 90% pollution abatement have been allowed for, and there is not much hope at present of discovering infinite natural resources.

3. Natural resource usage is zero, as in the previous policy. Pollution has been reduced to 10% of its original value, as in the previous policy. In addition, we now also suppress the effect in 1970 of crowding on the birthrates and deathrates. Now the population rises still further, becom-

ing 10.8 billion rather than 9.7 billion with "policy" 2. The quality of life from crowding is worse than before, and we now have even less food. So the overall quality of life is now substantially lower — by 70%.

4. Increased industrialization by higher capital investment generation is next attempted as a single policy. The pollution crisis reappears.

5. The normal birthrate is next lowered by 30% in 1970 as a policy. The resulting increase in quality of life and capital investment reduces some internal pressures limiting population, and the population actually rises. Natural resource depletion is a larger problem with the increased standard of living. The quality of life, which rises momentarily after reduction in the birthrate, again declines significantly.

6. The normal birthrate is lowered by 30% in 1970, and normal natural resource usage is reduced to 25% of its 1970 value. The quality of life rises even higher immediately after introduction of the policies and then declines dramatically as pollution rises and causes a dramatic decrease in population.

7. A technological innovation reduces normal pollution generation by 30% in 1970. This allows the population to grow even further, resulting in even more pollution than if the technological innovation had not occurred. Forrester views this as another example of proposed solutions producing counterintuitive and counterintended results.

8. In 1970 the food ratio for the world is increased by 25%. The instantaneous availability of more food causes the quality of life to rise, and the net effect of the resultant growth in population is to bring the quality of life back to its original baseline in about 20 years. Increased food production thus accomplishes virtually nothing.

9. Several combination policies are suggested. From these the "Forrester's suggested" strategy that emerges is as follows:
   a. Reduce normal natural resource usage rate 75%.
   b. Reduce normal pollution rate 50%.
   c. Reduce capital generation rate 40%.
   d. Reduce normal food production rate 20%.
   e. Reduce normal birthrate 30%.

The result of this hybrid policy is to decrease population slightly below its 1970 level and increase the quality of life. Natural resources are still slowly declining, however, and will create problems sometime after 2100 unless sufficient recycling and material substitution can occur prior to that time.

This last result, assuming validity of the model, suggests that an acceptable global equilibrium is possible but requires policies that will be very difficult to implement from both a political and a social viewpoint. Forrester presents the following issues as being raised by the world dynamics model. The gist of all of these appears to be that population and industrial growth must be checked if the world is to survive. Some specific conclusions drawn were as follows.

1. Industrialization may be more of a disturbing force to world ecology than population.
2. Shortages of natural resources and rapid population declines due to pollution, food shortage, or crowding will create serious problems over the next century.
3. There may be no way to increase the quality of life to its value in the immediate past.
4. Birth and population control programs may be inherently self-defeating.
5. As agriculture reaches a space and pollution limit and industrialization a natural resource and pollution limit, the quality of life will fall and will stabilize population.
6. Third World nations may have no hope of reaching the standard of living of industrial society. The population in Third World countries is quadruple that in industrial society, but industrial society places a 20 to 40 times greater burden on pollution generation and natural resources consumption systems. The world system would collapse before the Third World population could "catch up."
7. Highly industrialized society may not be able to sustain itself, and new international strife over the ecosystem could reduce the world standard of living to that of a century ago.
8. Thus, present efforts of Third World countries to industrialize may be ultimately unwise from the standpoint of their own long-range interests.

As might be imagined, this study resulted in considerable praise and concern. Some of the many criticisms are important. Lack of a technology sector to give weight to the human ability to solve environmental and resource problems is very unrealistic. There is potentially much that industrial ecology [64–66] and design for environment [67] can hope to accomplish. Also, pollution is not increasing exponentially on a worldwide basis. It is not possible at the present time to construct a model explicitly relating pollution and health. The world dynamics model is too highly aggregated. No real-world data have been used to validate the model.

Regardless of the criticisms of the world model, there appears to be little doubt that it provides a relatively clear exposition of the assumptions on which it is based. These can then be argued, and hopefully a "better" model is a result. A text by Clark and Cole [68] and a more recent paper by Bremer [69] present detailed surveys and comparative studies of existing global simulation models. System dynamics modeling has been used for a plethora of other application areas. These range from software project dynamics [70] to organizational learning [71,72], and they also include the limits to growth situations we discussed in our last two examples [73]. The DYNAMO simulation language is described in reference 74, and a personal computer-based language, STELLA, is discussed in reference 75. Even though these initial efforts were first described more than two decades ago, the challenges posed relative to sustainable development needs are quite important at this time.

4.4.3 Workshop Dynamic Models

In this subsection we develop a deterministic dynamic simulation methodology, referred to as Kane Simulation, or KSIM, originally postulated by Julius Kane [76,77]. In this cross-impact analysis-like approach, you assume that rather than estimating conditional probabilities for use in a cross-impact analysis, the cross-impact of various state variables upon the rate variables of a set of first-order differential equations is obtained. While it is suitable for use by an individual, it is particularly appropriate for use by a group.

Use of the KSIM approach results in a simple but powerful set of differential equations that can be used for forecasting and planning purposes. The KSIM process involves first the selection of a set of state variables, x_i, and then identification of a set of impact parameters, for the continuous-time nonlinear differential equation

$$\frac{dx_i}{dt} = -\sum_{j=1}^{N} \left(a_{ij}x_j + b_{ij}\frac{dx_j}{dt} \right) x_i \ln x_i \tag{4.4.15}$$

In this equation, we define

x_i = the ith state variable
N = the total number of state variables
x_j = the cross-impacting variables
a_{ij} = the long-term impact of x_j on x_i
b_{ij} = the short-term impact of x_j on x_i

The steps and associated assumptions made in the development of KSIM include to following:

1. *We Identify State Variable Elements, or x_i Variables for the Model.* All state variables are bounded, since variables of human and physical significance cannot increase indefinitely. With scaling of state variables, all state variables, can be bounded between 0 and 1.

2. *We Specify the Interactions Between These Variables.* A rate variable will increase or decrease depending upon whether the net impact of all state variables interacting with the rate variables is positive and enhancing or negative and inhibiting. This results in identification of the set of parameters a_{ij} and b_{ij} in Eq. (4.4.15). Complex interactions are described by a pair of interaction tables or matrices **A** and **B**.

3. *The Model is Simulated on a Computer.* Bounded growth and decay of state variables exhibit the familiar sigmoid shape. When a state variable is near its bounds of 0 or 1, the influence of impacting variables is less than when the state variable is not at either extreme of the operating region. For all state variables but one constant, increases in the one state variable will produce an increase in impact on the system.

4. *The Model Is Validated.* The results of the model simulation are analyzed and the various model parameters are adjusted until it is felt that the various responses are appropriate.

KSIM is designed to incorporate a feeling for the linkages that interconnect the elements of a complex system. These elements and an appropriate structure can often be obtained from sound application of the requirements and issue formulation methods you studied in the previous chapter. These procedures may be used to identify essential elements and a structure for a complex system. The KSIM methodology provides further insight into the implications of system structure by associating with the structure a set of interconnected first-order differential equations with unspecified parameters. If these parameters are specified, in much the same way that cross-impact probabilities are specified, computer simulation of the resulting equations will impart an appreciation of the influence of system structure and parameters within the structure on system behavior.

Rather than attempting to determine specific physical relations for all rate variables, each is assumed, in KSIM, to be computable from the set of equations (4.4.15). Inspection of this equation shows that for $x_i = 1$ the derivative, dx_i/dt, is 0. It is also straightforward for us to show that in the limit as x_i becomes 0, we also have $x_i \ln x_i = 0$. Thus Assumption 3 is satisfied by Eq. (4.4.15) in that the state variable will be bounded between 0 and 1 and the impacts will be very small near these bounds. The expression $x_i \ln x_i$ modulates the summed impacts in Eq. (4.4.15), and because it forces the derivative to go to 0 for $x_i = 0$ or $x_i = 1$, we see that Assumption 1 is satisfied and the state variables are indeed bounded between 0 and 1. The impact parameters a_{ij} and b_{ij} are adjusted to enforce satisfaction of Assumption 2.

Example 4.4.8. To gain further insight into this rather unusual differential equation, (4.4.15), let us consider the simplest case where $N = 1$ and $b = 0$ such that we have the "simple" system differential equation

$$\frac{dx}{dt} = -ax^2 \ln x \tag{1}$$

At first glance it might appear that it would be difficult to even simulate a first-order exponential decay model response $x(t) = x_0 \exp[-at]$, because we have a rather complex nonlinear first-order equation. If the term $-x \ln x$ is constant for an appreciable range of values of x, then the solution to the nonlinear first-order differential equation is essentially that of a first-order linear differential equation if the solution values of x do not exceed the range over which $-x \ln x$ is nearly constant. The expression $-x \ln x$ is indeed nearly constant and approximately equal to 0.3 for $0.15 < x < 0.65$, as you can easily verify. Thus if the state variable x remains within this region, then the response is essentially that of a linear system during the time that x remains within this region 0.15 to 0.65.

We can develop a difference equation solution to Eq. (1). To do this we propose the integration

$$\int_{x(t)}^{x(t+\Delta t)} \frac{dx}{x} = -\int_{t}^{t+\Delta t} a[x \ln x] \, dt \tag{2}$$

which is equivalent to Eq. (1). We note that $x \ln x$ is essentially constant over a wide range of x. Thus we approximate the $x \ln x$ term as a constant, specifically $x(t) \ln x(t)$, over the entire time range of integration, and we integrate Eq. (2) to obtain

$$\ln x(t + \Delta t) - \ln x(t) = -ax(t)[\ln x(t)] \, \Delta t \tag{3}$$

which is equivalent to

$$x(t + \Delta t) = [x(t)]^{1 - a\Delta t x(t)} \tag{4}$$

This will be an especially accurate difference equation, correct as $\Delta t \to 0$, and may be solved iteratively for an approximate $x(t)$. The accuracy of Eqs. (3) and (4) will suffer for positive a. Because x is less than 1, raising it to a power greater than 1 will not result in loss of numerical accuracy. For positive a the derivative dx/dt is positive and the solution is increasing in time. For negative a the solution decreases in time, and it is better to use for the constant $x(t) \ln x(t)$ the value $x(t) \ln x(t + t)$. The solution to Eq. (2) is then approximately

$$\ln x(t + \Delta t) - \ln x(t) = -ax(t) \ln x(t + \Delta t) \Delta t$$

which is equivalent to

$$x(t + \Delta t) = [x(t)]^{\frac{1}{1 + a\Delta t x(t)}} \tag{5}$$

When a is negative, we say that we have a negative or inhibiting impact of x on dx/dt through feedback (negative feedback). When a is positive, we say that we have a positive or enhancing impact of x upon dx/dt (positive feedback). Equations (3) and (5) may be combined as

$$x(d + \Delta t) = [x(t)]^{q(t)} \tag{6}$$

where

$$q(t) = \frac{1 + 0.5\Delta t[|a| - a]x(t)}{1 + 0.5\Delta t[|a| + a]x(t)} \qquad (7)$$

If you now approximate $-x(t) \ln x$ by 0.3, then Eq. (1) becomes $dx/dt = 0.3ax$, and this has the solution

$$x(t + \Delta t) = x(t)e^{0.3a\Delta t} \qquad (8)$$

which is an exact solution to Eq. (6). Solution of the difference equation (3) and the linear differential equation $dx/dt = 0.3ax$ for differing positive or negative values of a shows that, for times sufficiently small, the saturation effect does not occur and the solutions to the two equations are essentially the same.

We note that selection of Eq. (1) is fairly arbitrary. It is but one equation among many that has the desired property $dx/dt = 0$ for $x = 0$ and $x = 1$. The equations

$$\frac{dx}{dt} = ax^2(1 - x) \quad \text{and} \quad \frac{dx}{dt} = axf(x) \qquad \text{with } f(x) = 0 \text{ for } x = 0 \text{ and } x = 1$$

also have this property. Using the $x(1 - x)$ expression would allow generation of a more symmetric sigmoid curve. However, $x(1 - x)$ is not as constant for intermediate values of x as is $x \ln x$. Specifying the arbitrary function $f(x)$ could allow us to specify the precise nature of the saturation that occurs for extreme values of $x(t)$. However, this would represent an added complication to already complex problems. In his initial works, Kane has presented arguments to show that the growth rates near birth ($x = 0$) are generally faster than growth rates near maturation ($x = 1$), and actual sigmoid curves are often more like those generated from the $x \ln x$ term in the rate modification expression than they are like those in the $x(1 - x)$ rate modification expression. Obviously there will be processes for which a term other than the $x \ln x$ term will be more appropriate. However, we will use this term in all of our following developments concerning KSIM.

In Eq. (4.4.15), we see that the expression $x_i \ln x_i$ is convenient to force a birth–maturation-type sigmoid effect. Here, the a_{ij} represent the impact of x_j upon the rate variable dx_i/dt, and the b_{ij} represent the impact of changes in x_j, dx_j/dt, upon the rate variable dx_i/dt. In a similar way we might seek to incorporate a term $c_{ij}d^2x_j/dt^2$ into Eq. (4.4.15). However, this would convert the set of first-order equations into a set of second-order equations, which would greatly increase solution complexity as well as specification complexity. In fact, Kane reports that many groups have difficulty understanding and estimating the b_{ij} term. Thus there would appear to be little gained by incorporating this extra term. In fact we will often be able to set $b_{ij} = 0$ and obtain reasonable models from Eq. (4.4.15). We should note that there is no fundamental reason why the impact variables a_{ij} and b_{ij} cannot be functions of time, but this added complexity would be very difficult to deal with.

Rather than attempting to solve Eq. (4.4.15) as a continuous time differential equation, it is more convenient to obtain a difference equation that can easily be processed on a digital computer. You may rewrite this equation as

$$\frac{dx_i}{x_i} = -\sum_{j=1}^{N}\left(a_{ij}x_j + b_{ij}\frac{dx_j}{dt}\right)\ln x_i\, dt$$

integrate from t to $t + \Delta t$, and then regard as constant all the functions of x_j on the right-hand side of this equation such that you obtain

$$\ln x_i(t + \Delta t) - \ln x_i(t) = -\sum_{j=1}^{N}\left[a_{ij}x_j(t) + b_{ij}\frac{dx_j}{dt}\right]\ln x_i(t)\,\Delta t$$

This may be rewritten as

$$x_i(t + \Delta t) = [x_i(t)]^{q_i(t)} \qquad (4.4.16)$$

where

$$q_i(t) = 1 - \Delta t \sum_{j=1}^{N}\left[a_{ij} + \frac{b_{ij}}{x_j(t)}\frac{dx_j(t)}{dt}\right]x_j(t) \qquad (4.4.17)$$

A valid approximate solution to the original differential equation (4.4.15) may be obtained by solving the difference equations (4.4.16) and (4.4.17). We may regard $q_i(t)$ as 1 minus the sum of the impacts on x_i. As was the case in the previous example, this form of $q_i(t)$ may lead to computational inaccuracy for the positive or enhancing impacts on x_i, for which the signs of the a_{ij} and b_{ij} are positive. To avoid this potential problem, we use the same procedure that was used in the previous example. We rewrite the $q_i(t)$ term as

$$q_i(t) = \frac{1 + \Delta t(\text{magnitude of sum of inhibiting impacts on } x_j)}{1 + \Delta t(\text{magnitude of sum of enhancing impacts on } x_j)}$$

or

$$q_i(t) = \frac{1 + \dfrac{1}{2}\Delta t \sum_{j=1}^{N}[|I_{ij}(t)| - I_{ij}(t)]x_j(t)}{1 + \dfrac{1}{2}\Delta t \sum_{j=1}^{N}[|I_{ij}(t)| + I_{ij}(t)]x_j(t)} \qquad (4.4.18)$$

where $I_{ij}(t)$ is the total impact of state variable j on the ith rate variable,

$$I_{ij}(t) = a_{ij} + \frac{b_{ij}}{x_j(t)} \frac{dx_j(t)}{dt} \qquad (4.4.19)$$

We recommend use of Eqs. (4.4.16) through (4.4.19) to conduct a KSIM exercise. Appropriate initial conditions for Eq. (4.4.16) must be selected in order to start the computation. We suggest the following procedure for establishing a KSIM model:

1. Identify fundamental problem elements. The approaches described in our previous chapter may be used to formulate the effort.
2. Determine appropriate scaling such that each state variable can reasonably be expected to vary between 0 and 1.
3. Determine appropriate initial conditions for each state variable.
4. Determine the cross-impact relationships. A group unfamiliar with quantitative techniques and differential equations may well wish to begin this by assigning interaction impacts to a matrix with numbers chosen to represent zero, low, moderate, or intense interaction of an enhancing or inhibiting nature.
5. Determine the time response by computer simulation of Eqs. (4.4.16) through (4.4.19).
6. Iterate through steps 2 to 4 until the group accepts the model response as appropriate. By so doing, information supplied by the group is made explicit, and structural information is enhanced with numerical information.
7. Apply different proposed activities, or policies and policy interventions, to the model. You may accomplish this in a convenient manner by using step functions for the a_{jj} and b_{jj} terms to switch in different policies as a function of time.
8. Verify and validate the results of using the model.
9. Use the numerical results from the model as analysis inputs for the interpretation step of the particular systems engineering phase under consideration.

Example 4.4.9. Let us consider a deterministic dynamic simulation based upon cross-impact of potential growth and decay of passenger railroad service. We assume that the essential elements for the problem are as follows:

S = a qualitative variable called quality of service which will include numerous factors: train comfort, frequency of travel, adherence to schedules, courtesy of employees, etc.

I = a qualitative variable called management innovation

T = a quantitative variable called traveled passenger miles per month

It would appear that the cross-impact matrix

if	S	I	T	then
	0	+	− −	S
	+	+	+	I
	+	+ +	−	T

is valid for the **A** matrix. We shall assume $b_{ij} = 0$. The following reasoning is used to construct this cross-impact matrix:

$a_{11} = 0$ because improvement of quality of service has little impact upon improvement of quality of service

$a_{12} = +$ because increasing management innovation is enhancing to quality of service

$a_{13} = - -$ because volume of passengers is inhibiting to quality of service

$a_{21} = 0$ because quality of service is essentially unrelated to management innovation

$a_{22} = +$ because management innovation will enhance further management innovation

$a_{23} = - -$ because increased passenger volume will inhibit management innovation. This is the battered management hypothesis.

$a_{31} = +$ because quality of service enhances passenger volume

$a_{32} = + +$ because increased management innovation will enhance passenger volume

$a_{33} = -$ because increased passenger volume will inhibit further increases in passenger volume

You may easily write the explicit differential equations for the KSIM model as

$$\frac{dS}{dt} = (a_{12}I + a_{13}T)S \ln S$$

$$\frac{dI}{dt} = (a_{21}I + a_{23}T)I \ln I$$

$$\frac{dT}{dt} = (a_{31}S + a_{32}I + a_{33}T)T \ln T$$

You now have the task of choosing specific numerical quantities for the several values of a_{ij} and the initial conditions for S, I, and T.

You could attempt to solve these differential equations directly. Alternatively, you could use the discretization scheme discussed earlier and obtain, from

Eq. (4.4.16), the approximate difference equations

$$S(t + \Delta t) = [S(t)]^{q_1(t)}$$

$$I(t + \Delta t) = [I(t)]^{q_2(t)}$$

$$T(t + \Delta t) = [T(t)]^{q_3(t)}$$

The $q_j(t)$ terms may be obtained from Eq. (4.4.17). Because some of the impacts are enhancing and some are inhibiting, it is somewhat better from a numerical accuracy viewpoint to use Eqs. (4.4.18) and (4.4.19). While these six equations may appear more complicated than the original differential equations, they are in a form suitable for direct iteration, whereas a numerical algorithm, such as Runge–Kutta or Taylor series, will have to be used to obtain computable equations from the differential equations.

If you run several simulations with differing a_{ij} parameters and initial conditions, you should be able to test various hypotheses, including the following:

1. The *public is doomed* hypothesis, in which management deliberately downgrades service to drive off passengers. You may accomplish this by changing a_{12} from $+$ to $-$, so that the nature of the management innovation is designed to reduce the quality of service to passengers.

2. The *battered management* hypothesis, in which management is the scapegoat of government and other policy. The sign of a_{23} is therefore $-$, because increased passenger volume results in government policy to impose restrictions that decrease management innovation.

3. The *American dream car* hypothesis, which assumes that the public's love for the automobile drives it away from the railroad. Improvements in service and innovation do not act to increase passenger volume. To obtain this, you set $a_{31} = a_{32} = 0$.

Figures 4.41 and 4.42 illustrate simulations using this model for the specific parameters and initial conditions indicated in the figures. The first curves in Figure 4.41 might be regarded as an idealized nominal set of curves in which a_{13} and a_{33} are set equal to 0. In this idealized case, increased passenger volume will not directly impact or affect either quality of service or further passenger volume. The parameter a_{23} is initially positive $(+0.25)$ to indicate that increased passenger volume enhances management innovation. As seen in Figure 4.41, quality of service (S), management innovation (I), and passenger volume (T) all increase under the nominal idealized condition. Figure 4.42 represents the "battered management" hypothesis, in which a_{23} is changed to a negative value $(-0.25,$ whereas it was $+0.25$ in Figure 4.41) to simulate imposition of policy to decrease management innovation as passenger volume increases. The initial conditions used in Figure 4.41 are those of Figure 4.40 at

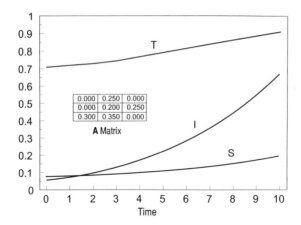

Figure 4.41 Idealized nominal conditions.

time $t = 7.5$. The results are dramatic in that quality of service (S) and management innovation (I) decrease significantly over time.

There have been a number of extensions of the basic KSIM modeling approach, and a relatively comprehensive summary of these may be found in reference 78. This approach is particularly well-suited to group modeling in workshops and, for this reason, is sometimes called a "workshop dynamic model." It is one of the few dynamic modeling approaches that lead to construction of a working model within a rather short time period. This fact suggests caution in that much time and effort are often needed in order to model a system adequately.

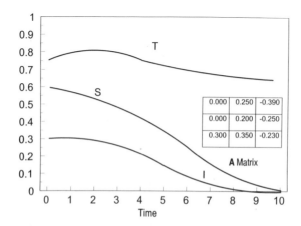

Figure 4.42 Battered management theory.

4.4.4 Summary

In this section we have examined three approaches that are potentially useful for the analysis of dynamic systems. The first approach is based on population models. In using this approach, you attempt to model a system based on the precise microlevel elements, or "physics," of the system you are considering. In the second approach, called system dynamics, you look at the macrostructure of a system. The third approach is based on even greater simplifying assumptions in that the form of the differential equation used to model the system is fixed. As we discussed, the KSIM approach is related to the cross-impact approach. The first step in construction of either a cross-impact or a KSIM model is specification of relevant events. This process can be enhanced using various formulation methods, such as those discussed in our last chapter. The most valuable inputs to a cross-impact or KSIM model are identification of the relevant state variables. These determine the structure of the system for either approach. Parameters are associated with this structure, either probabilistic parameters as in the cross-impact methodology or deterministic parameters as in the KSIM methodology. Joint probabilities for the occurrence of future events are determined in the case of cross-impact analysis. Deterministic responses are determined for the KSIM model output indicating levels of use, acceptability, and other pertinent factors.

Several simple simulations have been described in this section. These simulations were chosen for their usefulness in explaining significant features of the approaches rather than because they were indicative of the complex nature of actual problems for which solution has been attempted using them. We mentioned some of the software available to support modeling efforts. There is a variety of general-purpose simulation software. Matlab is one such general purpose package that contains a simulation application module that is called SimuLink. Matlab is available from The MathWorks Inc., 24 Prime Park Way, Natik, MA 01760-1500, telephone 508-647-7000.

4.5 ECONOMIC MODELS AND ECONOMIC SYSTEMS ANALYSIS

The accurate prediction of the cost, including effort and schedule, needed to accomplish activities associated with various life cycles in systems engineering efforts is a major need. In a similar manner, it is necessary to know the effectiveness that is likely to result from given expenditures of effort. When organizations underestimate the costs of activities to be undertaken, systems engineering programs may and typically do encounter large-cost overruns. This may lead to any of several possible embarrassments for the systems engineering organization, and perhaps even for the sponsor of or client for the effort. For example, cost overruns may lead to delivery delays and user dissatisfaction. When costs are overestimated, there may be much reluctance on the part of the sponsoring organization or its potential customers to undertake systems engineering activities that could provide many beneficial results. Thus, cost is an important ingredient in risk management, as we will mention briefly in our next chapter. In this section we will examine a number of issues that involve costing and cost estimation, as well as associated effectiveness estimation.

It is very important for you to note that a cost estimate is desired in order to predict or forecast the actual cost of developing a product or of delivering a service. Thus prediction and forecasting is the need and not after-the-fact accounting of costs. There are many variables that will affect cost:

1. The product or service scope, size, structure, and complexity will obviously affect the costs to produce it.

2. The newness of the product or service to those responsible for its production, and to the system development community in general, will be another factor.

3. The stability of the requirements for the product will be an issue because high requirements volatility over time may lead to continual changes in the specifications that the product must satisfy throughout an entire systems acquisition life cycle.

The integration and maintenance efforts that will be needed near the end of the product acquisition or fielding life cycle are surely factors that influence costs. There are a number of operating environment factors, such as the machine configuration and operating system on which a software system must run, that influence costs. The research and development costs to deliver an emerging technology capable of being incorporated into an operational system is another factor. Something about each of these factors must necessarily be known in order to obtain a cost estimate. The more detailed and specific the *definition* of the product is, the more accurate we should expect to be relative to the estimated costs for *development* and *deployment* of the product.

One guideline we might espouse is to delay cost estimation until as late in the product or service definition phase as possible, and perhaps even postpone it until some development efforts and their costs have become known. If we wait until after deployment, we should have an error-free estimate of the costs incurred to produce a product or system. This approach is hardly feasible, because estimates need to be known early in the definition phase in order to determine whether it is realistic to undertake detailed production. So, you must generally estimate costs and effectiveness of a system very early in the life cycle; this is not at all easy to do, especially when there are new and emerging technologies involved or when the requirements for the system to be developed are volatile.

We also see that there is merit in attempting to decompose an acquisition or production effort into a number of distinct components. A natural model for this is the life-cycle phases for production. We might, for example, consider the three- or seven-phase development life cycle, or any of the life-cycle models described in Chapter 2, and attempt to develop a work breakdown structure (WBS) of the activities to be accomplished at each of these phases. The work elements for each of these phases can be obtained from a description of the effort at each phase.

Figure 4.43 illustrates a hypothetical WBS for a software development life cycle. We could attempt to base our estimates of cost upon the detailed activities represented by this WBS, and a realistic WBS would be much more detailed than this, or we could attempt to develop a macrolevel model of development cost that is influenced by a number of primary and secondary drivers. The unstated assumption in the development of Figure 4.43 is that we are going to undertake a *grand design*-type approach for system production. This figure and the associated discussion need some modification for other then the grand design type approach, such as to enable us to consider iterative development or incremental development [32].

The broad goals of cost–benefit analysis (CBA) are to provide procedures for the estimation and evaluation of the costs and benefits associated with alternative courses of action. In many cases, it is not possible to obtain a completely economic evaluation of the benefits of proposed alternatives. In such cases, the word benefit is replaced by the term effectiveness, and a multiattribute effectiveness evaluation is used. We may use cost–benefit analysis (CBA) or cost–effectiveness analysis (CEA) to help choose among potential new projects or to evaluate existing systems for various purposes, such as identifying potential system modification needs. Oftentimes, the words benefit and effectiveness are used interchangeably, sometimes the distinctions made here are associated with use of the terms.

There are several steps in a cost–benefit analysis which correspond to steps in the systems engineering process. The iterative nature of these steps are very similar to the expanded logical steps of systems engineering. Figure 4.44 shows

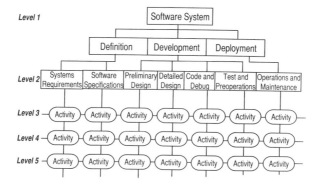

Figure 4.43 Breakdown structure for a grand design waterfall software development life cycle.

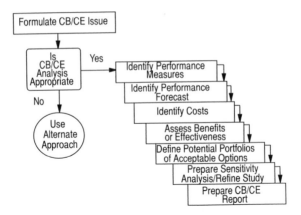

Figure 4.44 Typical cost–benefit and cost–effectiveness assessment process.

this relationship. It represents the steps involved in a typical cost–effectiveness assessment.

Let us now describe some of the basic techniques and principles needed to accomplish these steps. One of the efforts needed in a cost–effectiveness assessment is that of referring all costs to some particular point in time, such that it then becomes possible to compare different cost streams. We will briefly examine this subject here. Much more detail is available in any of the many excellent engineering economic and economic analysis texts that discuss this subject [79, 80], as well as those that take a more general approach [81]. Then we will return to our discussions of work breakdown structure and cost–effectiveness analysis.

4.5.1 Present Value Analysis

Present value analysis is a technique for evaluating the economic merit of a potential project. For example, suppose the purchase by a firm of a $60,000 machine would generate annual returns of $24,000 for 4 years. The firm estimates that an annual percentage return of 18% is required to justify the investment, given its risks. In present-value analysis, the firm calculates what amount would need to be invested at 18% to yield the four-year income flow of $24,000. A fairly simple calculation indicates that the figure is $64,560. This is the amount that is, at the present time, able to generate the income stream produced from the machine. Because this amount is more than the $60,000 cost of the machine, the investment is therefore worthwhile from the perspective of this simple analysis. Most economic models of costs and benefits use the basic concepts underpinning present value analysis. Therefore, let us review some of these basic algebraic concepts.

When we deposit an amount P_0 into a savings account at the beginning of year 0 and the annual constant interest rate is i, then we earn iP_0 in interest in one year, and we have available at the end of the first year the principal plus interest, or $F_1 = P_0 + iP_0(1 + i)$, where F_1 is the future value that we will have at the beginning of year 1. If this amount is invested for the next year, we then have an amount $F_2 = P_1(1 + i) = F_1(1 + i) = P_0(1 + i)^2$ at the end of the second year. At the end of N years we will have accumulated a grand total amount of

$$F_N = P_0(1 + i)^N \qquad (4.5.1)$$

Stated in present value terms, that is to say the present value of a future amount at year n, which we denote by $P_{0,n}$, this may be written as

$$P_{0,n} = A_n(1 + i)^{-n} \qquad (4.5.2)$$

where A_n is the amount of money invested at different points in time and the constant interest rate over the time interval is i. We simply add these amounts for all n under consideration and then obtain for the present value of a series of investments

$$P_0 = \sum_{n=0}^{N} P_{0,n} = \sum_{n=0}^{N} A_n(1 + i)^{-n} \qquad (4.5.3)$$

Of course there is no need for the interest to remain constant from year to year, just as there is no need for the amount invested to remain constant. Should the interest rate from year k to year $(k + 1)$ be i_k, we then have

$$P_{0,n} = A_n \prod_{k=0}^{n-1} (1 + i_k)^{-1}$$

$$P_0 = \sum_{n=0}^{N} P_{0,n} \qquad (4.5.4)$$

This expression represents the present worth of the several amounts A_n invested for n years, where the interest rate varies from year to year. Figure 4.45 illustrates three typical cash flow situations. First, it shows the equation you might use to calculate the future value of investing a payment for several time periods at a constant interest rate. It also shows the equation for calculating the present value of a series of future uniform amounts with a constant interest rate over the time interval. Finally, it depicts how to calculate the future worth of several uniform amounts at the end of a number of time periods, again assuming a constant interest rate.

In some cases the annual amounts of a benefit or a cost are constant over time. Then, if the discount rate is constant over the period under consideration, relatively simple closed-form expressions for present worth result as shown in Figure 4.44. Many ways can be used to derive these expressions. One simple

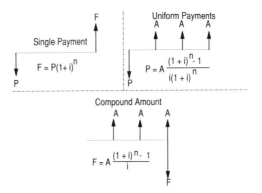

Figure 4.45   Simple net present and future value calculations.

way is to note that the present worth, at time 0 of an amount 1 that is initially invested at annual interest rate i beginning at year 0, can be obtained from the future worth equations

$$(1 + i)^{-1} + (1 + i)^{-2} + \cdots + (1 + i)^{-n} + \cdots = i^{-1}$$

This follows in a very simple manner from the identity

$$\frac{a}{1 - a} = a + a^2 + a^3 + \cdots$$

which will converge for $a < 1$. Here, we let $a = (1 + i) - 1$. The present value at time N of an amount 1 invested annually from time $N + 1$ is also $i - 1$ and the present value at time 0 of this amount is $i - 1(1 + i) - n$. We subtract this from the expression $i - 1$ and obtain the expression for the present worth, at time 0, of an amount 1 invested annually for N years.

We find that the present discounted worth P_0 of a constant amount A invested at each of N periods from $n = 1$ to N is given by

$$P = A[(1 + i)^N - 1]i^{-1}(1 + i)^{-N} \qquad (4.5.5)$$

We may easily calculate the future worth, at the Nth period, of annual payments from this relation as the expression

$$F = P(1 + i)^N = A[(1 + i)^N - 1]i^{-1} \qquad (4.5.6)$$

where you will notice that in Eqs. (4.5.5) and (4.5.6) we have dropped the subscripts on P and F for convenience.

Relations such as these are especially useful for calculating the amount of annuity payments that an initial principal investment will purchase or for calculating constant-amount mortgage payments. We must, however, remember that other factors such as inflation and depreciation, or appreciation, need to be considered when evaluating the present worth of an alternative. For the most part these other factors can be considered just as if they were interest. For example, the worth at year $n + 1$ of an investment at year n of P_n that is subject to interest i, inflation r, and depreciation d is given by the expression

$$P_{n+1,n} = P_{n,n}(1 + i_n)(1 - r_n)(1 - d_n) \qquad (4.5.7)$$

where i_n is the interest in year n, r_n is the inflation in year n, and d_n is the depreciation in year n. If we assume that this is augmented by an amount $A_n + 1$ at the beginning of year $n + 1$, then the worth of the investment at the start of the next period is

$$P_{n+1,n+1} = P_{n+1,n} + A_{n+1}$$

or

$$P_{n+1,n+1} = P_{n,n}(1 + i_n)(1 - r_n)(1 - d_n) + A_{n+1}$$

We can solve this difference equation with arbitrary i_n, r_n, d_n, and A_{n+1} to yield the future worth of any given investment path and annual investment over time.

In the simplest case, the initial investment $P_{n,n}$ is zero and the annual investment and rates are constant. Then it is a simple matter for you to show that the result of making an investment of amount A over N years is given by the future monetary amount at year N:

$$F = P_{N,N} = [(1 + I)^N - 1]AI^{-1}$$

where I is the effective interest rate which is given by

$$I = i - r - d + rd - ir - id - ird$$

If the percentage rates are all quite small, little error results from using the approximation obtained by dropping the products of small terms. We then can express the effective interest rate as

$$I \approx i - r - d$$

Hence we use an effective interest rate that is simply the true interest rate less inflation and depreciation. Actual calculation using these relations is quite simple conceptually, but can become tedious. Use of a spreadsheet is recommended.

4.5.2   Economic Appraisal Methods for Benefits and Costs Over Time

Several methods can be used to determine the economic value of alternative projects or systems. One is present-value analysis, as described above. Another method, payback period calculation, does not consider the time value of money at all. It simply determines the time required from the start of a project until total revenues flowing from the project start to exceed the total cost of the project. As can be easily shown, the payback period is a rather naive criterion to use in evaluating alternatives.

Another common method of project evaluation is called the internal rate of return (IRR). This approach determines the interest rate implied by anticipated returns on an investment, and then it compares that rate with some predetermined standard of measure. It is that interest rate that will result in economic benefits from the project being equal to economic costs, assuming that all cash flows can be invested at the internal rate of return. The assumption that it is possible to invest cash flows from the project at the internal rate of return will often be impossible to fulfill. Thus the number that results from the IRR calculation may give an unfortunate impression of the actual return on investment. It tends to favor short-term projects that yield benefits quickly, as contrasted with projects that yield long-term benefits only slowly. The reinvestment at the constant internal rate of return assumption is inherent in the IRR equation.

In some cases, the IRR is simply the inverse relation of net-present-value analysis (NPV). NPV starts with an assumed interest rate, solves for the present value of the income stream, and compares this figure with up-front cost. IRR starts with an estimated income flow, solves for the interest rate this flow implies, and compares this interest rate with a predetermined standard of comparison. Assuming the forecasts are done honestly, the calculations are done accurately, and that monetary returns received during the period that the investment is active can be reinvested at the IRR, which is to be calculated over the remaining life of the project, the two approaches represent equally legitimate tools of analysis. The choice between them should depend on the peculiar circumstances of the firm and the investment it is considering and whether this reinvestment at the IRR assumption is acceptable.

The IRR may be calculated fairly easily. If we assume that a project, of duration N years, has benefits B_n and costs C_n in year n, then the present value of the benefits and costs are, assuming that the interest rate is constant,

$$PB_0 = \sum_{n=0}^{N} B_n(1 + i)^{-n}$$

and

$$PC_0 = \sum_{n=0}^{N} C_n(1 + i)^{-n}$$

The net present value of the project is given by

$$NPV = PB_0 - PC_0 = \sum_{n=0}^{N} (B_n - C_n)(1 + i)^{-n}$$

To obtain the internal rate of return, we set the net present worth equal to zero and solve the resulting Nth-order algebraic equation for the IRR. These equations lead naturally to a consideration of the benefit–cost ratio (BCR), which is the ratio of benefits to cost and is given by

$$BCR = \frac{PB_0}{PC_0} = \frac{\sum\limits_{n=0}^{N} B_n(1 + i)^{-n}}{\sum\limits_{n=0}^{N} C_n(1 + i)^{-n}}$$

The return on investment (ROI) is the ratio of the net present value to the net present costs and is therefore given by

$$ROI = \frac{PB_0 - PC_0}{PC_0} = BCR - 1$$

Sometimes the ROI criterion is called the net benefit–cost ratio (NCBR), the net present value, or net present-worth–cost ratio.

It is important to contrast and compare these concepts to determine those most appropriate in particular circumstances. We first consider the selection of a single project or alternative from several. An immediate problem that arises is that NPV and IRR may lead to conflicting results. The BCR criterion may result in a different ranking also. Unless costs are constrained, there is no real reason to use a BCR criterion, however.

Two projects, A and B, may have the NPV-versus-interest rate curves shown in Figure 4.46. Here we see that if the actual interest rate i is less than I, the rate at which the NPV of the two investments are the same, we prefer

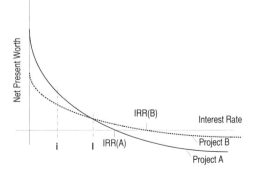

Figure 4.46   Illustration of potential difficulties with IRR as a criterion.

investment A over investment B if we use the NPW criterion. If the interest rate i is greater than I, we prefer investment B. However, if the actual interest rate is greater than the IRR for each investment, we would prefer not to invest at all, if this is possible, because the NPW of each investment is negative. The IRR of project A is IRRA and that of project B is IRRB. Because IRRB is greater than IRRA, we would prefer project B to project A by the simplest IRR criterion. This simple example illustrates some of the potential difficulties with careless use of IRR as an investment criterion.

There are several essential properties that any rational investment rule or criterion should satisfy. All cash flows should be considered. The cash flows should be discounted at the opportunity cost of capital — that is, the interest rate that can be obtained on an investment or some specified discount rate. The rule should select, from a set of mutually exclusive projects, the one that maximizes benefit or value as appropriately defined. Also, we should be able to valuate one project independently of all others. This requires that the investment criterion be complete. By complete, we require that we can always say that x is not preferred to y, or that y is not preferred to x on the basis of any two sequences of cash flows x and y. We also require that the criterion be transitive. With three cash flow vectors x, y, and z, if x is preferred to y, which we write as $x \rightarrow y$ and if $y \rightarrow z$, then transitivity requires that $x \rightarrow z$.

There are five desirable properties that should be possessed by any reasonable preference ordering of cash flow vectors.

1. *Continuity.* If $x \rightarrow y$, then for sufficiently small ε, cash flow vector x is also preferred to y, or $x - \varepsilon \rightarrow y$. This continuity property is very reasonable in that it assures us that the preference criterion and associated preferences are not schizophrenic for small arbitrary changes in the return of the investments.

2. *Dominance.* If cash flow x is at least as large as y at every period in the overall investment interval $(0, N)$ and is strictly greater in at least one period, then investment x is preferred to y, or $x \rightarrow y$. This dominance property is equivalent to greed, because more is always preferred to less, or to customer satiation.

3. *Time Value of Money.* This suggests that if two investments x and y are identical except that an incremental cash flow obtained by investment x at some period n does not result until period $n + 1$ for investment y, then investment x is preferred to investment y. The time value of money property is equivalent to impatience. We prefer to have money now than this same amount of money at some time in the future.

4. *Consistency at the Margin.* We prefer cash flow x to y if and only if the differential cash flow $x - y$ is preferred to a cash flow of 0. This property is called consistency at the margin. It is relatively easy to see the reasonableness of this criterion from the satiation of the consumer property.

5. *Consistency Over Time.* This simply says that the preference order over investments is unchanged if we shift all investment returns being considered by exactly the same number of periods. In making this shift, we must be careful that we do not lose any investments. The consistency over time property simply states that if $x = [x_0, x_1, \ldots, x_{N-1}, 0]^T$ and $y = [y_0, y_1, \ldots, y_{N-1}, 0]^T$ and if $x \to y$, then we must have $x' \to y'$, where $x' = [0, x_0, x_1, \ldots, x_{N-1}]^T$ and $y' = [0, y_0, y_1, \ldots, v_{N-1}]^T$. The consistency over time requirement is based on the interest or discount rate being constant throughout the investment time interval 0 to N.

If we accept these five requirements as reasonable, then it turns out that we can only accept the NPV criterion, or one that would yield the same preference ordering as the NPV criterion, as truly reasonable. It is possible to show that only preferences that satisfy these requirements are those given by the net present value criterion in which interest or discount rates, which may vary from period to period, are positive. Surely, it would be difficult to argue that any of these requirements are unreasonable. Although this discussion and the five requirements, which could be posed as axioms, appear quite abstract, it is just this sort of formulation that leads to normative or axiomatic theories.

Based on these requirements and the conclusion that follows from them, we should value investment x according to a net present value criterion, or decision rule, in which we obtain the present value of the investment component at the nth period x_n by use of the standard discounting relation

$$NPV(x_n) = x_n \prod_{j=1}^{n} (1 + i_{j-1})^{-1}$$

and then add together all of these present values over all of the interest bearing periods to obtain

$$NPV(x) = \sum_{n=0}^{N} NPV(x_n)$$

or

$$NPV(x) = \sum_{n=0}^{N} x_n \prod_{j=1}^{n} (1 + i_{j-1})^{-1}$$

where $ij - 1$ is the interest rate that is valid in the time interval from the period $j - 1$ to the period j.

Despite this argument, there are instances where it is justifiable to use the IRR criterion. In practice, the major potential problem with it is that it assumes that interest payouts during the time of the project can be reinvested at the internal rate of return. Often, this is not the case at all.

4.5.3 Systematic Measurements of Effort and Schedule

At first glance, it might appear that efforts to minimize cost, effort, and time required to produce a product or service are exclusively focused on the desire to become a low-cost producer of a potentially mediocre product. While this may well be the case, it is not at all necessary that this be the focus of efforts in this direction. An organization desirous of high product differentiation, and perhaps such other features as rapid production cycle time, is necessarily concerned with knowing the cost and time required to produce a product or service. In reality, it should wish to maximize the benefit or effectiveness of a product or service for a given cost (and effort and time), or it should minimize the costs to produce the product or service of a given fixed benefit or effectiveness. Alternately, it might try to maximize the difference between benefits and costs. These are sometimes, but not always, equivalent statements.

Concerns with organizational profits naturally turn to the desire to maximize the benefit–cost ratio for a product. This allows an organization to cope with competitive pressures through the development of new and improved products with simultaneous control on costs and schedule. You can then produce a superior product in a minimal amount of time that can be marketed at a lower price than might otherwise be the case. An exclusive focus to cost and schedule reduction can lead to poor results if other efforts are not also included. These include attention to product and process quality and defects, as well as attention to these effectiveness issues.

In this section we will comment on approaches for the minimization of effort, or cost, and schedule. These process-related approaches are closely related to approaches to maximize quality and minimize defects. In order to minimize effort, costs, and schedules necessary to produce a product or service, you need to know a good bit about the activities associated with production of a product or service. We interpret these activities in terms of effort, and then in terms of the associated costs and schedule. We might imagine that we could project the effort required for each of the life-cycle phases associated with a specific system engineering effort, perhaps as a fraction of total required effort within the steps associated with each of these phases, such as illustrated in Figure 4.47.

Several pragmatic difficulties emerge. A three-phase life cycle is hardly sufficient to describe a realistic systems engineering effort in sufficient detail for detailed costing. Often, you will need to consider more than three steps within each phase. There are interactions across the various life-cycle phases. Thus, process dynamics need to be considered. An increase in the effort associated with definition may well lead to a lowering of the effort required for system development. An increase in the development costs might be associated with a reduced effort required at subsequent maintenance. Thus, an increase in development costs might be associated with a reduction in the costs associated with deployment. The choice of a particular life cycle is a very important matter also. Some acquisition programs, or projects, may well be more

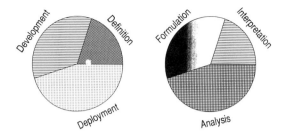

Figure 4.47 Hypothetical distribution of effort across systems engineering life-cycle phases and steps within these phases.

appropriate for a series of evolutionary or incremental builds than they are for a grand design life cycle, for example. The use of prototyping, in any of several forms, will influence the systems engineering effort considerably.

There are several approaches that we might use to estimate costs of a systems engineering project, or of an organizational activity within an overall project or program:

1. *Analogy* is an approach in which we identify a program with similar activities that has occurred in the past and use, perhaps with appropriate perturbations, the actual activity costs for that program.

2. A *bottom-up approach* to estimation would be based upon a detailed analysis of the activities or work involved and the subsequent decomposition of these into a number of tasks, the cost for which is then estimated.

3. An *expert opinion approach* would be based upon soliciting the wholistic judgment of one or mode individuals experientially familiar with the program under consideration, or closely associated programs.

4. A *parametric model approach* would be comprised of identification of a structural model for the program cost. Parameters within this structure would be adjusted to match the specific developments to be undertaken.

5. A *top down*, or *design to cost approach* would be based upon beginning with a fixed cost for the program or set of activities under consideration. This cost would be allocated to various activities and phases of development.

6. A *price to win approach* to costing is based not upon what it actually may cost to complete a set of activities, but upon a price that is believed as the maximum price that will win a proposed competition for a contract.

These approaches are not mutually exclusive and may be used in combination with one another. Each is reasonable in particular circumstances. Each may be more appropriate for some phases of the life cycle than are other approaches. The more analytically based approaches, for example, would generally be more useful for development than they would be for definition, for which analogy and expert opinion might be quite appropriate. There are a variety of other names used to denote these approaches. There are also a number of issues relating to support of the direct costs of labor, the accounting of these, and differences between direct costs and indirect costs. These are discussed in detail in more advanced efforts concerning systems management. Here, we provide only an overview of this very important area.

To estimate the cost of a set of activities, you need to know a number of direct cost rates for the individuals involved and a variety of nondirect costs. There are a number of approaches that you may use to estimate effort or cost rates. These include:

- Direct labor rate
- Labor overhead rate, or burden
- General and administrative rate
- Inflation rate
- Profit

These are the direct costs associated with labor. Costs may also be associated with materials, parts, and supplies, including subsystems incorporated into a product but which are procured elsewhere. A total cost equation will contain each of these elements.

The amount in the total cost equation that does not depend upon the number of products produced, or customers served in the case of a service, is the fixed cost, and the cost that does vary is called the variable cost. The total revenue, R, that we are going to obtain for sale of a product is just

$$R = PQ$$

where P is the price of each product and C is the number of products sold. The total profit, TP, is then the difference between the total revenue and the production cost. We can obtain a greater profit by increasing the price of the product, P, or by increasing the number of products sold, Q. Of course, there must be a consumer demand function for the product that suggests that the demand for the product will decrease as the price for them increases. There are many texts that discuss microeconomic analysis [81], including supply demand relations for issues such as this.

We may be constrained to sell at a price that is determined by the market situation that results when there are a large number of other sellers who are selling the product at a given price. If there is no special product differentiation associated with our product, then any competitive advantage [82] depends upon being a low-cost producer and selling at, or perhaps somewhat below, the market price. Suppose here that the market price for the product is $1200 and that we have the total cost and total revenue relations depicted in

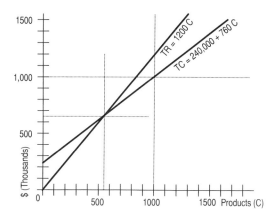

Figure 4.48 Total revenue and total cost for a simple single product firm.

Figure 4.48. If we can truly market 1000 products, our profit is $200,000 on the basis of the $1,000,000 in total costs and $1,200,000 in total revenue. As sales decline, we continue to make a profit until the sales volume drops to 545 units. Below this level of production and sales, the profit is negative. At a production of 1000 units, the return on investment (ROI) is $\text{ROI} = (\text{TR} - \text{TC})/\text{TC} = 0.20$, or 20%. Often, you will need to obtain curves such as these for realistic situations in efforts that involve technical direction and systems management.

There are a number of ways in which the fixed costs and variable costs may be defined in terms of how the total costs change as a function of production levels, or activities, associated with the cost allocation base. The usual way you would generally think of this is in terms of the sort of relationship given by the foregoing total cost relationship. It is possible, however, to attempt to define indirect overhead in terms of the direct labor costs only. Then you would need to consider an indirect general administration charge and apply it to the direct labor and materials costs. The fringe benefits associated with the direct labor would be included in the indirect overhead. We will not deal with these issues here and only point them out because they are present and do need to be considered in realistic situations. The results of considering various approaches to overhead illustrate that overhead may be determined in different ways and that the concept of cost needs to be very carefully considered and explained in accordance with these ways. The answer to the question, "What is the cost?" depends very much on the judgment and decision problem being considered and how that decision issue has been framed.

Among the many types of costs that can be defined are the following:

• Fixed costs and variable costs
• Direct costs and indirect costs
• Functional and nonfunctional costs
• Recurring and nonrecurring costs

There are many others such as incremental and marginal costs. So, while there is an answer to the cost question, it is perhaps better to say that there may be many answers, and each of them may be correct. The extent to which a given approach to costing, including the resulting cost, has value depends entirely on the purpose for which the information obtained is to be used.

There are a number of approaches to cost as pricing strategies. These include full-cost pricing, investment pricing, and promotional pricing. Another approach to costs is determining the cost required to achieve functional worth — that is, to fulfill all functional requirements that have been established for the system. While this is easily stated, it is not so easily measured. A major difficulty is that there are essential and primary functions that a system must fulfill, as well as ancillary and secondary functions that, while desirable, are not absolutely necessary for proper system functioning.

After the functional worth of a system has been established in terms of operational effectiveness, it is necessary to estimate the costs of bringing a system to operational readiness. If this cost estimate is to be useful, it must be made before a system has been produced. It is helpful to think of three different costs:

1. Could cost — the lowest reasonable cost estimate to bring all the essential functional features of a system to an operational condition
2. Should cost — the most likely cost to bring a system into a condition of operational readiness
3. Would cost — the highest cost estimated that might have to be paid for the operational system if significant difficulties and risks eventuate

Each of these types of costs — minimum reasonable, expected, and maximum reasonable — should be estimated because this provides a valuable estimate not only of the anticipated program costs but also of the amount of divergence from this cost that might possibly occur.

Quite obviously, it is very difficult to estimate each of these costs. The "should" cost estimate is the most likely cost that results from meeting all essential functional requirements in a timely manner. "Could cost" is the cost that will result if no potential risks materialize and all nonfunctional value adding costs are avoided. "Would cost" is the cost that will result if risks of functional operationalization materialize.

4.5.4 Work Breakdown Structure and Cost Breakdown Structure

As we have often noted, there are three fundamental phases in a systems engineering life cycle:

- System definition
- System development
- System deployment

These phases may be used as the basis for a work breakdown structure (WBS) or a cost breakdown structure (CBS), for depicting cost element structures. A work breakdown structure is mandated for proposals and contracts for work with the federal government in efforts that involve either, or both, of the first two phases of the systems life cycle. Military Standard STD 881A governs this. In general, we would expect that the total acquisition cost for a system would represent the initial investment cost to the customer for the system. This would be the aggregate cost of designing, developing, manufacturing, or otherwise producing the system, along with the costs of the support items that are necessary for it to be initially deployed. The MIl-STD 881A does not cover extended deployment efforts, but the WBS approach could easily be extended to cover these. The extension is simple from a conceptual perspective. Actually doing it in practice may call for information that is difficult to obtain.

To initiate a WBS, the initial system concept should be displayed as a number of component development issues. Figure 4.49 illustrates a hypothetical structure for an aircraft systems engineering acquisition effort. The particular subsystem, or system, level for which a WBS is to be determined is selected. Three levels are defined in the DoD literature.

1. Level 1 represents the entire program scope of work for the system to be delivered to the customer. This level deals with overall system or program management. Program authorization occurs at this level.
2. Level 2 is associated with various projects and associated activities that must be completed to satisfy the program requirements. Program budgets are generally prepared at the project levels.
3. Level 3 represents those activities, functions, and subsystem components that are directly subordinate to the level 2 projects. Program schedules are usually prepared at this level. It is at this level that various detailed systems engineering efforts are described.

We may also identify a fourth level as the level for the various life cycle

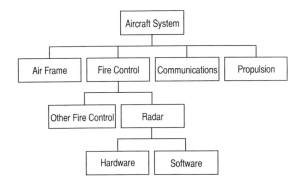

Figure 4.49 Work breakdown structure of subsystem elements in an aircraft system.

phases associated with the overall acquisition effort. It is at this level that detailed WBS estimates are obtained.

There are many components that comprise a WBS or CBS for a systems engineering life cycle. There are a number of related questions that, when answered, provide the basis for reliable cost estimation for the work breakdown structure. One DoD publication describes these in terms of organizational questions, planning and budgeting questions, accounting questions, analysis questions, and program and project revision questions. There are a number of questions that can be posed, and the responses to these provide valuable input for estimating WBS and costs. Details of WBS- and CBS-type approaches may be found in references 83–87.

In many cases a WBS is provided as a part of a request for a proposal (RFP), and proposers are expected to provide this detailed costing information as a portion of their contract bidding information. It then becomes a part of the statement of work (SOW) for the contractor selected for the program effort. A completed WBS displays the complete system acquisition effort and the costs for major activities associated with the effort in terms of schedule and costs, provides for subsequent performance measurement, and assists in the identification of risk issues.

4.5.5   Cost–Benefit and Cost–Effectiveness Analysis

Cost–benefit and cost–effectiveness analysis are methods used by systems engineers and others to aid decision makers in the interpretation and comparison of proposed alternative plans or projects. They are based on the premise that a choice should be made by comparing the benefits or effectiveness of each alternative with the monetary costs of full implementation of the alternative. Benefit is an economic term that is generally understood to be a monetary unit. Effectiveness is a multiattribute term used when the consequences of project implementation are not reduced to dollar terms. Often, the terms benefit and effectiveness are used as if they were synonyms. We will describe an approach for CBA and then indicate modifications to adapt it to COEA.

First, objectives for the project must be identified, and alternatives must be generated and defined carefully. Then the costs as well as the benefits, or effectiveness facets or attributes, of proposed projects are identified. These costs and benefits are next quantified, and they are expressed in common economic units whenever this is possible. Discounting is used to compare costs and/or benefits at different times. Present worth is the usual discount criterion of choice. Overall performance measures, such as the total costs and benefits, are computed for each alternative. In addition to this quantitative analysis, an account is made of qualitative impacts, such as social aesthetic and environmental effects. Equity considerations regarding the distribution of costs and benefits across various societal groups may be considered. The cost–benefit analysis method is based on the principle that a proposed economic condition is superior to the present state if the total benefits of a proposed project exceed the total costs.

Among the results of a cost–benefit or cost–effectiveness analysis are the following:

1. tables containing a detailed explanation of the costs and benefits, over time, of each alternative project, along with present value of costs or benefits of each alternative project.
2. Computation and comparison of overall performance measures, in terms of benefits and costs, for each of the alternative projects. Alternatively, effectiveness and costs may be used if a cost–effectiveness analysis is desired.
3. An accounting of intangible and secondary costs, such as social or environmental or even aesthetic, and the set of multiattributed benefits or effectiveness value that is associated with each alternative project.

The following major steps are generally carried out in a cost–benefit analysis.

1. Issue Formulation
   1.1 Problem definition
   1.2. Value system design
   1.3. System synthesis

These efforts are accomplished using techniques specifically suited to issue formulation, such as identification of objectives to be achieved by projects, some bounding of the issue in terms of constraints and alterables, and generation of alternative projects. The results of this formulation of the issue include a number of clearly defined alternatives, the time horizon for and scope of the study, a list of affected individuals or groups, and perhaps some general knowledge of the impacts of each alternative.

2. Issue Analysis
   2.1. Systems analysis — Identification of costs (negative impacts) and benefits (positive impacts) of effectiveness measures for each alternative is accomplished here. A list is made of the costs and benefits for each project. Measures for different types of costs and benefits are specified, and, if possible, conversion factors are developed to express different types of costs or benefits in the same economic units. For example, one of the benefits of a proposed highway project might be reduced travel time between two cities. In order to make this comparable to monetary costs, we have to determine how many dollars per time unit are gained by the reduction of the travel time. Determination of such conversion factors can be a sensitive issue, because the worth of various attributes can be dramatically different

for different stakeholders. For example, consider the difficulties involved in transforming additional safety benefits of a proposed project, measured in human lives saved, into monetary benefit units. Further complicating economic benefit evaluation are equity considerations, which may require the costs and benefits of the project to be allocated in different amounts to different groups.

2.2. Refinement of the alternatives is accomplished through more detailed quantitative analysis of costs and benefits. Costs and benefits are expressed in common economic units in so far as possible. Comparisons of projects may also be made with respect to two different quantified units, such as dollars (of cost) and human lives (for benefits). Economic discounting is used to convert costs and benefits at various times to values at the same time, and net value analysis is then used to compare benefits and costs. The impacts that cannot easily be quantified are assessed for each alternative project. These usually include intangible and secondary or indirect effects, such as social and environmental impacts, legal considerations, safety, aesthetic aspects, and equity considerations.

3. Interpretation
   3.1. Decision making is accomplished through selection of a preferred project or alternative course of action. There are a variety of criteria that can be used. We may select the project that maximizes benefits for a given cost. We may select the project that has minimum costs for a given level of benefits. We may maximize the cost–benefit ratio. We may maximize the net benefits, or benefit minus cost. Which is the most appropriate depends upon the number of alternatives being considered.

   3.2. Planning for and communication of results is the last effort in a CBA. This often takes the form of a report on both the quantitative and the qualitative parts of the study. The report may include a ranking or prioritization of alternative projects, or a recommended course of action. It is important that all assumptions made in the study are clearly stated in the report. The report should be especially clear with respect to costs and benefits that have been included in the study, costs and benefits that have been excluded from the study, approaches used to quantify costs and benefits, the discount rates that have been used, and relevant constraints and assumptions used to bound the CBA.

Cost–effectiveness analysis is accomplished through use of a very similar approach. Methods such as decision analysis and multiattribute utility theory can be used to evaluate effectiveness. We are able to use the resulting effectiveness indices in conjunction with cost analysis to assist in making the tradeoffs between quantitative and qualitative attributes of the alternative projects. There are a number of studies available addressing this important subject, and references 88–96 are particularly recommended.

4.5.6 Summary

In this section we have examined a number of issues surrounding the economic systems analysis and costing of products and services. We also made some comments concerning valuation, effectiveness, and other important issues. Modeling and estimation of costs, effort, and schedule are relatively new endeavors. Often, these models are validated from databases of cost factors that reflect characteristics of a particular systems engineering organization. Such models are useful when developer organizations are stable and continue to produce systems similar to those developed in the past for users who have similarly stable environments and needs. However, models have little utility if the mixture of personnel experience or expertise changes or if the development organization attempts to develop a new type of system or a user organization specifies a new system type with which the systems engineering organization is very unfamiliar.

It is vitally important to develop cost (including effort and schedule) estimation models that can be used very early in the systems engineering life-cycle process. The purpose of these models are to predict product or service life-cycle costs and efforts as soon as possible, as well as to provide information about the costs and effectiveness of various approaches to the life cycle and its management. An interesting study of cost estimation [97] provides nine guidelines for cost estimation.

1. Assign the initial cost estimation task to the final system developers.
2. Delay finalizing the initial cost estimates until the end of a thorough study of the conceptual system design.
3. Anticipate and control user changes to the system functionality and purpose.
4. Carefully monitor the progress of the project under development.
5. Evaluate progress on the project under development through use of independent auditors.
6. Use cost estimates to evaluate project personnel on their performance.
7. Management should carefully study and appraise cost estimates.
8. Rely on documented facts, standards, and simple arithmetic formulas rather than guessing, intuition, personal memory, and complex formulas.
9. Do not rely on cost estimation software for an accurate estimate.

While these guidelines were established specifically for information system software, there is every reason to believe that they have more general applicability. This study identifies the use of cost estimates for selecting projects for implementation, staffing projects, controlling and monitoring project implementations, scheduling projects, auditing project progress and success, and evaluating project developers and estimators. Associated with this, we suggest, is evaluating the costs of quality, or the benefits of quality, and the costs of poor quality.

4.6 RELIABILITY, AVAILABILITY, MAINTAINABILITY, AND SUPPORTABILITY MODELS

As anyone who has driven a car, traveled by plane, or worked with a computer can appreciate, people worry about system failures. Failures cost time and money and can even be dangerous. From a life-cycle viewpoint, much of the cost of any system depends on how frequently failures occur, how likely they are to occur, how difficult and expensive it is to prevent or repair failures, and how many resources must be dedicated to keep the system working. Reliability, availability, maintainability, and supportability (RAMS) are quantitative concepts that you can use to model and analyze these concerns.

The reliability of a system, from a mathematical perspective, may be defined as the probability that the system will perform in a satisfactory manner for a specified period of time when used under prescribed operating conditions. From an individual component perspective, reliability is the probability that a component will not fail during the specified time period of operation. A system "fails" when it violates one or more of its technical specifications. We can think of two different types of failure:

- Catastrophic failure, which refers to a failure that is sudden, complete, and generally irreversible

- Drift failure, or degradation failure, which refers to a failure associated with the gradual change in the system response over time, usually due to component drift or degradation, such that it no longer meets specifications after some point in time

Each of these is important, of course. Reliability, for most components and systems, deteriorates with time and is very dependent on the operational environment. Extremes in hot or cold weather, exposure to sand, dust, or water, damage from shock or vibration can all, over time, adversely impact reliability both before and during operation. Failures can be sudden and catastrophic, or gradual, in their appearance. There are many alternative words used for failure or fault. Among these are error, defect, miss, slip, and bug. Sometimes these terms are used interchangeably. Sometimes they are used to imply different types of failings. There are human reliability concerns as well as machine reliability concerns.

Blanchard and Fabrycky [98] note that an acceptable definition of reliability is based on four important ingredients:

- Probability
- Satisfactory performance
- Time
- Prescribed operating conditions or technical specifications

as contained in the system technical requirements specifications part of system definition. An understanding of probability concepts is basic to reliability modeling. Satisfactory performance is determined on the basis of requirements for the system, as is time, and prescribed operating conditions. Again, we see the importance of system requirements and formulation efforts in systems engineering.

As you know, probability is a quantitative term that represents the long-term chance of an event happening in a number of trials. For example, in basketball, coaches rate players on their ability to successfully make free-throw shots using probability basics. Coaches simply count the number of successful shots out of the total number of shots attempted to compute the free-throw percentage or probability of making free-throw shots. A 90% shooter, will, over many attempts, average nine successful shots for every ten attempted. The relative frequency of occurrence of successful free throws is 0.9 or 90%. This same idea is used in the reliability definition. If a component of a system is 0.8 or 80% reliable for at least 100 hours under standard operating conditions, you know that, over the course of many trial runs, identical components will operate successfully for at least 100 hours eight times for every ten attempts. You also know that these components fail an average of two times before 100 hours for every ten attempts. Still, you don't know when a specific component will fail. You only know that, ideally, components will operate successfully more often than they will fail. Every time you put one of these components into operation, there is still a 20% chance that it will be unreliable or fail sometime before the first 100 hours of operation. Hence, reliability is a quantitative term that depends on describing failures using probability concepts.

As you can imagine, a very important parameter in reliability models is the failure rate. The failure rate is the fraction of components failing per unit time. Formally, the failure rate, denoted by lambda, is defined by

$$\lambda = \frac{f}{T} \qquad (4.6.1)$$

where f is the number of system, or component, failures, in some specified operating time, T.

But how is the failure rate determined? Gathering field data from maintenance and service records can help determine the failure rate. Another way is to conduct tests before fielding in order to calculate the failure rate. You can expect that only a sample from some total number of identical components will be tested because it would be wasteful to wear out more items than necessary. Also, you can expect to observe that failures occur at different points of time during operation. Hence, you will usually have to approximate the true probability distribution of failures by using a standard known distribution that fits well with the sample data. The more observed failure data that are available, the more accurately you can select the type of distribution that best represents how the population of identical components performs. The Poisson, binomial, normal, exponential, and other probability density functions are often used depending upon the theoretical basis for one of these being appropriate failure and reliability density functions.

Consider the situation shown in Figure 4.50. The distribution of failures of a system as a function of time is modeled by a probability density function, $f(t)$. Initially, there is a large failure rate and this is corrected through replacing initially failed system components. This is sometimes called the period of "infant mortality," and failures here are usually those associated with poor quality. Following this period, the system enters a "useful life" period where the failure rate is essentially constant. This is associated with an exponential failure rate probability density function. This exponential is a reasonable density function in that it assumes that the rate of change of the failure rate is proportional to the failure rate, or

$$\frac{df(t)}{dt} = -\lambda f(t), \qquad f(0) = \lambda$$

such that we obtain

$$f(t) = \lambda e^{-\lambda t} \tag{4.6.2}$$

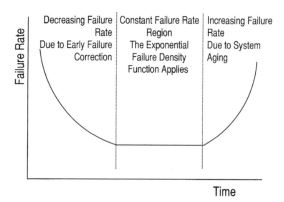

Figure 4.50   Traditionally assumed "bathtub" time-dependency of failure rate.

as the failure density function. The parameter λ, in the above relation is usually called the failure rate or, sometimes, hazard rate. Failures during this useful life period are generally random and are due to sudden catastrophic events, such as electrical failures or storms. This exponential failure distribution is generally valid for systems, such as modern VLSI systems, that do not show significant deterioration over time. The probability of such a system failing in a small time interval is independent of whether the system has been in operation for a long or a short period of time. We can define the inverse of the failure rate as the mean life, or as more commonly called the mean time between failures (MTBF). In equation form, we have

$$\theta = \frac{1}{\lambda} = \text{MTBF} \qquad (4.6.3)$$

The event–probability relations for failure and reliability, at time t, are given by

$$F(t) = \int_0^t f(t)\, dt = 1 - e^{-\lambda t} \qquad (4.6.4)$$

$$R(t) = \int_t^\infty f(t)\, dt = e^{-\lambda t} \qquad (4.6.5)$$

$$R(t) + F(t) = 1 \qquad (4.6.6)$$

If a system has a constant failure rate, $\lambda(t) = \lambda$, then the reliability of that system at its mean life is, with $t = \theta = 1/\lambda$, given by $R(t = 1/\lambda) = e^{-1} = 0.37$. Often, the failure rate per hour, or mean time to failure (MTTF), is used and we may therefore define, as we also indicated earlier,

$$\lambda = \frac{\text{number of failures}}{\text{total operating hours}} = \text{MTTF}$$

Clearly, we could also express failure rate or MTTF in terms of the number of failures per thousand operating hours.

By the definition of a probability density function, the entire area under the curve is equal to unity when integrated over all time. Now, suppose that you examine some time interval of interest called mission time, or operating time, from $t = 0$ to $t = T$. The probability that a failure will occur in time T equals the ratio of the area under the failure rate curve between 0 and T to the total area under the curve. Because the total area under the curve is, by definition, equal to 1, the area under the curve from $t = 0$ to $t = T$ is the probability that the system will fail in time T. This probability is the failure function and is usually called unreliability or $F(t)$. Because the total area under the curve equals 1, you can easily calculate reliability, $R(t)$, as $1 - F(t)$, as we have indicated.

Using the reliability function, $R(t)$, you can derive another very important function for reliability analysis, the hazard function $H(t)$. The hazard function is the instantaneous failure rate of the system and is usually denoted as $\lambda(t)$. You can derive an expression for $\lambda(t)$ rather easily. Consider the situation of N identical components that comprise a system and whose failures over time are represented by some probability density function, $f(t)$. When a component fails, it is repaired and the system is again functional. If N_t is the number of components operating at time t, then the number of times the system fails in a time increment is $N_t - N_{t+\Delta t}$. The fraction of components failing in the time increment of interest is the number of components that failed during the time period divided by the total number of operating components at the start of the time periods or $(N_t - N_{t+\Delta t})/N_t$. To calculate the fraction of components failing per unit time, the average failure rate, you just divide by the time period, Δt, to yield $(N_t - N_{t+\Delta t})/N_t \Delta t$. But you want this in reliability terms so divide the numerator and denominator by N, the total number of components, to get the expression $(N_t/N - N_{t+\Delta t}/N)/(N_t/N)\,\Delta t$. Now, we recall that we can approximate the differential reliability of a set of identical components, which is the probability that any one of the components will not fail by some time t, by dividing the number of operational components after some time t by the total number of components that were set into operation. From the last expression, this means that $R(t) = N_t/N$; $R(t + \Delta t) = N_{t+\Delta t}/N$. From these last two expressions, we see that we can now rewrite the average fraction of components failing per unit time as $-[R(t + \Delta t) - R(t)]/R(t)\,\Delta t$. Because you want an expression for the exact or instantaneous failure rate, you can find this by taking the limit of this average failure rate as the increment of time approaches zero. Thus, you obtain

$$\lambda(t) = \lim \Delta t \to 0 \; \frac{R(t + \Delta t) - R(t)}{R(t)\,\Delta t}$$

You recognize the definition of a derivative in the expression above, and this allows you to write

$$\lambda(t) = H(t) = -\frac{1}{R(t)}\frac{dR(t)}{dt} \qquad (4.6.7)$$

From this expression, you can easily show that the hazard or failure rate is constant for the exponential failure density function. Because you know that

$$R(t) = \int_t^{\infty} f(t)\,dt = 1 - \int_0^t f(t)\,dt$$

you can solve for the first derivative to find that

$$\frac{dR(t)}{dt} = -f(t) \qquad (4.6.8)$$

Therefore, by substitution, you now have a very useful expression for the instantaneous failure rate in terms of any probability density function and its associated reliability function as follows:

$$H(t) = \lambda(t) = \frac{f(t)}{R(t)} \qquad (4.6.9)$$

This instantaneous failure rate is a measure of the fractional number of system failures per unit time. You can express the failure rate in a number of different units depending on the components or system you are analyzing. For example, you can use failures per component-hour for a mechanical system, failures per cycle of operation, errors per line item of code for software, errors per page of documentation, or errors per job instruction.

Another important parameter in reliability modeling is the mean time to failure, or MTTF. It is the reciprocal of the mean time between failure, MTBF. The MTTF is simply the expected value of t for the probability density function $f(t)$. Using the definition of expectation, you can write an expression for MTTF:

$$\text{MTTF} = \text{MTBF}^{-1} = E[t] = \int_0^\infty t f(t)\, dt \qquad (4.6.10)$$

Example 4.6.1. In your professional effort as a systems engineer, you manage an electronics assembly plant that uses a large number of a particular component. A new supply source can potentially provide these components at a reduced price. The new source advertises extensive quality control and initial testing that they claim has a low constant failure rate of 0.00067 failures per component hour with an MTBF of 1500 hours. To compare the new source with your current supplier, you ask your plant operators to collect reliability data on a sample of 100 of the current components in use for 1000 hours of operation. You want to answer the following two questions relative to the components you are currently having tested:

1. What is their MTBF?
2. What reliability do they exhibit for a mission duration of 70 hours?

As your plant operators are collecting test data, you begin your analysis. Because the manufacturer of the current components also accomplishes debugging before delivery to your plant, you decide to assume that the current components will fail randomly in time due to chance at some constant failure rate that can be approximated by the exponential distribution. Using the formulas developed for reliability, failure rate, and mean time between failure, you derive the following results. You calculate the reliability from

$$R(t) = \int_t^\infty f(t)\, dt = \int_t^\infty \lambda e^{-\lambda t}\, dt = [-e^{-\lambda t}]_t^\infty = e^{-\lambda t}$$

Where you assume an exponential distribution

$$f(t) = \lambda e^{-\lambda t}$$

For the failure rate, you obtain

$$\lambda(t) = \frac{f(t)}{R(t)} = \frac{\lambda e^{-\lambda t}}{e^{-\lambda t}} = \lambda$$

And for the mean time to failure, you obtain

$$\text{MTBF} = E(t) = \int_0^\infty tf(t) = \int_0^\infty t\lambda e^{-\lambda t}\,dt = \left[-\frac{1}{\lambda}e^{-\lambda t}\right]_0^\infty = \frac{1}{\lambda}$$

Solution a. Using these results and the data collected from the reliability test, you create the spreadsheet shown in Figure 4.51. The data shown in Figure 4.51 was collected at discrete time intervals of 100 hours. The entire previous 100-hour period is considered as part of the successful operation time for any newly failed component. As depicted in Figure 4.51, the MTBF for the currently used components is only 1000 hours, which is less than that advertised by the new supply source (MTBF = 1500 hours). Therefore you decide to change to the new supplier to improve reliability and also to reduce costs.

Solution b. Because the failures are exponentially distributed, you can determine the reliability for a mission duration of 70 hours where you assume a constant failure rate of 0.0010 failures per component hour. This result is given by the expression $R = e^{-\lambda t} = e^{-(0.0010)(70)} = e^{-0.07} = 0.932$.

Time of Check (Hours)	Number of New Failures	Number of Working Components	Hours of Operation of Failed Components	Hours of Operation of Working Components	Total Component Hours of Operation	Total Number of Failures	Estimated Failures Per Component Hour	MTBF (hours)
0	0	100	0	0	0	0		
100	10	90	1000	9000	10000	10	0.00100	1000.00
200	9	81	1800	16200	18000	19	0.00106	947.37
300	6	75	1800	22500	24300	25	0.00103	972.00
400	7	68	2800	27200	30000	32	0.00107	937.50
500	5	63	2500	31500	34000	37	0.00109	918.92
600	3	60	1800	36000	37800	40	0.00106	945.00
700	2	58	1400	40600	42000	42	0.00100	1000.00
800	4	54	3200	43200	46400	46	0.00099	1008.70
900	2	52	1800	46800	48600	48	0.00099	1012.50
1000	4	48	4000	48000	52000	52	0.00100	1000.00

Figure 4.51   Reliability data and spreadsheet calculations for exponential failures.

For the new supplier, the components have a constant failure rate of 0.00067 failures per component hour so that for the same mission that lasts 70 hours, we have the expression

$$R = e^{-\lambda t} = e^{-(0.00067)(70)} = e^{-0.0469} = 0.954$$

Therefore there is a 95.4% chance that any individual component selected will survive a mission duration of 70 hours. From this result, you reach the important conclusion that useful life reliability is independent of component age so long as wearout does not begin. This means that if the component lasts throughout the 70-hour mission, it will also have a 95.4% probability of surviving the next 70-hour mission. It turns out that this same reliability result is also constant for any subsequent 70-hour period of operation as long as the failure rate remains exponentially distributed.

When a component enters wearout, its failure rate is no longer constant as in the case for useful life. Figure 4.50 has illustrated a characteristic failure rate curve that depicts the typical life cycle for a component that consists of three regions: debugging, a useful life period of constant failure rate due to chance only, and a wearout region of increasing failure rate due to chance and wearout. Also indicated are the appropriate failure rate distributions for calculating reliability. Because a component in wearout usually has an increasing failure rate, reliability in this region can be modeled by using the normal distribution. For any given normal distribution with a mean and standard deviation, you can calculate all the typical reliability formulas for any mission duration that includes wearout considerations.

Continuing with our example of components with a chance failure rate of 0.001 failures per component hour, we make further tests of the current components and find that the mean of their wearout distribution is 1000 hours with a standard deviation of 200 hours and ask the following questions.

1. What is the reliability of the component for 100 hours of continuous operation?
2. What is the reliability of the component for 900 hours of continuous operation?
3. What is the reliability of the component for a mission duration of 70-hours given that it has been operating continuously for 300 hours?

Answers to these questions can be quickly found using a graphical depiction of these three missions or time periods of interest. Figure 4.52 and the density function obtained for the case where wearout failures follow a normal distribution given by

$$f(t) = \frac{1}{\sigma\sqrt{2\pi}} \exp\left[-\frac{1}{2}\left(\frac{T-M}{\sigma}\right)^2 \right]$$

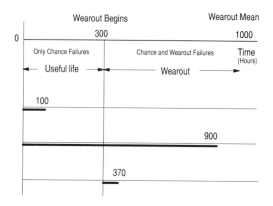

Figure 4.52   Graph of reliability mission time lines.

and this provide a basis for the calculation. We observe that the mean and standard deviation can be calculated using the time of each component failure, T_i, and the number of components, N, in the sample, if the failure data follow the normal distribution. These results are given by

$$M = \frac{\sum_{i=1}^{N} T_i}{N}$$

$$\sigma = \sqrt{\frac{\sum_{i=1}^{N} (T_i - M)^2}{N - 1}}$$

We proceed in the following manner. We first note that the standard normal variable Z, which gives the position of T as a number of standard deviations to the left or right of the mean, can be calculated as follows: $Z = (T - M)/\sigma$. This value is used to enter a table of values for the cumulative normal distribution. Using Z for any time T and a cumulative normal distribution table, the area representing the value of unreliability, Q, up to any time T can be found. Then, the reliability is obtained from

$$R = 1 - Q$$

Another convenient way to find the value of reliability for any time T is to use a spreadsheet's normal distribution function with appropriate specified parameters. The cell equation for reliability would be

$$R = 1 - \text{NORMDIST}(T, M, S, \text{TRUE})$$

where T = time, M = mean, S = standard deviation, and TRUE is a parameter that returns a cumulative distribution function.

To determine the time when wearout begins given a mean and standard deviation, the shape of the normal curve allows you to use

$$T_w = M - 3.5\sigma$$

Now that we have obtained information about wearout failures following a normal distribution, we can calculate answers to questions 3, 4, and 5 as follows.

Solution c. We may obtain the time when the components first enter wearout as

$$T_w = M - 3.5\sigma = 1000 - 3.5(200) = 300 \text{ hours}$$

Therefore, for a mission of 100 continuous hours of operation, none of the mission is during the wearout period. Hence we can just assume a constant failure rate and use the exponential distribution to find the reliability for this 100 hours as

$$R(t) = e^{-\lambda t} = e^{-0.001(100)} = 0.9048$$

We note that this reliability is less than what we calculated for a mission duration of 70-hours. This result is reasonable because, in the useful life region where failures are only due to chance and a constant failure rate applies, reliability depends only on mission duration.

Solution d. We now calculate the reliability for a continuous 900 hours of operation from time $t = 0$ until time $t = 900$. We note that 600 hours of this mission is under wearout from beginning of wearout at $t = 300$ hours until end of mission at $t = 900$ hours. Therefore we have to consider the reliability due to chance failures during useful life and the reliability due to wearout and chance in the wearout region so the total reliability is just the product of the two failure modes:

$$R_{total} = R_{chance} \times R_{wearout}$$
$$R_{chance}(900) = e^{-\lambda(900)} = e^{-(0.001)(900)} = e^{-0.9} = 0.04065$$

To continue our computations, we find the reliability due to chance for the entire 900 hours of operation. Then we find the reliability due to wearout for 600 hours from $T_w = 300$ until $t = 900$.

$$R_{wearout}(600) = 1 - \text{NORMDIST}(600, 1000, 200, \text{TRUE}) = 1 - 0.3085 = 0.6915$$

Finally, we calculate the total reliability considering both chance and wearout reliability. This result is

$$R_{total} = R_c \times R_w = 0.4065 \times 0.6915 = 0.2810$$

Solution e. Now, given that a component has been in operation for 300 hours, we want to find its reliability for the next 70 hours. Using conditional probability because the component must be capable of surviving to the beginning of the time period of interest and because the entire mission or period of interest is within wearout, we use the following formula to calculate total reliability:

$$R_{total} = R_c \times \frac{R_w(T + t)}{R_w(T)}$$

where T is the age of the component at beginning of the period of interest or mission and t is the length of the period of interest or mission time. R_c is the reliability associated with chance failures and R_w is the reliability associated with wearout failures. In this case, we compute the total reliability as

$$R_{total} = e^{-(0.001)(70)} \times \frac{R_w(300 + 70)}{R_w(300)}$$

$$R_{total} = 0.9320 \times \frac{(1 - \text{NORMDIST}(370,1000,200,\text{TRUE}))}{(1 - \text{NORMDIST}(300,1000,200,\text{TRUE}))} = 0.9320 \left(\frac{0.9992}{0.9998}\right)$$

$$= 0.9264$$

This result is quite reasonable because 70 hours is a relatively short time of operation given the failure rate of 0.001 failures per component hour. Even though this entire mission takes place during wearout, the mission starts just as wearout begins, so that the component has only just barely begun to experience the increasing failure rate in the wearout region. We note that these example calculations are for a situation where we have components whose reliability data indicate an exponential distribution of chance failures and a normal distribution of wearout failures. The results will not necessarily apply to other situations.

Advanced Reliability Models. Computing the probabilities of catastrophic failure and drift failure is needed in order to fully develop a reliability model. The probability of drift failure depends very much on the type of system that is being considered. The approaches that are taken usually involve Monte Carlo methods or methods based on an assumed m-variate normal density for failure rates of individual components. The first approach is a simulation-based approach that requires simulation of many actual systems. The second approach is based on assumed theoretical failure rate densities. For large systems, catastrophic failures are usually more of a problem than drift failures because drift failures can generally be prevented through routine maintenance.

Catastrophic failures analysis is often approached through the use of block diagrams. In these, a system is disaggregated into a number of subsystems or components associated with particular functional relationships. For each block, we determine the possible functionality conditions that may exist. We may, through use of this approach, describe systems as comprised of series elements, parallel elements, or a combination of series and parallel elements.

Figure 4.53 illustrates several simple systems comprised of elements — or subsystems — in series, parallel, and series–parallel form. The system is assumed to undergo a (catastrophic) failure if any of the individual subsystems fail. For all cases, we assume that the failure probability for system j is independent of that for a different system k. We easily see that the overall reliability for the series connection case is given by the product of the individual reliabilities. We easily obtain this result when we recall that the failure probability cumulative distribution function, or unreliability probability, is just one minus the probability that there is no failure in any of the subsystems. Mathematically, we have

$$F(t) = 1 - \prod_{i=1}^{N} [1 - F_i(t)] = 1 - \prod_{i=1}^{N} R_i(t) = 1 - R(t) \qquad (4.6.11)$$

For the particular case of exponential failure probabilities and constant failure rates, we have

$$R(t) = 1 - e \sum_{i=1}^{N} \lambda_i t \qquad (4.6.12)$$

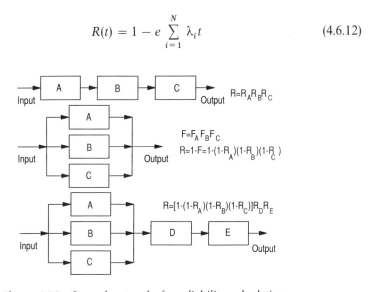

Figure 4.53 Several networks for reliability calculations.

For the parallel case, we actually can consider that all parallel elements are initially armed. The system fails if all subsystems fail. Thus, the overall system failure, or unreliability, distribution function for a system of N parallel elements is given by

$$F(t) = \prod_{i=1}^{N} F_i(t) = 1 - R(t) \tag{4.6.13}$$

where we again assume that the subsystems have independent failure probabilities. For the particular case where we have a constant and equal MTBF for each subsystem and an exponential failure density, we have

$$F(t) = [1 - e^{-\lambda t}]^N \tag{4.6.14}$$

Alternately, we consider a parallel standby system in which subsystem 2 is only switched in if system 1 fails, system 3 is only switched in if system 2 fails, and so forth. The overall system fails if the first system fails at time t_1 having been started at time 0, the second system fails at time t_2 having been started at time t_1, and so forth, and the last system failure at time t_N occurs at a time t. For the case where $N = 2$, we can show that the failure and reliability are given by $R(t) = (1 + \lambda t)e^{-\lambda t}$ and that this is always greater than the reliability of the redundant parallel structure. We can show that this is the case for all N, as is intuitively obvious because we only switch one system on after the lower-order system has failed. You will find it interesting to derive this result. You may show that the reliability of a set of N parallel switched systems, each with the same failure rate, where a single system is switched on only after the then operating system develops a fault, is

$$R(t) = e^{-\lambda t} \left[\sum_{i=0}^{N-1} \frac{(\lambda t)^i}{i!} \right] \tag{4.6.15}$$

Many other configurations can be established. For example, you could utilize a parallel redundant system where failure will occur if less than K components out of N are working or, equivalently, if more than $N - K$ of the N components are not working. We obtain

$$R(t) = \sum_{i=K}^{N} \binom{K}{i} [R(t)]^K [1 - R(t)]^{N-K} \tag{4.6.16}$$

as the overall reliability, $R(t)$, for this system of parallel elements with reliability $R(t)$.

Often, when a component or system fails, it is then subject to failure detection, diagnosis, and correction in order to restore the system to operational functionality. This leads us to consider systems that have a failure density or rate and a repair density. If we assume that these are constant in mean value and that the appropriate densities are exponential, then we can show that the event probabilities for the system being operational at time t and of being failed at time t are given by the expressions

$$A\ (t) = \frac{\mu + \lambda e^{-(\lambda + \mu)t}}{\lambda + \mu} \qquad (4.6.17)$$

and

$$F(t) = \frac{\lambda[1 - e^{-(\lambda + \mu)t}]}{\lambda + \mu} \qquad (4.6.18)$$

The term $A(t)$ is generally called the system availability, and it represents the probability that the system is operational or available at time t. This is really just the reliability of this composite system that is comprised of a single system and a single repair facility. We could obtain similar results for two subsystems and a single repair facility, two subsystems with two repair facilities, and so forth. Ultimately, things become rather involved, as you might well imagine. These issues form a part of the mathematical theory of operations research and have been explored for quite some time now. A relatively good source of reliability information is reference 99.

It turns out that there are at least three types of availability.

1. Inherent availability refers to the probability that a system, when used under prescribed operating conditions in an ideal environment, will operate satisfactorily according to specifications up to some point in time as required and without maintenance. Preventative or scheduled maintenance is *not* included. The inherent availability can be written as AI = MTBF(MTBF + MTTR) − 1, where MTBF is the mean time between failures and MTTR is the mean time to repair. The mean down time (MDT) is just the MTTR.

2. Achieved availability represents the probability that the system, when used under prescribed operating conditions in an ideal environment, will operate satisfactorily according to specifications up to some point in time. Preventative or scheduled maintenance is included in achieved availability. The achieved MDT is now the MTTR plus the planned preventative or scheduled maintenance. The availability can be written as AA = MTBM(MTBM + MAMT) − 1, where MTBM is the mean time between maintenance and MAMT is the mean active maintenance time.

3. Operational availability is the probability that a system, when used under prescribed operating conditions in an actual environment, will operate satisfactorily according to specifications up to some point in time when it is called upon. Preventative or scheduled maintenance is included. The achieved MDT is the MTTR plus the planned maintenance plus logistics time. This availability can be written as AO = MTBM(MTBM +

MMDT) − 1, where MTBM is the mean time between maintenance and MMDT is the mean maintenance down time.

4. It is often difficult to estimate the MTTR. A useful relation for the expected MTTR of N different systems is given by

$$E(\text{MTTR}) = \frac{\sum_{i=1}^{N} \text{MTTR}_i \lambda_i}{\sum_{i=1}^{N} \lambda_i} \qquad (4.6.19)$$

and we can compute this if we know the MTTR of each of the N units.

Reliability and availability are important needs that must be addressed in the definition, or requirements, phase of the systems engineering life cycle. Each of the three availability definitions is important. One of the outcomes of considering reliability during the requirements phase of the life cycle is specification of such important parameters as $R(t)$, $F(t)$, MTBF, MTTR, MTBM, MAMT, and MMDT in order to meet operational needs.

One of the major difficulties in performing reliability, availability, and maintainability (RAM) studies is that of obtaining reliable data for the models that may be constructed and in verifying the models themselves. This generally requires that we monitor probabilistic uncertainties associated with the various facets of the reliability, or RAM, model throughout the definition, development, and deployment phases of the system life cycle. Often reliability, or quality control specialists are selected as part of a systems engineering team for just this purpose. Part of an integrated logistics support (ILS) in the U.S. military concerns development of a system maintenance plan (SMP) as part of an overall systems engineering management plan (SEMP).

A failure mode and effect analysis (FMEA), sometimes called a failure mode, effect, and criticality analysis (FMECA), is often performed during the latter phases of the system definition phase in order to identify possible problems that could result from system failure. An FMEA will generally include the following:

1. Identification of each system or component that is likely to fail, or item identification
2. Description of most probable failure modes
3. Identification of possible failure effects
4. Identification of probability of occurrence of failures
5. Identification of criticality of failures
6. Diagnosis of failure causes
7. Identification of corrective actions and preventative measures

This follows the detection, diagnosis, and correction trilogy generally used for risk planning and error correction. A reliability, or RAM, planning effort generally integrates reliability, availability, and maintainability efforts with other operational requirements. Determination of maintenance–resource requirements is a need. This involves tradeoffs across reliability and maintainability. Our efforts in the next chapter are very useful in making these tradeoffs. The objectives of a RAM plan, which may be converted into a set of activities, are as follows:

1. To ensure compatibility between operational requirements and RAM requirements
2. To associate RAM requirements with appropriate organizational levels through use of the functional analysis results of the definition phase of the systems life cycle
3. To perform a failure mode and effect analysis (FMEA)
4. To estimate RAM needs for the various system elements
5. To enable detection, diagnosis, and correction of faults as they occur

In this section, we have discussed models for reliability, availability, and maintainability. There are a number of other important "ilities" such as usability not discussed here. References [33], [98], and [99] provide additional details concerning this important subject.

4.7 DISCRETE EVENT MODELS, NETWORKS, AND GRAPHS

There are many situations in the modeling and analysis of systems engineering efforts when it is desired to be able to represent the flow of materials, people, or ideas from one "state" to another. Flow is necessarily used in a very broad sense. Interest could be in the assignment or transportation of supplies, inventory, repair parts, or other physical quantities as needed to accomplish some objective. Alternately, concern might be with the flow of people, or telephone calls, in some waiting lines or queues. In a management decision situation, one might desire to know the structure of some decision situation, or the structure of an organizational plan that might enable accomplishment of some desired goal. In each of these situations, there is need for use of one or more of the many tools that fit under the general "network and graph" category.

There are several categories of methods that might be discussed and several taxonomies that might be used to partition these categories into more easily comprehended subunits. One particular taxonomy concerns whether or not there is probabilistic uncertainty associated with the particular state that will follow the present state. This is the taxonomy that will be used here. It allows us to consider two principal subtopics:

- Deterministic graphs, structured modeling, and network flows
- Stochastic graphs and network flows

This topic is not mutually exclusive and independent of other topics considered in our other chapters.

4.7.1 Network Flows

A good systems engineering analyst should have the capability of modeling systems using graph-theoretic concepts. As we saw in our earlier discussions in Section 4.3, there are four primitives or fundamental entities associated with a net or graph:

1. A set, P, of elements called points p
2. A set, R, of elements called directed lines, or more briefly, lines r
3. A function, f, whose domain is R and whose range is contained in P
4. A function, s, whose domain is R and whose range is contained in P

The set P contains the elements or points, p, which are to be structured, and the set R contains the elements or lines, r, which imply the structure. The functions f and s serve to identify the "first" and "second" points on each line.

Two axioms, which are very unrestrictive, are assumed when developing a theory of graphs.

1. It is always assumed that the set of points P is finite and nonempty.
2. Also it is assumed that the set of lines, R, is finite.

These axioms make it impossible to describe a network with no points; it is possible, however, to have a set of isolated points with no lines between them. There are a number of "relations" and important properties that a relation must possess. Among the important relations for graph-theoretic developments are: reflexive, irreflexive, symmetric, asymmetric, transitive, nontransitive, intransitive, and complete. In general, a relation is a network with no parallel lines.

Associated with these concepts, it is possible to develop the concept of a digraph. A digraph satisfies the same four primitives and two axioms as a network. Also, it is necessary to add two additional axioms for completeness.

1. No two distinct lines are parallel.
2. There are no loops.

Thus, a digraph is seen to be an irreflexive relation. While this may seem rather abstract, it is these assumptions that enable development of the important concept of a minimum edge digraph, a concept that greatly facilitates the analysis and interpretation of graphs.

Matrix representation of digraphs now becomes meaningful. In any digraph, consider two points, p_i and p_j; we can show that either p_i reaches, or does not reach, $p_j (p_i R p_j$ or $p_i \bar{R} p_j)$. The first relation will be true if there is a directed line from point i to point j, and the second will hold if there is not. This characteristic suggests a binary relation, and it is possible to represent a digraph by a binary matrix A with the set of points P serving as both the vertical and horizontal index sets. An entry in the matrix A is defined a $a_{ij} = 1$ if $p_i R p_j$ and $a_{ij} = 0$ if $p_i \bar{R} p_j$. This matrix A is typically called an adjacency matrix.

Example 4.6.1. A simple prototype of a minimum line adjacency matrix and the associated digraph is shown in Fig. 4.54. You may find it helpful to verify that the digraph corresponds to the adjacency matrix shown in the figure. You may also wish to use this for reference as we discuss some graph-theoretic concepts.

If there is a path from p_i to p_j, we say that p_j is reachable from p_i, or that p_i reaches p_j. The number of lines in the path p_i to p_j is called the length of the path. In Fig. 4.54, for example, p_4 is reachable from p_5 by a path of length 2. Reachability is a very intuitive concept, and a very useful one as well, if the relation being considered is transitive in the sense that $p_i R p_j$ and $p_j R p_k$ infer $p_i R p_k$.

The minimum-line or minimum-edge adjacency matrix is the matrix that describes reachability for all paths of length 1. If we add the identity matrix, I, to A, we obtain the matrix that describes reachability for all paths of length 0 and length 1. For the matrix in Fig. 4.54, we obtain

$$A = \begin{bmatrix} 1 & 1 & 0 & 0 & 0 \\ 0 & 1 & 1 & 0 & 0 \\ 0 & 0 & 1 & 1 & 0 \\ 0 & 0 & 0 & 1 & 0 \\ 0 & 0 & 1 & 0 & 1 \end{bmatrix}$$

It is easily shown that if we multiply the above matrix by itself, we obtain a matrix that describes reachability for paths of length 2 or less. This is

$$(A + I)^2 = \begin{bmatrix} 1 & 1 & 1 & 0 & 1 \\ 0 & 1 & 1 & 1 & 0 \\ 0 & 0 & 1 & 1 & 0 \\ 0 & 0 & 0 & 1 & 0 \\ 0 & 0 & 1 & 1 & 1 \end{bmatrix}$$

We continue to multiply this expression by $(A + I)$ until successive powers produce identical matrices. Then we have for sufficiently large r

$$(A + I)^{r-2} = (A + I)^{r-1} = (A + I)^r = P$$

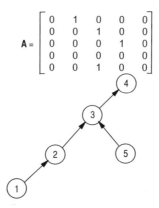

$$A = \begin{bmatrix} 0 & 1 & 0 & 0 & 0 \\ 0 & 0 & 1 & 0 & 0 \\ 0 & 0 & 0 & 1 & 0 \\ 0 & 0 & 0 & 0 & 0 \\ 0 & 0 & 1 & 0 & 0 \end{bmatrix}$$

Figure 4.54 Adjacency matrix and associated digraph.

where P is defined as the reachability matrix, or the transitive closure of A. Here, we see that

$$P = \begin{bmatrix} 1 & 1 & 1 & 1 & 0 \\ 0 & 1 & 1 & 1 & 0 \\ 0 & 0 & 1 & 1 & 0 \\ 0 & 0 & 0 & 1 & 0 \\ 0 & 0 & 1 & 1 & 1 \end{bmatrix}$$

It turns out that there is only one minimum-edge transitive adjacency matrix that corresponds to a given reachability matrix. There are, however, many intransitive adjacency matrices that will result in the same reachability matrix. It is for this reason that it is generally essential to consider transitive relations only when working with directed graphs, or digraphs. If a digraph is not transitive, many of the commonly used graph-theoretic and structural modeling techniques are no longer valid.

There are many uses where graph-theoretic notions are potentially useful. The process of structural modeling is one of the major results of graph theory that is especially useful in practice. The process of structural modeling generally proceeds as follows:

1. A meaningful contextual relation, with which to structure a set of elements, is selected.

2. A set of elements relevant to an issue is identified.

3. Data collected relevant to the existence or nonexistence of the relation between every pair of elements are explored [several options are available here: a strict binary relation with entries 0 and 1 may be used — in some

cases where there are potential transitivity difficulties with use of a strict binary relation, such as often arise when negatively transitive contextual relations are used, it is more reasonable to obtain a "signed digraph" by using the three level relations $+1$, 0, and -1, where 0 indicates no relation, $+1$ indicates an enhancing relation, and -1 indicates an inhibiting relation; the process of data collection may be computer-directed or may be interactively directed by the humans seeking to establish a structural model].

4. A reachability matrix of the resulting structural model is constructed using responses concerning relations between elements and transitive inference to fill in entries in the reachability matrix [this may be done in a computer-directed-and-controlled way after all questions concerning relatedness have been posed by the computer and answered, or software may be written such that the reachability matrix is constructed after each data entry or set of data entries].

5. The computer determines the minimum-edge adjacency matrix and displays the resulting structure for possible iteration and modification [this can be done in a totally computer-controlled-and-directed way or it can be accomplished in an iterative fashion after each data entry is obtained using one of the available algorithms for determining an adjacency matrix from a reachability matrix].

6. Often, it will be desirable to iterate and refine the structural model that is obtained such that there results a final structural model in the form of a digraph that can then be annotated with whatever signs and symbols are desired in order to convey the desired interpretation of the structural model.

There are various end uses to which the results of graph theory may be put. These include development of a variety of charts for project planning and scheduling, such as Gantt, DELTA, PERT, and CPM charts. It is also possible to obtain representations in the forms of trees or more general hierarchical networks. These can be useful in representing the structure of objectives, or the command structure of an organization, or the organization of a complex piece of equipment. It is also possible to represent various routing and transportation issues by a series of directed graphs. The resulting structural model of network routings may then serve as the basis for an optimization problem in which it is desired to extremize some cost function, such as the time required to go between two points in the network or the distance covered in servicing some given number of nodes. These result in zero–one integer linear programming problems [occasionally mixed integer linear programming problems] that can, in principal, be solved through a combination of the algorithms for graph theory noted here and an integer programming package. These are generally considered to be "transportation" problems or "network modeling" problems.

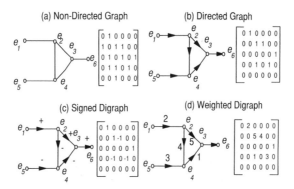

Figure 4.55 Minimum-edge adjacency matrix for several interaction matrix types.

One of the major uses to which graph-theoretic and structural modeling concepts may be put is that of obtaining interaction matrices. Interaction matrices provide a framework for identification and description of patterns of relations between sets of elements, such as needs, alterables, objectives, and constraints. Self-interaction matrices are used to explore and describe relations or interactions between elements of the same set. Cross-interaction matrices provide a framework for study and representation of relations between the elements of two different sets. Basically, a table is set up in which each entry represents a possible linkage between two elements. Figures 4.55 through 4.57 illustrate concepts useful in constructing interaction matrices. Each entry is

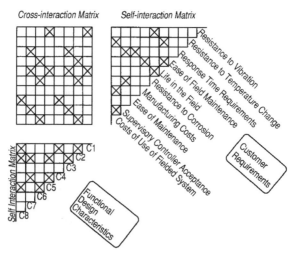

Figure 4.56 Sample self- and cross-interaction matrices for quality function deployment.

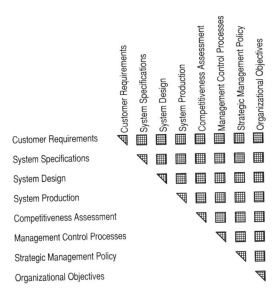

Figure 4.57 Interaction matrices for organizational policy assessment and quality function deployment.

considered, and the existence of a relationship is indicated in the table, such as in the adjacency matrices shown in Figure 4.55.

Development of an interaction matrix encourages us to consider every possible linkage, and it significantly enhances insight in the connectivity of problem elements. The resulting matrix may be helpful in identifying clusters of related elements, or elements that appear to be quite isolated from the others. Also, the table may be used to construct a graph or map showing a structure of elements. Interaction matrices are generally applicable in all circumstances where the structure of a set of elements has to be explored.

The theory of digraphs and structural modeling is several of the works we discussed earlier in Section 4.3. Geoffrion [100] has noted that a modeling environment needs five quality- and productivity-related properties.

1. A modeling environment should nurture the entire modeling lifecycle, not just a part of it.

2. A modeling environment should be hospitable to decision and policy makers, as well as to modeling professionals.

3. A modeling environment should facilitate the maintenance and ongoing evolution of those models and systems that are contained therein.

4. A modeling environment should encourage and support those who use it to speak the same paradigm neutral language, in order to best support the development of modeling applications.

5. A modeling environment should facilitate management of all the resources contained therein.

Structural modeling is a general conceptual framework for modeling and a prelude to many other forms of modeling. These references should be consulted for much greater and in-depth discussions of structured modeling. Here we provide only an overview of this important topic. We assume that one, two, or more sets of elements are given. These may have been obtained by an idea generation method or by using some other approach. Also, a decision has been made as to whether the interactions between the elements of one set (such as needs, alterables, stakeholders) or the linkages between elements of two or more different sets are to be explored. The following activities are typically followed in setting up an interaction matrix, including a graph-theory-based structural model.

1. *Determination of the Type of Relationship.* Directed or nondirected graph relationships might be considered, depending on the type of matrix, the type of elements, and the degree of specificity desired. For example, alterables might be related in that they belong to the same subsystem, or in that they tend to change in a similar nondirected way, or it might be that change in one alterable directly affects another alterable. For a cross-interaction matrix, nondirected relations seem appropriate, whereas when the self-interactions are considered, appropriated directed relations might be used. In general, any type of relationship may be used, as long as the meaning of the relation considered is clear to all those involved in setting up the interaction matrix.

2. *Setting Up the Interaction Matrix Framework.* The elements are listed and lines are drawn in such a way that the result is a table in which there is a box for each possible combination of elements. Nondirected self-interaction matrices have a triangular form, while directed self-interaction matrices are square and cross-interaction matrices are rectangular.

3. *Completion of Each Entry in the Interaction Matrix.* The entries in the matrix are completed one by one, each time asking the question, "Are these two elements directly linked according to the specified relation, or not?" A positive answer may be indicated by writing a "1" or an "x" in the appropriate box or by blackening the box, and a negative answer may be indicated by writing a "0" or by leaving the box blank. Information about the strength of the relation may be included, as noted in Figure 4.55.

4. *Revision of the List of Elements and Scanning the Pattern of the Relations That Are Displayed.* The process of completing the matrix entries may lead to the identification of important intermediate elements that are missing from the original list. If desired, such an element is added to the matrix. Also, elements may be redefined, requiring revision of related

matrix entries. Certain patterns may appear in the matrix, revealing either (a) clusters of interrelated elements or (b) elements appearing to be quite isolated from others. In an alternatives–objectives cross-interaction matrix, it is important to check whether, for each objective, there is at least one alternative that helps achieving it.

5. *Translation of the Matrix into an Appropriate Graph.* Because a graph conveys both direct and indirect linkages, it is generally a better way to communicate structure than a matrix. It may be worthwhile to construct such a graph from a self-interaction matrix, provided that the matrix is not too large and there are not too many interactions. The graph, essentially containing the same information as the matrix, can be directed or nondirected, depending on the relation used.

6. *Analysis and Formulation of Conclusions.* On the basis of the matrix and possibly the corresponding graph, conclusions may be drawn concerning the major subsets of the problem. Also, graphical analysis methods might be used to reorder the matrix or graph to show the structure more clearly. The consistency of several interaction matrices of elements relating to the same problem may be checked.

Interaction matrices are generally useful in the definition phases and issue formulation steps of a systems effort. They provide much assistance in the determination of function from structure. They provide a simple aid to exploring the structure of all the problem elements, such as needs, stakeholders, alterables, constraints, agencies involved, and so on. They may also be used to a considerable extent in the analysis effort throughout all of the phases of the life cycle.

Computer assistance may be helpful for setting up and displaying large matrices. Also, computers may be used to perform analysis of (binary) matrices and to help structure a matrix in such a way that an informative graph may easily be drawn. The usual approach is to first represent a mental model of a situation by identifying a number of elements and an appropriate contextual relationship. Questions are posed and answered concerning the binary relations among the elements according to the selected contextual relationship. This enables construction of a reachability matrix or a minimum-edge adjacency matrix according to the approach chosen.

If transitivity can be assumed, it is possible to uniquely determine a reachability matrix by asking a limited number of contextual relatedness questions about the elements and use the transitivity property to infer many responses. This approach is described in considerable detail in the structured modeling references given here. Many structural models take the form of a tree, or slightly more complex but hierarchical structure. Also, many structural models take the form of influence diagrams, which we discussed earlier in Section 4.3. The IDEF approach discussed briefly in Chapter 3 represents one approach to the development of system structure.

4.7.2   Stochastic Networks and Markov Models

The mathematical modeling approach we have described up to this point is deterministic in that it can be used to represent processes that are determined without an intervention from a chance mechanism. Indeterminism, which results from uncertainty, often exists. Most of the tools of graph theory, which deals essentially with deterministic models, can be modified to enable consideration of probabilistic and stochastic effects.

A Markov model is just a digraph that is weighted such that the weights associated with each edge correspond to the probabilities of a change in state between adjacent nodes in the directed graph. A Markov decision model consists of the following:

1. A statistical description of the typically not totally controllable dynamics of a process
2. A specification of the effects of each action alternative on process dynamics
3. A specification of a performance index

Markov processes are the finite-state stochastic counterpart of optimum systems control that typically deals with the infinite state value problem. While it would be possible to provide much additional detail concerning stochastic optimization, this is not our intent here.

Steps in the solution of the typical Markov optimization problem generally involve the following:

1. Definition of the problem and the determination of a digraph that represents the problem structure, and the encoding of a consistent set of state transition probabilities onto the various edges of the digraph
2. Determination of alternative courses of action and their impact upon the transition probabilities
3. Specification of the (typically single scalar-valued) objective performance index
4. Verification and validation of the resulting optimization model
5. Selection of the optimum alternative course of action using one of the existing algorithms for this purpose
6. Sensitivity analysis of the results

Determination of machine repair and replacement strategies, inventory and stock management, and the management of queuing systems are among the many problems that can be treated as Markov decision processes. Mathematical analysis of simple Markov process models involves just the solution of systems of linear equations and associated matrix–vector multiplication. There are many available extensions to the simple Markov process model discussed here. It is, for example, possible to introduce the concept of a "stochastic shortest route model" in which the availability of routings at each stage is probabilistic and where these probabilities depend upon the states that precede a particular stage. This results in a dynamic programming-type formulation of a Markov process and solution of the resulting optimization problem. This level of sophistication should probably be sought only after experience has been gained with operational concerns associated with use of simpler analysis activities. In a great many cases, simulation will be the preferred course of action to analytical solution of Markov process models. Section 4.6.4, which discusses discrete event digital simulation, should be referred to for a discussion of this topic.

Example 4.6.2. Let us consider a very simple repair or replacement exercise in which a production machine is inspected at the end of each day's effort. It can be found in one of four states:

1. As good as new, no faults
2. Operable, but with minor faults
3. Operable, but with major faults
4. Inoperable

If nothing relative to maintenance is accomplished, the state of a new machine will generally evolve from state 1 to state 2 to state 3 to state 4, where it will remain indefinitely. Data are obtained relative to the transition from one state to the other. These results are described in Figure 4.58. We note that it is not possible to go from a state of higher deterioration to one of lower deterioration in that the associated probabilities, not shown on the graph, are each 0.

We can calculate the probability of being in any of the four possible states on any chosen day, given the state on the previous day and that nothing is

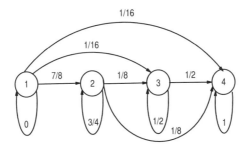

Figure 4.58 State transition diagram for Example 4.6.2—with no repair or replacement.

done relative to maintenance or replacement. We easily obtain

$$P(1, 0) = 1, \ P(1, N) = 0, \ N > 0$$

$$P(2, \ N + 1) = (\tfrac{7}{8})P(1, \ N) + (\tfrac{3}{4})P(2, \ N)$$

$$P(3, \ N + 1) = (\tfrac{1}{16})P(1, \ N) + (\tfrac{1}{8})P(2, \ N) + (\tfrac{1}{2})P(3, \ N)$$

$$P(4, \ N + 1) = (\tfrac{1}{16})P(1, \ N) + (\tfrac{1}{8})P(2, \ N) + (\tfrac{1}{2})P(3, \ N) + P(4, \ N)$$

and recognize this as a set of first-order linear difference equations. We start the solution at day zero, with $N = 0$. It is easy to show that we obtain solutions like

$$P(2, \ N) = (\tfrac{7}{8})(\tfrac{3}{4})^{N-1}$$

$$P(3, \ N) = \left(\frac{1}{16}\right)\left(\frac{1}{2}\right)^{N-1} + \frac{7}{64} \sum_{i=2}^{N} \left(\frac{3}{4}\right)^{N-i}\left(\frac{1}{2}\right)^{i-2}$$

Ultimately, we see that we obtain $P(K, N) = 0$ for $K = 1$, 2, and 3; and $P(4, N) = 1$. The system becomes unrepairable and remains there.

After inspecting a machine and detection and diagnosis of a fault, the machine operator can choose from among combinations of three possible courses of action:

A Do nothing at all.

B Overhaul the machine such as to return it to state 2 with probability 1.

C Replace the machine such as to return it to state 1 with probability 1.

The production costs increase as the condition of the machine deteriorates, presumably due to increased production difficulties or the need to correct low quality production. Suppose that the production cost increases are as follows:

State 1 No costs
State 2 Expected cost increase of $1000 per day
State 3 Expected cost increase of $3000 per day.

There are costs associated with maintenance and replacement. Suppose that maintenance, in either state 2 or 3, costs $2000. Suppose that the replacement costs are $4000. In each case a production delay of one day is introduced, and this costs an additional $2000. Thus, the costs in state 3 become $4000 when we repair the machine and $6000 if we replace it. The costs in state 4 are $6000 for a replacement.

The problem is to find the general repair or replacement strategy for which the expected long-term average costs per day of operation are minimum. Clearly, the only option if the machine gets to state 4 is to replace it. This must

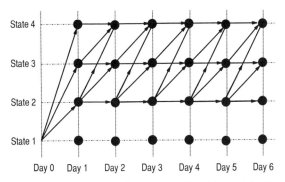

Figure 4.59   State transition diagram for the "do nothing" policy.

be done or we will have no production at all from that point on. If we are in state 1, we surely would not wish to replace the machine because there is not even a minor fault. So, the only question is which of the three options should be chosen if we are in states 2 or 3.

One policy that we might adopt is to do nothing until the machine becomes unreparable and then repair it. This will alter the state transition diagram of Figure 4.58 to produce that shown in Figure 4.59. The difference equations expressing the state transition probabilities now become

$$P(1, N + 1) = P(4, N)$$

$$P(2, N + 1) = (\tfrac{7}{8})P(1, N) + (\tfrac{3}{4})P(2, N)$$

$$P(3, N + 1) = (\tfrac{1}{16})P(1, N) + (\tfrac{1}{8})P(2, N) + (\tfrac{1}{2})P(3, N)$$

$$P(4, N + 1) = (\tfrac{1}{16})P(1, N) + (\tfrac{1}{8})P(2, N) + (\tfrac{1}{2})P(3, N)$$

We can solve this difference equation if we wish with the initial condition $P(1, 0) = 1$, $P(K, 0) = 0$ for $K = 2, 3, 4$. The solution for large N is of particular interest. We know that $P(1, N) + P(2, N) + P(3, N) + P(4, N) = 1$. Thus, we obtain, for large N, $P(1, N) = P(3, N) = P(4, N) = 2/11$, $P(2, N) = 7/11$. The expected cost of implementing a strategy is

$$E(C) = C(1)P(1) + C(2)P(2) + C(3)P(3) + C(4)P(4) = 1000P(2)$$
$$+ 4000P(3) + 6000P(4)$$

and so we obtain for this strategy, $E(C) = \$2454.54$.

We could adopt the strategy of replacing the machine if it is in state 4 and repairing it if it is in state 3. Figures 4.60 and 4.61 illustrate the resulting state transition diagram. We can easily write the difference equation for the

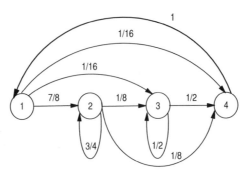

Figure 4.60   State transition diagram for Example 4.6.2. There is replacement when machine becomes inoperable.

probability evolution from this diagram and obtain

$$P(1, N) = P(4, N)$$

$$P(2, N + 1) = (\tfrac{7}{8})P(1, N) + (\tfrac{3}{4})P(2, N) + P(3, N)$$

$$P(3, N + 1) = (\tfrac{1}{16})P(1, N) + (\tfrac{1}{8})P(2, N)$$
$$P(4, N + 1) = (\tfrac{1}{16})P(1, N) + (\tfrac{1}{8})P(2, N)$$

Recalling that $P(1, N) + P(2, N) + P(3, N) + P(4, N) = 1$, we obtain for large N the steady-state conditions $P(1, N) = P(3, N) = P(4, N) = 2/21$ and $P(2, N) = 5/7$. We see that this strategy results in the machine being in state 2 for a much longer fraction of time than the strategy of only replacing the machine when it is inoperable. The expected costs of this strategy are \$1666.67 per day. It turns out that this is the best possible strategy. We can compute others if we desire. Here we have enumerated all strategies and selected the one with minimum cost. While possible for this simple example, this would be very

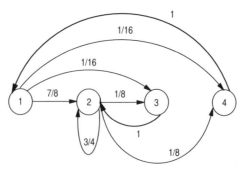

Figure 4.61   State transition diagram for Example 4.6.2. There is replacement when machine becomes inoperable, and there is renovation when it enters state 3.

computationally intensive for more complex problems and mathematical programming techniques of operations research may be used to advantage.

We have presented a brief overview of Markov decision processes here. These have been applied to a number of application areas. It is generally necessary to use dynamic programming to obtain solutions over a finite-time horizon or use linear programming in the case of infinite-time horizon problems. Much more detail is available in reference 101.

4.7.3 Queuing Models and Queuing Network Optimization

Queuing theory is the applied mathematical analysis of probabilistic concerns. In particular, it has been developed to study problems related to queues or waiting lines. Typically, queuing phenomena arise when there are, at certain times, more customers desiring a service than can be served at once. A typical, elementary queuing system consists of customers, waiting and being served, and a service facility with one or more servers. The state of the system changes with the arrival of a customer or with the departure of one or more customers that have been served. Both customer arrivals and service time for each customer are governed by a chance process.

Application of queuing theory requires a proper statistical description of customer arrival and service time distribution, plus information on queuing and service disciplines. These include such things as personnel behavior when faced with long waiting lines, and the effects of policies, such as possible priority rules on the behavior of personnel in these lines. Queuing theory enables the analyst to compute expected values and distributions of queue length, waiting times, and server occupancy. Given reliable data, an analysis based on queuing theory can yield general insight and precise forecasts that are much more veridical than comparable results obtained from use of more intuitive methods.

There are six primary defining characteristics of queuing processes:

1. Arrival pattern of customers
2. Service pattern of servers
3. Queue discipline
4. System capacity
5. Number of service channels
6. Number of service stages

In most cases, knowledge of these six characteristics provides an adequate description of a queuing system.

The arrival pattern or input to a queuing system is often measured in terms of the average number of arrivals per some unit of time, the mean arrival rate, or by the average time between successive arrivals, the mean interarrival time.

These quantities are clearly related, and knowledge of either one is usually sufficient to enable description of the system input. In the event that the stream of input is deterministic—that is to say, completely known and thus void of uncertainty—then the arrival pattern is fully determined by either the mean arrival rate or the mean interarrival time. On the other hand, if there is uncertainty in the arrival pattern, then these mean values provide only measures of central tendency for the input process, and further characterization is required in the form of the probability distribution associated with this random process.

Another factor of interest concerning the input process is the possibility that arrivals come in batches instead of one at a time. In the event that more than one arrival can enter the system simultaneously, the input is said to occur in bulk or in batches. It should be noted that entering the system does not necessarily mean entering into service, but may instead require joining the line when immediate service is not available. In the bulk-arrival situation, not only may the time between successive arrivals of the batches be probabilistic, but also the number of elements in a batch.

It is also necessary to know the reaction of a system element, such as personnel for example, upon entering the system. A person may decide to wait no matter how long the queue becomes, or if the queue is too long, may decide not to enter it. If a person decides not to enter the queue upon arrival, the person is said to have "balked." On the other hand, an arrival may enter the queue, but after a time lose patience and decide to leave. In this case the arrival is said to have "reneged." In the event that there are two or more parallel waiting lines, people may switch from one to another, that is, they jockey for position. These three situations are all examples of queues with impatient customers.

Much of the discussion concerning the arrival pattern is also appropriate in discussing service patterns. For example, service patterns can also be described by a rate or as the time required to service a customer. One important difference exists, however, between service and arrivals. When one speaks about service rate or service time, these terms are conditioned on the fact that the system is not empty, that is, that there is someone in the system requiring service. If the system is empty, the service facility is idle. Service may also be single or batch. One generally thinks of one person being served at a time by a given server, but there are many situations where customers may be served simultaneously by the same server, such as by a computer with parallel processing capabilities.

Even if the service rate is high, it is most likely that some customers will be delayed by waiting in the line. In general, customers arrive and depart at irregular intervals. Hence, the queue length will assume no definitive pattern unless arrivals and service are deterministic. Thus it follows that a probability distribution for queue lengths would be the result of two separate processes, namely, arrivals and services. These are generally, though by no means always, assumed to be mutually independent phenomena.

Queue discipline refers to the process through which customers are selected for service when a queue has formed. The most common discipline that can be observed in everyday life is first come, first served, or a first-in first-out (FIFO) queue. This is certainly not the only possible queue discipline. Some others in common usage are the last-in first-out (LIFO) queue, which is applicable to many inventory systems when there is no obsolescence of stored units. In this case, it is easier to reach the nearest items that are the last in the queue. Another queue selection is service in random order independent of the time of arrival to the queue (SIRO). There exist a variety of priority schemes, where arrivals are given priorities upon entering the system, and the ones with higher priorities are selected for service ahead of those with lower priorities, regardless of their time of arrival to the system.

In some queuing processes there is a physical limitation to the amount of waiting room, so that when the line reaches a certain length, no further customers are allowed to enter until space becomes available through a service completion. These are referred to as finite queuing situations. In these systems, there is a finite limit to the maximum queue size. A queue with limited waiting room can be viewed as one with forced balking where a customer is forced to balk if the customer arrives at a time when queue size is at its limit. In more complex cases, the circumstances under which arriving customers will balk may not be known exactly.

The number of service channels refers to the number of parallel service stations that can service customers simultaneously. A queuing system might have only a single stage of service or it may have several stages. An example of a multistage queuing system would be a routine equipment maintenance procedure, where each equipment undergoing maintenance must proceed through several stages in order to complete the maintenance process. In some multistage queuing processes, recycling may occur. Recycling is common in manufacturing processes where quality control inspections are performed after certain stages, and parts that do not meet quality standards are sent back for reprocessing.

The six characteristics of queuing systems are generally sufficient to describe completely a process under study. One can see from the discussion thus far that there exist a wide variety of queuing systems that can be encountered. Before performing any mathematical analysis, and in particular a queuing analysis, it is absolutely necessary to describe adequately the process being modeled. Knowledge of the six characteristics of a queuing process is essential in this task.

To accomplish a queuing analysis, a minimum of three distinct subsystems must be identified:

1. The set of people or things that require service — that is to say, the calling population

2. One or more implicit or explicit waiting facilities, known as queues

3. A service facility that services the calling population

The calling population may be finite or infinite, and it can also be homogeneous or nonhomogeneous depending upon the type of service that is required. We can also consider single arrival systems in which all customers arrive one at a time, or batch arrival systems in which all customers arrive at one time. The most critical aspect in modeling the calling population is the probability function that determines the arrival of customers from the population to a queuing subsystem. The simplest calling population to deal with is one that is infinite, homogeneous, and single arrival and which has a well-behaved probabilistic arrival pattern.

Two distinct aspects of a queuing subsystem need to be considered: the queuing configuration and the queue discipline. A queuing configuration may involve single versus multiple queues. Also, we may have serial or parallel queues when the service that is being sought involves more than a single activity. A queue discipline may involve first-in first-served protocols, last-in first-served protocols, or random service protocols. Balking, which gives a person the option of not joining a queue if it is too long, or reneging, which allows a person to leave a queue after once joining it, are among the many possible queue discipline options.

As already noted, we can have single or parallel service facilities. There are many variations of this. We could have a single server, multiple identical servers in parallel, or parallel sets of servers that are in series. Servers in a series may or may not have queues between each server. There may be a single queue for parallel servers, or individual queues for each. Service rates for multiple servers may be homogeneous, in the sense of identical service time for the same service activity, or nonhomogeneous. A most critical aspect of a service system is the probability function that determines the service rates. The simplest service facility subsystem is that of a single server with a very well behaved probabilistic service time. These three subsystems are very highly interdependent in operation, and it is not possible to meaningfully analyze one aspect of queuing without considering the other two.

Most solutions obtained through use of queuing theory apply to steady-state queuing phenomena. The forms of arrival and service time distributions are assumed to be constant, and the system is supposed to have stabilized. Queuing theory has been applied successfully in a variety of fields, for the design of new systems as well as for improving existing systems. While it can be an extraordinarily theoretical and complicated subject, most of the successful applications involve only a small portion of the available theory. Discrete event simulation may be used as an alternative approach to obtaining the solution to queuing problems that become analytically intractable.

Several steps are generally associated with solution of a queuing problem:

1. *Specification of the Structure of the Three Subsystems of a Queuing System.* Characteristics of the queuing phenomenon are phrased, and restrictions imposed on the system by its environment are identified. Major elements and events are structured. These include customer arrivals, queues, and serving facilities. Alternative solutions are specified,

and criteria to evaluate the merits of alternative solutions are formulated.

2. *Collection of Information.* Data on the statistical properties of events are collected from theory or by direct observation. When the system does not exist yet, observation of comparable events or a comparable solution may be possible. Behavior is observed closely to determine whether and when the personnel, or other elements being studied, shift queues or leave the service facility when they perceive long waiting lines. Possible priority rules and other alternative policies are specified.

3. *Identification of Appropriate Mathematical Descriptions.* A statistical description of random events is chosen for each event, generally on the basis of available data. Hypothesis testing techniques may be used to identify the description that best fits the data. Also, other parameters are determined to specify the equations describing the queuing system. A general classification scheme with three parameters is used to identify the type of problem. The first parameter refers to the type of arrival distribution, the second one refers to the departure (or service time) distribution, and the third one refers to the number of parallel service channels. This notation takes the form $U/V/W$, where specific symbols or letters substituted in the three positions describe standard models. M is used to represent a Poisson arrival or exponential distribution. D is used for a deterministic arrival or service distribution, and G is used for some other general distribution of arrival or service times. For example, an $M/M/2$ queue means that both interarrival times (times between individual arrivals) and service times can be described using an exponential distribution (indicated by "M"), while there are two parallel service channels. Sometimes three additional parameters are used to specify the service discipline or priorities, maximum number of customers allowed in the system, and the size of the customer population. Then, we have a classification system $U/V/W/X/Y/Z$, where the meaning of U, V, and W is as before, and where X indicates queue capacity, Y indicates queue priority system, and Z indicates the number of series servers. This list of symbols is not complete. For example, there is no indication of a symbol to represent bulk arrivals, nothing to represent series queues, and no symbol to denote any state dependence. Some notation does exist for unmentioned models, but many of these are not in standard use. In addition, there are queuing models for which no symbolism has ever been developed. This is often true for those problems less frequently analyzed in the literature. At this point in the analysis and modeling effort, past experience allows the systems analyst to decide whether the complexity of the queuing problem and the resulting model allows analytical treatment. If not, discrete event digital simulation approaches may be used. If it is elected to continue with a formal queuing analysis, step 4 is accomplished. If this is not appropriate, then attention should shift to use of a digital simulation approach.

4. *Solution, Verification, and Validation of the Queuing Model and Analysis.* The systems analyst uses the results of queuing theory to compute expected queue length, waiting times, and other performance indices for the situation(s) on which data are available. The results are compared with actual observations. If serious discrepancies appear, the model is adjusted appropriately and the analysis is repeated.

5. *Solution of the Queuing Model for Each Alternative.* Models for each alternative course of action, solution, plan, or design are specified and results are obtained for each of these.

6. *Evaluation and Refinement of Results.* The results of the queuing analysis are used to compare alternatives with respect to their merits. When clear criteria have been specified, a performance index for each alternative solution can be computed. For example, waiting costs and server costs may be computed for a system with a variable number of servers, and the minimum overall cost solution is identified. When closed-form analytical solutions have been obtained, a general expression giving total cost as a function of controllable parameters can be derived. Mathematical programming may be used to find the minimum cost solution within the allowable range of the variable parameters.

7. *Selection of the Best Alternative.* The alternative course of action with the largest payoff (or smallest if a minimization-type extremization is sought), is selected, and this resolves the queuing analysis issue. While it would generally always be desirable to perform a sensitivity analysis, this can be quite a complex undertaking.

The presence of several conditions makes use of queuing models and the associated queuing analysis quite appropriate:

1. There is a problem related to queues or waiting lines in an existing system.
2. Various design options have to be compared with respect to their queuing related aspects.
3. A minimum total cost solution for a queuing system has to be identified.
4. The problem is well-structured and relatively simple.
5. The primary interest is in steady-state queuing behavior.
6. The results, or a portion of the results, of a discrete event digital simulation have to be analyzed and understood.
7. Sufficient data to conduct the analysis can be obtained, and precise results are desirable.

Queueing theory is most appropriate for analysis and refinement of alternative designs, or to study the effects of possible changes in an existing system. Hypothesis testing might be used to select the most appropriate statistical descriptions for the various random elements in the system to be analyzed. Discrete event digital simulation is an alternative for queuing problems that are too complicated to be handled easily. Mathematical programming approaches might be used to determine the optimum design or solution in terms of a given performance function. When there is a need for determining state-dependent control laws in a queuing system, a Markov decision model also might be useful.

There are several validity concerns that should be addressed when you use queuing models:

1. Application of analytical results of queuing theory is restricted to relatively simple problems because of the difficulties in establishing an appropriate model and in conducting a queuing analysis for other than simple models. Discrete event digital simulation is a more appropriate, and less time-intensive and expensive, method for complicated problems.

2. Necessary simplifications and adaptations of queuing events to fit standard mathematical descriptions may affect the validity of results.

3. Costs of waiting may be very difficult or impossible to estimate, and they generally involve subjective value judgments. A sensitivity analysis that allows variation of the various assumptions concerning waiting costs should be included in the analysis if this is possible.

4. Most analytical queuing studies are restricted to equilibrium or steady-state conditions.

5. The mathematics needed for a queuing analysis may easily render the procedure, and quite likely the results of the analysis as well, incomprehensible for other than queuing specialists.

6. As already noted, there are three system responses of interest: some measure of the waiting time that a customer might be forced to endure; an indication of the manner in which customers may accumulate; and a measure of the idle time of the servers. Because most queuing systems have stochastic elements, these measures are often random variables; thus, their probability distributions or their expected values need to be found.

The task of the queuing analyst is to accomplish two analytic efforts. The analyst must determine the values of appropriate measures of effectiveness for a given process or must design an "optimal" system according to some criterion. To accomplish the first task, the analyst must relate waiting delays and queue lengths to the given properties of the input stream and the service procedures. To accomplish system design or system evaluation, the analyst would probably want to balance customer waiting time against the idle time of servers according to some inherent cost structure, or to measure these quantities for existing systems. If the cost of waiting and idle service can be obtained directly, they could be used to determine (a) the optimum number of channels to maintain and (b) the service rates at which to operate these channels. Also, to design the waiting facility it is necessary to have information regarding the possible size of the queue. There may also be a space cost that should be considered along with customer-waiting and idle-server costs to obtain the optimal system design. In each case, the analyst would strive to solve the problem by analytical means. When this fails, which will often occur when complex systems are considered, discrete event digital simulation is used. Some appropriate references for additional study of queuing include references 102 and 103.

4.7.4   Discrete Event Digital Simulation

In no way can simulation be considered as a tool of last resort by the systems analyst. Simulation models allow systems engineers to engage in "what-if" and "if-then" type exercises to a degree not possible using totally analytical methods. There are many other potential benefits to simulation, but the ability to explore the implication of many alternative courses of action is perhaps the principal one that allows more effective planning and decision making. There are a variety of simulation types that are possible, and we have already examined several of these. In a later section of this chapter, we will discuss time series modeling and simulation. Our concern in this subsection is with probabilistic simulations that involve a finite number of state values. This is generally known as discrete event digital simulation [104].

Central to methods of discrete, or finite, event digital simulation is the notion of an activity that is an elementary task that requires time to complete. Activities are always preceded and proceeded by other activities. The collection of activities in a given situation is known as a process. An event is an instant in time at which a given activity begins or ends. Events, therefore, are ordered instants in time at which the system undergoes perceptible change. These changes could represent customers being serviced, equipment being repaired, or any of a multitude of changes of potential interest.

Of importance also are the notions of attributes and entities. An entity is simply an item of interest in the process itself. These entities could be temporary or permanent. Examples of the former are telephone calls and parts in an inventory. Examples of the latter include telephone networks and repair depots. Entities, events, and activities all have characteristics that are of interest. These characteristics are called attributes. Average repair time for a particular procedure would be one example of an attribute.

Discrete event digital simulations may be based or focused on activities or events. Thus the use of the word "event" may be somewhat inappropriate, although it is very common. Existing simulation languages such as GASP and SIMSCRIPT accomplish modeling by focusing on events. Languages such as GPSS and CSL are block oriented process approaches which focus on activities. A particular subclass of the activity-focused, discrete-event simulation language is the entity progress approach that allows focus on the evolution over time of the progress of an entity in the system. SIMULA is a particular entity progress simulation language that enables one to follow the interactions of an entity in the system, from initial arrival event through to final departure event. Clearly both are appropriate representation methods and newer languages such as SLAM and GERT allow implementation by either approach.

The steps that are involved in the typical discrete event simulation study include the following:

1. *Problem Formulation.* The problem is defined and the various activities, events, entities, and attributes are specified such that we obtain the problem formulation elements.
2. *Data Collection and Analysis.* Data relevant to the activities, events, entities, and attributes are obtained. To do this meaningfully, we will generally have to first establish a structural model of the system and an identification of possible alternative courses of action.
3. *Choice of Modeling Method and Interpretation of the Results from the Study.* Based on the decision situation structural model, a discrete event simulation approach is chosen and the simulation is conducted.

The event scheduling approach to discrete event simulation emphasizes detailed consideration of occurrences that result when individual events occur. The discrete event structural model is basically a model of events and potential times of occurrence for these events. Activities are decomposed into their respective events, and an event generation diagram is constructed. Determination of times between recurrent events is needed. These times are generally random and are often described in terms of a probability distribution. Each event is processed by transferring control to a specially written event processing routine that accomplishes necessary housekeeping in order to record event occurrence and to create other events caused by the occurrence of the now current event. The activity scanning approach emphasizes study of the activities in a simulation to determine which can be begun or terminated each time that an event occurs. In a process interaction approach or block-oriented approach, the processes represented are a chronological sequences of activities. The progress of an entity through a system, from an arrival event to its departure event, is tracked. SIMULA and GPSS are examples of this approach.

The method for advancing time in a discrete event simulation model is the same, regardless of whether an event-oriented or block-oriented approach is used. Discrete event simulation languages keep track of the current simulated time through use of a specifically defined variable that allows time to be passed from the time of the event currently being processed to the time associated with the next event. It is important to note that only the times where events occur need to be recorded. These are the only time instants at which it is possible for the outputs of the simulation model to undergo changes. It is this execution of the model through the discrete advances in time, from one event to the next succeeding event, that gives the method "discrete event simulation" (model) its name.

1. *Model Validation.* The actual simulation model is constructed and the simulation is run. Suspected errors in the structure of the model are corrected and the modeling process iterated until the model is judged to be a good representation of the real system.
2. *Evaluation of Alternatives.* Each of the alternative courses of action are programmed into the discrete event digital simulation model and the model is exercised, such as to allow evaluation of each potential alternative course of action.
3. *Selection of Best Alternative* If the intent of the simulation is to enable selection of a best alternative policy, this is accomplished. If the intent is to describe an existing situation without selection of a appropriate new alternative, this and the preceding step are unnecessary.
4. *Sensitivity Analysis.* Parameters within the discrete event simulation model that are subject to change are changed, and the sensitivity effects of these changes upon recommended courses of action are determined.

In order to develop an appropriate set of algorithms to enable modeling in terms of graphs, queuing networks, and associated discrete event simulation techniques, careful consideration must be devoted to basic data structures. Much experience has indicated that the analysis that can be accomplished using graph and queuing concepts becomes extraordinarily difficult for other than simple models. Discrete event digital simulation is, in these situations, the preferred technique. There are a number of useful references that describe modeling and simulation, including discrete event digital simulation. Among these are references 105–107.

4.7.5   Time-Series and Regression Models

There are many occasions when people wish to make forecasts of possible future states and events. There are several methods that might be used to accomplish this. Those based on forecasting models are of interest here. Some forecasting models are very crude, and some are sophisticated. Some are based upon the information that constitutes the expert judgment of an individual or a group. Others are based upon mathematical approaches and formal reasoning. Here, we will be concerned with quantitative forecasting approaches based on ordered time series of observations.

Many issues that involve information processing in humans and organizations can be associated with the analysis of change over time based on ordered sequences of observations. By definition, these constitute what is called a time series. Many time series are not well-behaved, and any persistent regularities and variances are hidden from all except perhaps the most experienced observer. Often these provide early clues to some impending crisis; consequently, their detection is very important. Many techniques are useful for the analysis of change over time. Regression analysis techniques are one of the most commonly employed and useful techniques. In the design of a system to support human information processing, one important ingredient is a library of software tools that will support the user in determining how important variables will likely behave in the future based upon their behavior in the past and the assumption of a model structure to relate system inputs and outputs.

Representation of many real-world phenomena in terms of time series is a very practical and useful approach for understanding and predicting system behavior. A time-series analysis takes into account the nature of an observed process as it evolves through time. A time series is constructed so as to reflect the way a system behaves over time due to changes in input to the system. Usually, time-series analysis is considered to be an area of statistics, or statistical estimation theory. It is rich in the choice of models potentially offered the user; for this reason, modeling is a particularly important component of a realistic time-series analysis. It is important that an information system provide several ways of encoding dynamic behavior. There are at least five approaches to modeling observed phenomena:

1. Pictorial or graphical representation
2. Verbal representation
3. Flow diagram or graphic representation
4. Control theoretic modeling in the form of differential equations, perhaps with unspecified parameters that are to be determined as part of the modeling and analysis effort
5. Finite-state modeling, in which there exist a limited number of states or events, one of which can be used to characterize each of a system's state variables

The first two of these relate primarily to artificial intelligence and expert system based approaches where various forms of knowledge representation may be used — for example, production rules, scripts, cognitive maps, and schema. The last of these relate to the discrete event and queuing representations we discussed earlier in this chapter. Methods 3 and 4 are representations appropriate for the use of time-series analysis.

There are at least two ways of representing a time series of an observed variable:

1. If we choose serial dependencies in discrete time as the way of representing observations, we obtain what is called an autoregressive model and what are called input–output transfer functions. This is the common

method used in time-series analysis and in control theory and is the approach emphasized in this section. The major potential problem with this representation is that there is no general way in which we can become familiar with other than the input–output behavior of a system through use of input–output data, spaced at regular time intervals, only.

2. When it is needed to know various internal and structural aspects of system behavior, a state-space model representation is appropriate. This approach is more powerful that the input–output analysis approach but will often require more complex mathematics and a more detailed knowledge of the structure and interactions of the system being modeled.

Statistical procedures can vary from the drawing and assessment of a few simple graphs, with perhaps "eyeball" estimations of fit and average values to very complex mathematical analysis that use very large computers and sophisticated models. In any area that is appropriate for statistical analysis, there is an essential random nature to the observations that are taken. In fact, statistics may be defined as the collection and analysis of data from random observations. Probability theory, which provides the mathematical basis for the models that describe random phenomena, is a necessary ingredient in any statistical analysis. Many statistical methods are designed to use data to identify a parameter within a system or perhaps even the underlying probability functions that are responsible for some random phenomena. Problems in this area are known as system identification problems.

One purpose of a statistical analysis is the summarization, or standardized representation in terms of various norms, of data or information. There are a variety of ways in which this might be accomplished. You may obtain a representation of the *"central value"* of a random phenomenon, such as a time series, by using the average value of the observations. It turns out that there are a variety of measures of "average" such as mean, median, and mode of the observation. Another very important average measure of an observation concerns the variability of one piece of data from others. An often-used measure of this "spread" is known as the *variance*. By definition, *this is the ensemble or time average of the squared difference between the values of the observations and the average value.* The average of the square of the observations is known as the mean square value. It can be shown that the variance of an observation is the mean square value of the observation minus the square of the mean of the observation. The square root of the variance is commonly known as the *standard deviation.* This is often a very useful measure of "spread" in a set of observations. It is sometimes useful to describe a probability function by a mathematical relation that contains a few unknown parameters. Various approaches can be used to "identify" these parameters.

The most-used form of probability density function is the normal or Gaussian density function which, in the single-variable case, is described by and depends on two parameters only, namely, the mean and the variance. In the multivariable case, it is necessary to also use various variance, correlation, and covariance terms to describe a Gaussian process. While the variance of a set of observations is a measure of their spread or dispersion, the *covariance* is a statistical measure of the association between the variables. *The covariance of two variables x and y is the average of the product of the deviations of corresponding x and y values from their respective means.* Thus, we see that a variance function is a particular case of a covariance function. The correlation function is often used also; and this is just the average of the product of the values of the variables x and y themselves. In most of the applications of interest here, we will be concerned with processes that evolve over time. In this case the definition of the *autocorrelation function* of a variable $x(t)$ is

$$\phi_x(t_1, t_2) = E\{x(t_1)x(t_2)\}$$

where the symbol E denotes *probabilistic expectation* and is defined by

$$E\{x(t)\} = \int_{-\infty}^{\infty} x(t)p[x(t)] \, dx(t)$$

Often it is necessary to compute the cross-correlation function of two time variables $x(t)$ and $y(t)$. We define this as

$$\phi_{xy}(t_1, t_2) = E\{x(t_1)y(t_2)\}$$

In general, this expectation must be taken over an ensemble of records such that only a probabilistic definition and interpretation of this relation can be given. Many physical processes are stationary in the sense that the time average of the product of two random processes is a function only of the time difference in the age variable of the two processes. A condition that is stronger than stationarity is *ergodicity. An ergodic process is one for which the time average moments of the process and the ensemble average moments are the same.* An ergodic process is always stationary. The converse is not necessarily true, however. When random processes are ergodic, then the expectation operator in the foregoing two relations can be replaced by a time average over a sufficiently long period of time. Measurement error is associated with a noninfinite time interval, and there exists a large body of knowledge concerning measurement of correlation functions. For an ergodic random process, we have

$$\phi_{xy}(t_1 - t_2) = \phi_{xy}(t_1, t_2)$$

The time difference variable $t_1 - t_2$ is generally replaced by a single variable such as τ. An ergodic autocorrelation function is symmetrical in this variable, and it is consequently an even function. The variance function is always nonnegative. Generally, it will have a smaller value for a random variable x

whose observations are always close to the mean value. The covariance may have any numerical value. It is positive when increases in x are generally associated with increases in y. The autocorrelation and cross-correlation functions may take on value, positive or negative. Essentially the only restriction is that the autocorrelation function must be nonnegative for zero difference in the age variable. At zero age variable, the autocorrelation function, is just the variance plus the square of the mean value. The statistical technique known as *analysis of variance* is generally concerned with disaggregation of the components of variance, covariance, and correlation functions into components that arise from specific causes.

Regression. Regression techniques are used to obtain a mathematical model that specifies the relations between a set of variables. The input to the model is data that represent observations of those variables. Generally, *a regression analysis equation describes the value of one variable, the dependent variable, as a function of other independent variables.* Regression analysis equations may be helpful for interpolation or extrapolation, or forecasting of events of interest. Alternately, they may be used as a part of a more complicated mathematical description of some problem. The result of a regression analysis may also be useful as evidence to support or reject hypothetical theories about the existence of relations between variables in a system.

Estimation theory, which is closely related to regression analysis, is concerned with the determination of those parameter values in a given equation, such that use of the equation results in the best possible fit to observed data. Regression analysis and estimation theory also include the search for an appropriate structural equation or, alternately, an input–output model that best replicates observed data. This aspect of regression analysis is often not emphasized to the extent appropriate for identification of useful models.

The following activities are associated with the solution of a typical regression problem:

1. *Determination of Candidate Variables and Data Collection.* The dependent variables that need to be described as a function of other variables are defined. This is usually guided by intuition and existing theory and knowledge. Then, it should be ascertained that a sufficient number of joint observations of the values of all the variables considered are available. Usually, the number of data points should be not smaller than 10 times the number of variables. Often, it is not possible to directly observe the values of variables; because of this, noise-corrupted observations must be made.

2. *Postulation of a Mathematical Model or Structure.* The form of the postulated equation may be linear, multiplicative, logarithmic, exponential, or some other appropriate mathematical form. An initial postulate is made, and, generally when possible, transformations are performed such

that a linear relationship between the transformed variables results. If, for example, the assumed model structure has the form $y = ax^b z^c$, then the logarithm is taken on both sides to yield $Y = \ln(y) = \ln(a) + b\ln(x) + c\ln(z)$ or $Y = A + bX + CZ$ where $X = \ln(x)$ and $Z = \ln(z)$. It is important to note here that while the resulting equation is now linear in the transformed variables X, Y, and Z, it is not linear in the original variables x, y, and z. It should be remarked that even though the logarithm of all data is taken so that the postulated relationship between the transformed data becomes linear, the values of a, b, and c that best fit the nonlinear equation are not generally the values that best fit the linear logarithmic equation.

3. *Choice of Estimation and Selection Method.* The most widely used estimation method is generally referred to as "least squares," to indicate that it determines those coefficient values that will yield the smallest possible value for the sum of the squares of the differences between observed values of the dependent variable and values computed from the estimated relationship. In mathematical notation, if the function $f(x)$ is to be determined such as to best express y as a function of the set of state variables $x = [x_1, x_2, \ldots, x_n]$, and we have N observed values of $y = [y_1, y_2, \ldots, y_N]$, then we determine the unspecified coefficients in $f(x)$ such that the sum from $i = 1$ to $i = N$ of the squared error expression $e_i^2 = [y_i - f(x)]^2$ is minimal. There are a number of generalizations on the basic least squares estimation criterion, especially to include the dynamic evolution of observations over time and the notion of weights. These extensions make regression analysis and estimation theory problems virtually indistinguishable.

In regression analysis and estimation theory, one needs to determine which of the candidate independent variables needs to be taken into account in order to obtain a good description of variations in the dependent variable. One approach to this calls for first taking all of the candidate variables into account and then estimating the associated coefficient values and their uncertainty. Then, through use of hypothesis testing techniques, it is determined which of the coefficients is most likely to represent no relation at all between the dependent and independent variables. The state variable corresponding to this is then dropped from further consideration in the analysis, and the process is repeated until the likelihood that any of the remaining coefficients actually represents no relation at all is smaller than some preset value or level of significance. The end result is the appropriate regression equation.

Another approach is based on a procedure that is inverse to the one just described. After all state variables to be considered have been included in the proposed regression equation, a ranking of the levels of significance of the respective coefficients in the regression equation is determined. Then, an equation is estimated using the most significant variables only. One state variable at a time is added to this regression equation, in decreasing order of initial significance, until it is observed that the addition of one more additional state variable does not lead to an "appreciable" improvement in the goodness of fit of the resulting regression equation.

Clearly, there are many variations of these basic approaches that are possible and potentially desirable for many applications. For example, it is possible to add more than one state variable at a time. As noted earlier, these structural aspects of regression analysis, and systems engineering in general, are under explored relative to areas more subject to complete analytical exploration. The success of a modeling effort is critically dependent, in most cases, upon success in choosing an appropriate structural model. Here are the steps to follow:

1. *Determination of the Regression Curve.* This step involves obtaining the needed data, along with the use of a subportion of the regression analysis program in which algorithms for parameter estimation have been encoded.

2. *Iteration and Sensitivity Analysis.* Depending on the criticality of obtaining a good regression equation, various iterative and sensitivity forms of testing should be performed in which, for example, other structural models or different selection procedures are used.

Figure 4.62 illustrates the flow of these steps in regression analysis.

The following are conditions under which the use of regression analysis and estimation theory techniques may be appropriate:

1. Data and theory need to be combined in order to determine an equation that best expresses the relations between and among a set of observed variables.

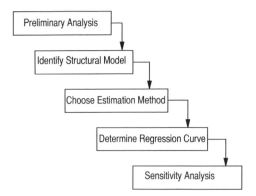

Figure 4.62 Systems engineering phased efforts in regression analysis and system estimation.

2. A mathematical model of observed phenomena is desired and there exists no established theory to explain variations in a variable; as a consequence, the determined model must be primarily data-based as contrasted with theory-based.

3. Data need to be critically examined to test the validity of a postulated hypothetical theory or assumption.

4. It is desired to use data as a basis for suggesting theoretical relations between variables.

5. Extrapolation of historical data into the future, or an interpolation of likely values occurring between data points, is needed.

6. Parameter values for an assumed structural model need to be determined on the basis of empirical evidence.

7. Data are corrupted by observation noise, and it is desired to "filter" this observed data in order to best separate data from noise.

8. Regression analysis and estimation theory are often used in conjunction with other forms of mathematical modeling in order to attain extrapolations of likely futures or trends. Hypothesis testing is generally used as part of the regression analysis process. The methods of regression analysis and estimation theory are closely related methodologically to optimization methods because, in each case, parameters are determined such as to lead to extreme values of a performance index.

In order to determine the completeness and usefulness of a regression analysis, it is important that the answers to the following questions be "yes":

1. Have you taken all the important explanatory variables into account?

2. Do the obtained results make sense to you, and can this assertion be validated in some manner?

3. Are the results of the analysis useful to you in clarifying the structure of the problem and in leading to enhanced wisdom relative to its resolution?

The user of regression analysis and estimation theory techniques should be concerned with several observations that affect model validity:

1. The results of regression analysis and estimation theory will be unreliable if they are based on an insufficient number of observations.

2. Results that are obtained through use of these approaches in situations in which there exists little theoretical knowledge should be examined very carefully because there is no guarantee that a regression relation that displays an excellent fit to observed data will really have any predictive power at all. Causation is required here, and a good fit obtained using regression approaches only assures high correlation. In practice, this caveat often seems overlooked!

3. In a similar way, the results of a regression analysis do not necessarily provide evidence or proof of causal relations among events.

4. Poor data quality may make even the optimum fit a very poor one. There is no automatic assurance that the "best" is necessarily very good.

5. The criteria for inclusion or exclusion of variables in an estimation or regression algorithm must necessarily be strongly dependent upon the purpose to which the resulting model is ultimately to be used.

Trend extrapolation and *time-series forecasting*, which are very closely related to estimation and regression theory, are widely used as the basis for projection of the future in terms of a series of historical observations of one or more observed variables over time. *The essential difference between time-series forecasting and regression analysis is that time is necessarily considered as the independent variable in the former case, but not necessarily in the latter.* In this sense, time series forecasting is a subset of regression analysis. However, there have been many studies in this area, and the subject doubtlessly deserves a separate treatment.

As in regression analysis, the *basic objectives* in *time-series forecasting* are as follows:

1. To identify a mathematical structural relation that might potentially explain observed phenomena.

2. To best identify unspecified parameters within this structure, such that some statistical error measure is minimized, and then

3. To best estimate a parameter within the time interval of observation, and/or

4. To use the determined relation in order to extend the observed data into the future based on the assumption that past trends will continue.

The typical *results* or final products of a *time-series analysis* include the following:

- A projection or forecast of one or more future values of one or more variables that are of interest

- An identified mathematical function or structural model describing past observations which is potentially useful as an aid in forecasting

- Evidence that can be used to support or reject assumptions or theories about the mechanisms that govern the past behavior of one or more variables that are of interest

In the actual use of time-series analysis algorithms, it is generally assumed that a time-series of observations of sufficient length and quality is available. This length and quality depends upon the purpose of the forecast, in particular the time length of the extrapolation into the future that is needed and the dynamics of the process being modeled. The time interval over which observations are available should be long enough to allow detection of trends of potential interest, and the time interval between observations should be sufficiently small to enable isolation of phenomena of interest.

The following steps provide an elementary description of the process of time-series analysis:

1. *Preliminary Review of Observed Data and the Environment in Which the Process Evolves.* A rough sketch of all, or a portion of, observed data as they evolve over time is an initial and very helpful approach leading to identification of readily apparent characteristics of a time series. Often cyclic components, general trends, and the random fluctuations can, at least in a preliminary way, be identified. This may be very helpful in the selection of appropriate structural forms or characteristics for an initial time-series model. Information on the physical process, and the environment into which it is embedded, may be of much assistance in enabling the selection of an appropriate mathematical model for the time-series representation.

2. *Identification of a Structural Model for the Time Series.* There is no general systemic approach for finding the appropriate structural model to use to represent a time series. When only a very crude analysis is contemplated, parameters for a linear differential equation of low order may be identified in order to provide a best fit to observed data. Both the functional form and order of the structural model and the criterion used to best fit the data to the model output are subject to change. The general form of a time-series model is $Y_t = f(Y_{t-1}, Y_{t-2}, Y_{t-3}, \ldots, Y_{t-n}) + W_t$, where Y_t indicates the potentially transformed observation of the time series at time t. The expression W_t represents a white noise driving term. Other statistical analysis methods are used to characterize the error term associated with this. Once a tentative structural model has been identified, the undetermined coefficients or parameters of this model are estimated in order to make the output of the model best fit the observed data. Statistical tests are conducted to judge the adequacy of an assumed time-series model. Another candidate structural model is subjected to experimentation if the one under test is shown to be inadequate.

3. *Verification and Validation of an Assumed Model.* Often, a time-series model is purposely based only on part of the available data. The remaining portion of the data is then used to verify the model. For example, a model might be identified based on 30 days worth of data, and then the model may be used to predict behavior for the next 5 days where data are available but where the data have not been used to identify the model. After the model has been verified in the way described, the data previously unused to identify model parameters may be used to refine the parameters of the model.

4. *Actual Forecasting.* The appropriate time-series model is used to forecast values for variables of interest. Often, this is accompanied by an error analysis in which statistical uncertainties of the random functions are taken into account to enable computation of moments, perhaps even probability density functions, of appropriate variables such as forecast values. It is possible to obtain adaptive estimation and adaptive time-series analysis algorithms in which these estimated values are used to tune the parameters of the estimation or time-series analysis algorithms.

Examination of Figure 4.62 indicates that the time-series analysis process is essentially the same, generically, as the regression analysis process.

The presence of several conditions make use of time-series analysis appropriate:

1. There is a need for forecasting future values of critical variables.
2. There exist sufficient past data about these variables.
3. It is not necessary or not possible to fully specify an appropriate model on the basis of accepted theories about the causes of change in important variables.
4. It is reasonable to assume that the information contained in historical data is the best and most reliable source of future predictions.

As we noted before, regression analysis, estimation theory, curve fitting, hypothesis testing, and time-series analysis methods have a lot in common. For almost all intents and purposes, all of these describe similar methods that are used for essentially similar purposes. Earlier comments concerning validity cautions regarding use of regression analysis and estimation theory are applicable here as well. With each of these, it should also be cautioned that:

1. The environment may change over the forecasting time interval and unless this is accommodated, poor results will often occur.
2. It may be difficult to cope with information that is not easily quantified, and a consequence of this may be to ignore such information to the detriment of the analysis.
3. Time-series and regression models may not be fully useful for predicting the effect of options that were in effect over the time period for which data is obtained, especially if implementation of these options change the structure of the model used for prediction and this change is not recognized.

It would be impossible to provide details on all of the methods appropriate for time-series forecasting here. There are three basic types of models used for time-series forecasting: *autoregressive models, moving average models,* and *autoregressive moving average models* that represent a hybrid approach. There are also variations of these. Simply stated, an autoregressive model [AR] is one represented by the first-order difference equation $x_t = a_1 x_t - 1 + w_t$, where w_t is a zero mean white noise forcing function. The process x_t generated by this model is known as a first-order autoregressive process, and is also a Markov process. The general zero mean autoregressive process of order N is that process which evolves from the model $x_t = a_1 x_t - 1 + a_2 x_{t-2} + \cdots + aN x_{t-N} + w_t$. A first order moving average [MA] process is one that evolves over time from the model $x_t = w_t + b_1 w_{t-1}$, where w_t is zero mean white noise. For an Nth-order MA process, we have the model $x_t = w_t + b_1 w_{t-1} + b_2 w_{t-2} + \cdots + b_N w_{t-N}$. The autoregressive moving average [ARMA] is just a combination of the AR and the MA process and can be written as $x_t = a_1 x_{t-1} + a_2 x_{t-2} + \cdots + a_N x_{t-N} + w_t + b_1 w_{t-1} + b_2 w_{t-2} + \cdots + bN w_{t-N}$.

The tools useful for implementation of approaches to time-series analysis are just computer implementations of algorithms associated with the methods discussed here. These include:

- Least-squares curve fitting
- Regression analysis
- Estimation theory
- Hypothesis testing
- System identification
- MR, AR, and ARMA time-series analysis

The models to be used are the assumed models for the physical or organizational process being represented. Generally, these will be difference or differential equations of appropriate order.

The general problem of constructing a mathematical representation for observed phenomena is fundamental to *system identification*, which is *the process of constructing a model that describes observed system behavior*. This also describes the subject of statistical estimation theory and regression analysis. Although observed descriptive behavior is generally used as the basis for identification of a system, it is very important to note that the uses for system identification are primarily normative. That is to say that we need to identify, or estimate, the characteristics of systems at some future time in order to evolve optimal policies for these systems over this future time horizon. The purpose of the optimal policy is to accomplish some meaningful goal. The ultimate goal of systems control is to provide a certain function, product, or service within reasonable cost and other constraint conditions such as to enable the fulfillment of some desired performance goals. These overall performance goals or objectives are typically translated, through use of the systems engineering process, into a set of expected values of performance, reliability, and safety. On the basis of this, it is the task of the system planner attempting conceptual specification of system architecture, the task of the system designer attempting concept realization in the form of operational system specifications, and the task of a man–machine intelligent system to examine future issues in such a way as to be able to do the following:

1. Identify task requirements to determine issues to be examined further and those not to be considered further
2. Identify a set of hypotheses or alternative courses of action that may resolve the identified future issues
3. Identify the impacts of the alternative courses of action
4. Interpret the impacts in terms of the objectives for the task at hand
5. Select an alternative for implementation and implement the resulting control or policy
6. Monitor performance to enable determination of how well the integrated system is functioning

These are just the fundamental formulation, analysis, and interpretation steps of systems engineering as we have discussed so often here.

System identification needs abound in many of the above six activities. The identification of task requirements involves an effort to determine what is often called the *contingency task structure* — that is to say, *the specific task at hand* and *the general objectives for the system, the environment into which the task is embedded,* and *the experiential familiarity of the problem solver with the task and the environment.* Identification of the impacts of alternative courses of action can only be accomplished through use of some model of system operation. If that model has unspecified parameters associated with it, then there is a fundamental requirement for what is generally called system (parameter) identification, or generalized system estimation.

There are many modeling issues that arise in the use of time-series algorithms, as well as in the use of related approaches in systems identification. Among them are the following:

1. *Nature of the Input–Output Relations Involved.* In this characterization, we determine whether the system with which we are dealing is causal or noncausal, dynamic or static, finite state or infinite state, discrete or continuous event, and so on.
2. *Nature of the Process Involved.* In this characterization, we determine the basic process involved in the form of a set of structural laws that are assumed correct together with a set of behavioral structural assumptions, the parameters of which need estimation or identification.
3. *Information Imprecision and Uncertainties Involved.* In this characterization, we determine the nature of the uncertainties involved and the degree of precision and completeness that is associated with the process.

Representation of time-series analysis and system identification problems in terms of the nature of the input-output relations involved is usually not a conceptually difficult task, although it may be very tedious to accomplish, once the nature of the process that is involved has been characterized. There are a number of potential ways that may be used for process characterization, and the choice of one of these depends upon the method chosen to represent uncertainty and imprecision.

There appears to be three fundamental activities associated with any given time-series analysis and system identification effort:

1. Characterization of the generic type of effort involved
2. Determination of the structure of the specific system to be identified
3. Identification of parameters within this structure

Each of these three activities are related. The results of the first activity clearly influences the second, which in turn influences the third activity. In a similar way, what is presently available in terms of software implementation for structural representation and parameter determination influences the types of estimation and identification issue characterization that is used.

As with other topics in this section, and indeed in this chapter, we have only scratched the surface of a number of important issues in systems analysis and modeling. With respect to the topics discussed here, much additional information can be found in references 108–113.

4.8   EVALUATION OF LARGE-SCALE MODELS

The past several sections have examined various approaches to the determination of models of various aspects of system behavior. Our intent has been primarily to describe various methodologies for model making; this section will discuss the very important subject of appraisal of system models. We will not discuss statistical evaluation of models here. Chapter 22 of reference 33 contains a description of approaches for statistical evaluation in systems engineering.

In a very real sense this entire text concerns the model-building process. Our last chapter was concerned with issue formulation and this includes such elements as problem definition, needs, constraints, and alterables. We also need to determine who wants a model and what the model is wanted for, along with value system design and system synthesis, considering objectives and their measures as well as policies and their measures and thus specifying or elaborating the problem so that we can begin formalizing a model and contextual relations to determine a structure for the model. This chapter has discussed specification of model structure and parameters within the structure. Here, valuation and validation of system models will be discussed. This will generally require data and perhaps an optimization process to best fit the model to the data. Finally, the model, to be of ultimate value, should be useful in the decision-making process, which will be discussed in our next chapter.

Model usefulness cannot be determined by objective truth criteria, but only with respect to well-defined and stated functions and purposes. Recall our earlier comments that a model has structure, function, and purpose. There are no objective criteria for model validity, and thus there is little likelihood of the development of a general-purpose context-free simulation model.

Model credibility depends to a considerable extent upon the interaction between the model and the model user. It is, at best, difficult to build a model that does not reflect the outlook and bias of the modeler. For this reason the earlier chapters placed considerable emphasis on techniques appropriate for the model specification steps of problem definition, value system design, and system synthesis. To determine model credibility we must examine the evidence that will be required before the model user can use the model as an aid to decision making, as opposed to using some other method such as judgment or the result of a coin toss. A credible model is one that has been verified and validated.

Parameter estimation is a very important subject with respect to model validation. It would be impossible to present an in-depth discussion of this topic without first presenting considerable material concerning optimization, estimation, and system identification. Such a presentation would not be consistent with the goals of this text. Nevertheless, a qualitative understanding of what can be accomplished with respect to this phase of model validation is quite important. If we are given a sample from a population whose specification involves one or more unknown parameters, we may form estimates of these parameters. There are quite a number of different types of estimators of a given parameter. These types are often remarkably similar, and estimation theory is concerned with properties of different estimators.

Observation of basic data and estimation or identification of parameters are essential steps in the construction and validation of system models. The simplest estimation procedure, both conceptually and from the point of view of implementation, is the *least square error estimator*. Many advanced estimation algorithms are available and in actual use. In estimating or identifying parameters such as those needed to validate and calibrate system models, two subprocesses may immediately be recognized. One is the *formulation of the system structure*, and the second is the *determination of system parameters* which determine behavior within the system structure. Observed data may be used to help formulate the system structure, and then unknown system parameters may be determined within that structure to minimize a given error criterion. In a complex, large-scale system, this process is very difficult because of the large number of individual subsystems that need to be structured to form the overall system model, and also because of inherent noise and observation error. There are no one-to-one relationships between the structure of a complex nonlinear feedback system and its behavior, and it is not true that only one model will produce given observed data. An essential complicating problem in a large-scale system is the need to correctly represent the structure of a system rather than just to accurately reproduce observed data. Thus we want to postulate correctly the forces operating between various subsystems of a complex system. In this way we are able to show how problems are created so that corrective actions may be taken and control policies established, in addition to the simple but important problem of explaining behavior. Only by obtaining proper system structure can there be a proper understanding of the underlying cause-and effect relationships. Selection of a poor structure will complicate system parameter identification and design and inhibit or prohibit proper system operation.

Verification of a model is needed to ensure that the model behaves in a gross fashion as the model builder intends. This can be achieved if we can determine that the structure of the model corresponds to the structure of the elements obtained in the problem definition, value system design. and system synthesis steps. Even if a model is verified, there is no assurance that it is valid in the sense that predictions made from the model will occur. Because data concerning the results of policies not implemented are generally not available, there is no way, as indicated in the preceding paragraphs, to completely validate a model. Nevertheless, there are several steps that can be used to validate a model, at least with respect to policies that have been implemented. These include a *reasonableness test*, in which we determine that the overall model as well as model subsystems respond to inputs in a reasonable way, as determined by knowledgeable people. The model should also be valid accord-ing to statistical time series used to determine parameters within the model. It is much more likely that relevant data will be available for these analyses in operational situations than for strategic issues. Finally, the model should be epistomologically valid in that the policy interpretations of the various model parameters, structure, and recommendations are consistent with the profes-sional, ethical, and moral standards of the group affected by the model.

A simulation model is generally the most appropriate when the following four conditions are satisfied:

1. It is either impossible or very costly to observe the effects of implementing certain policies in the real world.
2. The system under study is too complex to be described by a simple analytical model.
3. There is no reasonable straightforward analytical technique for solution of the system model.
4. The process under investigation has many state variables and/or is highly nonlinear in its behavioral patterns.

The majority of large-scale and large-scope systems, organizational and tech-nical, certainly satisfy these four conditions. However, there are also conditions under which it may be questionable whether a simulation model can be developed which will be appropriate. These conditions are as follows:

1. More appropriate techniques exist.
2. The elements related to the policy questions to be asked are not readily accessible and measurable at acceptable costs.
3. The needed databases are inconsistent and inadequate, or inaccessible.
4. The issue context is not reasonably well specified. The problem definition, value system design, and system synthesis steps have not been adequately completed or, for some reason, cannot be completed.
5. There are short-term deadlines or the cost of modeling outweighs potential benefits.

Brewer [114], who initially discussed many of these factors, also suggests that the following evaluation scorecard of potential models be used to evaluate their usefulness:

1. Intention, problem definition, and context demarcation
   a. Is the major purpose for which the model is to be built described?
   b. Are model appraisal criteria appropriate for the model intentions? Have we connected the intended use and criteria?
2. Specification and problem elaboration
   a. Have logicotheoretical strengths and weaknesses of the model been considered? Are the underlying structural assumptions well-treated?
   b. Have data aspects of the model received explicit consideration?
3. Control, information collection, and management
   a. Have data collection and management procedures been carefully considered and scrutinized?
   b. Has technical evaluation of the database been considered?
4. Validation and performance appraisal
   a. Have we assessed the model validation techniques we propose to use?
   b. Will we assess model validity with respect to the intended purpose of the model?
5. Overall appraisal function
   a. Theoretical appraisal: Is the model structurally sound? Have structure, function, and purpose been considered?
   b. Technical appraisal: Does the model reproduce historical data? How accurately have the parameters within the structure been tuned?
   c. Ethical appraisal: Does the model or the results from it offend moral codes, duty, or principles?
   d. Pragmatic appraisal: Is the model realistic and applicable with respect to policy or forecasting questions?

4.9 SUMMARY

This extensive chapter has examined many types of systems analysis and modeling methodologies. The relation of systems analysis and modeling to the other steps of systems engineering has been indicated previously, and a brief discussion has been devoted to questions of model evaluation. This is a particularly important topic, because the mere reproduction of a historical set of data is not in itself sufficient to make the model useful for explaining behavior. Almost any simple model with two or three constant parameters can be tuned to fit a curve. We must consider structure, function, and purpose when devising a model if it is to be a truly useful one.

There are limitations and problems associated with current systems analysis and modeling approaches. These include documentation problems, data source adequacy, error propagation, sensitivity considerations for variations in internal model parameters as well as exogenous variables, and model transferability from one setting and in one environment to another. There may be problems associated with building the proper connecting links between the formulation elements, including identification of these, and their use in intelligently constructing a model and linking it to the decision maker. Insightful, responsible, and appropriate model development should result in models that are both relevant to large-scale and large-scope issues.

We have presented many lists of steps to support systems modeling and have provided a number of detailed examples of some of the modeling approaches. We have also discussed cautions concerning their use. The problems at the end of the chapter provide the opportunity to construct and evaluate a number of simulation models. We will discuss some aspects of sensitivity analysis in our next chapter. In this chapter we wish to consider an analysis-related problem that can be used to model decisions under uncertainty. A simple illustration of such a situation is represented in Figure 4.63. Here the selection of one or the courses of action leads to a chance situation in which the outcome is uncertain. This is the sort of structure that the efforts in the latter portions of Chapter 5 are most useful.

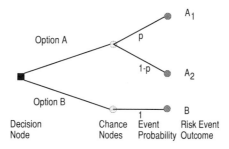

Figure 4.63 A decision tree model of actions and risk event outcomes.

PROBLEMS

4.1 Determine a feedback control system model for the epidemic propagation problem. Contrast and compare this with the system dynamics model of an epidemic.

4.2 Consider development of a system dynamics model to predict the percentage of children in public schools and the percentage in private school in a city. Both schools exist and will accept all who apply. The private school is very short of money and has high tuition. As tuition rates are increased in the private school, enrollments decrease. What other important factors do you believe should be incorporated in the model?

4.3 A system dynamics model for energy resource utilization might consist of the following level variables: level of reserves (NR), average usage rate (URA), level of technology (T), and substitute fraction (SF). A representative model would appear to consist of a basic supply–demand loop consisting of actual cost (AC), demand (DE), usage rate (URA), and reserves (NR). The reserves to cost link may actually be indirectly implemented through two other loops representing distribution (D) and pollution abatement (PA). This would be accomplished in such a way that as reserves are depleted, the increased cost of distribution and pollution abatement is reflected in the total cost of the resource. Provision should also be made for the discovery and development of new reserves on nonrenewable resources. This may be accomplished by means of exploration (EX), which may account for the effects of higher prices in stimulating the search for new reserves. Sales revenue from business income may be used to finance the search for new energy resources. Research and development may also lead to exploration for new resources, and newly developed techniques from this research and development may be so used. Technology (T) is basically the result of research and development financed through revenue (R) and demand (DE). Also, a competitive substitution variable (S) may determine changes in demand resulting from the use of alternative energy resources. The exogenous variable for this model may be assumed to be normal demand (ND) for a given energy source. Normal demand might be assumed to be an exponentially increasing function of time. This normal demand may be modified by actual developments, such as the ratio of cost to average cost or substitution, in order to generate actual demand (DE). Determine a system dynamics model for energy supply and demand with respect to a single energy source. Indicate how this basic model could be expanded to incorporate a variety of energy sources by cascading several simple models of the sort you have just developed and coupling them through some average cost and availability relationships.

4.4   Determine a system dynamics diagram for pertinent elements in your educational institution. Be sure to specify your assumed problem boundary and elements (faculty, research support, faculty promotions and discharges, student enrollments, etc.) used to construct your system dynamics diagram.

4.5   Among the several uses to which a model may be put are the following: (a) To display the perception and belief of a person concerning what has happened and what will happen in the world, (b) Same as (a) but for a group, (c) to learn more about a particular subject, (d) to predict or forecast the future, (e) to determine policy. Please comment upon the potential abilities of systems dynamics modeling in accomplishing each of these.

4.6   Determine the bounds for the future probabilities in Example 4.1.1 for the case where event j is inhibiting. What do the results of Example 4.1.1 become for the case where event i precedes event j in time?

4.7   Show that there are N^2 probabilities that must be estimated to employ the classical cross-impact analysis approach and $2N - 1$ questions to be answered If we use the direct tree probabilities suggested here.

4.8   Determine a KSIM model for the four elements of a highway transportation problem: M, quality of maintenance by highway department; S, vehicle speed; C present road conditions; V, vehicle volume. Use the model to examine two theories: (a) Larger maintenance funds result in better roads that carry more people at higher speeds, and (b) Increased volume damages highways and results in high maintenance costs and low speed.

4.9   Prepare a brief paper in which you contrast and compare cross-impact analysis with KSIM. For what purposes would one method be applicable and the other inapplicable? For what purposes would they both be equally applicable?

4.10 Reexamine the solution to the KSIM exercise considered in the text in which you use $x(1 - x)$ for the $x \ln x$ expression in the KSIM differential equation.

4.11 Prepare a brief paper in which you discuss amplitude and time-scale considerations associated with determination of a KSIM model and its associated solution.

4.12 A person borrows \$20,000 at a bank to buy a car. The loan arrangement calls for a four-year loan and for a quoted interest rate of 6%. The total interest is computed as 4 times 0.06 times 20,000, or \$4,800. The total to be repaid is \$24,800 and, spread over 48 months, the monthly payment is \$516.67. What is the actual, or effective, rate of interest in this case?

4.13 Based on the time value of money relations developed in this chapter, derive the algorithm that expresses the relationship between annual interest rate, amount borrowed, duration in months of the loan, and constant monthly payments in order to fully amortize a mortgage. What do the results of this problem become if there is a certain amount of the principal, called the balloon amount, which is not retired over time and which becomes due upon expiration of the mortgage.

4.14 How much should you be willing to pay now for a promissory note to pay $100,000 thirty years from now if the discount rate is 6%? 10%?

4.15 Suppose that you need to decide whether to accept investment A or investment B as the best investment. Each investment requires an initial investment of $10,000. Investment A will return $30,000 in a period of two years. Investment B will return $20,000 in one year. Suppose that the opportunity cost of capital (OCC), or discount rate, is 20% per year. Please compare the net present worth of the two investments and the internal rate of return of the two. Which project is the best according to these criteria? Suppose that the opportunity cost of capital is considered to be a variable. Is there some OCC where the two investments have the same value? Explain the implications of this.

4.16 Please indicate how you might modify the net present worth criterion such that you could consider a borrowing interest rate that is different from a lending interest rate.

4.17 Costs and benefits can be calculated from the perspective of an individual or firm (private sector analysis), the economic system (public economic analysis), or a complete sociopoliticoeconomic system (social cost–benefit analysis). Market prices would be used for benefits and costs in the first analysis, and shadow prices would be used in the second. The third analysis would necessarily attempt to weigh various potentially noncommensurate factors into the analysis. Please write a brief paper on how we might go about each analysis.

4.18 Identify relevant factors that will enable you to bring about a cost–effectiveness analysis of a make or buy decision for (a). database management software, (b) a complete office automation system, and (c) a decision support system [115].

4.19 Prepare for and conduct a cost–effectiveness analysis of the potential decision to purchase a personal computer. Please do this from the perspective of an individual, the organization in which the individual is employed, and society.

4.20 Two projects each cost $300,000 and are directed at technologies that will supply the same service. The technology resulting from one project has a 0.6 probability of supplying the service at a cost of $1 per unit and a 0.40 probability of supplying it at a cost of $2 per unit. Comparable risks and outcomes result from project B. The worth to the project developer of being able to supply the service at $1 per unit is estimated at $2M, and the worth of supplying it at $2 per unit is $1M. What is the expected worth of the two projects, and which should be selected if only one project is to be developed? What is the probability that the technology produced by the nonfunded project is at least as good as that developed by the funded project? What does this say in terms of risk management?

4.21 A company wishes to explore the possibilities of implementing a computer-integrated manufacturing process for its production efforts. Please prepare a cost–benefit analysis or cost–effectiveness analysis of a CIM effort in terms of productivity, profitability, and competitiveness. There are tangible and intangible benefits to a CIM effort, and you should discuss these.

4.22 The most elementary approaches to cost–effectiveness analysis are intended for application to those situations where the following conditions exist. (a) Enough is known about the technologies under development to develop credible forecasts and estimates of their performance. (b) There is no transfer of information between projects. (c) The performance of each product does not depend on the performance of the others. These criteria are generally met by large-scale projects directed at the final stages of technology development but are not met by many basic research efforts. Discuss issues associated with expanding the approach such that it is applicable to each of these situations.

4.23 You are buying a Sport Utility Vehicle (SUV) which costs $25,000. You put $5000 down on it, which will leave you with a $460/month payment for 48 months. You also estimate that it will cost you $100/month to operate, $200/month to insure, and $50/month to maintain. The manufacturer projects the resale value of the truck to be $12,000 in four years. (You plan to sell the car after four years.) Based on this information, draw the cash flow diagram (CFD) for the period you will own your SUV. What is the value, in present-day dollars, of how much you will lose or gain by buying this SUV?

4.24 Build a cost breakdown structure for a project based on the following figures:

    a. Research and development costs: 3.2 million during FY93–94
    b. Factory modifications: 2.2 million in FY94
    c. Advertising: 1.2 million in FY94, 0.8 million annually after that.
    d. Distribution costs: 0.5 million per year
    e. Replacement model research: 0.5 million during FY95–97, 1.5 million in FY98
    f. Required service/maintenance support: 1.2 million per year
    g. Production costs: 180 million per year

4.25   The ACME Corporation faces a decision whether to lease or purchase a piece of manufacturing equipment. **Purchase:** The purchase price is $400,000 and installation costs will be $40,000. The equipment can be expected to have a useful life of 5 years, after which it can be sold as scrap with a net realization of $30,000. To maintain the equipment is expected to require $3000 per year; taxes and insurance, $1200. Assume that the annual outlays occur at the beginning of each year. **Lease:** The machine can be leased for 5 years at an annual rental fee of $58,000, payable in five equal annual installments at the beginning of each year. The lessor agrees to install the equipment ready to use, to remove it at the end of five years, and to see that it is kept in working order. The ACME Corporation will have to spend about $1000/year for ordinary maintenance and $500/year for insurance. Alternative investment opportunities would produce a 5% return to the ACME Corporation at this time. What should the corporation management do? Use present values comparison.

4.26   You are the systems engineer working on the design of a new subway system. One part of your system is automatic token vending machines. Each machine operates for an average of 9000 hours between chance failures (MTBF). The mean time for wearout failures on these vending machines is 27,000 hours with a standard deviation of 3500 hours. What is the chance failure rate (λ)? You want a maintenance plan whereby the machines are overhauled before any wearout failures occur. When is the latest possible time you can do this? A vending machine has been in operation for 14,000 hours without failure. Determine the reliability of the machine for the next 5000 hours.

REFERENCES

[1]   Gass, S. I., and Sisson, R. L., *A Guide to Models in Government Planning and Operations*, Sauger Books, Potomac, MD, 1975.

[2]   Atherton, G., and Borne, P. (Eds), *Concise Encyclopedia of Modeling and Simulation*, Pergamon Press, Oxford, UK, 1992.

[3]   Sage, A. P., and Botta, R., On Human Information Processing and Its Enhancement Using Knowledge-Based Systems, *Large Scale Systems*, Vol. 5, 1983, pp. 208–223.

[4]   Gordon, T.J., and Hayward, H., Initial Experiments with the Cross-Impact Matrix Method of Forecasting, *Futures*, Vol. 1, No. 2, December 1968, pp. 100–116.

[5]   Gordon, T. J., Cross Impact Matrices — An Illustration of Their Use for Policy Analysis, *Futures*, Vol. 2, No. 4, December 1969, pp. 527–531.

[6]   Enzer, S., A Case Study Using Forecasting as a Decision-Making Aid, *Futures*, Vol. 2, No. 4, December 1970, pp. 341–362.

[7]   Enzer, S., Cross Impact Techniques in Technology Assessment, *Futures*, Vol. 4, No. 1, March 1972, p. 3051.

[8]   Duval, A. E., Fontella, E., and Gabus, A., Cross Impact Analysis: A Handbook of Concepts and Applications, in Baldwin, M. (Ed.) *Portraits of Complexity*, Battelle Monograph 9, Battelle Memorial Institute, Columbus, OH, 1975, pp. 202–222.

[9]   Helmer, O., *Looking Forward — A Guide to Futures Research*, Sage Publishers, Beverly Hills CA, 1983.

[10]   Porter, A. L., Roper, A. T., Mason, T. W., Rossini, F. A., and Banks, J., *Forecasting and Management of Technology*, Wiley, New York, 1991.

[11]   Enzer, S., Delphi and Cross-Impact Techniques: An Effective Combination for Systematic Futures Analysis, *Futures*, Vol. 3, No. 1, March 1971, pp. 48–61.

[12]   Enzer, S., and Alter, S., Cross Impact Analysis and Classical Probability: The Question of Consistency, *Futures*, Vol. 10, No. 2, March 1978, pp. 227–239.

[13]   Quinlan, J. R., "Inferno: A Cautious Approach to Uncertain Inference," *The Computer Journal*, Vol. 26, No. 3, 1983, pp. 255–266.

[14]   Sage, A. P. (Ed.), *Concise Encyclopedia of Information Processing in Systems and Organizations*, Pergamon Press, Oxford, UK, 1990.

[15]   Gettys, C., and Willke, T. A., The Application of Bayes' Theorem When the True Data State is Uncertain, *Organizational Behavior and Human Performance*, Vol. 4, 1969, pp. 125–141.

[16]   Edwards, W., Phillips, L. D., Hays, W. L, and Goodman, B. C., Probabilistic Information Processing Systems: Design and Evaluation, *IEEE Transactions on System Science and Cybernetics*, Vol. SMC-4, 1968, pp. 248–265.

[17]   Pearl, J., *Heuristics: Partially Informed Strategies for Computer Problem Solving*, Addison-Wesley, Reading, MA, 1983.

[18]   Pearl, J., *Probabilistic Reasoning in Intelligent Systems: Networks of Plausible Inference*, Morgan Kaufman, San Mateo, CA, 1988.

[19]   Schum, D. A., *Evidential Foundations of Probabilistic Reasoning*, Wiley, New York, 1994.

[20]   Dempster, A. P., Upper and Lower Probabilities Induced by a Multivalued Mapping, *Annals of Mathematical Statistics*, Vol. 38, 1967, pp. 325–339.

[21]   Shafer, G., *A Mathematical Theory of Evidence*, Princeton University Press, Princeton, NJ, 1976.

[22]   Shafer, G., The Combination of Evidence, *International Journal of Intelligent Systems*, Vol. 1, No. 2, 1986, pp. 155–179.

[23]   Toulmin, S., Rieke, R., and Janik, A. *An Introduction to Reasoning*, MacMillan, New York, 1979.

[24]   Suppes, P., *A Probabilistic Theory of Causality*, North-Holland, Amsterdam, 1970.

[25]   Ottes, R., A Critique of Suppes' Theory of Causality, *Synthese*, Vol. 48, No. 2, 1981, pp. 167–189.

[26]   Lagomasino, A., and Sage, A. P., An Interactive Inquiry System, *Large Scale Systems*, Vol. 9, No. 3, 1985, pp. 231–244.

[27] Lagomasino, A., and Sage, A. P., Imprecise Knowledge Representation in Inferential Activities, in M. Gupta, (Ed.), *Approximate Reasoning in Expert Systems*, North-Holland Elsevier, Amsterdam, 1985, pp. 473–497.

[28] Janssen, T., and Sage, A. P., Toulmin-Based Logic in a Group Decision Support System for Pest Management, *Proceedings, IEEE Systems Man and Cybernetics Annual Conference*, Beijing, China, October 1996, pp. 2704–2709.

[29] Janssen, T. J., and Sage, A. P., A Group Decision Support System for Science Policy Conflict Resolution, in Beroggi, E. G. (Ed.), Special Issue on Public Policy Engineering Management, *International Journal of Technology Management*, 1999, in press.

[30] Kahneman, D., Slovic, P., and Tversky, A. (Eds.), *Judgment Under Uncertainty: Heuristics and Biases*, Cambridge University Press, New York, 1982.

[31] Sage, A. P., *Systems Engineering*, Wiley, New York, 1992.

[32] Sage, A. P., *Systems Management for Information Technology and Software Engineering*, Wiley, New York, 1995.

[33] Sage, A. P., and Rouse, W. B. (Eds.), *Handbook of Systems Engineering and Management*, Wiley, New York, 1999.

[34] Harary, F., Norman, R. Z., and Cartwright, D., *Structural Models: An Introduction to the Theory of Directed Graphs*, Wiley, 1965.

[35] Warfield, J. N., *Societal Systems: Planning, Policy, and Complexity*, Wiley, New York, 1976.

[36] Sage, A. P., *Methodology for Large Scale Systems*, McGraw-Hill, New York, 1977.

[37] Steward, D. V., *Systems Analysis and Management: Structure, Strategy, and Design*, Petrocelli Books, New York, 1981.

[38] Lendaris, G. G., Structural Modeling: A Tutorial Guide, *IEEE Transactions on Systems, Man, and Cybernetics*, Vol. SMC 10, No. 12, December 1980, pp. 807–840.

[39] Eden, C., Jones, S., and Sims, D., *Messing About in Problems*, Pergamon Press, Oxford, UK, 1983.

[40] Roberts, F. M., *Discrete Mathematical Models*, Prentice-Hall, Englewood Cliffs, NJ, 1976.

[41] Geoffrion, A. M., An Introduction to Structured Modeling, *Management Science*, Vol. 33, No. 5, May 1987, pp. 547–588.

[42] Geoffrion, A. M., The Formal Aspects of Structured Modeling, *Operations Research*, Vol. 37, No. 1, January 1989, pp. 30–51.

[43] Geoffrion, A. M. Computer Based Modeling Environments, *European Journal of Operations Research*, Vol. 41, No. 1, July 1989, pp. 33–43.

[44] Forrester, J. W., *Urban Dynamics*, MIT Press, Cambridge, MA, 1969.

[45] Howard, R. A., and Matheson, J. E., Influence Diagrams, *Readings on the Principles and Applications of Decision Analysis*, Vol. II, Strategic Decisions Group, Palo Alto, CA, 1984, pp. 719–762.

[46] Shachter, R. D., Evaluating Influence Diagrams, *Operations Research*, Vol. 34, No. 6, 1986, pp. 871–882.

[47] Howard, R. A., Knowledge Maps, *Management Science*, Vol. 35, No. 8, August 1989, pp. 903–922.

[48] Genest, C., and Zidek, J. V., Combining Probability Distributions: A Critique and an Annotated Bibliography, *Statistical Science*, Vol. 1, No. 1, 1986, pp. 114–148.

[49] Shachter, R. D., Probabilistic Influence and Influence Diagrams, *Operations Research*, Vol. 36, No. 4, July 1988, pp. 589–604.

[50] Shachter, R. D., An Ordered Examination of Influence Diagrams, *Networks*, Vol. 20, 1990, pp. 535–563.

[51] Call, H. J., and Miller, W. A., A Comparison of Approaches and Implementations for Automating Decision Analysis, *Reliability Engineering and System Safety*, Vol. 30, 1990, pp. 115–162.

[52] Tatman, J. A., and Shachter, R. D., Dynamic Programming and Influence Diagrams, *IEEE Transactions on Systems, Man, and Cybernetics*, Vol. 20, No. 2, March 1990, pp. 365–379.

[53] Buede, D. M., and Ferrell, D. O., Convergence in Problem Solving: A Prelude to Quantitative Analysis, *IEEE Transactions on Systems, Man, and Cybernetics*, Vol. 22, No. 6, December 1992, pp. ???–???.

[54] Tatman, J. A., and Shachter, R. D., Dynamic Programming and Influence Diagrams. *IEEE Transactions on Systems, Man, and Cybernetics*, Vol. 20, No. 2, April 1990, pp. 365–379.

[55] Forrester, J. W., *Principles of Systems*, Wright Allen Press, Cambridge, MA, 1971.

[56] Forrester, J. W., *World Dynamics*, Wright Allen Press, Cambridge, MA, 1971.

[57] Roberts, N. D. F., Anderson, R. M., Deal, R. M., Garet, M. S., and Shaffer, W. A., *Introduction to Computer Simulation: A Systems Dynamics Modeling Approach*, Addison-Wesley, Reading MA, 1983.

[58] Meadows, D. L., and Meadows, D. H. (Eds.), *Towards Global Equilibrium*, Wright Allen Press, Cambridge, MA, 1973, pp. 155–186.

[59] Meadows, D. H., Richardson, J., and Bruckmann, G., *Groping in the Dark: The First Decade of Global Modeling*, Wiley, New York, 1981.

[60] Meadows, D. H., Meadows, D. L., and Randers, J., *Beyond the Limits*, Chelsea Green, White River Junction, VT, 1992.

[61] Saeed, K., *Development Planning and Policy Design: A Systems Dynamics Approach*, Ashgate, Brookfield, VT, 1994.

[62] Chen, K. (Ed.), Special Issue on Urban Modeling, *IEEE Transactions on Systems, Man, and Cybernetics*, Vol. 2, No. 2, April 1972.

[63] Sage, A. P. (Ed.), *Systems Engineering: Methodology and Applications*, IEEE Press, New York, 1977.

[64] Graedel, T. E., and Allenby, B. R., *Industrial Ecology*, Prentice-Hall, Englewood Cliffs, NJ 1995.

[65] Graedel, T. E., *Streamlined Life-Cycle Assessment*, Prentice-Hall, Upper Saddle River, NJ, 1998.

[66] Allenby, B. R., *Industrial Ecology: Policy Framework and Implementation*, Prentice-Hall, Upper Saddle River, NJ, 1999.

[67]   Kiksel, J., *Design for Environment*, McGraw-Hill, New York, 1996.

[68]   Clark, J., and Cole, S., *Global Simulation Models: A Comparative Study*, Wiley, New York, 1974.

[69]   Bremer, S. A., Computer Modeling in Global and International Relations: The State of the Art, *Social Science Computer Review*, Vol. 7, No. 4, July 1989, pp. 459–478.

[70]   Abdel-Hamid, T., and Madnik, S. E., *Software Project Dynamics: An Integrated Approach*, Prentice-Hall, Englewood Cliffs, NJ, 1991.

[71]   Senge, P., Kleiner, A., Roberts, C., Smith, B., and Ross, R. (Eds.), *The Fifth Dimension Fieldbook*, Doubleday, New York, 1994.

[72]   Morecroft, J., and Sterman, J. (Eds.), *Modeling for Learning*, Productivity Press, Portland, OR, 1994.

[73]   Meadows, D. H., Meadows, D. L., and Randers, J., *Beyond the Limits*, Chelsea Green, Post Hills, VT, 1992.

[74]   Richardson, G. P., and Pugh, A. III, *Introduction to System Dynamics Modeling with DYNAMO*, Productivity Press, Portland, OR, 1981.

[75]   Richmond, B., and Peterson, S., *An Introduction to Systems Thinking*, High Performance Systems, Hanover, NH, 1992.

[76]   Kane, J., A Primer for a New Cross Impact Language—KSIM, *Technological Forecasting and Social Change*, Vol. 4, No. 1, 1972, pp. 129–142.

[77]   Kane, J., Thompson, W., and Vertinsky, I., Health Care Delivery: A Policy Simulator, *Socio-Economic Planning Sciences*, Vol. 6, No. 3, 1972, pp. 283–293.

[78]   White, K. P., Jr., Workshop Dynamic Modeling, in Sage, A. P., *Concise Encyclopedia of Information Processing in Systems and Organizations*, Pergamon Press, Oxford, UK, 1992., pp. 508–514.

[79]   Thuesen, G. J., and Fabrycky, W. J., *Engineering Economy*, Prentice-Hall, Englewood Cliffs, NJ, 1989.

[80]   Park, C. S., *Contemporary Engineering Economics*, Addison Wesley, Reading, MA, 1993.

[81]   Sage, A. P., *Economic Systems Analysis: Microeconomics for Systems Engineering, Engineering Management, and Project Selection*, North-Holland, New York, 1983.

[82]   Porter, M. E. *Competitive Advantage*, Free Press, New York, 1985.

[83]   Michaels, J. V., and Wood, W. P., *Design to Cost*, Wiley, New York, 1989.

[84]   Blanchard, B. S., *Systems Engineering Management*, Prentice-Hall, Upper Saddle River, NJ, 1998.

[85]   Fabrycky, W. J., and Blanchard, B. S., *Life-Cycle Cost and Economic Analysis*, Prentice-Hall, Englewood Cliffs, NJ, 1991.

[86]   Kerzner, H., *Program Management: A Systems Approach to Planning, Scheduling, and Controlling*, 4th edition, Van Nostrand Reinhold, New York, 1992.

[87]   Eisner, H., *Essentials of Project and Systems Engineering Management*, Wiley, New York, 1997.

[88]   Bussey, L. E., *The Economic Analysis of Industrial Projects*, Prentice-Hall, Englewood Cliffs, NJ, 1978.

[89]   Mishan, E. J., *Cost–Benefit Analysis*, Praeger, New York, 1976.

[90] Porter, A. L., Rossini, F. A., Carpenter, S. R., Roper, A. T., 1980 *A Guidebook for Technology Assessment and Impact Analysis*, North-Holland, New York, 1980.

[91] Sassone, P. G., Schaffer, W. A., *Cost–Benefit Analysis — A Handbook*, Academic Press, New York, 1978.

[92] Sugden, R., and Willliams, A., *The Principles of Practical Cost–Benefit Analysis*, Oxford University Press, Oxford, UK, 1978.

[93] King, J. L., and Schrems, E. L., Cost–Benefit Analysis in Information Systems Development and Operation, *Computing Surveys*, Vol. 10, No. 1, March 1978, pp. 20–34.

[94] Ewusi-Mensah, K., Evaluating Information Systems Projects: A Perspective on Cost–Benefit Analysis, *Information Systems*, Vol. 14, No. 3, 1989, pp. 205–217.

[95] Merkhofer, M. W., *Decision Science and Social Risk Management*, Reidel, Dordrecht, Holland, 1987.

[96] Layard, R., and Glaister, S. (Eds.), *Cost Benefit Analysis*, Cambridge University Press, New York, 1994.

[97] Lederer, A. L., and Prasad, J., Nine Management Guidelines for Better Cost Estimating, *Communications of the Association for Computing Machinery*, Vol. 35, No. 2, February 1992. pp. 51–59.

[98] Blanchard, B. S., and Fabrycky, W. J., *Systems Engineering and Analysis*, 3rd edition, Prentice-Hall, Upper Saddle River, NJ, 1998.

[99] Ireson, W. G., and Combs, C. F. (Eds.), *Handbook of Reliability Engineering and Management*, McGraw-Hill, New York, 1988.

[100] Geoffrion, A. M. Computer Based Modeling Environments, *European Journal of Operations Research*, Vol. 41, No. 1, July 1989, pp. 33–43.

[101] Bertsekas, D. P., *Dynamic Programming and Stochastic Control*, Academic Press, New York, 1976.

[102] Gross, D., and Harris, C. M., *Fundamentals of Queueing Theory*, 2nd Edition, Wiley, New York, 1985.

[103] Kleinrock, L., *Queueing Systems*, Wiley, New York, 1975.

[104] Fishwick, P. A., *Simulation Model Design and Execution: Building Digital Worlds*, Prentice-Hall, Upper Saddle River, NJ, 1995.

[105] Pritsker, A. A. B., *Introduction to Simulation and Slam II*, 2nd edition, Wiley, New York, 1984.

[106] Pritsker, A. A. B., and Sigal, C. E., *Management Decision Making: A Network Simulation Approach*, Prentice-Hall, Upper Saddle River, NJ, 1983.

[107] Cloud, D. J., and Rainey L. B. (Eds.), *Applied Modeling and Simulation: An Integrated Approach to Development and Operation*, McGraw-Hill, New York, 1998.

[108] Melsa, J. L., and Sage, A. P., *An Introduction to Probability and Stochastic Processes*, Prentice-Hall, Englewood Cliffs, NJ, 1973.

[109] Armstrong, J. S., *Long Range Forecasting: From Crystal Ball to Computer*, Wiley, New York, 1978.

[110] Box, G. E. P., *Time Series Analysis, Forecasting and Control*, Holden Day, Oaklands, CA, 1970.

[111] Kotz, S., and Johnson, N. L. (Eds) *Encyclopedia of Statistical Sciences*, Wiley, New York, 1983.

[112] Sage, A. P, and Melsa, J. L. *System Identification*, Academic Press, New York, 1971.

[113] Sage, A. P., and Melsa, J. L., *Estimation Theory: with Application to Communication and Control*, McGraw-Hill, New York, 1971.

[114] Brewer, G. D., Policy Analysis by Computer Simulation: The Need for Appraisal, Rand Corporation Memo P-4893, Santa Monica, CA, 1973.

[115] Gremillion, L. L., and Pyburn, P. J., Justifying Decision Support and Office Automation Systems, *Journal of Management Information Systems*, Vol. 2, No. 1, Summer 1985, pp. 5–17.

CHAPTER 5

Interpretation of Alternative Courses of Action and Decision Making

People make decisions. This is a central fact that must be remembered even in today's world of increasing computer automation where many times it seems that machines are making decisions for us. For example, your request for a cash advance at an automated teller kiosk is disapproved. But, you know better than to get mad at the machine, because the machine—unless it is malfunctioning—is just executing a policy that was decided by people and programmed into the automated teller's software logic.

This chapter is about helping people make better decisions in general and in helping systems engineers with the decision-making tasks they typically encounter in interpreting the impacts of alternative courses of action, or in working with clients in so doing. The systems engineering life cycle is intended to enable evolution of high-quality, trustworthy systems that have appropriate structure and which also provide functional support for identified purposeful client objectives. Many major efforts called for throughout all phases of a systems engineering life cycle involve the making of decisions, or *decision assessment*, or decision making, or decisionmaking, or decision-making, and even decision taking in the United Kingdom and elsewhere. All of these terms are used throughout the *decision* literature. Without question, decision-making is the most common term. Decision assessment is perhaps the least common of the several noted. We feel that it is an appropriate term because we are interested in accomplishing much more than an analysis of decisions that have been, or which could be, made. So, we need to make a decision concerning which term to use. It is not an easy decision. A relevant question is; Could we use the methods outlined in this chapter to assist in making, or taking, the decision? Would the results of the analysis differ across approaches used? Why? What are the implications of this difference?

This chapter continues with our discussions of methodological issues in systems engineering. The focus is, however, shifted to concerns that relate to normative, descriptive, and prescriptive aspects of decision making, or perhaps more appropriately stated *decision assessment*. The term *decision analysis* is more common than *decision assessment*. One of our steps in the effort involves analysis of the impacts of alternative decisions. For this and other reasons, we prefer the slightly more generic term, decision assessment. Clearly, this is a very small point. We will often use the terms interchangeably. There has been much said on this subject of decision analysis. We will attempt to present the salient features of some of the work in this area that is most relevant to systems engineering. Because our coverage is necessarily broad, it will be helpful to provide an overview and perspective. Then we will examine normative decision analysis and prescriptive decision analysis. There is much of importance that has been said concerning decision assessment. Although our chapter is, necessarily, lengthy, we can cover only a rather small part of this wealth of material.

5.1 INTRODUCTION: TYPES OF DECISIONS

A *decision* is an allocation of resources. It is generally irrevocable, although a new decision may reverse the effects of an initial decision. The *decision maker* is the person who has authority over the resources that are being allocated. The decision maker makes a decision, by allocation of resources, in order to further the achievement of some *objective* that is felt to be desirable. It is generally important to distinguish between decision and objective. For example "to lose weight" is an objective and not a decision. To exercise or diet (in order to lose weight) is a decision, in that it involves a course of action that will hopefully lead to attainment of an objective.

A decision maker makes, or at least should make, decisions that are consistent with *values*, or those things that are important to the decision maker or those represented by the decision maker. A very common value is economic, in an effort to increase wealth, but such personal or social values as happiness, security, and equity are also very important. Sometimes a distinction is made between a *goal* and an objective, where a goal refers to a specific degree of satisfaction of a given objective. For example, the objective of the decision might be to lose weight, and the goal might be to weigh 10 pounds less in a month. This goal is also an objectives measure.

Decision analysis, is a structured and formal viewpoint that relates how a course of action taken would lead to a result. Generally, there are three features of a decision situation that are of importance: a decision to be made and course of action to be taken, the unknown events and outcomes that can affect the result, and the obtained result itself. The decision analysis approach is based on construction of *models*, which represent logical, often mathematical, representations of the relationships among these three features of the decision situation. Use of these models leads to estimation of the possible implications of each alternative course of action that might be taken, thereby leading to understanding of the relationships between actions and objectives and to selection of the most appropriate decision.

At the time a decision is made, or taken, the decision maker has available at least two *alternatives*, or courses of action that might be taken. If there are not at least two alternatives, one of which is the do-nothing option, then no decision is possible! When an alternative is chosen and committed to, the decision has been made or taken and *uncertainties* generally arise. These are uncontrollable chance elements. Different alternatives may be associated with different uncertainties. In each case, it is the combination of alternatives with uncertainties that ultimately lead to an *outcome*. Thus, an outcome is a generally uncertain result that follows from the decision situation, the impacts of which should be measured on the decision- maker's value scale.

A *simple decision* is one in which there is only one decision to be made, even though there might be many alternatives. If there are two or more decisions to be made, we have a problem of *strategy*. Sometimes, several decisions need to be made at the same time and this special case of a strategy problem is the *portfolio* problem. Often, the various decisions associated with the strategy are of a similar nature, and there are not sufficient resources to enable selection of all alternatives. The prototypical example is an investment portfolio. Here, the decision maker is aware of a number of investments that might potentially be made, but there is a finite budget. In a situation like this, we approach the problem as one in which we accomplish a *prioritization* of the various opportunities. If one opportunity has a higher priority than another, then the decision maker would prefer to invest in the alternative that leads to this opportunity or outcome.

In other cases, *sequential decisions* are involved. Here, the decision maker is able to observe the outcome that results from one decision before having to take a subsequent decision.

Decision making would generally be easy if we could predict precisely the outcome that would result from selection of an alternative course of action. When the quantities to be forecasted are uncertain, we often try to describe the uncertainty using a *probability distribution* to capture what is known about uncertainties. When a decision maker contemplates a decision, the notion of *risk* often arises. For the most part, we are *risk averse* and will be willing to pay something to have risk reduced. Thus, we often value alternatives at less than their expected values. These are some of the most important considerations that affect important formal decisions, and we examine them in this section.

Decisions range from very simple to very complex, and from very narrow to very broad in scope. We make personal, and organizational or business, decisions every day. We make very simple decisions without much explicit consideration of the factors affecting the decision or those individuals who may be affected by the decision. At least we often think that this is what we do. In reality, a decision often seems simple because it concerns an issue, environment, and relevant considerations with which we are familiar. Intuition or other skill-based forms of reasoning may not only be perfectly acceptable in these situations, but perhaps actually preferred to a very formal and perhaps lengthy process. In the case of more complex decisions, we normally give much more thought and consider more of the factors involved, especially when we do not have a wealth of initial experiential familiarity with the decision situation. The term *decision situation* is a relatively broad term that includes the following:

- The objectives to be achieved
- The needs to be fulfilled
- Constraints and alterables associated with the decision
- Those affected by the decision
- The decision options, or alternative courses of action, themselves
- The environment in which all of these are imbedded
- The experiential familiarity of the decision maker with all of these

Thus, the *decision situation* is very dependent upon contingency variables. Figure 5.1 is a representation of some of these variables as they impact a decision situation. Decisions may also be taken by an individual, or by a team or a group of individuals.

As shown in Figure 5.1, there are at least seven important elements that make up the situation surrounding a decision situation. These seven elements

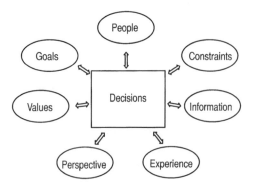

Figure 5.1 Surrounding and contingency elements that comprise the decision situation.

of a decision situation are *people, goals, constraints, values, perspective, experience,* and *information.* People are always a primary concern because they ultimately make the decisions and are impacted by them. There can be one or many people involved in a decision situation. Oftentimes a group of people are brought together to decide or advise on a tough decision problem. For example, many corporations and government bodies form ad hoc committees for this purpose. Something to remember about decision-making groups is that the roles that people are supposed to play in these groups are often unclear. As a result, strong personalities can sometimes dominate the thinking of the group even though others in the group have better ideas.

Teams are often formed to overcome the problems of groups by assigning well-defined roles for people to play. This structured team approach is often used in real-time decision-making situations where the cost of improper decisions is high and where the decision-making tasks are beyond the limitations of a single person. A good example of this is the airplane crew where the captain, co-pilot, and navigator all are selected and trained to perform coordinated decision-making tasks to safely fly planes.

In the final analysis, many decisions are a result of people acting as individuals. Sometimes decisions are mainly a matter of personal style and taste. This can be true not just of deciding on a fine wine to accompany dinner. It can even be true of large corporations where a single, powerful individual imposes their taste and style on key organizational decisions. Other times, even a consensus-building leader must retreat to make an individual decision because, even though they are open to listening to lots of advice from others, the advice they get is often conflicting. A leadership position often requires an individual decision. Perhaps that is one reason why it is "often lonely at the top."

As a systems engineer working with people as individuals or organized into groups or teams, you must be aware of these influences on the decision situation. What are the groups or teams involved? Is there a dominant individual? Is the dominant individual also the most knowledgeable individual? What is the personal taste and style of key decision makers? Will there be conflicting advice? If so, who will give it and will the decision maker retreat to make the decision alone? If so, when will the decision be made? Of course, another important aspect that people bring into decision situations is their values.

In considering decisions, especially social decisions, we must exercise great care to incorporate the *values* of those affected by the decision into the decision-making process. Incorporating these values into the process often results in making some very basic decisions. These decisions may be made consciously or they may be made unconsciously. Suppose, for example, that a city is considering the construction of a monorail mass transit system. Very likely, the questions asked by the city planners concern right of way, scheduling, and other difficulties. To ask such questions, they must have decided that a monorail system is the best of several alternative methods of moving people from one location to another. Such a decision could very well come from the systems engineering approach considered in earlier chapters.

However, before even beginning the systems engineering approach, some person or some group has made a decision that it is good to be able to move large numbers of people from one location to another with great speed. Thus, the decision makers have incorporated into their value system the concept that a mobile society is desirable and even preferred to some other form of society that would not permit such mobility. A decision analyst who has been called upon to help solve a problem must accept the fundamental decisions and values of those stakeholders whose brokerage has funded the problem study. It is essential that those fundamental values that act as constraints upon problem solution be explicitly enumerated as such when presenting the results of the decision analysis.

Goals and constraints are important aspects of any decision situation because they help determine feasible options. A goal or objective of a stakeholder is a statement of their intent. It is what they want to accomplish or achieve. If ignored, then the entire systems engineering effort can be judged a failure. For example, in designing investment options you need to know the ultimate goals and objectives of the investor. Otherwise you have no way of evaluating the suitability of alternative investment portfolios.

Constraints on possible decisions or outcomes or objectives are almost always present. Someone may have a goal of a lavish retirement lifestyle but may be constrained in reaching that objective by their modest income. So money or budget constraints are often important in decision situations. Another important constraint is time. The time that you have in making a decision is itself an important consideration. Sitting around a conference table in a boardroom and contemplating market strategy for next year is a crucial decision. Yet contrast that situation with trying to decide how to maneuver a commercial airliner to evade a thunderhead of wind sheer that just erupted in front of your flight path. The time pressure of events is a critical aspect of the decision-making situation and can cause significant stress on decision makers. They may even be in fear for their own safety in some situations that can create emotional responses to decision situations.

Perspectives that people have about decision situations are sometimes vitally important to what ultimately is decided. Four human rationality perspectives that you must always consider in almost every decision situation are emotional, organizational, political, and technoeconomic. People some-times respond to a decision situation on the basis of their emotions. Because of previous personal or other experiences, they may decide in a particular situation out of anger, sympathy, jealousy, or fear. Other times people, as we mentioned before, will be largely influenced by their position and role in an organization. Political motivation is another reality of decision-making situ-ations. People sometimes make decisions not just for the long-term good of the organization, but, at times, for the purpose of gaining political or organi-zational power. The fourth perspective is the technoeconomic rationality perspective that is the main topic of this chapter. The formal-reasoning-based technoeconomic rationality approach to decision-making is based on making the best possible use of the information available for a decision situation from the perspective of maximizing cost-effectiveness of the resulting solution. A much expanded discussion of rationality perspectives may be found in Chapter 9 of reference 1, which contains much more material concerning descriptive decision assessment than presented here.

Experience is a crucial aspect in decision-making situations. People who have extensive experience in dealing with the same or nearly the same decision situation over and over again can develop expert knowledge that helps them make these decisions much better than an inexperienced or novice decision maker could. Also, the "style" of decision making is often different as also indicated in some detail in reference 1 and in a number of recent efforts that provide discussions of how people make decisions [2] and how to make better decisions through formal approaches [3].

Information, and associated knowledge, is the last of the important seven elements of a decision situation. Hopefully, all decisions are based on information about the decision problem and the situation in which the decision problem is embedded. People also have biases that affect how they process information, so it is important for a decision analyst to understand and account for these biases.

Depending on the complexity and scope involved, the thought first given to an impending decision may include a brief mental comparison of experientially familiar alternatives, or it may be a thorough analysis appropriate to a complex, less-understood situation in which there are significant differences in the impacts of various alternative courses of action.

The *decision assessment process* may be described as follows. The decision maker is presented with a problem that requires a decision. Certain objectives may be provided by those to whom the decision maker is responsible. Also available, or identified as part of the effort, are certain alternative courses of action, each of which satisfies the objectives in some way and to some extent. The problem is to choose the alternative course of action that best meets or satisfies the objectives and that is fully responsive to any constraints. If there is more than one factor that contributes to the satisfaction of the objectives, the decision maker should generally find some way to combine the effects of these factors in some "best" or most appropriate way.

Descriptive, or *behavioral*, *decision-making*, or *assessment* is concerned with the way in which human beings in real situations actually make decisions. The term "way" is intended to include both the process used to make a decision and the actual decision that is made. *Normative decision assessment* is concerned with the decisions that should be made if the decision maker wishes to satisfy some, presumably desirable, set of axioms of rationality. Normative decision analysis is often called rational decision analysis or axiomatic decision analysis. In the more formal approaches, a set of *axioms* that a *rational* person would surely agree with is postulated. From this results the *normative*, or most desirable, or optimum, behavior which the rational decision maker should seek to follow in order to conform to the accepted axioms.

Prescriptive decision assessment involves suggestions of appropriate decision behavior that tempers the formal normative approaches to ensure that the process reflects the needs of real people in real decision situations. For the most part, our concern here is with prescriptive approaches. Because the prescriptive includes some blend of the normative and the descriptive, as indicated in Figure 5.2, we must really be aware of all three approaches. Experimental and other evidence indicate that the some of the prescriptions of the normative theory are often unacceptable as a descriptive theory of decision assessment. The normative theory is intended, of course, as a standard guide to what people would have to do to accommodate the intent of what appear to be very desirable axioms for judgment. Because unmodified normative theory does not describe unaided descriptive decision assessment, considerable caution in the use of the normative theory for aided prescriptive behavior should be exercised. Otherwise, very serious cognitive stress may result from potential acceptance of a normative theory that may be at variance with unaided descriptive behavior. The fact that unaided decision behavior (what people do) differs from the normative (what people ideally might do) and the prescriptive (what real people should do) appears not to be a criticism of the normative theory but, rather, a strong indication of need for the theory.

A number of years ago, Janis and Mann developed what is often called a *cognitive mode model of judgment and choice* [4]. Judgment and decision-making efforts are often characterized by intense emotion, stress, and conflict, especially when there are significant consequences likely to follow from decisions. As the decision maker becomes aware of various risks and uncertainties that may be associated with a course of action, this stress becomes all the more acute. Janis and Mann developed a *conflict model of decision-making* that reflects this thinking. Here, conflict refers to "simultaneous and opposing tendencies within the individual to accept and reject a given course of action." The most frequent of such symptoms of conflicts are hesitation, feelings of uncertainty, vacillation, and acute emotional stress with an unpleasant feeling

Figure 5.2   Prescriptive decision support as the confluence of the normative and the descriptive.

of distress. For this reason, this model is often called a *stress-based model*. The term time–stress-based model would seem more appropriate. The major elements associated with the conflict model of Janis and Mann are the concept of vigilant information processing, the distinction between "hot" and "cold" cognitions, and several coping patterns associated with judgments.

"Cold" cognitions are defined to be those cognitions made in a calm, detached state of mind. The changes in utility possible due to different decisions are small and easy to determine. "Hot" cognitions are those associated with vital issues and concerns and are associated with a high level of time–stress. Whether a cognition should be hot or cold is dependent upon the task at hand and the experiential familiarity and expertness of the decision maker with respect to the task. The symptoms of stress include feelings of apprehensiveness, a desire to escape from the distressing choice dilemma, and self-blame for having allowed oneself to get into a predicament where one is forced to choose between unsatisfactory alternatives. Janis and Mann state that "*psychological stress*" is used as a generic term to designate unpleasant emotional states evoked by threatening environmental events or stimuli. They define a "*stressful*" event as "any change in the environment that typically induces a high degree of unpleasant emotion, such as anxiety, guilt, or shame, and that affects normal patterns of information processing."

Janis and Mann describe several functional relationships between psychological time–stress and decision conflict:

1. The degree of time–stress generated by decision conflict is a function of those objectives that the decision maker expects to remain unsatisfied after implementing a decision. If implementation of a course of action is expected to produce a timely high-quality result, there is little stress associated with the decision.

2. Often a person encounters new threats or opportunities that motivate consideration of a new course of action. The degree of decision stress is a function of the degree of commitment to adhere to the present course of action.

3. When the degree of time–stress is low, and there is satisfaction with the present course of action, *unconflicted adherence to the present course of action* will be the chosen decision. When the time–stress is low, and there is dissatisfaction with the present course of action and a single satisfactory alternative can be identified, the decision maker will implement a decision involving *unconflicted change to a new course of action*.

4. When decision conflict is severe because all identified alternatives pose serious risks, failure to identify a better decision than the least objectionable one may lead to *defensive avoidance*, or undue procrastination.

5. In severe decision conflict when the decision maker anticipates having insufficient time to identify an adequate alternative that will avoid serious losses, the level of stress remains extremely high. The likelihood that the dominant pattern of response will be *hypervigilance*, or panic, increases.

6. A moderate degree of stress, which results when there is sufficient time to identify acceptable alternatives, in response to a challenging situation, induces a *vigilant effort to carefully scrutinize all identified alternative courses of action and to select a good decision.*

Based upon these relationships or propositions, Janis and Mann present five coping patterns that a decision maker would use as a function of the level of stress: unconflicted adherence or inertia, unconflicted change to a new course of action, defensive avoidance, hypervigilance or panic, and vigilance.

Figure 5.3 presents an interpretation of this conflict model of decision-making in terms of the contingency model discussed here. This model points to a number of markedly different tendencies that become dominant under particular conditions of time–stress. These include open-mindedness, indifference, active evasion of potentially disconfirming information, failure to assimilate new information, and other cognitive information processing biases. There are a number of cognitive information processing preferences and decision modes potentially generated by this conflict model. Particularly evident is the striking complexity entailed by the vigilant information processing pattern in comparison to the other coping patterns. This *vigilance pattern* is characterized by formulation, analysis, and interpretation of a number of alternative courses of action and implementation of the most preferred course. Selection of a

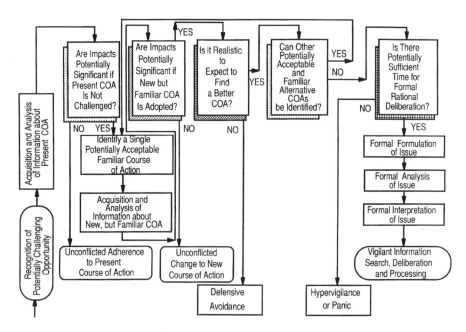

Figure 5.3 Interpretation of the contingency model of judgement and choice due to Janis and Mann.

coping pattern may be made properly or unwisely, just as selection of a decision style may be proper or improper. *We note that there are three proper coping strategies and two improper coping strategies in this figure.* Unconflicted adherence to a present course of action, unconflicted change to a new course of action, and vigilant search and deliberation are all appropriate *if* the contingency task structure is assessed correctly. The first two of these coping patterns involve basically intuitive cognition, and the latter involves analytical cognition. Defensive avoidance, or decidophobia, and hypervigilance are inappropriate modes of coping with a decision situation, and humans and organizations should proactively seek to avoid the creation of conditions that lead to them.

Janis and Mann present a decision balance sheet, an adaptation of the moral algebra of Benjamin Franklin, on which to construct a profile of the identified options together with various cost and benefit attributes of possible decision outcomes. They have shown that decision regret reduction and increased adherence to the adopted decision results from use of this balance sheet. Strategies for challenging outworn decisions and improving decision quality are also developed in this seminal work.

In this time–stress-based conflict-theory model, there are many activities that are associated with information processing and associated situation assessment. These include obtaining information, or additional information:

1. To enable recognition of a potentially challenging opportunity
2. To enable determination of potential losses if the present course of action is continued
3. To enable determination of potential losses if a change is made to a new but familiar course of action
4. To enable determination of whether it is reasonable to find a better course of action than the familiar ones already considered and initially dismissed as improper
5. To ascertain if familiar courses of action not previously considered are acceptable
6. To ascertain whether the remaining time, until the decision must be made, is appropriate to formal rational deliberation
7. To support a formal formulation, analysis, and interpretation of the issues and the resulting vigilant search, processing and deliberation

In Figure 5.3, there are five possible decision types. Three of them are inherently appropriate if the contingency task structure and situation are assessed appropriately. Two of these are the result of skill-based or intuitive reasoning: unconflicted adherence to the present course of action, and unconflicted change to a new course of action. One is based on formal analytical thought, which is appropriate in situations when we do not have sufficient expertise to enable skill-based intuitive thought and results in vigilant information search, deliberation, and processing of information and knowledge about the decision situation. Two are defective modes of choice: (a) defensive avoidance and (b) hypervigilance or panic. We should always attempt to avoid being placed in situations where these modes of choice prevail. This is a very realistic model of judgment and choice. We discuss it here in terms of individual decision making. Janis has examined the use of this time–stress-based model, as well as others, in group settings that encourage *groupthink* [5] and in prescriptive settings for leadership in policymaking and crisis management [6].

In providing prescriptive support for decision assessment, great care must be taken to understand the decision situation extant. Unless there is a good assessment of the decision situation, efforts on decision selection may not produce the intended results. The primary objectives of a prescriptive or formal decision assessment are as follows:

1. To formulate the issue in terms of objectives to be obtained, needs to be satisfied, and the identification of potential alternative courses of action
2. To analyze the impacts of the alternatives upon the needs and objectives of the appropriate group of stakeholders
3. To interpret these impacts in terms of the objectives so as to enable selection of a most appropriate alternative course of action, given all of the realities of the decision situation

These are just the formal steps of the systems engineering process, as described earlier, and there appears to be little difference between decision assessment and a general systems engineering approach to problem resolution. Often the term *decision analysis* is used for the complete effort. In the way that we describe it, analysis is one of the three fundamental steps of an assessment effort. Many would argue that these are precisely the objectives of normative, or formal–rational, technoeconomically based decision analysis. Then, the identification of a prescriptive approach that is distinct from the normative approach would be unneeded.

We may subdivide decision assessment efforts into five types:

1. *Decision under certainty issues* are those in which each alternative action results in one and only one outcome and where that outcome is sure to occur. The decision situation structural model is established and is correct. There are no parametric uncertainties.
2. *Decision under probabilistic uncertainty issues* are those in which one of several outcomes can result from a given action depending on the state of nature, and these states occur with known probabilities. The decision situation structural model is established and is correct. There are outcome uncertainties, and the probabilities associated with these are known precisely. The utility of the decision maker for the various event outcomes can be quite precisely established.

3. *Decision under probabilistic imprecision issues* are those in which one of several outcomes can result from a given action depending on the state of nature, and these states occur with unknown or imprecisely specified probabilities. The decision situation structural model is established and is correct. There are outcome uncertainties, and the probabilities associated with the uncertainty parameters are not all known precisely. The utility of the decision maker for the events outcomes can be quite precisely established.

4. *Decision under information imperfection* issues are those in which one of several outcomes can result from a given action depending on the state of nature, and these states occur with imperfectly specified possibilities. The decision situation model is established but may not be fully specified. There are outcome uncertainties, and the possibilities associated with these are not all known precisely. Imperfections in knowledge of the utility of the decision maker for the various event outcomes may exist as well.

5. *Decision under conflict and cooperation issues* are those in which there is more than a single decision maker, and where the objectives and activities of one decision maker are not necessarily known to all decision makers. Also, the objectives of the decision makers may differ.

Problems in any of these groupings may be approached from a normative, descriptive, or prescriptive perspective. Problems in category 1 are those for which deterministic principles may be applied. This condition of known states of nature ignores the overwhelming majority of issues faced in a typical private or public sector decision assessment, including such factors as: How will my customer base or constituency be affected? What will happen to the cost or quality of my product or service? How will the morale and organization of my workforce be affected? How does my ever-changing institutional and societal environment bear on this decision? What are the implications of this decision on other decisions I must make elsewhere in my organization? and so on.

Problems in category 5 are game-theoretic-based problems, and they will not be considered in this text. The majority of decision assessment efforts have been applied to issues in grouping or category 2, although current approaches to decisions under information imperfections allows solution to some problems in categories 3 and 4 as well. We will initially concentrate on a description of problems in category 2, and we will then provide an overview of some solution methods for problems in categories 3 and 4.

Figure 5.4 illustrates a simple decision tree structure that can be used to illustrate the differences among these 5 categories. For category 1, a decision under certainty problem, all of the probabilities are equal to either 1.0 or 0.0, such that there is a single known outcome from each alternative. In general, Figure 5.4 is actually most suitable to a category 2 problem. It is also valid for a category 3 problem, except that some of the probabilities are, themselves,

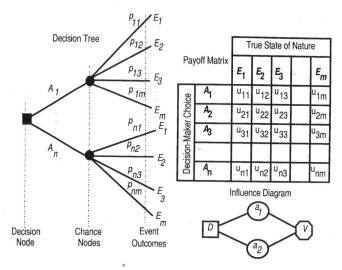

Figure 5.4 Decision analysis tree structure for simple, single decision lottery and associated payoff matrix.

unknown. In a category 4 problem, there may be event outcomes that have not been identified. In addition, probabilities of the outcomes may not necessarily be specified. In a category 5 problem, the probabilities take on the nature of variables that are generally capable of being altered by a competitor or collaborator in a game situation. Decision assessment provides us with a framework for describing how people do choose, or ideally could best choose, or should choose, among alternative courses of action when the outcomes resulting from these alternatives are clouded by uncertainties and information imperfections.

It is helpful to examine these five generic decision types in greater detail. Figure 5.5 depicts a typical *decision under certainty*. As shown in Figure 5.5 there are three options and each option has just one outcome or state of nature. Note that the probability associated with each state of nature equals one. Decisions may be made under certainty whenever each option results in one and only one known outcome or payoff. It seems easy to select the best option in this type of decision situation. For the problem shown in the example payoff matrix of Figure 5.5, you would simply choose the option with the best payoff. In the example shown, if the payoffs represent dollars, then the best answer is option *b*, with a payoff value of $90, if your objective is to maximize the payoff value. If your objective is to minimize the payoff value, then your best answer is option *a* with a minimum cost of $24.

In real-world problems, selecting the best alternative course of action often isn't this easy. In part, this is because there are a very large number of possible options in many decision situations. Therefore analytical approaches are still needed for decisions under certainty. Deterministic optimization problems that

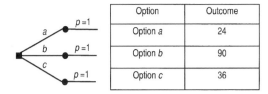

Figure 5.5 Decision tree and payoff matrix for decision under certainty illustration.

can be solved using linear programming are one example of appropriate solution techniques for decisions under certainty.

The next category is *decision under risk*. As shown in Figure 5.6, you know the structure of the decision tree, as well as the states of nature and their associated probabilities or likelihood of occurrence. Decisions may be made under risk when each option may result in more than one outcome and when you know the probabilities of each outcome or state of nature. Across the top horizontal row, the probability of each outcome or state of nature is listed. The sum of these probabilities must equal one. If the probability for each of the three outcome states were different for each option, the payoff matrix representation becomes cumbersome. To solve this decision problem, you calculate the *expected value* of each option and enter it in the expected value column as shown. To find the expected value of an option, you simply sum across an option row the products of the probability of each state of nature times the corresponding payoffs. For option *a*, the expected value is $0.10(24) + 0.40(12) + 0.50(60) = 37.20$. If the best option is the option with the highest expected value, in this case we obtain option *c* as the best option. It has an expected

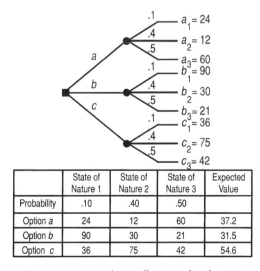

Figure 5.6 Decision tree and payoff matrix for decision under risk.

value of 54.6. If your objective is to minimize the expected value, then option *b* with an expected value of 31.5 is best.

At this point, you might have some concerns about recommending option *b* to a decision maker who wants to minimize cost. Why? Well, because option *b* also has a 10% chance of resulting in the highest outcome or cost of 90, which you note is the highest outcome or payoff in the table. What if your decision maker really wanted very much to avoid the possibility of such an adverse outcome? How could you account for this in analyzing this decision? The answer lies ahead in the use of the appropriate criterion to make sure that the recommended choice reflects the decision maker's true preferences, including risk preferences. If we had selected option *a*, the worst possible, in this case largest, outcome that we could have obtained is 60. In this case, we might have decided to select the option that minimizes the maximum possible outcome. If we really have great aversion to a possible high outcome, option *a* might well be better than option *b*, even though it has a larger expected value. Because we know all of the probabilities with precision here, we might try to develop a method that would allow us to consider our significant risk averse personal characteristics. This brings about the need for what is called utility theory. Before we introduce utility theory, let's examine the other types of decisions and the opportunities and challenges they present.

Decisions under uncertainty is the third type of decision. Similar to decisions under risk, the options result in more than one outcome. The payoff matrix in Figure 5.7 illustrates how much information you might have in a specific instance when making a decision under uncertainty. Although you know the

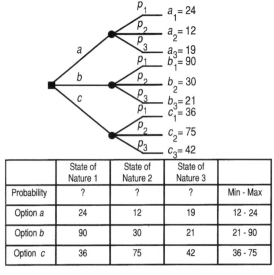

	State of Nature 1	State of Nature 2	State of Nature 3	
Probability	?	?	?	Min - Max
Option *a*	24	12	19	12 - 24
Option *b*	90	30	21	21 - 90
Option *c*	36	75	42	36 - 75

Figure 5.7 Decision tree and payoff matrix for decisions under probabilistic uncertainty.

different options and their associated payoffs, you do not know the probabilities or likelihood of occurrence of the outcomes that result from selecting a given option. Figure 5.7 has a question mark for the probability of states 1, 2, and 3 because you really don't have any idea which state of nature is likely to occur in these situations. There are several ways you can analyze and solve the decision under uncertainty problem shown in Figure 5.7. You could try first using the idea of *dominance.* Is there one option you can choose because every outcome for each state of nature is better than every other outcome of any other alternative? Or is there one option we can eliminate because every payoff for each state of nature is worse? If the objective is to maximize the payoff value, then option *a* can be eliminated because it is dominated by each of the other two options. That is, option *a* has the lowest payoff value in each column. However, if your objective is to minimize the payoff value, then option *a* would dominate the other two options in that the outcome regardless of the state of nature that prevails is always lower than the outcome that would have resulted from selecting either option *b* or option *c.* Thus, you could confidently recommend option *a* as the best decision.

We continue with the problem of deciding between option *b* and option *c* when you want to maximize your payoff value. Note the right hand column shows the minimum and maximum payoff for each option. If you are optimistic, you could recommend option *b* because it has the highest possible payoff value of 90. However, it also has the worst payoff value of 21 among options *b* and *c*! So if you think nature may be unkind, you could recommend option *c* because it avoids the worst outcome. You will learn more detailed ways to deal with these issues soon, but first there are two more types of decisions to learn about.

A fourth category called *decision under information imprecision* has also been defined [1]. In this type of decision the probabilities of the outcomes may be unknown, or not fully known, and there may also be outcomes that are unknown, or not fully known. In the first three types of decisions, you can usually fully specify a correct decision model. But, in this fourth type of decision, the imprecision of the information available to the analyst makes it very difficult to build a fully specified decision model. Figure 5.8 represents a decision tree and payoff matrix for decision under imprecision. Note that there are many more question marks in the matrix to show that more unknown or imprecise information exists in this type of decision. Also, we may pose more generic representations of decision under information imprecision.

There are many different ways to try to resolve a decision issue under information imprecision. For example, you could assign some specific values for the unknown payoffs. One way to assign missing values is to use the average of the known outcome payoffs for an option. Or you could use the lowest or highest known outcome value depending on whether or not you were pessimistic or optimistic about a particular option. In either case, once you assume some values for the missing or imprecise information, you could use strategies similar to decisions under uncertainty. We will not explore these

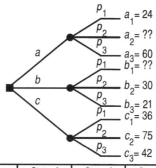

	State of Nature 1	State of Nature 2	State of Nature 3	
Probability	?	?	?	Min - Max
Option a	24	?	60	24? - 60?
Option b	?	30	21	21? -30?
Option c	36	75	42	36 - 75

Figure 5.8 Decision tree and payoff matrix for information imprecision.

issues here, but refer the interested reader to reference 1 and the references contained therein for additional discussions.

The last type of decision type is decision under conflict and cooperation. In this type of decision, there is more than one decision maker. The states of nature are affected or replaced by the actions of these other decision makers. Some or all of the other decision makers may be either hostile or cooperative such as in warfare or a business situation where there is competition for market share. This type of decision making is often modeled using game theory. Figure 5.9 shows a simple two-person decision under conflict situation. We assume that each player must exercise their options at the same time, and without knowledge of what exact option the other player has chosen. Here, the payoffs in the matrix are assumed to be those for player *A* wins, who hopes to maximize the outcome value. The outcomes for player *B* are the negative of the outcomes for player *A*, and player *B* desires to also obtain the maximum possible result. Looking at the table in Figure 5.9, if *A* chooses Option 1 and *B* chooses Option 1, then *A* wins with an outcome of 7. *B* obtains an outcome of -7. Negative payoffs represent an outcome situation where player *B* wins or has a net gain. As you can imagine, the strategies for making decisions in competitive or cooperative situations can become quite complex. This is especially so because most zero sum games are not fully realistic and nonzero sum games are generally even more complicated to analyze than are zero-sum games. It turns out in the zero-sum case that player *A* should usually choose the row that yields the maximum of row value for the minimum column results.

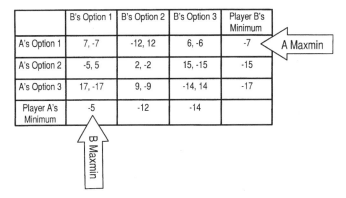

	B's Option 1	B's Option 2	B's Option 3	Player B's Minimum	
A's Option 1	7, -7	-12, 12	6, -6	-7	◁ A Maxmin
A's Option 2	-5, 5	2, -2	15, -15	-15	
A's Option 3	17, -17	9, -9	-14, 14	-17	
Player A's Minimum	-5	-12	-14		

Figure 5.9   Two-person zero sum game situation.

Player *B* should generally choose the column that leads to the maximum of column value for the minimum score across rows. These turn out to be the optimum solutions when the game has what is called a stable equilibrium point. Here, each player should select option 1. An excellent discussion of modern game theory and associated strategies is presented in reference 7.

There are many needs that we should consider for decision making, policy formulation, resource allocation, and planning. Four other complicating factors that you need to recognize and learn to deal with are usually present. These are as follows:

1. *Outcome Uncertainty.* Decisions must be made and policies formulated in an atmosphere of uncertainty; that is, the outcome resulting from any action is seldom known with certainty.

2. *Multiple Decision Makers.* Often the process involves a large number of decision makers who act according to their varied preferences and the often diverse data available to them.

3. *Sequential Decisions.* Decisions sometimes affect other decisions. For example, before making a risky decision such as making an investment or buying an expensive product or service, you may want to know if you should first purchase advice from a consultant or some other information. A decision to focus your business on a certain market segment today may impact your ability to penetrate other market segments in the future.

4. *Multiple Attributes.* The outcomes resulting from an action can often be adequately characterized only by using several different attributes to describe an outcome, thus making comparison of these outcomes difficult. For example, when considering your decision to buy a car, you usually consider at least three attributes: safety, performance, and cost. In systems engineering projects, cost, schedule, and performance are three important

attributes of many decisions. For example, when considering whether to buy or make a new component for a system, the systems engineer will ask three questions. How much will it cost? Can it perform to our specifications? Will it be able to be delivered or made soon enough to meet our project schedule?

We will consider these issues in our discussions in this chapter.

Many foundations for decision assessment are available. Behavioral psychology provides us with information concerning both the process and the product of judgment and choice efforts. Systems modeling, at least in principle, provides us with methods that can be used to represent decision situations. Probability theory allows, at least in principle, the decision maker to make maximum use of the uncertain information that is available. Utility theory, at least in principle, guarantees that the choice will reflect the decision maker's preferences if all aspects of the decision situation structural model have been modeled correctly.

In order to develop a theory that agrees with how decisions could be made in a normative sense, we need to assume the following:

1. Past preferences are valid indicators of present and future preferences.
2. People correctly perceive the values of the uncertainties that are associated with the outcomes of decision alternatives.
3. People are able to assess decision situations correctly, and the resulting decision situation structural model is well-formed and complete.
4. People make decisions that accurately reflect their true preferences over alternatives that have uncertain outcomes.
5. People are able to process decision information correctly.
6. Real decision situations provide people with decision alternatives that allow them to express their true preferences.
7. People accept the axioms that are assumed to develop the various normative theories.
8. People make decisions without being so overwhelmed by the complexity of actual decision situations that they must necessarily use suboptimal decision strategies.

Given these necessary assumptions for a valid normative theory, it is not surprising that there will exist departures between normative and descriptive theories of decision assessment. A principal task and responsibility of those systems engineering professionals who seek to aid others in decision assessment and decision aiding is to retain those features from the descriptive approach which enable an acceptable transition from normative approaches to prescriptive approaches. The prescriptive features should eliminate potentially undesirable features of descriptive approaches, such as flawed judgment heuristics and information processing biases, while retaining acceptable features of the normative approaches.

There are a variety of approaches to normative, prescriptive, and descriptive decision assessment. Most of these rely on what is called utility theory. Peter Fishburn [8,9] has identified 25 different variations, or generalizations, of what we will call expected utility theory. Most of the variations are minor, and most lead to relatively similar normative recommendations. We will discuss some of the major generic approaches here.

5.2 FORMAL DECISIONS

In this section, we will examine approaches that are intended to aid in the process of making decisions, or, more precisely, making effective decisions.

5.2.1 Prescriptive and Normative Decision Assessments

A prescriptive decision assessment provides a systematic framework of search, deliberation, evaluation, and selection that facilitates selection of a most preferred course of action for a decision maker in a complex decision situation. In a formal sense, this may be accomplished through use of an eight-step process that comprises formulation, analysis, and interpretation of the decision situation.

    I. *Formulation.* This will result in understanding and knowledge of the decision issue to be resolved.

        1. Assess the decision situation. This will include the identification of objectives, needs, constraints, alterables, and potential alternative courses of action. These are the ingredients of the decision situation structural model.

    II. *Analysis.* This will result in a complete decision tree model of the structure and parameters of the decision situation.

        2. Structure the relationship between the decision alternatives and outcomes, typically in the form of a decision tree or some other appropriate decision situation structural model, such as an influence diagram.

        3. Encode the information known by the decision maker, or by others, on the decision situation structural model in order to describe the probability of occurrence of the outcomes of uncertain situations. With steps 2 and 3 completed, we have an impact analysis model of the decision situation that enables us to predict the likelihood of various outcomes resulting from alternative courses of action.

        4. Determine the impacts of the identified alternative courses of action.

    III. *Interpretation.* This will result in identification of the best decision alternative.

        5. Assess the decision maker's preferences for the various attributes that characterize the possible outcomes, and then formally include the decision maker's attitudes toward the risk associated with uncertain event outcomes. This will generally result in a number of different utility functions.

        6. Use the utility functions, or scores, associated with decision outcomes and relevant probabilistic information to evaluate an overall expected utility score for the set of event outcomes that are associated with each alternative.

        7. Rank the alternatives in terms of the expected utility of each alternative course of action. Discuss the implication of these results with the decision maker.

        8. Use sensitivity analysis and other approaches to refine the decision assessment process and product. This is, ideally, done such that the decision maker more fully appreciates, accepts, and is supportive of the final results obtained. Communicate these as appropriate. Select the alternative with the maximum expected utility as the "best."

Ideally, what is prescribed here is that of the normative theory. Our description here is actually that of a typical normative decision analysis process that evolved in the 1960s from the efforts of a number of investigators [10–13]. A very interesting and readable overview of the historical foundations of normative and prescriptive decision analysis is provided by Fishburn [14]. An overview of descriptive decision making may be found in the introductory chapter of reference 15.

    The fact that the process just described is suggested for implementation qualifies it as a prescriptive process. It also happens to be the product of a normative theory. Thus, this is a prescriptive–normative process. Even if a formal normative approach is optimal under hypothetical circumstances, modifications may be needed to make it more appropriate for a specific decision situation. For example, the decision tree may be extraordinarily complex, and it may be appropriate to simplify it in any number of ways, such as by modeling some subsequent acts as if they were events. This would replace uncertain act–event combinations by uncertain events only. In a very complex situation, obtaining solutions might be more realistic of accomplishment than if such an approximation were not made. An approximate and useful answer is generally much better than a potentially more correct answer that is not obtained until the time for decision enactment is past. This suggests a major role for experiential familiarity with the decision situation extant in making appropriate decisions.

The typical final products of formal normative decision assessment generally involve the following:

1. A description of the decision situation structural model
2. A decision tree that represents the real decision situation and that indicates probabilities assigned to outcomes
3. A utility function that describes the decision makers' preferences for relevant attributes.
4. An evaluation and ranking of alternative courses of action according to their utility to the decision maker
5. A sensitivity analysis and interpretation of how the results might change with changes in inputs, such as subjective probabilities, personal preferences, and attribute measures

The specific steps of the decision analysis and assessment effort, including the decision situation structural model, are configured to enable successful completion of the decision assessment effort and achievement of these five products. It is worth noting that these steps of a formal decision situation do not always correspond to descriptive approaches to decision making. There are situations when other than a formal approach is best, as suggested in Figure 5.3. There have been a number of efforts to bring the formal approaches into more widespread usage [2–22] when they are the appropriate decision styles. One of the tasks of those doing decision assessments, therefore, is to be sure that the customer or client for the effort is aware that these formal descriptions of decisions are in fact quite appropriate in many decision situations. In particular, the Janis–Mann model of Figure 5.3 indicates the information and experience situations in which the formal approach presented here is appropriate.

Quite a bit of information is needed to accomplish the decision assessment process just outlined. This includes information about the following:

1. The structure of the decision situation, including an identification of needs, constraints, alterables, objectives and measures by which outcomes shall be evaluated
2. Alternative courses of action that are available to the decision maker
3. Possible outcomes that may be realized from implementation of each alternative course of action
4. The probability of occurrence of each outcome
5. The attributes or objectives measures of all possible outcomes.
6. The costs of purchasing additional information about the decision situation

Quite clearly, this information will not always be completely known with precision. If we have all of this information, we can incorporate it into a structural model of a decision situation, use this model to analyze the impact of the alternative decisions that have been identified, and interpret these according to the value system of the decision maker in order to enable formal rational choice.

There are a number of questions that should be answered to appraise the worth of a projected decision assessment effort, including the following:

1. Was the decision situation identified in a satisfactory manner?
2. Did the decision situation structural model satisfactorily capture the reality of the decision situation such as to give the decision maker confidence in the indicated results?
3. Were a sufficiently robust set of alternative courses of action identified?
4. Were the possible outcomes of the alternatives made explicit, and was knowledge about the probability of occurrence of each sufficient for the intended purpose?
5. Were the attributes of the outcomes identified and explained?
6. Did the decision assessment produce decisions of sufficient quality to justify its cost?
7. Did the decision maker have confidence in the decision situation structural model and the validity of the utility function?

Good decisions do not necessarily lead to good outcomes, and bad decisions may result in good outcomes. For example, the decision to pay $1 for a lottery ticket in which one might win $100 with probability 0.99 and receive nothing with probability 0.01 is, from almost all perspectives, a good decision. We are able to purchase something with an expected return of $99 for only $1. But, even though the decision may be a good one, we may lose our $1 in that we obtain an unlucky outcome. So, even though you have made a good decision, you receive a bad outcome due to the inherent uncertainties. There is no totally objective method of directly assessing the wisdom of a given decision effort by observing only the outcome! Often, people are judged only on the basis of outcomes, and this is not always a wise approach to evaluate their worth or that of the advice that they have prescribed. An indirect evaluation approach to assess the quality of a decision and not just the quality of the outcome may be more appropriate, Often, we should be interested in valuing the decision support process itself and not just the outcome or product of the decision support process, important as that may be [23, 24].

5.2.2 A Formal Normative Model for Decision Assessment

Figure 5.4 illustrates representations of alternatives, event outcomes, and utilities. This forms the basis for the model that we will use for our study of normative decision analysis, and it leads to the decision analysis process represented in Figure 5.10.

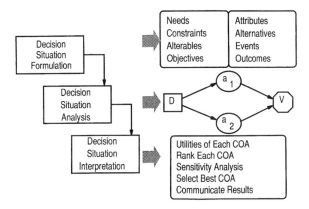

Figure 5.10 Fundamental steps in prescriptive decision analysis and assessment.

A formal decision analysis model is generally comprised of six elements:

1. The set of n alternative actions $A = \{a(1),\ a(2),\ \ldots,\ a(n)\}$.
2. The set of m states of nature $E = \{e(1),\ e(2),\ \ldots,\ e(m)\}$.
3. The set of nm outcomes $Q = \{q(11),\ \ldots,\ q(ij),\ q(nm)\}$, where $q(ij)$ corresponds to the pair of action–event possibilities $\{a(i)e(j)\}$.
4. A set of probabilities of these outcomes, generally represented as a set of conditional probabilities $P[e(j)|a(i)]$.
5. A utility function $U = \{u(11),\ u(ij),\ \ldots,\ u(nm)\} = u_{ij}$, where $u(ij) = U(q(ij))$. This utility function expresses the decision maker's utility at having selected alternative i and receiving state of nature j as a result.
6. The objective function which represents what is desired to be accomplished and which we seek to extremize or, most often, maximize.

The term decision analysis, or decision assessment, has come to be a generic term that refers to a wide variety of problems in such decision-related areas as systems engineering. The one common bond among all these problems is that they involve some kind of optimization, in terms of selecting a best alternative. For example, in the model just described, the decision maker seeks to choose the alternative a_i that makes the resulting utility, u_{ij}, "best" in some sense. Often "best" will mean "maximum." But the resulting u_{ij} will also depend upon the particular value of the random variable e_j, so the best the decision maker can do here is to maximize some function such as the expected value of the random variable u_{ij}. This is given by the maximum expected utility expression

$$\mathrm{MEU} = \max_i \mathrm{EU}\{a_i\} \sum_{j=1}^{\infty} u_{ij} P(e_j | a_i)$$

where $P(ej|a_i)$ is the probability mass function that the state of nature that results or follows from selection of alternative a_i is e_j. More precisely, this is the probability of state e_j occurring, given that alternative a_i is selected. We use notation $EU\{a_i\}$ to mean the expected utility of taking the alternative course of action a_i.

5.2.3   Decision Assessment with No Prior Information on Uncertainties

To solve decision problems where there is information concerning the probability of various outcomes e_j that may result from selection of an alternative a_i, we need to know the associated conditional probability mass function $P(e_j|a_i)$. Generally, we deal with finite event situations in decision analysis. Thus, the probabilities are probabilities of events, or probability distributions, or probability mass functions. It is not uncommon to find terms such as probability density function used to describe these probabilities.

Let us suppose for the moment that nothing is known about this probability density function. Although we may argue that there are few, if any, practical problems where we know absolutely nothing about the probabilities of the states of nature, there is nothing to prevent us from hypothesizing this total lack of knowledge. We now turn our attention to several approaches, described in many early operations research books [25], which may be taken in such a situation.

1. *The Laplace Criterion.*   Assuming all $P(e_j|a_i)$ are equal — that is, where $P(e_j) = P(e_j|a_i) = 1/m$ — we obtain $EU\{a_i\} \sum_{j=1}^{\infty} u_{ij} P(e_j|a_i) = u_{ij}/m$. Because the common factor $1/m$ is invariant over the summation, it is not significant in influencing the optimization result. The optimal decision rule is to pick that a_i which maximizes $\sum_{j=1}^{\infty} u_{ij}$. We should note that the Laplace criterion is not a very good decision rule in general, and unless there is some reason for believing that all $P(e_j|a_i)$ are equal, or nearly so, the application of this rule may potentially yield very undesirable results.

2. *The Max–Min, or Pessimist, Criterion.* This is an extremely pessimistic rule which holds that regardless of which action a_i is chosen, nature will somehow malevolently cause the state e_j, which yields the very smallest utility possible for the action taken. We partially compensate for this pessimism by choosing the alternative course of action a_i which has the largest possible value of the smallest such utility. Thus, we have a criterion that allows us to minimize the maximum damage that can be done. This criterion will allow us to refuse to pay $1 for the lottery noted earlier in which one receives a $100 outcome with probability 0.99 and a $0 outcome with probability 0.01. One might extend the decision situation structural model such that refusal to pay is the desirable alternative. We are stuck on an island that is scheduled to be destroyed tomorrow.

All we have on the island is our clothing and \$1. Someone offers to take us off of the island for \$1. Before deciding whether to buy the ticket off the island, we have the option of paying our \$1 and playing the lottery. Clearly, we would keep the \$1, reject the lottery, and buy the escape ticket. Obviously, however, we have altered the initially stated decision situation by inserting uses for the money, and this introduces a multiattribute decision situation.

3. *The Max–Max, or Optimist, Criterion.* Here we assume that nature takes on a benevolent character such that regardless of which a_i is chosen, nature will act kindly to produce the state e_j that yields the largest utility, or u_i^+. We then choose the action that results in the largest such u_i^+, where $u_j^+ = \max_i u_{ij}$. An alternative explanation of this criterion is that the decision maker chooses that action whose row in the payoff matrix contains the largest element, and he or she hopes for the best. Such an approach would result in paying \$10 for a lottery in which the payoff is \$100 with probability 0.01 and \$0 with probability 0.99. This bad decision might result in a good outcome, but you would likely not express much hope of obtaining it. Clearly there is a potential problem if we receive advice that leads to accepting and paying for the lottery, and we win and think that we received good judgmental advice. Again, we could consider being stuck on an island that is going to be destroyed tomorrow. It costs \$100 to buy a ticket on the last boat off of the island and we presently only have \$10. We have the option of playing the lottery before possibly buying the ticket off the island. Now, the lottery becomes quite desirable as we face certain death unless we win the lottery. But, as in the illustration used for the pessimist criterion, we have altered the initially stated decision situation structural model and have introduced some multiattribute considerations.

4. *The Optimism Index a.* In this approach we combine the max–min criterion with the max–max criterion. We let $u_i^- = \min_j u_{ij}$ and $u_i^+ = \max_j u_{ij}$. Next we define a function of the optimism index $u_i(a) = au_i^+ + (1-a)u_i^-$ for $(0 \leqslant a \leqslant 1)$. This describes a linear function of a which can be plotted for each value of i. The maximum over i of these functions defines a convex function of a that gives the optimal utility according to this rule. For the two extreme cases of $a = 0$ and $a = 1$, this rule reduces to the max–min rule and the max–max rule, respectively.

5. *The Min–Max Regret Criterion.* From the payoff matrix we generate a regret matrix R according to the rule $R_{ij} = u_j^+ - u_{ij}$, where $u_j^+ = \max_i u_{ij}$. Regret is the sorrow we experience about not getting a better outcome. For each alternative a_i we find the largest element in each row of the R matrix. This is $R_i^+ = \max_j R_{ij}$. Finally, we choose that alternative course of action a_i which yields the smallest such regret, which is $R_i^+ = \min_i R_i^+ = \min_i \max_j R_{ij}$.

Example 5.2.1 To observe how these methods may be applied in a practical decision problem, we consider a simple example. The payoff matrix, decision tree, and influence diagram for this simple decision problem are shown in Figure 5.11.

First, we apply the Laplace criterion. Course of action a1 results in an expected utility of 13. Course of action a_2 results in an expected utility of 21. For course of action a_3, we have an expected utility of 17. All of these expected utilities are based on the assumption that the probability of each outcome state being realized, $P(e_j|a_i) = 1/m$, is 0.333. Because $21 > 17 > 13$, we see that the course of action a_2 is the best course of action from a Laplace criterion perspective. The expected utility of this course of action is 21 if all probabilities are equal.

Next we apply the max–min criterion and obtain $u_1^- = 3$, $u_2^- = 6$, $u_3^- = 9$ as the minimum utility value associated with each course of action. Because $u_3^- > u_2^- > u_1^-$, we see that a_3 is the best course of action because it yields the maximum value of all of the minimum return values.

Applying the max–max criterion results in $u_1^+ = 24$, $u_1^+ = 36$, $u_1^+ = 27$, because we assume that nature will give us the maximum utility for each action. From this we see that the best course of action is a_2, because this yields the maximum utility.

Combining the max–min and max–max techniques using the optimism index a, we obtain

$$\hat{U}_1(a) = 24a + 3(1 - a) = 3 + 21a$$
$$\hat{U}_2(a) = 36a + 6(1 - a) = 6 + 30a$$
$$\hat{U}_3(a) = 27a + 9(1 - a) = 9 + 18a$$

It would be easy to sketch the utility versus a curves. Such an illustration

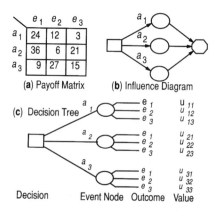

Figure 5.11 Payoff matrix and decision situation structural models.

would show that for $a < 1/4$, course of action a_3 is the best course of action. For $a > 1/4$ the optimal course of action alternative, or decision, is a_2.

Finally, applying the min–max regret criterion yields $u_1^+ = 36$, $u_2^+ = 27$, $u_3^+ = 21$, and thus we have for the regret matrix

$$R = \begin{bmatrix} 12 & 15 & 18 \\ 0 & 21 & 0 \\ 27 & 0 & 6 \end{bmatrix}$$

We now compute the maximum regret for each alternative. These are obtained as $R_1^+ = 18$, $R_2^+ = 21$, $R_3^+ = 27$. Thus the optimal decision is the course of action a_1, because we obtain the minimum of the maximum regret by this selection.

In the absence of any prior knowledge of the event probability describing occurrence of the various events, e_j, we find that the optimal decision may, and generally will, depend upon the particular criterion selected. For the example considered we can, by judicious choice of the decision criterion, cause any action to be optimal. There are other approaches that might also be used. For example, we might observe that for two of the three outcomes, alternative a_1 is better than alternative a_2. Also, we observe that for two of the three outcomes, alternative a_2 is better than alternative a_3. If we stop here, we might be tempted to conclude that because $a_1 \to a_2$ and $a_2 \to a_3$, with each preference on two out of three outcomes, then we should have $a_1 \to a_3$. Sadly, however, it turns out that a_3 is better than alternative a_1 for two out of three outcomes. It turns out that we will often be in deep trouble in trying to make pairwise comparisons of alternatives, as we are here. Intransitive preferences may result, as they do here, if we are not very careful in formulating the comparison approach wisely.

Actually, this example is relatively academic in the sense that it would be rare that we had no knowledge whatever about the probabilities of outcome occurrence. Usually, we do have some information. Nevertheless, this somewhat academic example does provide some insight concerning how we might obtain solution to problems using less-than-complete information. Other approaches to decision under information imprecision and imperfection are discussed in reference 26.

5.2.4 Decision Assessment Under Conditions of Event Outcome Uncertainty

We now turn our attention to the case where the probability density function of the states of nature is known. For many decades, basic notions of probability have intuitively come to be associated with events that could be repeated exactly many times, such as flipping a coin or rolling a die, thereby giving the term a relative frequency interpretation. However, there are events that may occur only once, or less. In these situations, a relative frequency interpretation may be less than fully meaningful. Subjective factors will need to be included in such estimates. Often, different people will associate different personal probability assessments associated with the same action. The early works of von Neumann and Morgenstern [27] and Savage [28] were concerned with notions of subjective probability and subjective utility, and we will soon discuss some of these findings.

In a standard decision situation, once the appropriate probability density functions and outcome values have been determined such that the decision tree is well-structured, all the information necessary for determining the expected utility resulting from any decision action is then potentially available. This is the decision under probabilistic uncertainty situation described earlier. Most efforts in normative decision analysis have been concerned with this problem. In such a situation, a techno-economic rational decision maker attempts to maximize subjective expected utility. How this is done can perhaps best be shown by an example. Numerous pedagogical examples of this type decision problem exist in the literature. The "anniversary" problem described by North [29] in 1968, and the "wildcatter" problem outlined by Raiffa [12] are especially good. A number of very useful examples are provided by Holloway [30]. In order to keep the discussion relatively context-free, our approach in this subsection roughly parallels the "ball–urn" approaches that were initially described by Raiffa.

We assume a collection of opaque urns, each identical in external appearance and each containing a number of red balls and black balls. The decision maker knows in advance that there are n_1 urns, called type Ξ_1, containing r_1 red balls and b_1 black balls. There are n_2 urns that we call type Ξ_2. They contain r_2 red balls and b_2 black balls. The person conducting the experiment chooses an urn at random and gives it to the decision maker, who must, on the basis of the information given above, decide whether the urn is of type Ξ_1 or of type Ξ_2. The reward or return to the decision maker will depend upon the type of urn that is actually obtained by the decision maker, called the true state of nature, and the choice of decision alternative a_1 or a_2. The payoff matrix is given in Figure 5.12, where we use W and L as mnemonics for win and lose, respectively. Any of the four numbers W_1, W_2, L_1, and L_2 could be positive or negative, but it is assumed that $W_1 > L_1$ and that $W_2 > L_2$. Also it is assumed that $W_1 > L_2$ and that $W_2 > L_1$; otherwise, one decision would always dominate the other, and the decision choice would be obvious: We should select the nondominated decision alternative.

For the moment let us assume that our decision maker's utility curve for money is a linear function of the amount of money involved. Then the utilities W_1, W_2, L_1, and L_2 may be represented numerically by the equivalent amount of money. Such a decision maker is called an expected monetary value player, or EMVer. More will be said about the nonexpected monetary value player. or non-EMVer, in Sections 5.2.5 and 5.2.6, which specifically discuss utility theory.

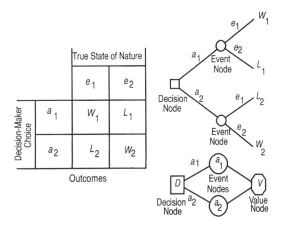

Figure 5.12 Payoff matrix, decision tree, and influence diagram for simple decision options and outcomes.

The decision maker has two choices. Either of two decisions, a_1 or a_2, may be chosen. If the decision maker chooses a_1, the expected reward will be given by $E\{a_1\} = W_1 P(e_1) + L_1 P(e_2)$. On the other hand, if the decision maker chooses a_2, the expected reward will be given by $E\{a_2\} = L_2 P(e_1) + W_2 P(e_2)$, where $P(e_i)$ is the probability of being given an urn of type Ξ_i. These probabilities are easily determined, using the relative frequency concept, as $P(e_1) = n_1/(n_1 + n_2)$ and $P(e_2) = n_2/(n_1 + n_2)$. These probabilities are called *a priori* probabilities because they represent the prior knowledge, or the knowledge before sampling that exists concerning the relative proportions of type e_1 and type e_2 urns.

In this case, the appropriate decision rule is simply stated:

If $EU\{a_1\} > EU\{a_2\}$, select a_1.
If $EU\{a_2\} > EU\{a_1\}$, select a_1.
If $EU\{a_1\} = EU\{a_2\}$, either a_1 or a_2 may be selected.

There are two basic approaches that can be taken to actually determine the best decision, given a well-structured decision tree. These are described as the extensive form and the normal form approaches. Decision assessment in the extensive form has just been described. Solution of problems in extensive form is comprised of six steps, which will be illustrated in the next example:

1. Structure the decision tree diagram.
2. Model the deterministic outcome payoffs.
3. Model the probabilistic outcomes by identifying probabilities at all chance forks.

4. Determine the utilities of the actual outcomes that are possible.
5. Analyze the problem, generally through averaging out and folding, or rolling, back.
6. Determine the optimum decision and communicate the results.

The extensive form is often called the "roll-back form." Figure 5.13 illustrates these steps and the associated decision analysis process. The extensive or roll-back form is generally used when all probabilities are known and the problem involves decision assessment under risk.

Example 5.2.2 Suppose a decision maker is presented with the following decision situation model. An experimenter owns 1000 urns, 750 of which are called type 1 and 250 of which are type 2. Each type 1 urn contains seven red and three black balls. Each type 2 contains two red and eight black balls. The experimenter will select an urn at random, and the decision maker must guess whether the urn is type 1 or type 2. If the DM guesses type 1 and the urn is actually a type 1 urn, a \$500 return is received. But if the urn is actually a type 2 urn, the DM must pay \$100. If the DM says it is a type 2 urn and the urn is actually a type 2 urn, an \$800 return is received. But if the urn is actually a type 1 urn, the DM must pay \$150. Thus, the DM has three alternative courses of action, or decisions.

a_0 Refuse to play
a_1 Guess urn type 1
a_2 Guess urn type 2

If the DM refuses to play, nothing is gained and nothing is lost; the payoff is zero. If the DM plays this little game, the payoff or consequence of play is

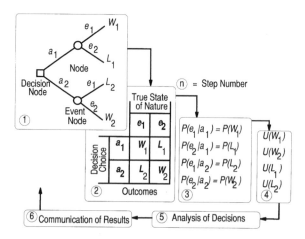

Figure 5.13   The decision assessment process.

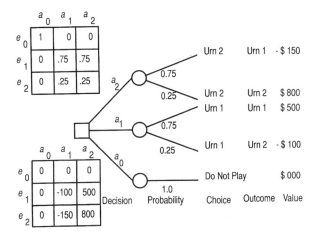

Figure 5.14   Payoff matrix, probability matrix, and decision tree for simple urn model example.

depicted by the payoff matrix shown in Figure 5.14, which also illustrates the complete decision tree and influence diagram that represents the decision situation structural model up to this point.

We can easily compute the expected return from each decision. In general, we should obtain the expected utility of the various decision options. In this case, the utility of a given amount of money is just that amount of money. Thus, $EU(A_i) = E(a_i)$. Therefore, we obtain $E\{a_0\} = 0$, $E\{a_1\} = 500(0.75) + (-100)(0.25) = \$350$, and $E\{a_3\} = -150(0.75) + 800(0.25) = \$87.50$. Because the expected return is the greatest if decision a_1 is chosen, the DM should play the game and should choose a_1 for an expected return of $350. For this simple example, there is no real opportunity to average out and fold back in the process of making the optimum decision. In more complicated problems, which we will now consider, the operations of averaging out and folding back are quite useful.

Now suppose that before the decision maker decides to select option a_0, a_1, or a_2, it is possible, for a price S, to draw one ball out of one of the urns, to be selected at random, and observe its color. The rational decision maker still seeks to maximize the expected utility resulting from the decision. Also, the decision maker should desire to use the information gained from sampling one ball from an urn to update the assessment of $P(e_1)$ and $P(e_2)$. Thus, the DM needs to determine $P(e_1|R)$ and $P(e_2|R)$ if a red ball is drawn and $P(e_1|B)$ and $P(e_2|B)$ if a black ball is drawn. These are calculated from Bayes' rule as

$$P(e_i|R) = P(R|e_i)P(e_i)/P(R)$$

$$P(e_i|B) = P(B|e_i)P(e_i)/P(B)$$

where

$$P(R) = P(R|e_1)P(e_1) + P(R|e_2)P(e_2)$$

$$P(B) = P(B|e_1)P(e_1) + P(B|e_2)P(e_2)$$

$$P(R|e_i) = r_i/(r_i + b_i)$$

$$P(B|e_i) = b_i/(r_i + b_i)$$

$$P(e_i) = n_i/(n_1 + n_2)$$

The probabilities $P(e_i|R)$ and $P(e_i|B)$ are called *a posteriori* probabilities, because they represent the probabilities of the outcome states of nature, e_i, that result after the random sampling that occurs by picking a ball from an urn.

Using $E\{a_i|R\}$ to indicate the expected reward from choosing option a_i, given that a red ball was ball was drawn from the urn, the appropriate expected values are

$$E\{a_1|R\} = (W_1 - S)P(e_1|R) + (L_1 - S)P(e_2|R)$$

$$E\{a_2|R\} = (L_2 - S)P(e_1|R) + (W_2 - S)P(e_2|R)$$

$$E\{a_1|B\} = (W_1 - S)P(e_1|B) + (L_1 - S)P(e_2|B)$$

$$E\{a_2|B\} = (L_2 - S)P(e_1|B) + (W_2 - S)P(e_2|B)$$

It is important that we note that the cost of sampling, S, has been subtracted from each reward. Thus we see that after sampling, the optimal decision rule is as follows:

If a red ball is drawn and

If $E\{a_1|R\} > E\{a_2|R\}$ Select alternative a_1

If $E\{a_2|R\} > E\{a_1|R\}$ Select alternative a_2

If a black ball is drawn, and

If $E\{a_1|B\} > E\{a_2|B\}$ Select alternative a_1

If $E\{a_2|B\} > E\{a_1|B\}$ Select alternative a_2

These expected values provide us with the expected reward after sampling. The expected reward before we sample, and given that we do sample, is given by

$$E_s\{a_i\} = [\max_i E\{a_i|R\}]P(R) + [\max_i E\{a_i|B\}]P(B)$$

Now we need to determine whether it is worthwhile to sample. To determine this we compare the expected reward without sampling to the expected reward with, but before, sampling. The optimal decision rule here is simply

If $\max_i E\{a_i\} > E_s\{a_i\}$ Do not sample

If $E_s\{a_i\} > \max_i E\{a_i\}$ Sample

It is important to note that the decision maker (DM) must now first decide whether to sample or not to sample. If the DM decides to sample, then based upon the result of the sampling outcome as a red ball or a black ball, the DM must make another decision. This concerns whether to choose option a or to choose option a_2. Thus, we see that two decisions must be made, with an intervening chance event. This type of decision problem is generally called a sequential decision problem. The large majority of realistic decision problems are sequential in nature. The decision situation structure for the problem with sampling as described here is represented by the decision tree as shown in Fig. 5.15. For problems of any complexity, the decision tree can become quiet lengthy and the analysis effort can be quite laborious and fraught with difficulties associated with obtaining the relevant probabilities.

If we use the extensive, or fold-back, approach, the optimal policy is developed by starting at the terminal nodes on the right-hand side of the tree and working back through the tree by assigning to each decision node the maximum of all the expected utilities emanating from that node. This approach is called by Raiffa [12] averaging out and folding back. It is often a tedious approach to track through because we must have all the information regarding probabilities, sampling costs, and rewards before we can begin working our way through the tree. This makes appropriate software for decision analysis very desirable.

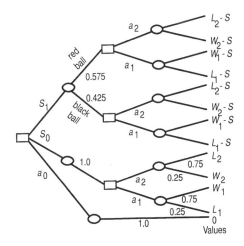

Figure 5.15 Decision tree, with sampling, for the simple urn model example.

Fortunately, there are a number of appropriate software packages available, along with a noteworthy text and related software that discusses the utilization of spreadsheets to support decision analysis and assessment [31].

Example 5.2.3   Let us consider an extension to our previous example. Suppose further that we are permitted to sample the selected urn by drawing a ball and observing its color before making our decision. We are granted that privilege in return for a sampling fee of $60. Of course we may elect not to sample and then would pay no sampling fee. The question is, "Should the DM choose to sample or not to sample?" And then, "Should the DM select urn 1 or urn 2?" Decision analysis or decision assessment provides the answer for this well-structured problem with no information imprecision. We again assume that the decision maker is an expected monetary value player. That is to say, the DM will select the course of action that yields the highest expected value monetary payoff.

In this example, the decision situation has already been explicitly stated. Thus, the initial step in the decision assessment associated with this example, as we see from Figure 5.10 or Figure 5.13, is to structure the decision flow diagram. This result of doing this is illustrated in Fig. 5.16. This figure indicates all the possible courses of action the decision maker may take. The square nodes are those points at which the decision maker selects the branch to follow, while the round nodes are those points at which the branch to follow is selected by the chance outcome of an experiment that occurs as the decision situation evolves in real time.

Our next step is to assign payoffs at the tips of the branches. These payoffs are those indicated in the payoff matrix of the previous example. The decision flow diagram with payoffs is shown in Figure 5.16. The $60 fee for sampling is

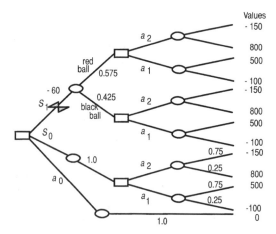

Figure 5.16   Decision tree outcome values, with sampling, for the simple urn model example.

indicated in this figure. We note again that the payoff for alternative a_0, the option of refusing to play, is $0.00.

The ensuing step calls in the decision assessment effort for the assignment of probabilities at the chance forks. These follow each uncertainty node. In the branch that follows decision a_0, there are no chance events. Hence, there are no probabilities to be assigned because we have encountered a deterministic node. The branches for decisions a_1 and a_2 involve only the chance node outcome of whether the urn is a type 1 urn or a type 2 urn. The only information available to the DM is that there are 750 type 1 urns and 250 type 2 urns. Because the random selection of an urn implies that an individual type 1 or type 2 urn is equally likely to be selected, we assess probabilities at the chance nodes in the appropriate branches as

$$P(R) = P(R|\text{urn } 1)P(\text{urn } 1) + P(R|\text{urn } 2)P(\text{urn } 2)$$

$$= (0.7)(0.75) + (0.2)(0.25) = 0.575$$

$$P(B) = P(B|\text{urn } 1)P(\text{urn } 1) + P(B|\text{urn } 2)P(\text{urn } 2)$$

$$= (0.3)(0.75) + (0.8)(0.25) = 0.425$$

Figure 5.17 illustrates the completed decision tree showing these probabilities. We have deducted the $60 sampling cost from the payoff in each case in which we sample, as also indicated in Figure 5.17. The information from the sample can now be used to calculate the probability of each type of urn. We use *Bayes' rule* to find the probability that the urn is type 1, given that a red ball was drawn. This is given by the expression

$$P(\text{urn } 1|R) = P(R|\text{urn } 1)P(\text{urn } 1)/P(R) = (0.7)(0.75)/0.575 = 0.913.$$

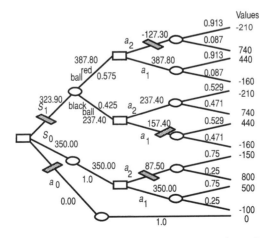

Figure 5.17   Decision tree outcome values, with sampling, for simple urn model example.

We determine the other probabilities in the same manner and have indicated them in Figure 5.17.

The next step involves the *averaging-out* and *folding-back effort. In it, the DM evaluates the expected value or mean of the payoff at each chance node and selects the alternative that leads to the largest expected value. This averaging process begins at the points closest to the end of the tree and works back to the start to indicate the alternative with the highest expected value.* This is why an appropriate name for this portion of the effort is "averaging out and folding back".

The expected value of the events at a chance node is the sum of the payoffs at that chance node multiplied by the respective probabilities. If we elect not to sample, the expected value of decision a_1 is, as was obtained previously, $E(a_1) = (0.75)(500) + (0.25)(-100) = \$350$. Similarly, $E(a_2) = \$87.50$. The same kind of computation must be made at the tips of nodes that spread out from the branch associated with the sample decision. For the decision node following the situation where the DM obtains a red ball sample, we have $E(a_1) = (0.913)(440) + (0.087)(-160) = \$387.80$ and $E(a_2) = -\$127.30$. For the decision branch that follows a black ball sample, we obtain $E(a_1) = \$157.40$ and $E(a_2) = \$237.40$.

Each of these expected values are shown in Figure 5.17. If we examine the red ball sample obtained branch, we see that decision a_1 has a higher expected monetary value outcome than does decision a_2. The implication of this is that a DM who has drawn a red ball should decide to accept option a_1. Because decision option a_2 is less desirable in this situation, a slashed box is associated with this alternative option a_2 in order to indicate its removal from any further consideration. The decision node on the red sample obtained branch then has an expected value of \$387.80, provided that the decision maker chooses a_1 after drawing a red ball. Our efforts in this example have just led us to a determination of a part of the optimal decision.

In a similar manner, the DM observes that a_2 has the greatest expected value on the black ball sample obtained branch. Thus, the DM should remove a_1 from any further consideration there. The DM assigns an expected value to the decision node on the black ball sample obtained outcome node of \$237.40. The DM finally folds back one more time and is then able to find the expected value of the sample chance node e_1 as $E(e_1) = (0.575)(387.80) + (0.425)(237.40) = \$322.90$. *Now each of the alternatives has an expected monetary value assigned to it, and the decision maker can select the one with the largest expected monetary value.* For this problem, the alternative with the greatest expected value is the no sample alternative a_1. That is to say, we should simply select urn type 1 without a sample. If the sampling cost were lower, say \$30 rather than \$60, the sample branch would have the higher expected value, and a different decision would then be indicated. We would then sample and choose a_1 if we draw a red ball and a_2 if we drew a black ball. This concludes this relatively simple example. While nothing at all is difficult, either mathematically or conceptually, the example is tedious to go through. The tree structure assists greatly in tracking the several conditional probabilities.

In this simple example, we assumed that the DM made judgments on the basis of the expected value of money. This expected monetary value approach is open to some criticism. Not all decision makers would be at all content or comfortable with an expected monetary value approach. Concepts of utility theory will be introduced in the next subsection in order to permit risk attitudes to be considered, such that the DM may then be other-than-expected monetary value-oriented.

The *expected value of perfect information* and the *expected value of sampling information* are two important concepts that arise in examples of the sort we have just considered. To continue in the context of the ball–urn problem, suppose that the decision maker knows in advance, perhaps by consulting an oracle of some sort, which type of urn will be given for purposes of taking the sample. For each state of nature, e_1 or e_2, the DM should choose that decision alternative, a_1 or a_2, which maximizes the resulting reward. Each such reward is multiplied by the corresponding *a priori* probability of obtaining e_1 or e_2. These weighted rewards are then summed over all the states, and from this sum is subtracted the optimal expected outcome without sampling. Thus for the problem posed here, we obtain the *expected value of perfect information* as $\text{EVPI} = W_1 P(e_1) + W_2 P(e_2) - \max E\{e_i\}$. A related term, the *expected value of sampling information*, is simply the difference between the optimal expected reward with sampling and the optimal expected reward without sampling. We easily obtain this value as $\text{EVSI} = E_S(a_i) - \max E(a_i)$.

The important point here is that perfect information represents an upper limit to the amount of knowledge that can be gained by sampling. Consequently, *we should never pay more than EVPI, the expected value of perfect information, for sampling information.* The expected value of sampling information must be no greater than EVPI, and the amount by which it is less is indicative of the quality of sampled information.

Example 5.2.4 We consider now the determination of the expected value of perfect information and the expected value of actual sampled information for our previous example. Application of the relevant equations easily yields, with the probabilities and returns shown in Figure 5.17, $\text{EVPI} = 500(0.75) + 800(0.25) - 350 = \$225.00$. Also, we obtain $\text{EVSI} = 323.90 - 350.00 = -\$26.10$. Thus we see that perfect information would be worth quite a bit here. The expected return without sampling is \$350.00. We could increase this return by \$225.00 to \$575.00, for a 64 percent increase in the anticipated return, if we had perfect information. Actual sampled information would increase our expected return by only \$33.90, even when this information is free. It costs the DM \$60 to sample, however, and it is not worth it for the DM to pay \$60 for information worth \$33.90, because there is a loss of \$26.10 by doing this.

This sort of episode can go on and on! Suppose that, instead of drawing the ball from the unknown urn and observing its color, the DM decides to hire an assistant to perform this sampling task and report back the color of the ball to the DM. Unfortunately, many assistants are not completely reliable and will occasionally report results inaccurately. The conditional probabilities describing such a situation would then be given by an appropriate matrix of probabilities. For example, we would have the conditional probability that the assistant reports red when the true color is black. Ideally the elements along the diagonal should be much larger than the off-diagonal elements. In the case of such "noisy" measurements, the *posteriori* probabilities of interest are of the form $P(e_i|r)$ — that is, the probability that the urn is of type e_i given that the assistant reported that a red ball was drawn from it. Also, the DM will need the probabilities $P(r)$ and $P(b)$, the probabilities of what the assistant reports concerning the color of the balls, rather than the previously needed $P(R)$ and $P(B)$. These needed expressions may be obtained from

$$P(r) = P(r|R)P(R) + P(r|B)P(B)$$

$$P(e_1|r) = P(r|e_1)P(e_1)/P(r)$$

$$P(r|e_1) = P(r|R)P(R|e_1) + P(r|B)P(B|e_1)$$

Similar expressions result for the other needed probabilities. We could now specify all of these probabilities for and continue the example of this section. We will leave this as an example for the interested reader (See Problem 5.4).

What has been accomplished in this section may appear, at least at first glance, not especially related to systems engineering problems, and decision assessment issues. It may appear more attuned to gambling strategy. However, the urn model example used here may easily be used as a model for a large number of important decision-related issues. It is especially useful as a basic probability model in choice under uncertainty situations.

5.2.5   Utility Theory

When choosing among alternatives, we must be able, somehow, to indicate preferences among decisions that may result in a diversity of outcomes. With the monetary rewards in the just considered ball-in-the-urn examples, this was relatively straightforward. We assumed an expected value approach and preferred a larger amount of money to a smaller amount. There are many situations when the value associated with money is not a linear function of the amount of money involved. Utility theory allows us to address this issue.

Utility theory is needed for at least two reasons. First, the decision maker may have an aversion toward risk. On the other hand, a DM may be risk-prone and make decisions that are associated with high-impact unfavorable consequences simply for the "fun" of it, or because of very large rewards if the chance outcome does turn out to be favorable. Second, not all consequences are easily quantified in terms of monetary value, especially when there are multiple and potentially incommensurate objectives to be achieved. A scalar utility function makes it, in principle, possible to attempt to resolve questions that involve uncertain outcomes with multiple attributes such as these. But before discussing utility theory, we need to define and discuss a lottery, because this provides a basis for the utility concept.

It is desirable to reflect briefly on what we mean by the concept of utility in decision analysis and decision assessment. Formally, a *utility function* is a mathematical transformation that maps the set of outcomes of a decision problem into some interval of the real line. In other words, a utility function assigns a numerical value to each outcome of a decision problem. The basic assumptions, or axioms, needed to establish utility theory may be described as follows.

1. Any two outcomes resulting from a decision may be compared and placed in a preference order. If A and B are two such outcomes, then one must either prefer A to B ($A > B$ or $A \mathscr{P} B$), prefer B to A ($B > A$ or $B \mathscr{P} A$), or be indifferent between A and B ($A \sim B$ or $A \mathscr{I} B$). An extension of this assumption leads to the concept of transitivity. This requires that if a decision maker prefers A to B and also prefers B to C, then this decision maker must also prefer A to C. If a decision maker is not transitive, all sorts of maladies can result, as can easily be demonstrated.* In the simple way posed here, every individual should normatively seek to be transitive relative to preferential expressions. What applies to individuals does not apply to groups, because group preferences may well be intransitive, even though all individuals in the group have transitive preferences. We will discuss this reality in our next section.

2. Utilities may be assigned to lotteries involving outcomes as well as to the outcomes themselves. The term *lottery* is defined as a chance mechanism that yields outcomes $e_1, e_2, \ldots, e_n$ with probabilities $P_1, P_2, \ldots, P_n$, respectively, and where each $P_i \geq 0$, and where the sum over all P_i must be equal to 1. This will insure that we have a valid probability function; it will not assure us that the probabilities are appropriate for the problem at hand. A lottery, as described here, is just an alternative outcome pair that is intended to describe the alternatives that may be selected and their possible outcomes. It is denoted by $L = (e_1, P_1; e_2, P_2; \ldots; e_n, P_n)$ and is easily modeled by a tree-like diagram, such as the many we have illustrated in this chapter. From this it follows that if one prefers outcome

*For example, one can become a money pump. If, for example, you have the preference structure $A > B$, then there should be some amount of money, say \$1, such that $A \sim B + \$1$. Now suppose that $B > C$ and $B \sim C$ and $B \sim C + \$1$. If I have alternative C to start with, then I should be willing to trade in that alternative and \$1 for alternative B. If I have alternative B, then I should be willing to trade it and \$1 for alternative A. This new relation, $A \sim C + \$2$ is quite consistent with a transitive preference structure $A \sim C$, but very much a problem if you are intransitive and have $C > A$. If this were your preference structure, then you should be willing to pay someone to take away A and give you back C. Then the cycle can start all over again and continue ad infinitum!

A to outcome B, then one should also prefer A to the lottery $(A, P; B, 1 - P)$, and one should prefer this lottery to outcome B. Furthermore, if we prefer A to B, then we should prefer the lottery $L = (A, P; B, 1 - P)$ to the lottery $L' = (A, P'; B, 1 - P')$ if and only if $P > P'$.

3. If A is preferred to C and C is preferred to B, then for sufficiently large P, the lottery $(A, P; B, 1 - P)$ is preferred to C. Similarly for sufficiently small P, C is preferred to the lottery $(A, P; B, 1 - P)$. Thus there exists some P $(0 \leqslant P \leqslant 1)$, such that one is indifferent between receiving risk free outcome C and the lottery $(A, P; B, 1 - P)$. This insures continuity of the utility function. It is an especially important condition in that it establishes the certain money equivalent (CME). This is the lottery that is believed to be quite exactly equivalent to a specific sum of money. It is obtained by letting $B = 0$.

4. There is no intrinsic reward in the lotteries themselves, or as it is more commonly expressed, there is "no fun in gambling." In other words the decision maker's preferences should not be affected by the particular means chosen to resolve the uncertainty. If there is indeed "fun" in gambling, then this should be considered as one of the attributes of the alternative outcomes and multiattribute evaluation methods used. This is easily accomplished, and the result is a situation in which there is no reward associated with the lotteries themselves. The DM is ambivalent between lotteries of comparable value, and complex lotteries may be, appropriately, simplified.

Condition 1 is often called the *orderability* condition, or axiom. It ensures that the preferences of the DM impose a complete ordering and a transitive preference ordering across outcomes. Condition 2 is a *monotonicity* condition. It ensures that the preference function is smooth and that increasing preferences are associated with increasing rewards. Condition 3 is a *continuity* condition that ensures comparability across outcomes with equivalent preferences. Condition 4 is a *decomposition* condition that ensures transformation of lotteries and substitutability of equivalent lotteries.

Condition 3 contains the key to resolving a decision-with-uncertain-outcomes problem through use of decision analysis techniques. In actual practice, a situation like the one described in condition 4 is used to measure the utility that a decision maker has for various outcomes. These four assumptions form a basic set for the establishment of a utility theory. While this theory is not at all complicated, putting the theory into practice successfully is very complicated. There are a number of important theoretical issues that can easily be overlooked.

It is important, for example, to keep in mind that the concept of preference precedes the concept of utility. In other words, A may be preferred to B $(A > B)$ but not because the utility of A is greater than that of B, or $U(A) > U(B)$; on the contrary, we have the relation $U(A) > U(B)$ simply and directly because of the fact that A is preferred to B. Also, it is important to note that the numerical values assigned by the utility function are not unique. Given a utility function $U_1(x)$, it is possible to generate a new utility function $U_2(x) = aU_1(x) + b$, where a is a positive constant, and b is any constant, such that the preferences are not altered. Thus in a sense, a utility function may be viewed as a kind of preference thermometer. The utility function must also be monotonically nondecreasing in the sense that the highest utility is assigned to the most preferred outcome and the lowest utility is assigned to the least preferred outcome. Intermediate utility values and preferences are assigned in an analogous ordering. There are many potential behavioral pitfalls as well. Many of them are more insidious than the theoretical and quantitative pitfalls. Some of these we mention briefly in this text and are discussed in detail in references 1 and 15 and the references contained therein.

According to expected utility theory, the decision maker should seek to choose the alternative a_i which makes the resulting expected utility the largest possible. The utility, $u(ij)$, of choosing decision a_i and obtaining outcome event e_j will also depend upon the particular value of the probabilistically uncertain random variable e_j as conditioned on the decision path that is selected. Thus, the best the decision maker can and should do here is to maximize some function, such as the expected value of utility

$$\max_i \text{EU}\{a_i\} = \max_i \sum_{j=1}^{n} u_{ij} P(e_j | a_i)$$

where the maximization is carried out over all alternative decisions, and $P[e_j | a_i]$ is the probability that the state of nature is e_j. Often, we will use the abbreviated notation $\text{EU}\{a_i\}$ to mean the expected utility of taking action a_i. Generally, this is also called the *subjective expected utility* and written SEU. The expression subjective is used to denote the fact that the probabilities and utilities may be personal or subjective—that is to say, belief probabilities and personal utilities for obtaining outcomes from selecting alternatives.

A lottery is a chance mechanism that results in an outcome with a prize or consequence $e(1)$ with probability $P(1)$, an outcome with prize or consequence $e(2)$ with probability $P(2), \ldots$, and an outcome with prize or consequence $e(r)$ with probability $P(r)$. The probabilities $P(i)$ must be nonnegative and such that they sum to one, so that we have

$$\sum_{j=1}^{M} P(j) = P(1) + P(2) + \cdots + P(M) = 1$$

where the lottery is denoted as

$$L = [e_1, P_1; e_2, P_2; \ldots, \ldots; e_N, P_N]$$

We can now formally define a utility function as a transformation which maps the set of consequences into an interval on the real line. We will denote the utility of a consequence $a(i)$ as $U[a(i)]$, even though this will usually require that we calculate expected utility or subjective expected utility and should therefore be written as $EU[a(i)]$ or $SEU[a(i)]$. We now utilize the consequences of the four utility relevant assumptions just made. The number of assumptions varies from four to six, depending on which approach is taken. The basic approach taken by Luce and Raiffa [32], for example, in a classic and seminal work, utilized six assumptions. We have just described four assumptions, conditions, or axioms, which enable establishment of subjective expected utility as the criterion of choice for rational decision assessment. This very strong statement is necessarily true if we accept the four axioms or assumptions stated earlier.

The result of these assumptions is that the utility function $U(ij)$ satisfies several properties. These utilities turn out to be indicators of preference and not absolute measurements. They are unique only up to a general linear transformation or, more properly stated, affine transformation. A considerable motivation for introducing these concepts is to be able to deal with concepts of risk aversion and relative risk aversion [33]. Let us examine some of these concepts here.

A utility curve for money can be generated in the following manner. We arrange all the outcomes of a decision in order of preference. We denote the most preferred outcome W and the least preferred outcome L. We arbitrarily assign utility $U(W) = 1$ and $U(L) = 0$. Next, the decision maker is encouraged to answer the question, "Suppose I owned the rights to a lottery which pays W with probability $1/2$ and L with probability $1/2$. For what amount, say $X(0.5)$, would I be willing to sell the rights to this lottery?" Since the decision maker is indifferent between definitely receiving $X(0.5)$ and participating in the lottery $[W, 0.5; L, 0.5]$, then we must have $U[X(0.5)] = 0.5u(W) + 0.5u(L) = 0.5$. Figure 5.18 illustrates this lottery in decision tree format.

We could have picked any other lottery with outcomes W and L, such as, $[W, 0.01; L, 0.99]$. There are several analytical reasons, however, for choosing a lottery which gives W or L with equal probability, or $[W, 0.5; L, 0.5]$. It may be conceptually quite simple for the decision maker to imagine a lottery that gives W or L with equal probability. In this case, the decision maker need only be concerned with the relative preferences of W and L. Also, the value of the lottery $[W, 0.5; L, 0.5]$ occurs midway on the U axis between W and L and therefore facilitates plotting the utility curve. This choice of 0.5 probability for the assessment may, however, be quite unrealistic. For example, W might mean some enormous increase in profit to the firm and L might mean go bankrupt. It might be that a business person would so abhor a situation in which there is a 50% chance of going bankrupt that it becomes impossible to realistically think of utilities in these situations.*

---

*In a situation such as this, there would generally be a very significant effort to cause the decision situation to be changed such that the alternative options would be changed, especially the very significant probability of going bankrupt.

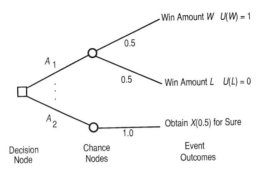

Figure 5.18 Decision analysis tree structure to determine decision-maker utility for *X(0.5)*.

Next the decision maker is encouraged to answer the question, "Suppose I owned the rights to a lottery that paid W with probability 1/2 and $X_{0.5}$, which has a utility of 0.5, with probability 1/2. For what amount, call it $X_{0.75}$, would I just be willing to sell the rights to this lottery?" Here, the decision maker is indifferent between receiving $X_{0.75}$ for certain and participating in the lotter $[W, 0.5; X_{0.5}, 0.5]$. Therefore, $U(X_{0.75}) = 0.5U(X_{0.5}) + 0.5U(L) = 0.75$. Next, we consider the lottery $[X_{0.5}, 0.5; L, 0.5]$, and the amount to be determined is $X_{0.25}$. In this case, we obtain $U(X_{0.25}) = 0.5U(X_{0.5}) + 0.5U(L) = 0.25$. This process could be continued indefinitely. However, the three points generated here and the two end points are often sufficient to give an idea of the shape of the utility curve, which will usually take on the general shapes shown in Figure 5.19. The utilities we have obtained are indicators of preference and are not absolute measurements. As a result, they are unique only up to a linear transformation of the form $U'(L) = aU(L) + \beta$ and where $a > 0$, as we have noted.

It is rare that a decision maker is risk prone and the majority of people are risk averse in almost all situations. One of the few exceptions to this would be

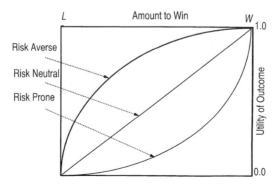

Figure 5.19 Hypothetical utility curve.

in situations where there is a very small probability of receiving a very large reward, such as a probability of 10^{-6} of receiving $\$10^6$ and a person is willing to pay \$5 for such a lottery. There, the typical utility curve is concave downward. There are some other notable exceptions to this generally risk averse nature of the DM, especially when the utility function is not required to be symmetric about losses and gains.

The first step in a utility assessment process generally, and ideally, results in a definition of the scope of the assessment procedure. Elements of the value system have been identified and often structured, at least partially, in the issue formulation step. Attributes, according to which alternatives are to be assessed, are identified and structured. The attributes are next related to measurable characteristics of the alternatives. Much of this effort is accomplished in the formulation and analysis steps, and the results should be organized to match the requirements of the utility assessment portion of the interpretation step. The final part of the utility assessment process concerns either (a) elicitation of the weights of each attribute and aggregation of the information to facilitate selecting the best alternative or (b) use of a number of holistic judgments and regression analysis to identify these weights.

There exist a number of *techniques for eliciting utility functions* in terms of attribute weights and alternative scores [34–37]. Four very common and useful approaches will be described here.

1. *Direct Elicitation.* The decision maker is asked to assign a relative, or ordinal, utility score for outcomes. Generally, these value scores would be based on experiential familiarity with previous situations. The utility might be assigned into categories such as fair, good, or excellent, or might be based on numerical values, perhaps anchored on a scale. The range of the scale might typically be 0–100, or it could be left indeterminate until after the elicitation is completed. Direct methods can provide precise numerical scores and therefore are very attractive for their speed in application. But providing precise, consistent, and meaningful numerical values is a very difficult task. It is also quite difficult to consider risk aversion in an approach of this sort.

2. *Ranking Methods.* In these methods, the decision maker orders the levels of a specific set of attributes from most preferred to least preferred. The ordering may consist of levels of a single attribute, or combinations of the levels of two or more attributes, as in the case of investigating value tradeoffs, or assessing the utility of interrelated attributes. This type of utility assessment measures the relative strength of preference of one preference relation with respect to another preference relation. Outcomes are assumed to occur with certainty for purposes of outcome preference ranking.

3. *Indifference Methods.* These methods consist of identifying indifference points in a decision space. Indifference methods may involve the assess-

ment of several possible combinations of levels of attributes and determining indifference among these combinations. Judgments related to "tradeoffs" between the utilities of various attribute level combinations must be made. When uncertainty is involved, indifference methods rely on the decision maker's ability to choose among uncertain outcomes with known probabilities. In this case, we obtain the certainty equivalent of a lottery — that is, the level of attributes for which the decision maker is indifferent to that lottery. Generally, indifference methods are preferred by most practicing decision analysts, especially those deeply committed to axiomatic approaches. There are a number of heuristic and approximate approaches to utility assessment using indifference methods. In principle, indifference judgments involve lotteries. Thus, they provide a combined assessment of strength of preference and risk attitude. It is important to note that indifference methods require prior knowledge of the decision situation model, including the outcome attribute tree. In standard multiattribute utility decision assessment, which we will soon discuss, ranking methods are often used prior to the use of indifference methods. This provides a rough, imprecise assessment of the possible values of the parameters of the model. It facilitates coordination of the subsequent precise assessments by means of indifference methods.

4. *Assessment Through Decision Observation.* This approach consists in observing decision behavior in real-world or simulated real-world situations and inferring from these observations the parameters of a prespecified model. The work of Ken Hammond and his colleagues on social judgment theory, or policy capture, makes use of this concept. Techniques based on this approach are sometimes called bootstrapping techniques. Use of regression-analysis-based approaches to estimate parameters in the assumed structural model is central to the decision observation approach. Social judgment theory can be used whenever there is a need to determine the relative importance that an individual or group adhere to different attributes of a decision situation. It can be particularly appropriate as an aid for evaluation and comparison of alternatives for decision assessment, and also for value system design in issue formulation. This approach has been used in a rather large number of applications [38–42], and it is described in reference 1. A single volume that provides a definitive overview of this approach, as well as some discussion of other approaches, is also available [43].

There are three generally used utility-scale types in common use. The *ordinal scale* is one where items are ranked in preference order. The statement $A_1 > A_2 > A_3 > A_4 > A_5 > A_6$ is a statement of ordinal preferences. The *interval scale* is one where preferences are entered according to their distance from some arbitrary end points. A classic example of an interval scale would represent a Fahrenheit temperature scale. 40°F is warmer than 20°F, but not twice as warm. We might say for example that an outstanding result on a test receives a score of 100 and a very poor result receives a score of 0. The statements that $A_1 = 90$, $A_2 = 30$ indicate that the first alternative is quite good and the second one rather poor. But it would not be meaningful to say that the first alternative is three times as good as the second one. A *ratio scale* is a scale in which alternatives are associated with a cardinal ranking according to their relative distance from some calibrated and nonarbitrary end points. The indifference methods just discussed lead to ratio scales. The assessment through direct observation approach may do this. The other approaches generally lead to ordinal or interval scale measurements.

Utility functions are only fully useful when they are intelligently obtained and thoughtfully applied. Structuring of a decision assessment issue is a crucial part of an overall decision assessment process. The final result that is obtained will, in practice, depend upon the veridicality of both the decision situation structural model and the parameters within this structure.

The process of eliciting probabilities and associated utilities such that they veridically represent individual judgments is one of the major tasks in decision analysis. The seminal text by von Winterfelt and Edwards [44] describes many of the approaches that have been taken to eliciting these values. The efforts of Merkhofer [45], Wallsten and Budescu [46], Borcherding et. al [47], and Keeney et al. [48] are also pertinent. Harvey has documented a number of useful findings in this area [49, 50]. Risk assessment and the identification of rare event probabilities is considered in Sampson and Smith [51]. A definitive three volume set provides a compendium of a great variety of approaches to judgment and decision-related measurement [52–54], including measurements of probabilities and utilities for decision assessment.

Example 5.2.5   We provide a simple example of utility determination. To give a specific numerical example, let $W = \$100$ and $L = -\$50$. Then we might find $X_{0.5} = \$10$, $X_{0.75} = \$47$, and $X_{0.25} = -\$25$. In other words, the decision maker owns the rights to the lottery $[\$100, 0.5; -\$50, 0.5]$ and would just be willing to sell the rights to this lottery for \$10. An EMVer, of course, would be willing to sell the rights to the lottery for $0.5(\$100) + 0.5(-\$50) = \$25$. This utility curve may be expressed graphically as shown in Figure 5.20. Here, the units on the abscissa $X_0, \ldots, X_1$ represent increments from the decision maker's present assets.

The concave shape of the utility curve in Figure 5.20 results from the risk-averse nature of the decision maker. An EMVer would have the utility curve shown by the straight line in Figure 5.20. Because of the concavity of the curve, the decision maker will place a value on a lottery somewhat lower than its expected monetary value. For example, suppose our decision maker owned the rights to a lottery which pays \$75 with probability 0.7 and \$30 with probability 0.3. The EMV of such a lottery is $0.7(\$75) + 0.3(\$30) = \$61.50$. However, the expected utility is given by $0.7u(\$75) + 0.3u(\$30) = 0.7(0.9) + 0.3(0.65) = 0.825$. This corresponds to a CME (certain monetary equivalent)

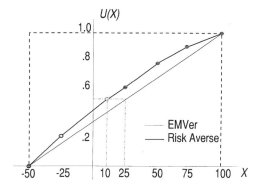

Figure 5.20   Curves representing the utility of money.

of $60. The fact that the CME is less than the EMV is indicative of the conservative or risk-averse nature of our decision maker.

Once the decision maker's utility curve has been defined, the standard tenets of axiomatic decision analysis suggest that it may be employed in any decision problem as long as all the outcomes fall between L and W in desirability. In principle, the decision maker can now delegate authority to an agent, who can employ this utility curve to arrive at the same decisions that the decision maker would make. The utility curve conveys all the information necessary for making decisions that reflect the decision maker's preferences for the problem under consideration. Clearly, all of this is very idealistic. While what is said may not be untrue, there are a lot of contingencies that need to be associated with these statements. Clearly, these are highly provocative statements. Substantively, there are several major problems. First, and probably most obvious, is that delegating authority has all sorts of ego, status, and authority issues that have nothing to do with utility curves and the substance of the decision in question. Second, delegation of authority would presume that the subordinate could be trusted fully to understand and adhere to the utility curves of someone else, and that the subordinate has equivalent knowledge.

At a more theoretical level, these statements assume that, once determined, a utility function is immutable. This is generally not the case. A person's utility curves, regarding any number of aspects of life, continually change as the person learns new things, has new experiences, or develops new perspectives. Similarly, a utility curve that applies to one kind of problem, say, preferences for apples versus oranges, may offer very little insight into another problem, say preferences for apples versus pears. Thus, it may simply make little sense to suggest that a subordinate (or anyone else) can learn someone's utility curves, and then operate as if these can be rigidly applied in changing circumstances.

Example 5.2.6   Thus far we have assumed that our decision maker owned the rights to the lottery $[W, P; L, 1 - P]$ and was trying to determine a selling price, or a CME, for the lottery. Now suppose that the DM does not own the rights to such a lottery, but instead is seeking to buy these rights. What should be the buying price? For an EMVer the buying price and the selling price would be the same, but for a non-EMVer they would not necessarily be the same, because we are, in reality, considering two different lotteries. This is true because in the case of buying a lottery, the buying price must be subtracted from the outcomes of the lottery. Thus if b is the buying price of a lottery, we are trying to find b such that $U(0) = PU(W - b) + (1 - P)U(L - b)$. This is also illustrated by the decision tree of Figure 5.21(a). Similarly, the selling price s can be determined either from the relevant equation or from the decision tree of Figure 5.21(b). Here, we assume that $W = \$100$, $L = 0$ and that the probability of winning is 0.25. Thus an EMVer should be willing to either buy or sell this lottery for $25. With the risk-averse behavior we have assumed in Figure 5.20, the buying price is such that $0.44 = 0.25U(\$100 - b) + 0.75U(-b)$. Thus, the buying price for the lottery is approximately $19. The selling price for a risk-averse owner is determined from $U(s) = 0.25(1) + 0.75(0.44) = 0.58$ and is, from the utility function of Figure 5.20, approximately $22.00.

Suppose we are faced with a gamble with an equal chance of winning or losing $10,000. Rather than risk losing $10,000, we might be willing to pay, say, $1000 to avoid having to take this gamble, especially if we did not have $10,000 to lose! This is, of course, the basis for the insurance industry. An example of this is described in Fishburn [55]. *The shape of an individual's utility curve will clearly depend upon that individual's present assets.* Because an individual with a high asset position would be less averse to taking risks that might result in a loss, we might expect that the utility curve would approach that of an EMVer (i.e., a straight line) as the present asset position increased. Such a decision maker exhibits what is called decreasing risk aversion. In other words, when faced with a risky gamble, say an even chance of winning or losing $100,000, the amount that a risk-averse decision maker would pay to avoid this gamble,

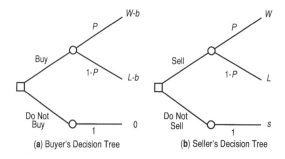

Figure 5.21   Decision trees for buyer and seller.

called the insurance premium, is a function of current assets and should therefore decrease to zero as assets increase indefinitely.

Example 5.2.7   Let us examine briefly the interesting result that life insurance policies may be good for both the buyer and the seller in that they increase the utility of both the seller and the buyer of the insurance. We will assume that we are considering a $100,000 term life insurance policy which costs $1000 per year. We will assume that the individual is risk-averse. We shall assume that the probability of death in a given year is $P_{Death} = 0.009$, the present assets of the buyer are $10,000, and the value of future income discounted to the present is $100,000. Figure 5.22 details the payoff matrix representations of this life insurance problem for both the buyer and the seller. The expected value of a buy decision for the buyer is $109,000, which is less than the expected value of the do-not-buy decision, which is $109,100. If this were not the case the insurance company average profit would be negative. Even if the utility curves for both buyer and seller are risk-averse and precisely the same, the regions of operation are so different that the buyer is much more risk-averse than the seller.

Consider, for example, a utility for money of the form $U = \ln$ (assets $\times 10^{-3}$) that is valid for both buyer and seller. The buyer is operating over the region $10,000 to $110,000 for determination of this utility curve, whereas the insurance company is operating over the region $99,901,000 to 100,001,000. We may use the relation $U' = aU + b$ to scale the utility curves for the buyer and seller over the range 0 to 1. The original and scaled utility curves are illustrated in Fig. 5.23.

The total utility curve is concave, because both buyer and seller are risk-averse; but the buyer is relatively more risk-averse, because the seller can afford to "play the averages" but the individual buyer cannot. They buyer is willing to reduce total assets to less than the EMV, $109,000 versus $109,100, of the "lottery" in order to prevent a financial hardship for the surviving

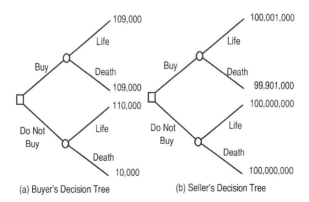

(a) Buyer's Decision Tree   (b) Seller's Decision Tree

Figure 5.22   Life insurance decision trees for buyer and seller.

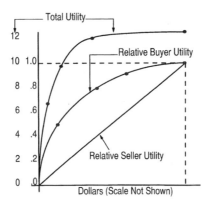

Figure 5.23   Utility curves for money.

members of the family in the event of death. The insurance company is willing to assume this risk for the buyer, because the EMV is now higher, $100,001,000 as contrasted with $100,000,000, than its assets before selling the policy. In the payoff matrices of Fig. 5.22, none of the outcomes for the seller differs significantly from the initial assets. This is not at all the case for the buyer. Thus, it is not too surprising that the insurance company is more of an EMVer than the individual buyer.

It is quite interesting to note that these results were obtained assuming that the utility curves for both the buyer and the seller were the same. However, the two parties operate over such different portions of the total utility curve that the buyer is quite risk-averse, whereas the seller is effectively an expected monetary value player.

Notions of risk aversion and utility are of much interest. Early works in decision analysis were much concerned with concepts of risk aversion incorporated in the von Neumann–Morgenstern utility functions. *A von Neumann– Morgenstern utility function for money* is a function of final wealth at the end of the decision problem, and not the gains and losses relative to a generally uncertain initial position. Thus $U(x)$ represents the utility that results when one's final wealth is exactly equal to x. If we let the random variable x denote a level of final wealth, then presumably there is some amount of money that, if we possessed it with certainty, would make us just as happy as having the uncertain wealth level x. This is called the *certainty equivalent of* x, often denoted by $CE(x)$. Most people would be willing to trade x for less than its actuarially fair value in order to be rid of the *risk* associated with x, and these people are said to be risk-averse. The *risk premium* is the amount we are willing to give up — that is, the "premium" we are willing to pay to an insurance company — to get rid of all the risk. The risk premium of x is usually written as RP(x) and is defined by the equation

$$RP(x) = E(x) - CE(x)$$

One is defined to be risk-averse if $RP(x) > 0$ for any wealth x that is uncertain. This is what is often referred to as "risk aversion in the large." It has been demonstrated that there is a measure of "risk aversion in the small" that is associated with risk aversion in the large. The function $ra(x)$ defined by

$$ra(x) = -U''(x)/U'(x)$$

measures one's local aversion to risk in the vicinity of wealth level x and is often called the "Arrow–Pratt measure of absolute risk aversion." Sometimes the reciprocal of $ra(x)$ is called the local risk tolerance, $rt(x)$.

It seems reasonable that one becomes less risk averse as their wealth level rises because, intuitively, a person who can afford to absorb larger losses should be willing to undertake greater risks. This suggests that the wealthier one is, the less they should be concerned about any specific wealth variance, and the smaller the premium one should pay to be rid of it. Utility functions are often classified in terms of whether, and in what way, risk aversion changes as a function of wealth. There are several utility functions in which risk aversion, $ra(x)$, or risk tolerance, $rt(x)$, as the inverse of risk aversion, is a linear function of wealth, and some of these are given in Table 5.1.

The exponential utility function is interesting in that it has the property of constant absolute risk aversion. This does not seem to be an appropriate utility function for representing risk attitudes over a potentially large range of final levels of wealth but may be very useful as an approximation to a utility function for evaluating small to moderate size changes in outcome level. For this reason, this utility function is often used in decision analysis applications. An exponential utility function has very convenient mathematical properties. It satisfies the "delta property" that if all the outcomes of a gamble are increased in value by the same amount Δ, then the certainty equivalent for the

TABLE 5.1 Four Simple and Interesting Utility Functions

	Utility Function	Local Risk Aversion	Local Risk Tolerance
Linear	$U(x) = x$	$ra(x) \equiv 0$	$rt(x) \equiv \infty$
Exponential	$U(x) = -e^{-ax}$	$ra(x) \equiv a$	$rt(x) \equiv 1/a$
Logarithmic	$U(x) = \ln(x + b)$	$ra(x) = 1/(x + b)$	$rt(x) = x + b$
Power	$U(x) = (1/c)(x + b)^c,$ $c < 1, c \neq 0$	$ra(x) = (1 - c)/(x + b)$	$rt(x) = (x + b)/(1 - c)$

gamble is also increased by exactly the amount Δ. With an exponential utility function, a group of agents with different beliefs and different risk tolerances behaves like one "representative agent" whose utility is a combination of the utilities of the individuals. Under these conditions, we can "aggregate" diverse beliefs and preferences of a number of agents in a convenient manner — for example, the risk tolerance of the single "representative agent." The logarithmic and power utility functions are such that these utilities are decreasingly risk-averse for increases in final wealth position.

5.2.6   Multiple Attribute Utility Theory

In this subsection, we describe an extension of the basic decision analysis paradigm to include situations where the outcomes have multiple attributes. We assume that a set of feasible alternatives $A = (a, b,...)$ and that a set $(X_1,..., X_n)$ of attributes or evaluators of the alternatives can each be identified. Initially, we assume that a single known outcome follows with certainty, from selection of an alternative. We will comment on the decision under outcome uncertainty case later. Associated with each alternative course of action a in A, there is a corresponding consequence $X_1(a), X_2(a), ..., X_n(a)$ in the n-dimensional attribute (consequence) space $X = X_1 X_2,..., X_n$.

The problem faced by the decision maker is to choose an alternative a in A so that the maximum pleasure with the payoff or consequence, $[X_1(a), X_2(a), ..., X_n(a)]$, results. It is always possible to compare the values of each $X_i(a)$ for different alternatives, but in most situations the values $X_i(a)$ and $X_j(a)$ for i not equal j cannot be easily compared because they may be measured in totally different units. It would be very convenient if we had a scalar utility value, just as in our earlier efforts in this section. Thus, a scalar-valued function defined on the attributes $(X_1,..., X_n)$ is sought that will allow comparison of the alternatives across the attributes. The interested reader is referred to the extended and thorough conceptual discussions in Keeney and Raiffa [56, 57] for additional information.

A primary interest in multiattribute utility theory (MAUT) is to structure and assess a utility function of the general form

$$U[X_1(a), ..., X_n(a)] = f\{U_1[X_1(a)], ..., U_n[X_n(a)]\}$$

where U_i is a utility function over the single attribute X_i, and f aggregates the values of the single attribute utility functions such as to enable one to compute the scalar utility of the alternatives. We assume that the utility functions U and U_i are continuous, monotonic, and bounded. Usually, they are scaled by $U(x^+) = 1$, $U(x^-) = 0$, $U_i(x_i^+) = 1$, and $U_i(x_i^-) = 0$ for all i. Here $x^+ = (x_1^+, x_2^+,..., x_n^+)$ designates the most desirable consequence and the expression $x = (x_1^-, x_2^-,..., x_n^-)$ denotes the least desirable consequence. The symbols x_i^+ and x_i^- refer to the best and worst consequence, respectively, for each attribute X_i. Thus, we have $x_i^+ = X_i(a^+)$, where a^+ is the best alternative for attribute *i*, and $x_i^- = X_i(a^-)$ where a^- is the worst alternative for attribute *i*. In the simplest situations, what is called *additive independence of attributes* [55, 56] exists such that the MAUT function may be written as

$$U(a_i) = w_1 U_1(a_i) + w_2 U_2(a_i) + \cdots + w_n U_n(a_i) = \sum_{j=1}^{n} w_j U_j(a_i)$$

Here, the w_j are the weights of the various attributes of the decision alternative a_i, and the U_j are the attribute scores for that alternative.

Mutual preference independence is required for this utility function to be valid in the case of certain outcomes. There are many cases where this independence relationship holds. It simply requires that preferences over specific outcomes on attribute $X_i(a)$ do not depend upon the outcomes on attribute $X_i(a)$ for all *i*, *j*. Somewhat stronger than the notion of preferential independence is that of utility independence. The *utility independence condition* is basically the same as that of preference independence, except that the preferences for uncertain choices involving different realizations of $x_i(a)$ do not depend on a fixed value that might be set for $X_j(a)$ for all *i*, *j*.

It turns out that precisely this same linear form of utility expression is valid in the case where decision outcome uncertainties are involved. In cases where there is probabilistic event outcome uncertainty, we determine the expected utility of the multiattributed outcomes as suggested by Figure 5.24.

We have just described very briefly the case of certainty in a multiattribute decision assessment framework. Associated with each alternative there is a known consequence that follows with certainty from implementation of the

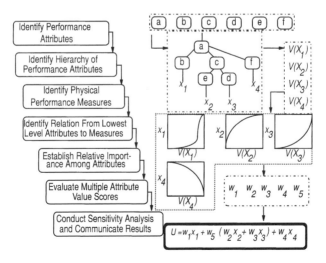

Figure 5.24 Simplified MAUT process.

alternative. These results also carry over to the case of decisions under uncertainty of event outcomes. The foundations for decision assessment under risk were discussed in an earlier portion of this section. *The implications of this work are that probabilities and utilities can be used to calculate the expected utility of each alternative, and that alternatives with higher expected utilities should be preferred over alternatives with lower ones. In the multiple attribute case, we simply calculate the scalar utility function of each multiattribute outcome and use this scalar utility function to calculate the subjective expected utility function as we have done previously.*

Multiattribute utility theory (MAUT) is an approach for evaluating, prioritizing, and ranking the outcomes associated with different action alternatives in complex decision situations. The outcomes of alternative policies may occur with either certainty or uncertainty, and they have multiple attributes that indicate the degree to which the outcomes meet prespecified objectives.

Even though the theory of multiattribute decision analysis is conceptually straightforward, especially when the linear additive independence of attributes condition holds, there are circumstances that make its implementation very complex. Putting the methodology into practice is much more involved than we might initially believe. Each of the foregoing decision analysis steps requires substantial interaction between the analyst and the decision maker if useful results are to be obtained. There are a number of subtleties associated with scaling and, of course, a number of simplified approaches that can be utilized [58]. We will describe one of these soon.

Use of the MAUT approach presumes the following:

- The decision situation has been structured, for example, in a decision tree.
- A hierarchy of objectives or attributes has been developed in the form of an attribute tree.
- The lowest-level attributes are capable of being observed and measured, as we indicate in Figure 5.24.

In the case of uncertainty, probabilities should be assigned to each of the possible outcomes of a decision alternative. Then, a utility or value function is derived, based on preference information provided by the decision maker. Using the multiple-attribute utility function, the probability estimates, and the utility values for the lowest-level attributes associated with each outcome, an overall expected utility is computed for each decision alternative. These utility values reflect the decision maker's preferences. Therefore, they may be used for prioritization of alternative decisions.

The use of multiple-attribute utility theory should be preceded by an effort in which objectives are defined, alternative decisions or policies are identified, and their impacts are investigated. Following this decision issue formulation effort, the decision situation is structured, usually in the form of a decision tree or inference diagram that illustrates the possible outcomes of each action alternative, along with its associated probability. The attribute tree should always be structured in such a way that the lowest-level attributes of the decision outcome events are measurable. Also, attributes should be defined in such a way that it is easily possible to elicit the decision maker's preferences over different possible outcomes. This requires that the attributes be defined in such a way that the linear independence conditions hold. Then, assessment of the decision maker's utility function results in a single number that represents the preference of the decision maker over the expected outcome of each decision alternative. The lowest-level attribute measures for each outcome, along with the probability estimates for the outcomes, are used as input data to this calculation. The resulting cardinal multiple-attribute utility function represents the decision maker's value system, assuming that we have been faithful to the tenets of the MAUT theory and that the theory is applicable in the decision situation at hand.

There are a number of desirable characteristics of multiple-attribute utility functions. These include the following [55, 56].

1. *Complete.* If a multiple-attribute utility tree or multiple-attribute value hierarchy is complete, the hierarchy of attributes taken together must adequately envelop all of the concerns necessary to evaluate the overall decision objective. Thus, if the attribute tree shown in Figure 5.25 is complete, you only need to know how well an alternative performs with respect to the lowest-layer attribute in order to know how well it performs with respect to the overall objective of "purchase the most appropriate software."

2. *Nonredundant.* In a nonredundant hierarchy of attributes, no two evaluation considerations have common characteristics in the same layer of the attribute hierarchy. In this way, each layer in the attribute hierarchy will have partitioned the layer above it into more detailed and descriptive attributes. In Figure 5.25, the software performance consideration is partitioned into three more descriptive attributes. If this is to be a nonredundant partitioning, a score can independently be assigned to one of the three lowest-level attributes shown.

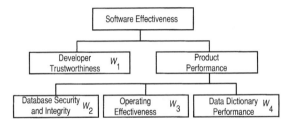

Figure 5.25 Attribute tree for software effectiveness.

3. *Decomposable.* In a decomposable hierarchy, the hierarchy is nonredundant. And also the value that is attached to changes in attribute measure scores at any lowest-level attribute does not depend on other lowest-level attribute considerations. When this property is satisfied, attribute values can be added across the hierarchy.

4. *Operable* (*Understandable*). An operable attribute hierarchy is one that is understandable for those who must use it. This is especially important in those situations where technical specialists must interact with the public in policy relevant matters.

5. *Small Size.* It is desirable to have a reasonably small attribute tree in that a smaller hierarchy is capable of being more easily communicated to others. Also, less resources are usually required to estimate performance of alternatives with respect to the attribute measures.

Most often, it will be highly desirable to obtain a sensitivity assessment of the effort, in order to investigate the effects of changes in probability estimates, elicited multiple attribute utility parametric coefficients, and other imprecise information that may have been used in the analysis. The results of a sensitivity assessment should give a good idea of the robustness of the alternative preference rankings that have been obtained and should indicate the critical sensitivity factors and areas of the decision assessment that are in need of further attention.

The following simplified approach, an adaptation of a procedure called worth assessment [59–61], is suggested as a heuristic for *construction of a multiple attribute utility function.* The process flow is also illustrated in Figure 5.24.

1. *Identify the Overall Performance Attributes.* The list of identified attributes should be restricted to those performance attributes with the highest degree of importance, and they should include all relevant attributes. Ideally, the attributes in the list should be mutually exclusive and independent so that no one attribute encompasses any other attribute. The attributes should also be linearly independent in the sense that the decision maker is willing to trade partial satisfaction on one attribute for reduced satisfaction on another attribute.

2. *Construct a Hierarchy of Performance Attributes.* Once the overall attributes have been established, they are subdivided until the decision maker feels there is enough detail in each attribute that its value can be measured. This dividing and subdividing process will result in a tree-type hierarchy.

3. *Select Appropriate Physical Performance Measures.* In constructing a hierarchy, the decision maker defined a set of lowest-level attributes that are combined in some fashion to define the overall performance attributes. Some characteristic of performance must be assigned to each

lowest-level attribute to measure the degree of criterion satisfaction. It is important to distinguish between an attribute measure and a performance objective. A performance objective reflects what a decision maker desires from a set of decision consequences, whereas attribute measures reflect what a decision consequence can actually deliver across the identified attributes. In considering a possible standard of measure, the following question can be asked: Would changes in the value of this measure bring about significant increases or decreases in the extent to which each lowest-level attribute is satisfied? If the answer is yes, the measure is appropriate.

4. *Define the Relationship Between Lowest-Level Attributes and Performance Measures, and Deal with the Scoring Problem.* Selecting the performance measures establishes the connection between the measure and the worth indicated by that measure, but it does not completely specify the connection. The connection is established by a scoring function, which assigns an attribute score to all possible values of a given performance measure. The domain of the scoring function is the set of all possible values of the performance measure, and its range is the closed interval $[0, 1]$. The scoring function may be stated explicitly or graphically.

5. *Establish Relative Importance Within Each Attribute Subcriteria Set.* In constructing the attribute hierarchy in step 2, each attribute is successively subdivided. At each point of division, an attribute criterion is defined by its subcriteria. Some of those subcriteria may be more important than others, and a weighting function is needed to indicate the relative importance of attributes. When this step is completed, the hierarchy can be used to create an overall attribute score for each alternative being considered. Attribute weights are assigned to each attribute set such that the sum of the attribute weights is unity. The first step in this process is to rank the subcriteria by relative importance of attribute satisfaction. Next, the most important criterion is assigned a temporary value of 1.0. Then the decision maker must be more precise in terms of a response to the concern "How much less important is the second criterion than the first?" If the answer is "half as important," the criterion is assigned a temporary value of 0.5. Next, the second and third most important are compared. If the third is one-fifth as important as the second, we assign it a temporary value of $1/5 \times 1/2 = 1/10$. Note that the third is also one-tenth as important as the first. This process is continued until all attributes have been assigned temporary weights. Then those temporary weights are scaled so that their sum is unity. The final weight for the ith subcriterion is determined from $w_i = \beta_i/(\beta_1 + \beta_2 + \cdots + \beta_p)$. It is important to note that the relative importance of any two attributes is reflected in the ratio of their assigned weights. Thus, we have obtained a ratio scale. The issues of attribute independence and subdivision of attributes are important. Clearly, if the overall attribute performance objectives are

independent of each other, the elements of an attribute set must also be independent, and they must be included within the meaning of the attribute they subdivide. Theoretically, it makes little difference whether we consider a single-level attribute tree or a multiple-level attribute tree. In practice, the effort involved and the results obtained may differ quite a bit across various approaches taken. There are really two problems here. The first is how to define the attribute sets, and the second is how to determine attribute independence. Let us examine these separately. In comparing an attribute with its related higher-level attributes, the following two statements should be examined to ascertain which better describes the relationship. A given attribute may be included within the meaning of, or be an integral part of, the higher-level criterion. Alternately, an attribute may represent one of several alternative means of satisfying the higher-level criterion and may be important only insofar as it contributes to it. If the first statement is the more accurate description, the candidate subcriterion may belong in the hierarchy under the higher-level criterion. If the second is more accurate, worth independence does not exist, and so the attribute does not belong in the hierarchy. *Attribute independence* can be defined by three interrelated statements:

- The relative importance of satisfying separate performance criteria does not depend on the degree to which each criterion has in fact been satisfied.
- The rate at which increased satisfaction of any given criterion contributes to the overall worth is independent of the levels of satisfaction already achieved on that and other criteria.
- The rate at which decision makers would be willing to trade off decreased satisfaction on one criterion for increased satisfaction on another is independent of the levels of satisfaction already achieved by the criteria.

6. *Calculate the Multiple-attribute Utility.* Using the attribute weights and performance scores on lowest-level attributes, the performance of each alternative being evaluated is calculated.
7. *Conduct Sensitivity Studies as Needed, and Communicate the Results of the MAUT Evaluation.*

The approach is sufficiently general that it can accommodate either single-level attribute tree structures or multiple-level attribute structures.

Example 5.2.8. We present below some sample criteria for database management system (DBMS) software selection. These will provide us with an appropriate summary of the critical and important factors in conducting a MAUT-based study. First, it is important to present the sequence of steps with which we approach DBMS software evaluation:

1. Identify user requirements for DBMS software.
2. Define critical attributes for the DBMS software. Identify the range of performance expected for each attribute, from best performance to worst performance. Also identify the importance of this performance range across each attribute.
3. Obtain literature describing potential software packages and arrange for an evaluation/demonstration of the software. If needed, iterate back to step 2 and refine attributes and attribute scoring measures;
4. Evaluate each software package according to its score on the various performance attributes.
5. Conduct a sensitivity analysis of the results of the analysis, if needed.
6. Select the highest scoring software package.

In the first step, we identify critical user needs to be fulfilled by the DBMS software. When we speak of user requirements, we refer to the purposes and associated functions that the user wishes or needs to perform. These user requirements can be converted into requirements specifications, which are the more technical and structural characteristics of the DBMS.

Two closely related, but different, approaches are possible. One results in a *ratio scale*, and the other results in an *interval scale*. The first approach, called the *relative weight* or *swing weight approach*, is anchored on the best performing and worst performing attribute scores for the actual alternatives and yields a ratio scale. The second approach, called the *absolute weight approach*, is based on definition of a hypothetical best and worst attribute score. In the way accomplished here, it results in an interval scale.

1. In the relative weight approach, we may list desired attributes in order of importance; evaluate the degree to which each candidate software package has each such attribute; weigh these evaluations according to relative importance of each attribute; sum to come up with a final evaluation of each candidate alternative.
2. In the absolute weight approach, we may list attributes in order of importance; establish maximum desired and minimum expected performance level for each attribute; evaluate each candidate alternate by the degree to which it satisfies this maximum value for each attribute; weigh these evaluations according to relative importance of each attribute; sum to come up with a final evaluation of each candidate alternative.

We will now describe these two multiattribute-based approaches to performance evaluation. The first of these is based on relative weights and relative performance scores across attributes of performance and performance alternatives.

In the *swing-weight-based approach*, we identify the maximum performance expected on each of the identified performance attributes, and we assign these performance levels a score of 1.0. The actual score used to indicate the maximum is arbitrary, and the overall evaluation will not depend upon this value. We could just as well use 100 as the maximum score possible. We identify the minimum performance expected on each attribute and assign these performance levels a score of zero.

Next we assign a level of importance to each attribute. Generally, some attributes will be very important and some unimportant. A possible way to identify weights for the attributes is to identify the most important performance attribute, in terms of the worst to best performance scores expected. It is necessary to identify this importance in terms of the performance ranges expected. If all software packages have virtually the same performance on some attribute, then the relative importance of this attribute is zero. Recall that there will be a performance score of 1.0 for the best-performing software package and a performance score of 0.0 for the worst-performing package.

The most important attribute is assigned a value of one. The next most important performance attribute, in terms of the range in performance expected, is identified. If the most important attribute has a weight of 1.0, then we should associate a lower weight, which is called a swing weight for this second most important attribute. This is continued until we have identified swing weights for all attributes of importance. These swing weights are then normalized such that they sum to one. The normalized swing weight, w_j, becomes $w_j = W_j/(\Sigma_{j=1}^N W_j)$, where N is the number of attributes considered and W_j is the unnormalized swing weight of the jth attribute.

Following this, we identify performance scores for each alternative on each attribute. We have already identified the 0.00 and 1.00 performance scores and their meaning, and so the performance evaluation task is just that of placing the performance of each DBMS alternative within this range for each of the performance attributes. The final performance score is then determined from the weighted sum $\S_i = \Sigma_{j=1}^N (w_j S_{ji})$, where w_j is the swing weight associated with the jth attribute and S_{ji} is the performance score of the ith software package on the jth attribute of performance.

The absolute-weight-based approach is based on identifying an ideal performance alternative, one whose performance is ideal across all performance attributes, and then associating a performance score of 1.0 with this, for all attributes. The following associations of numerical scores with performance characteristics are somewhat mercurial, but may be of value in identifying an ideally performing DBMS package and in setting the standards for judgment of actual DBMS packages.

1.00   Package completely meets performance requirements and expectations on this attribute.

0.75   Package is very acceptable on this performance attribute, in all but very minor ways.

0.50   Package generally satisfies maximum performance requirements on this attribute, but is deficient in some important aspects.

0.25   Package fails to meet performance specifications associated with this attribute.

0.00   Package is almost totally deficient in performance on this performance attribute.

Now that we have defined the performance of an ideal standard DBMS packages and have established performance scores across each attribute for a less-than-ideal performance, we need to identify the weight to assign with each attribute. Clearly, some attributes will be very important and some very unimportant. These importance weights can be expected to vary, perhaps considerably, across specific software evaluations. A possible way to identify the (absolute) attribute weights is to identify the most important performance attribute. This attribute is assigned a value of one. The next most important performance attribute, in terms of the range in performance expected, is identified. We might associate weights in accordance with the following associations:

1.00   The feature described by the attribute in question is mandatory and very important.

0.75   The feature described by the attribute in question is quite important.

0.50   The feature described by attribute in question is moderately.

0.25   Performance at the level specified by the ideal standard for this attribute is desirable, but not needed.

0.00   Performance of this attribute is unimportant, and this attribute can be disregarded in terms of performance evaluation.

The process of listing the attributes in terms of importance continues. After they are identified, the absolute weight values are identified. These must be veridical to the interval scale listing. A danger with this particular approach is identifying too many performance attributes and assigning most or all of them too high a worth. A worth of 0.35 many not seem very high. However, if there are 10 attributes with this weight, then it is possible to obtain a performance score of as much as 3.5 on these attributes alone. This is 3.5 times the maximum score of 1.0 that can be obtained due to performance on the most important attribute. For this reason, primarily, we prefer and suggest a relative, or swing weight, approach.

In the relative, or swing weight, approach, the absolute weights are normalized such that they sum to one. The normalized absolute weight, w_j, becomes $w_j = W_j/(\Sigma_{j=1}^{N} W_j)$, where N is the number of attributes considered and W_j is the unnormalized weight of the jth attribute. The final performance score is then determined from the weighted sum $\S_i = \Sigma_{j=1}^{N}(w_j S_{ji})$, where w_j is the absolute weight associated with the jth attribute and S_{ji} is the performance score of the ith software package on the jth attribute of performance.

Even though the formula to be used to compute the final software package worth appears to be the same for the two approaches, the interpretation placed on the attribute weights and alternate scores are quite different. Nevertheless, each method is both subjective and potentially appropriate. *The swing weight approach has the greatest theoretical basis for use.* It would be appropriate to evaluate the performance of candidate software packages using each approach; and if there is a discrepancy in the final result, then explore the causes for this, perhaps by means of a sensitivity analysis and more refined definition of terms. The critical difference between the two methods is whether one wants to think critically about tradeoffs among attributes, implying the value of the swing weight method on relative attributes, or whether one wants to focus on the overall performance of the system, implying the value of the absolute method, on some absolute standard.

As a specific numerical example of DBMS software package evaluation, let us consider that four lowest-level attributes are initially thought to be of importance, as represented by the multiple-attribute tree of Figure 5.25.

1. Software developer trustworthiness (W_1)

2. Database security and integrity (W_2)

3. Software functional effectiveness (W_3)

4. Quality of data dictionary (W_4)

We assume that three software packages are being considered for purchase. To use the swing weight approach, we first determine the best- and worst-performing software packages across each alternative. This might result in the following partially filled in performance scoring matrix:

$$
S_{ji} = \begin{bmatrix}
0.00 & 1.00 & \\
0.00 & 1.00 & \\
1.00 & & 0.00 \\
& 1.00 & 0.00
\end{bmatrix}
$$

This says that we have determined that package 1, represented by the matrix column S_{j1} is the worst performer on the first two attributes we are using in the evaluation and the best performer on the third attribute. On the other hand, package 2 is the best performer on attributes 1, 2, and 4. It is not a worst performer on anything. Package 3 is the worst-performing package on attributes 3 and 4. It is important to note that there is a single 0.00 and a single 1.00 in each row.

Based on these initial assessments, we proceed to evaluate the performance of the other alternative attribute pairs. Suppose that we obtain the following elicited results, perhaps based upon experimentation with the three packages:

$$S_{ji} = \begin{bmatrix} 0.00 & 1.00 & 0.60 \\ 0.00 & 1.00 & 0.90 \\ 1.00 & 0.00 & 0.00 \\ 0.40 & 1.00 & 0.00 \end{bmatrix}$$

Based on this definition of best and worst, we now need to determine the attribute swing weights. Suppose that we indicate that the difference between worst and best performance is such that attribute 4 is the most important. We then assign it an unnormalized weight of 1.00. Given this, we might say that attribute 3 is only 0.80 as important as attribute 1 and that attribute 1 is 0.40 and attribute 2 is 0.30 as important as attribute 4. We then calculate the normalized swing weights and obtain

$$W_1 = 0.40 \qquad W_2 = 0.30 \qquad W_3 = 0.80 \qquad W_4 = 1.00$$
$$w_1 = 0.16 \qquad w_2 = 0.12 \qquad w_3 = 0.32 \qquad w_4 = 0.40$$

The final evaluation scores are obtained using the equation $\S_i = \Sigma_{j=1}^{N}(w_j S_{ji})$ as $\S_1 = 0.46$, $\S_2 = 0.62$, and $\S_3 = 0.20$. Thus, we see that alternative 2 is the best alternative, actually by quite an amount.

To perform this same evaluation using the absolute weight approach, we might proceed as follows. We recognize that the order of assigning weights and performance scores is immaterial. We assign weights, perhaps the following:

$$W_1 = 0.60 \qquad W_2 = 0.50 \qquad W_3 = 1.00 \qquad W_4 = 1.00$$

and then calculate relative weights as

$$w_1 = 0.19 \qquad w_2 = 0.16 \qquad w_3 = 0.32 \qquad w_4 = 0.32$$

Next, we assess performance scores. Ideally, it would be best to do this across all alternatives at a single attribute, and then proceed to a new attribute. To make this assignment across attributes for a single alternative is generally more cognitively demanding and invites cognitive bias. These might be

$$S_{ji} = \begin{bmatrix} 0.50 & 1.00 & 0.70 \\ 0.50 & 1.00 & 0.90 \\ 1.00 & 0.60 & 0.50 \\ 0.60 & 1.00 & 0.30 \end{bmatrix}$$

such that for the evaluation scores we obtain $\S_1$ 0.56, $\S_2 = 0.86$, and $\S_3 = 0.48$.

Thus, we see again that alternative 2 is the best alternative, according to the criteria and approach used. The other two alternatives appear to be close competitors, however. In part this is caused by the fact that we are working over a smaller range of performance scores in that no performance score is now less than 0.30 and many are close to 1.00.

5.3 GROUP DECISION MAKING AND VOTING

Decisions often are either made by, or affect, groups of people instead of an isolated individual. In such cases managers, planners and decision makers are usually expected to consider the preferences of the individuals concerned. Approaches to the problem of amalgamating the individual preferences into a form that can be used to guide the decision maker are discussed in this section. Voting is one very common form of group decision making, and our initial emphasis is on voting.

In many presentations of the theory of decision making, there is not necessarily a clear distinction between decision theory for individuals and decision theory for groups: This lack of distinction is not a liability of the theory because the theory prescribes maximization of the expected value of utility as a guide to rational behavior. It does not specify whether that utility must describe (1) a single attribute for a single individual, (2) multiple attributes for a single individual, (3) a single attribute for a group, or (4) multiple attributes for a group. The utilities involved in a decision situation include attitudes toward risk. This incorporation of risk attitudes into utility makes interpersonal comparison of utilities, needed for group judgment, very difficult. Preference and utility are each required for decision making. The preference structure is, in fact, the input to the utility determination process, and the resulting utilities are such that if A is preferred to B, which we generally write* as $A > B$ or $A\mathscr{P}B$, then $U(A) > U(B)$ which indicates that the utility of A is greater than that of B. Stated another way, the cardinal utility numbers possess the same ordinal relationships as the preferences they describe.

Some of the methods used in the past to determine social or group preference include dictatorship, widely encompassing sets of traditional rules or customs, market mechanisms, and voting. With the possible exception of voting, it is reasonably clear that these methods do not necessarily use any "fair" scheme to determine group preferences for making decisions. As we shall indicate, voting is not nearly as perfect for determining group preferences as one might intuitively think. Thus we come to the question explored in this section: Just how do, or how should, we amalgamate individual preference structures into a group preference structure?

---

*We attempt to use the symbols $>$ and $\sim$ to indicate individual preference and indifference in a decision situation and use $\mathscr{P}$ and $\mathscr{I}$ to indicate preferences and indifferences of groups or individuals in a group decision situation. Unfortunately, it is not always possible to follow this convention.

Before continuing our development, we pause to define some of the notation we will use. Preference must be determined among a set of alternatives $\{a, b, \ldots, x, y, z\}$. The preferences are to be considered for a society of n individuals denoted by $1, \ldots, i, \ldots, n$. If individual i prefers x to y, we write $x\mathscr{P}_i y$, where the subscript indicates that this preference is true for individual i and not necessarily for any other individuals. If individual i is indifferent between x and y, we write $x\mathscr{I}_i y$. If individual i does not prefer x to y, we write $x\bar{\mathscr{P}}_i y$. The statement $x\bar{\mathscr{P}}_i y$ is equivalent to stating either $y\mathscr{P}_I x$ or $y\mathscr{I}_I x$. We use the unsubscripted forms $x\mathscr{P}y, x\mathscr{I}y$, and so on, to indicate preferences that are somehow ubiquitous across, or otherwise accepted by, the entire group. We also require that each individual be transitive in their preference structure. If $x\mathscr{P}_I y$ and $y\mathscr{P}_I z$, then we require that $x\mathscr{P}_I z$. It might seem intuitively apparent that a group, consisting of individually transitive people, will be transitive. Sadly, this is not necessarily so, and this fact causes great difficulties in determination of an appropriate voting system, as we shall soon discuss.

5.3.1   Voting Approaches

We assume that, by some means, a voting system is selected. Some of the most common voting rules, or systems, in use are as follows:

1. *Plurality.* Voters vote for one alternative, and the candidate receiving the most votes wins.

2. *Majority.* Voters vote for one alternative. A candidate alternative must receive more than 50% of the votes to win. If no alternative wins, runoff voting is held among alternatives who received the most votes and whose aggregate votes in the present iteration constituted a plurality.

3. *Weighted Voting.* Voters assign weighted votes for the candidate alternatives according to their strength of preference for each one. The total weighted votes for each alternative are counted, and the winner is the candidate with the largest total.

   Borda voting is an example of a weighted voting method. Each voter gives $N - 1$ votes for the most preferred alternative among N candidates, $N - 2$ for the second preferred alternative, and so on. The total votes for each candidate are counted by the formula: $(N - 1)M_1 + (N - 2)M_2 + \cdots + M_{N-1} + OM_N$, where M_1 is the number of voters having the candidate as first choice, M_2 is the number of voters having the candidate as second choice, and so on, and M_N is the number of voters having that candidate as last choice.

   There are many possible variations of weighted voting. One of these—that is, receiving much current acclaim—is known as *approval voting*. In approval voting that has been simultaneously discovered and advocated by several authors [62, 63], a person votes for, or approves of, as many candidates as desired. The winner is the candidate with the most votes. Approval voting collects much more information than does plurality voting in that a voter should vote for all acceptable candidates who are above a somehow set threshold of acceptability. It is a form of weighted voting in which the weights for each candidate are either zero or one. If a person assigns a weight of one to a candidate, the voter *approves* of that candidate. If a weight of zero is assigned, the person making the assignment *disapproves* of that particular candidate. Less information is available than in Borda voting, but the voting scheme is also much less cumbersome to administer.

4. *Binary Comparison Voting.* Voters vote in a binary fashion for all possible paired alternative combinations. One such approach to binary comparison voting is the method of *Condorcet voting*. Here, the alternative or candidate that wins by a simple majority over all other alternatives in pairwise contests is the winner in the election.

   Another variant of this type of voting is *Copeland voting*. Here, the score of each alternative is calculated by subtracting the number of its losses in pairwise contests with all other alternatives from the number of its wins. The alternative with the highest score is the winner.

In each and every voting system, the winner or winners are decided according to rules of the voting system. If there is no winner, a decision is made either to start another round of voting or to choose winners by some other method. It is of interest to examine several voting schemes.

5.3.2 Potential Problems with Voting: Some Illustrative Examples

We will illustrate some of the problems inherent in voting with two examples. The first example results in the well-known *"paradox of voting."* Suppose three individuals constitute a society, and they must establish a preference structure among three alternatives x, y, and z. Furthermore, let us suppose that the individual preference orderings are transitive preference orderings and are given by $x\mathcal{P}_1 y\mathcal{P}_1 z$, $y\mathcal{P}_2 z\mathcal{P}_2 x$, and $z\mathcal{P}_3 x\mathcal{P}_3 y$. If the society decides to use majority voting, we see that x is preferred to y on two out of three occasions, or by two of the three individuals; thus for this society, we say that $x\mathcal{P}y$. Similarly, y is preferred to z on two out of three occasions, thus $y\mathcal{P}z$. If we require the social preference ordering to be transitive, we would have $x\mathcal{P}z$. However, the majority rule for the aforenoted preference structures of the three individuals is easily shown to yield $z\mathcal{P}x$. Hence, we have a set of transitive individual preference orderings that lead to an intransitive group preference ordering.

Intransitive behavior, on the part of individuals or groups, is usually deemed quite undesirable. A society that generated a preference structure like the one mentioned should question its validity, and certainly must question its usefulness in any case. If valid, the preference structure is one that is likely to create considerable difficulties for the decision maker and for others. In this particular example, all *ordinal* preferences are known. In most voting systems, in particular plurality voting, nowhere near this much information is available. A candidate may be chosen as the winner, even if that candidate is not approved of as much as some other candidate by the majority of voters, and is strongly disapproved of by more voters than any other candidate. One can question the extent to which these possibilities are pathologic. Unfortunately, we do not know the real answer because sufficiently complete information to make this determination is generally unavailable in most plurality-type elections.

Example 5.3.1. This example considers the type of voting procedure used. Suppose a group of 60 individuals are voting for an office holder from a field of three candidates a, b, and c. Suppose that among the individuals, 23 have preference order $a \mathscr{P} c \mathscr{P} b$, 19 have preference order $b \mathscr{P} c \mathscr{P} a$, 16 have preference order $c \mathscr{P} b \mathscr{P} a$, and 2 have preference order $c \mathscr{P} a \mathscr{P} b$.

If a plurality is used to select the winner, candidate a has 23 first-place votes, b has 19 first-place votes, and c has 18 first-place votes. Thus a is the winner.

It could be that a majority is required to select the winner. If so, there must be a run-off election between a and b because these two candidates, taken together, have a majority of the total votes cast. The preference orders remain unchanged, and $23 + 2 = 25$ voters express a $\mathscr{P} b$ either directly or by transitivity. Similarly, $19 + 16 = 35$ voters express $b \mathscr{P} a$. In this case, candidate b is the winner.

One other method, the intensity method, is sometimes used to select a winner. In the intensity method, which is a Borda-weighted voting system, a weight of 2 might be given to the first-place vote, a weight of 1 is given to the second-place vote, and the third-place vote is not counted. Under this scheme, a would receive $(23 \times 2) + (2 \times 1) = 48$ votes. Candidate b would receive $(19 \times 2) + (16 \times 1) = 54$ votes, while candidate c would receive $(23 \times 1) + (19 \times 1) + (16 \times 2) + (2 \times 2) = 78$ votes. Thus candidate c is the winner. Even if a run-off election occurs between b and c, candidate c will still be the winner.

Let us now examine the situation that exists when approval voting is used. We see immediately that we cannot really determine the answer because we do not know the *approval threshold* that each voter will use. And, each voter is free to adjust it differently. It makes no sense for a voter to not approve of at least one alternative, unless, for some reason of principle, that voter wishes to throw their vote away. In a similar way, it is senseless to approve of all alternatives, because that has the same effect as voting for none. From the group of 23 people, we see that a will for sure receive 23 votes and b will receive 0 votes. Alternative c can receive anywhere from 0 votes to as many as 23 approval votes from this group. In a similar way, the group of 19 voters will approve of b a total of 19 times and will approve of a 0 times because a is the least preferred candidate by this group. Alternative c can receive anywhere from 0 to as many as 19 approval votes from the group of 19 voters. Alternative c will receive a total of 16 votes from the group of 16 people with preference structure $c \mathscr{P} b \mathscr{P} a$. In a similar way, alternative a will receive 0 votes from this group. Alternative b will receive at least 0 votes and as many as 16 votes. Finally, the small group of two voters will provide 2 votes for alternative a and 0 votes for alternative b. The votes for a will be at least 0 and no more than 2. The final tallies for the various alternatives, using the approval voting scheme, will be

$$23 \leqslant a \leqslant 25, \qquad 19 \leqslant b \leqslant 35, \qquad 18 \leqslant c \leqslant 60$$

and we see that any of the three candidates could be a winner! We do see that it is rather unlikely that a will be a winner unless the voters tend to approve of only one alternative — that is, the one best for them. If, for example, half of the people in each group *approve* of their mid-ranked alternative, then the approval votes will be $a = 24$, $b = 27$, and $c = 39$, and so c is the winner. On the other hand, if all voters approve of their two top choices, the final approval votes are $a = 25$, $b = 35$, and $c = 60$; we see that c wins by a "landslide." It is very interesting to note that c does not win when the plurality system is used, a does. Nor does c win when the majority plurality system is used; b wins because candidate c is eliminated in the first round of elections. Figure 5.26 presents some interesting graphical results for this example.

Had we used *Condorcet voting*, where we simply count the total number of votes that each alternative obtains in all possible binary comparisons, we obtain $a = 48$, $b = 54$, and $c = 78$. When we use *Copeland voting*, we obtain the same number of "wins" as in Condorcet voting. The number of "loses" that we have are $a = 72$, $b = 66$, and $c = 42$. Thus the final scores for the various alternatives is $a = -16$, $b = -12$, and $c = +36$. Again, candidate or alternative c is the winner.

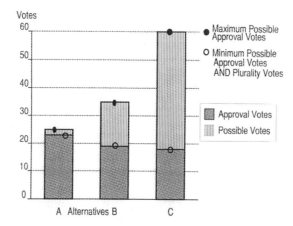

Figure 5.26 Approval voting possibilities and plurality voting results for a simple example.

It is, of course, curious that such a fundamental and very old subject as voting would be subject to present-day inquiry. Yet, it is. There are a number of questions that need to be posed relative to any voting scheme, such as following:

1. Is the system subject to *strategic voting* possibilities in which a person can steer the outcome to one more to their liking by voting other than their own preferences?

2. Will the choice of voting system really make a difference? The classic example of a case where all *ordinal* voting schemes are equivalent is that of a two-person election to select a single winner. In three-or-more-person elections, most anomalies are caused by one alternative or candidate being strongly favored by a minority, where a second alternative is highly regarded by almost everyone, but not sufficiently well regarded to be the first choice for hardly anyone. In a case like this, which is the case in our last example, almost any winner can result as a function of the voting system selected.

3. Is a particular voting system feasible for implementation? If a particular system is too costly to implement or too difficult to understand, then it is very doubtful it will have much value.

Doubtlessly, all voting systems based on *ordinal* preferences are vulnerable even if people correctly express their feelings. Even the *cardinal* schemes are vulnerable to strategic voting, as we will soon see.

After more than two centuries of efforts by many, quite useful results concerning voting procedures are now emerging [62–66]. The major difficulties occur when we have multiple candidate elections. These situations are easy to construct. Suppose that 400 voters have preferences $A\mathcal{P}B\mathcal{P}C$, 390 have preferences $B\mathcal{P}C\mathcal{P}A$, and 40 have preferences $C\mathcal{P}B\mathcal{P}A$. A plurality election in which we vote for our top-ranked candidate only is $A\mathcal{P}B\mathcal{P}C$ with a 400:390:40 vote tally. The extraordinary poor support for C suggests that this candidate, or alternative, because this could be a decision among alternatives issue, is of little or no interest. However, when we compare the candidates in pairs, then we find that $C\mathcal{P}A$ (430 to 400), $C\mathcal{P}B$ (440 to 390), and $B\mathcal{P}A$ (430 to 400). Thus, it seems like $C\mathcal{P}B\mathcal{P}A$, an outcome that is precisely the opposite of the plurality ranking. Simply by changing the procedure that voters use to select their second-ranked candidate, we can change the outcome of the election. Some now suggest that the Borda approach, which we described earlier, is a preferred approach. In this example, it results in a $C\mathcal{P}B\mathcal{P}A$ ranking of the candidates or alternatives with a 870:820:800 vote result, which is not a significant win for candidate c.

Several decades ago, Kenneth Arrow [67] proved an *Impossibility Theorem that demonstrated under what appears to be a very reasonable set of conditions that there is no system without these flaws, that is, no perfect voting system exists.* Thus the notion of "fair voting system" is subjective, and any voting system can be made to look unfair. One problem with many systems, including the Borda system, is that they are potentially very susceptible to forms of "strategic voting" in which voters may influence and manipulate outcomes through insincere voting. Many feel that a relatively fair system is "approval voting" in which voters cast votes for all candidates that they "approve" of and the candidate with the most votes wins. Because the alternative with the largest approval base wins, those in the "center" will often get elected. A difficulty is that many voters are very positive about those they like very much and those they dislike very much, but do not have strong feelings about candidates in the middle. If they choose randomly for or against approval for those in the middle, an entire election can become a random process.

The following advantages are represented to be characteristics of approval voting by one of the originators of this approach [68]:

1. *It gives voters flexible options*, under which plurality voting can allow choice of a single favorite, as well as voting for all candidates found acceptable. If a voter's preferred choice is felt to have little chance of winning, a vote for the first choice and an acceptable more possible choice should be preferred to voting for the very unpopular candidate only.

2. *It helps elect the strongest candidate.* A candidate that will be strong in a Condorcet election by defeating every other candidate in separate pairwise contests will generally win under approval voting. They often lose under plurality voting because votes are generally split with one or more other centrist candidates.

3. *It will reduce negative campaigning* by providing incentives for candidates to espouse the views of the majority of voters and not minorities whose votes might result in an edge in a plurality contest. Thus, candidates will have incentives to broaden their appeal to a larger audience and to thereby seek approval by voters who may have different first choices. At the same time, being negatively critical of other candidates will risk alienating that candidate's supporters and thus lose their approval.

4. *It will increase voter turnout* because voters are more likely to vote when they perceive that their vote will really count.

5. *It will provide proper support for minority candidates* because these candidates will not suffer through having voters vote for another and less appealing candidate who is felt to be a stronger contender. Such candidates will receive a true level of support, and the election returns will be a better indication of acceptability of a candidate, rather than a result distorted by strategic voting.

6. *It is practicable* in that it is simple for voters to understand and use.

There are a number of elections being conducted under "approval voting" today. This includes election of the Secretary-General of the United Nations and election by a number of professional societies, including Mathematical Association of America (MAA), Institute for Operations Research and Management Science (INFORMS), American Statistical Association (ASA), and Institute of Electrical and Electronics Engineers (IEEE).

Our discussions in this section indicate that any of the candidates in many elections could be elected, depending on the method of voting that is employed. This is clearly an undesirable situation because every individual would prefer the method of voting which elects their candidate. The "social welfare function" is intended to alleviate this problem and provide an acceptable method for the amalgamation of individual preferences. In order to be acceptable, the social welfare function must satisfy some conditions, and we discuss them in the next subsection. It is very necessary to distinguish between *ordinal* social welfare functions, which require measurements of weak preference orderings only, and *cardinal social* welfare functions, which require measurements of the interpersonal intensities of preference across individuals. We first discuss ordinal social welfare functions. This is the type of *social welfare function* that we have tacitly assumed in all of our voting examples.

5.3.3 Ordinal Social Welfare Functions

There are a number of advantages in having to specify preferences in an ordinal sense only. Thus, we give a fairly precise definition of *ordinal social welfare function*. A society of n individuals desires to determine a preference structure among the alternatives in the set $\{a, b, \ldots, x, y, z\}$. The preference structure for each individual is transitive. Also, for any two alternatives x and y, exactly one of the following is true for individual i: $x\mathscr{P}_I y$, $y\mathscr{P}_I x$, or $x\mathscr{I}_I y$. Kenneth Arrow, in a seminal work [67], states the requirements for acceptability of the individual preference orderings and the resultant social ordering in the form of two axioms, where $\mathscr{R}$ is either $\mathscr{P}$ or $\mathscr{I}$:

Axiom I: For all x and y, either $x\mathscr{R}y$ or $y\mathscr{R}x$.
Axiom II: For all x, y, and z, $x\mathscr{R}y$ and $y\mathscr{R}z$ imply $x\mathscr{R}z$.

A relation that satisfies Axiom I is said to be connected, and a relation that satisfies Axiom II is said to be transitive.

The "paradox-of-voting" examples that we have just considered in our previous section yielded intransitive social orderings from a set of transitive individual orderings. In our examples, Axiom II is not satisfied for the groups that are voting even though it is satisfied for every individual in the group. Thus, it turns out that ordinal voting could not yield an acceptable social ordering. Thus conventional voting *may* not be an acceptable social welfare function for these particular examples.

Together, these axioms imply that the alternatives form a linear order with respect to the relation $\mathscr{R}$ for group preferences. In general, $\mathscr{R}$ is a weak linear ordering, but if $\mathscr{R}$ is such that $x\mathscr{I}y$ cannot exist for any x and y, then $x\mathscr{R}y$ becomes $x\mathscr{P}y$, and we denote such an occurrence as a strong linear ordering.

Arrow's primary contribution to the problem of amalgamating individual preferences into a group preference structure is his definition of a set of five desirable conditions for which it seems reasonable that an ordinal social welfare function should satisfy.

Part of the difficulty in determining a social welfare function stems from the diversity of preferences held by members of the society. At one extreme, we could have all individuals sharing the same preference structure. In such a case of unanimity, the task would be trivial, because the social welfare function would simply identify this single preference structure as the societal preference structure. Because unanimity is the exception rather than the rule, we must consider a broader class of admissible individual preference orderings. In particular, we consider all possible transitive individual orderings as admissible. To do less would deny the existence of certain preference orderings. If the group consisted of a single individual, we would again have a trivial problem, and the social preference structure would be simply the individual preference structure. Hence, we consider groups with at least two individuals. If there were only one alternative under consideration, the choice would be trivial. If there were two alternatives, the problem would be more complex but still fairly easily treated. Thus, we consider the case of at least three alternatives. Let us pose Arrow's axioms in a relatively simple form. There are five conditions, and this discussion summarizes the requirements for Condition 1.

Condition 1

(a) The number of alternatives is greater than or equal to three.

(b) The social welfare function is defined for all possible profiles of individual orderings.

(c) There are at least two individuals.

To develop condition 2, we suppose that there is a social welfare function which asserts that society prefers a_1 to a_2, or $a_1 \mathscr{P} a_2$. Suppose that individual 1 changes their preference from $a_1 \mathscr{I} a_2$ to $a_1 \mathscr{P} a_2$. We would certainly expect that society would still prefer a_1 to a_2. Also suppose that society had initially been indifferent between a_1 and a_2, or $a_1 \mathscr{I} a_2$. Suppose that individual 1 changes their preference from $a_1 \mathscr{I} a_2$ to $a_1 \mathscr{P} a_2$. We would certainly expect that society would remain indifferent between a_1 and a_2 or would possibly prefer a_1 to a_2. Condition 2 is known as positive association of social and individual values.

Condition 2. If the social welfare function asserts that x is preferred to y for a given profile of individual preferences, it shall assert the same when the profile is modified as follows:

(i) The individual-paired comparisons between alternatives other than x are not changed.

(ii) Each individual-paired comparison between x and any other alternative either remains unchanged or is modified in favor of x.

To illustrate the requirements of condition 3, suppose a rank-order method is used to vote among four candidates w, x, y, and z. Let the method of voting require that the first place vote be given a rank of 4, the second place vote a rank of 3, the third place vote a rank of 2, and the fourth place vote a rank of 1. Suppose there are three individuals, and two of them express preferences $w\mathscr{P}x\mathscr{P}y\mathscr{P}z$, while the third expresses their preferences as $y\mathscr{P}z\mathscr{P}w\mathscr{P}x$. Under this system, w receives 10 votes, x receives 7 votes, y receives 8 votes, and z receives 5 votes. Thus w is the winner. If we delete x from consideration, we would expect the same results, especially because w is preferred to x for all the voters. However, if we use the same voting scheme on the remaining candidates, w receives 10 votes, y receives 10 votes, and z receives 7 votes. Hence, w and y are tied for first place with 10 votes each. The existence or nonexistence of x has made a difference in the social preference ordering. We refer to Condition 3 as independence of irrelevant alternatives.

Condition 3: Let H be a subset of alternatives from $\{a, b, \ldots, x, y, z\}$. If a profile of orderings is modified in such a way that the paired comparisons among the elements of H are unchanged, the social preference orderings resulting from the original and modified profiles should be identical for the elements in H.

Consider, for example, a social welfare function which asserts that $x\mathscr{P}y$ regardless of the preferences of any of the individuals in the society. Such an undesirable social welfare function is said to be imposed. To avoid imposed social preference orderings, we state Condition 4, which we call the condition of citizens' sovereignty.

Condition 4. For each pair of alternatives x and y, there is some profile of individual orderings in such that society prefers x to y.

We do not want the social welfare function to be biased so that one individual's preference ordering necessarily controls the social preference ordering for the entire group. To avoid this kind of dominance, we define Condition 5, which is known as the condition of nondictatorship.

Condition 5. There is no individual with the property that whenever they prefer x to y, for any x and y, society does likewise, regardless of the preferences of other individuals.

Three theorems result from these axioms and conditions. The first two are as follows:

Theorem I. Possibility Theorem for Two Alternatives. If the total number of alternatives is two, the method of majority decision is a social welfare function which satisfies Conditions 1–5 and yields a social preference ordering of the two alternatives for every set of individual orderings.

Theorem I supports the concept that majority rule is a desirable and "fair" social welfare function when there are only two alternatives. It could be viewed, somewhat simplistically perhaps, as the basis for the two-party political system.

Theorem II. Possibility Theorem for Three or More Alternatives. Unfortunately, for three or more alternatives, there does not exist a social welfare function that satisfies Conditions 1–5 by yielding a social ordering relation that is consistent with Axioms I and II. As a result, we obtain an Impossibility Theorem.

Before seeking ways to circumvent the implications of Arrow's Impossibility Theorem, we briefly examine two possible choices of social welfare functions and their drawbacks.

Majority Rule as a Social Welfare Function. The method of majority decision, or majority rule, satisfies Conditions 1–5 when there are only two alternatives. For more than two alternatives, intransitivities may result as shown in the "paradox-of-voting" examples. In such a case, the relation induced on the set of alternatives by the social welfare function is not consistent with Axioms I and II. While these are major drawbacks, it turns out to be almost the only ones that the majority rule is associated with.

Arrow presents his proof of the possibility theorem for two alternatives in such a way that Conditions 2, 4, and 5 are independent of the number of alternatives. He also shows that Condition 3 is satisfied by the method of majority decision. In doing so, he proves the following theorem.

Theorem III. For any space of alternatives, the method of majority decision is a social welfare function satisfying Conditions 2–5.

This theorem suggests that the method of circumventing the difficulties imposed by impossibility theorem must involve the transitivity problem. Alternatively, we may abandon the hope of finding "ordinal" social welfare functions and seek a "cardinal" welfare function. A possibility theorem for cardinal welfare functions has, in fact, been obtained by Keeney [69]. This necessarily involves interpersonal comparison of utilities, which is a very difficult task. Many excellent works on applied decision analysis [70] do not consider these group issues.

Unanimity as a Social Welfare Function. It is possible that a group could consist of individuals who all have the same preference orderings for the alternatives. This unanimity makes selection of a social welfare function trivial — the transitive social preference order is the same as the transitive individual preference orders. One might argue that this state of affairs permits a dictator in the group to establish their preference ordering. This is true, as we see from consideration of the definition of a dictator in Condition 5. In that case, everyone is a dictator, and the undesirability of a dictatorship becomes less important. While unanimous societies are the exception rather than the rule, unanimity is certainly an acceptable social welfare function in those few instances when it is applicable.

5.3.4. Modifications to Achieve a Social Welfare Function

Although Arrow's Impossibility Theorem ensures that we cannot find an ordinal social welfare function which satisfies Conditions 1–5, the problem of amalgamating individual preference orderings into some sort of a social preference ordering still exists, and it must be dealt with. Decisions which affect groups still must be made, and responsible decision makers still seek ways of incorporating individual preferences into an overall preference structure that is acceptable to those concerned.

All of Arrow's conditions seem reasonable, and the logical method of attack seems to be to relax one or more of those conditions and find a social welfare function which satisfies that modified set of conditions.

There have been a number of approaches that relax one, or more, of Arrow's conditions. Luce and Raiffa [32] and others have identified Condition 3 as the most vulnerable of Arrow's conditions. They claim that the irrelevant alternatives may not be irrelevant at all, but that they can and should be used to indicate strengths of preferences. The following example, presented by Goodman and Markowitz [71], illustrates this point. A host intends to serve refreshments to two guests. The host can serve them either coffee or tea, but not both. Guest 1 prefers coffee to tea and guest 2 prefers tea to coffee. Based on this information, one might conclude that the (ordinal) welfare function should indicate equal preference between tea and coffee. Suppose, however, that the host obtains additional information. It is discovered that guest 1 prefers coffee to tea, tea to cocoa, and cocoa to milk. On the other hand, guest 2 not only prefers tea to coffee, but also prefers cocoa to coffee, milk to coffee, and even water to coffee. With this additional information, it seems plausible to serve tea because it does not make "much difference" to guest 1 and it makes "a lot of difference" to guest 2. Although the (not truly considered) alternatives of cocoa and milk are irrelevant in the sense that they will not be served, they are relevant in indicating strength of preferences.

The preceding approach leads to consideration of the problem of interpersonal comparison of utilities. We have discussed the difference between utility and preference as measures of desire for an alternative. The work of Arrow concerned ways to combine preferences of individuals into a group preference structure. Utilities could have been used in that work, but the same results would have been obtained because the ordinal properties are the properties considered, and ordinal relationships for utilities and the preferences they describe are identical.

We may also attempt to achieve a social welfare function for various preference structure modification approaches. If we leave Condition 3 intact, the other likely candidate for modification is Arrow's Condition 1. We can seek some way to modify the admissible preference orderings so that intransitivity will not occur. The acceptable social welfare function for majority rule, the "paradox-of-voting" example, showed that $W^{(n)}$ was mapped into a social preference ordering that was not contained in W.

Either of two modifications appear to be appropriate:

1. We could leave $W^{(n)}$ unrestricted and require the mapping to yield a transitive social order contained in W. This is equivalent to saying that we will modify the range of the social welfare function.
2. We could restrict the set of admissible profiles in $W^{(n)}$ such that the mapping yields a social preference order contained in W. This is equivalent to saying that we will modify the domain of the social welfare function.

We consider only the second of these approaches here. This is based on the presence of single peaked preferences.

Some of the preference profiles in $W^{(n)}$ yield a transitive social order when considered by the method of majority decision. If we restrict the set of admissible orderings to those selected profiles, the method of majority decision will satisfy Conditions 1–5 and serve as a social welfare function. One way to identify a set of admissible orderings is with the concept of single-peakedness developed by Black [72].

Consider a set of alternatives $\{a, b, \ldots, x, y, z\}$ that have been ranked according to preference. Because the preference relation results in a linear order, we can represent it on a linear scale. As an example, consider the five alternatives v, w, x, y, and z. Suppose individual i ranks them as $z\mathscr{P}_I x\mathscr{P}_I v\mathscr{P}_I y\mathscr{I}_I w$. This ranking is purely an ordinal one. We may represent this preference order in any of several ways. We may use a two-dimensional plot with the order-of-preference scale as the vertical scale, and we may use some other ordering of the alternatives as the horizontal scale. For the preference order given, the plot in Figure 5.27 results.

We may rearrange the order of the points on the horizontal axis to obtain a different plot of the same information. Two possible horizontal orders and their corresponding plots are shown in Figure 5.28. The vertical scale is the same as that for Figure 5.27. Both of the curves in Figure 5.28 are said to be single-peaked. In Figure 5.28(b) the line segment from x to z is up-sloping, the

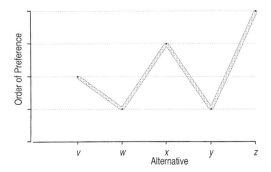

Figure 5.27 Two-dimensional plot of preference.

line segment from z through v to y is down-sloping, and line segment from y to w is horizontal. We define a change of direction as a change in the curve from an up-sloping segment to a down-sloping segment, or vice versa, but not from an up-sloping or down-sloping segment to a horizontal segment. Thus Figure 5.27 has three changes of direction, Figure 5.28(a) has no changes of direction, and Figure 5.28(b) has one change of direction. We formally define a single-peaked preference curve as a preference curve that has, at most, one change of direction.

An individual's preferences may always be ordered on the horizontal axis so that a single-peaked preference curve exists. In the cited reference, Black demonstrated that when a specific horizontal ordering of the alternatives exists such that all members of the society have single-peaked preference curves, the method of majority decision results in a transitive social preference order. However, lack of single-peaked preference curves for all members of the society does not guarantee intransitivity. Therefore, the existence of single-peaked preference curves for all individuals is a sufficient, but not necessary, requirement for a transitive social ordering.

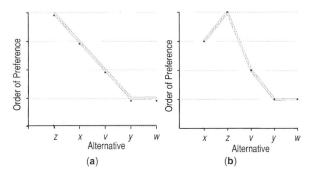

Figure 5.28 Alternative two-dimensional plots of single-peaked preference.

An example illustrates the determination of a social preference ordering when the members of the society have single-peaked preference curves. Suppose that a group with 15 members is selecting a new president from a field of three candidates, x, y, and z. Further suppose that the following preference orders hold: Five members rank $x \mathscr{P} y \mathscr{P} z$, four members rank $x \mathscr{P} z \mathscr{P} y$, and six members rank $y \mathscr{P} x \mathscr{P} z$. With three alternatives, there are $3! = 6$ possible rankings, but all six do not form a single-peaked group. There are three rankings that are single-peaked in ordering z, x, y. According to Black, the median preference curve will contain the winner as its most preferred alternative. In this example, there are two distinct curves with peaks at x, but the median curve (counting from either direction) is one of the three solid line curves, as shown in Figure 5.29. Thus x is declared the winner. If knowledge of the second place candidate is desired, x can be deleted from Figure 5.29 and the resulting single-peaked curves considered again. Thus y is the second choice, and the overall preference structure is $x \mathscr{P} y \mathscr{P} z$. To verify this result, we may examine the preference orders on a pairwise basis. Candidate x wins over candidate y by a nine to six vote, and $x \mathscr{P} y$. Candidate y wins over candidate z by an eleven to four vote, and $y \mathscr{P} z$. If the order is transitive, the pairwise comparisons should indicate $x \mathscr{P} z$. Indeed, Candidate x wins over Candidate z by a fifteen to zero vote. Again, we see that $x \mathscr{P} y \mathscr{P} z$.

We have just considered determination of a preference order among a finite set of alternatives. Both the individual preference orders and the resulting social preference order deal with the same finite set of alternatives. Single-peaked preference curves contain a distinct alternative that is preferred above all the rest for each individual. The method of majority decision will yield one of those alternatives as society's most preferred choice. If the society wants the remainder of the preference order, it can be determined by the method of majority decision, and it will be transitive. Such a system is certainly acceptable when the alternatives possess this distinct nature — for example, candidates in an election or brands of consumer products.

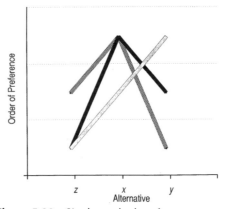

Figure 5.29   Single-peaked preference curve.

Suppose that the society is asked to indicate its assessment of some quantity that is measured on a continuous numerical scale. For discussion purposes, suppose the quantity is the probability of the occurrence of some event. Each individual indicates their preferences as to what that probability is in the interval [0,1]. Any probability different from an individual's assessment will be less preferred by that individual than their assessment. Furthermore, for two distinct probabilities that are both lower (or higher) than the individual's assessed value, the lowest (or highest) will be less preferred. Under these conditions, each individual's preference curve for assessment of the probability will be single-peaked, and the peak will be at the individual's assessed value. The set of individually assessed values (peaks) will be finite, and the method of majority decision will yield the social preference. In this case, however, the individually assessed peaks do not have the distinct nature that the candidates in an election have. In order for a particular value in the interval [0,1] to be selected, it must be the particular assessment of one of the individuals. Because there are many other possible values in [0,1], one might reasonably suggest that some "averaging" scheme be used instead.

To illustrate, suppose that 50 people assess 0.5 as the probability of the occurrence of an event and that 51 people assess 0.7 as the probability of the event. Under majority rule, 0.7 would be selected. The 50 people who assessed 0.5 as the probability may feel that they have been treated unfairly, but they would be perfectly willing to settle for some compromise value of about 0.6 because 0.6 is also an alternative, although it has not been suggested by anyone. In this illustration we have used the mean as the function for determining preferences. It turns out that the mean is acceptable as a social welfare function for single-peaked, continuous preference curves.

Consider the problem of selecting a single numerical quantity from a continuous interval. Any member of the uncountable infinite set of numbers may be selected. Arrow's five conditions were stated for determination of preferences among a finite set of alternatives, but the concepts he identified are desirable for an infinite number of alternatives; and we examine the mean to determine if it satisfies the intention of Arrow's conditions for social welfare functions. We note that this, in effect, assumes a quadratic social choice function, and this will, of course, justify selection of the mean. In a sense this choice is totally arbitrary, and a great many functions may satisfy the Arrow conditions if assumed arbitrarily. However, there are many desirable features inherent in quadratic cost functions; and this, at least in the past, justifies examination of the quadratic social choice function as one useful criterion reflecting strength of preference.

The mean value $\bar{x}$ of a set of numbers suggested by n individuals is defined as

$$\bar{x} = \frac{1}{n} \sum_{i=1}^{n} x_i$$

where x_i is the number suggested by individual i and x_i is contained in the interval in question. We experience no loss in generality if we consider the interval [0,1]. We must consider a modified version of Condition 1 wherein the set of admissible individual preference orders are continuous and single-peaked on the interval [0,1]. Condition 1 requires that the relation induced on the alternative set by the social welfare function be connected and transitive. The purpose of the connected and transitive requirement is to ensure a rational approach and to reflect the individual preferences in the social preference ordering as accurately as possible. The mean does not generate an order on the uncountably infinite number of points $I \in [0,1]$. Rather, it defines a single alternative that should represent the social preference for the quantity in $[0,1]$ being sought. Hence, the concepts of connectedness and transitivity lose meaning where the mean is the sought-after alternative. We can, however, indicate how accurately any choice identifies the social preference. We let c denote any choice selected as the social preference, and let

$$\text{DI} = \sum_{i=1}^{n} (x_i - c)^2$$

denote the "dissatisfaction index." Such an index is appropriate when the preference curves are single-peaked. It is easy to show that dissatisfaction is minimized when $c = \bar{x}$, where $\bar{x}$ is defined as the mean value. We conclude that the mean most accurately measures, in the sense defined, the social preference and therefore satisfies the intent of Condition 1.

Condition 2, positive association of social and individual values, is satisfied by the mean value. If individual i changes preference from x_i to x_i', where x_i' is a lesser value, the mean also changes to a lesser value. Likewise, if i changes preference from x_i to x_i'', where x_i'' is a larger value, $\bar{x}$ also changes to a larger value, as required.

To verify satisfaction of Condition 3, which is the independence of irrelevant alternative condition, we note that by asking for a numerical assessment from an interval, we automatically consider all values in the interval as relevant and all values outside the interval as irrelevant. If we prohibit consideration of those irrelevant alternatives, satisfaction of Condition 3 is trivial.

Clearly the social preference is determined by the individual preferences, and the mean satisfies Condition 4, the condition of citizens' sovereignty. Similarly, there is not a single individual such that their selection of a numerical quantity from [0,1] will determine the mean, regardless of the preferences of other individuals. Because there is not a dictator, Condition 5 is satisfied.

With Conditions 1–5 satisfied, we have established that the mean is one appropriate social welfare function for determining group preference when the group must select a numerical quantity from a continuous interval and when the individuals' preference curves are single-peaked.

5.3.5 Cardinal Social Welfare Functions

There have been many studies of group decision making which show that, under a very mild set of realistic axioms, there is no assuredly successful and meaningful way in which ordinal preference functions of individuals may be combined into a preference function for the entire society. Conflicting values are the major culprit preventing this combination. This has a number of implications that suggest much caution in using ordinal preference voting systems, and any systemic-approach-based only on ordinal preferences among alternatives. Among other possible debilitating occurrences are agenda dependent results which can, of course, be due to other effects.

Very definitive studies of the interpersonal comparison of utilities have been conducted by Harsanyi [73, 74]. He argues convincingly that we make interpersonal utility comparisons all the time whenever we make any allocation of resources to those to whom we feel the allocation will do the most good. The prescription against such comparisons is one of two key restrictions that lead to the Arrow impossibility theorem. By using cardinal utilities such that it becomes possible to determine preferences among utility differences, that is to say whether $U(a) - U(b) > U(b) - U(c)$, and interpersonal comparison of utilities, Harsanyi shows that Arrow's Impossibility Theorem becomes a possibility theorem. This is a major point in that it is generally not possible for a group to express meaningful transitive ordinal preferences for three or more alternatives, even though all individuals in the group have individually meaningful transitive ordinal preferences.

Harsanyi is concerned primarily with organizational design [75] — that is, how to design social decision making units so as to maximize attainment of social objectives or value criteria. He shows "that rational morality" is based on maximization of the average (cardinal) utility level for all individuals in society. The utilitarian criterion is applied first to moral rules, and then these moral rules are used to direct individual choices. Thus each utilitarian agent chooses a strategy to maximize social utility under the assumption that all other agents will follow the same strategy. Harsanyi recognizes a potential difficulty with this particular utilitarian theory of morality in that it is open to dangerous political abuses, as well as the numerous problems associated with information acquisition and analysis in a large centralized system. He posits a difference between moral rationality and game-theoretic rationality. He argues for the unavoidable use of interpersonal cardinal utility comparisons in moral rationality, as well as for the inadmissibility of such comparisons in game theory. Much of Harsanyi's efforts concern game situations in which outcomes depend on mutual interactions between morally rational individuals, each attempting to better their own interests.

Harsanyi's concept of utilitarianism has occasionally been criticized for making inadequate provision for equity, or equivalently for social group equality. John Rawls, a philosopher, has presented a theory of justice [76] that involves a difference principle in which decisions are made under uncertainty rather than under risk. This difference principle advocates selection of the alternative choice which is the best for the worst-off member of society and is, therefore, the direct social analog of the maximum principle for the problem of individual decisions under certainty. Rawls uses a "veil of ignorance" concept in which individuals must determine equitable distribution of societies' resources before they know their position in society. His argument is essentially that people will select a resource allocation rule that maximizes the utility of the worst-off member of society.

Other useful interpretations of cardinal utility and interpersonal utility comparisons have been made by Keeney and Kirkwood [77]. Their axioms allow development of a multiplicative group utility function in contrast to the additive utility function of Harsanyi. It is possible to more directly deal with equity considerations in a multiplicative group utility model than in an additive model. Papers by Bodily, Brock, Ulvila and Snider, and Keeney in Kirkwood [78] contain insightful discussions concerning group and individual utilities of a multiattribute nature. Other discussions of social choice and its relationship to voting and decisions from economic and social perspectives may be found in references 79 and 80.

5.4 SUMMARY

In this chapter we have examined approaches to decision analysis and decision assessment. A goal in this was to provide an appropriate basis for the selection of appropriate prescriptive approaches to aid people in evaluating alternatives and making decisions. There are a number of variants of the approaches that we have discussed here. A multiple-attribute utility-theory-like approach has, for example, been devised by Saaty [81] and is called the analytical hierarchy process (AHP). A substantial number of applications have been considered, and the process has been studied intently by a number of researchers [82–85]. There is much additional work to be found in the references cited in this chapter.

PROBLEMS

5.1 Suppose that a city planner must decide whether to do nothing A_1 or to build, in a given uninhabited area, a single-lane highway A_2, a dual-lane highway A_3, or a giant expressway A_4. The land use characteristic of the area in the future may be: uninhabited as it is now E_1, sparsely populated E_2, residential usage E_3, or heavily industrialized E_4. The planner will receive blame or praise for foresightedness in the decision made. The utilities of the outcomes are given by the payoff matrix:

	E_1	E_2	E_3	E_4
A_1	4	4	3	3
A_2	3	5	4	3
A_3	2	5	4	3
A_4	1	2	4	6

If the city planner uses (a) the max–max criterion, (b) the max–min criterion, (c) the minimum regret criterion, or (d) the Laplace criterion, what is the optimum strategy for each criterion? Maximum payoff from the decision taken is desirable.

5.2 Suppose that the probabilities of the various states of nature in Problem 5.1 are $P(E_1) = 0.1$, $P(E_2) = 0.2$, $P(E_3) = 0.5$, and $P(E_4) = 0.2$. Which alternative yields maximum expected utility?

5.3 Two medical doctors must decide whether to treat, T, a patient for a particular diagnosed disease or wait, W. The utilities for the outcomes cure, paralysis, or death may be assumed independent of the corrective strategy or decision adopted. They are

	$U(C)$	$U(P)$	$U(D)$
MD_1	1.0	0.4	0
MD_2	1.0	0.7	0

The conditional probabilities of the outcomes are given by $P(C|T) = 0.6$, $P(P|T) = 0.1$, $P(D|T) = 0.3$, $P(C|W) = 0.2$, $P(P|W) = 0.6$, and $P(D|W) = 0.2$. Each doctor desires to maximize the utility of their decision strategy for the patient. What will be each doctor's best strategy? It is reasonable for the utilities of the outcomes to be independent of the alternatives, so that these utilities are really utilities of the state of nature? Draw several models of the decision situation and analyze each. How could the doctors reach agreement if they attempted to arrive at a single utility curve for the two of them?

5.4 Please reconsider Example 5.2.4 discussed in the text on page 391. Suppose that the conditional probabilities associated with the judgment of the assistant are $P(R|r) = 0.7$, $P(B|r) = 0.3$, $P(R|b) = 0.2$, and $P(B|b) = 0.8$. What are the best decisions for this example?

5.5 Show that the buyer should indeed buy insurance in Example 5.2.7 on page 409.

5.6 Reconsider the insurance example on page 409, if the initial assets of the buyer are $500,000. What is the smallest value of the buyer's assets such that the best decision is not to buy the insurance? Does the value of the buyer's assets make any difference to the seller?

5.7 A manufacturer desires to ship goods worth $100,000 by truck. There is a probability P that the shipment may be lost, stolen, or otherwise destroyed in transit. It costs $5000 to insure the shipment. What must be the probability P such that the business executive should buy the insurance. Consider the cases where the manufacturer is an expected value operator, risk-averse, and a risk-prone gamble. Also consider several cases for total assets of the manufacturer, say $100,000, $1,000,000 and $100,000,000.

5.8 A city planner must decide whether to recommend formulated plans A_1 or A_2 to the mayor, who in turn may submit it to the city council for approval. One plan must be recommended, and both cannot be recommended. In order for the plan to become policy, both mayor and city council must prove it. The city planner believes A_1 to be the best plan but believes that there is a 20% chance the mayor will not approve it and pass it on to the city council. If the mayor approves it, there is a 50% possibility that the city council will not approve it. The city council will, in the opinion of the planner, surely approve plan A_2, but the probability of the mayor passing it on to them is only 0.5. The planner would prefer that rejection, if it occurs, occur at the hands of the city council. The planner's utilities are as follows:

Alternative	Mayor Outcome	Council Outcome	Utility
A_1	Accepts	Accepts	1.00
A_1	Accepts	Refects	0.50
A_1	Rejects		0.30
A_2	Accepts	Accepts	0.80
A_2	Rejects		0.00

What should the planner recommend? Please illustrate your solution according to the tenets of SEU theory.

5.9 An investor who is wealthy enough to be an expected value operator may buy shares of a stock for $25,000. The net return r_i on the investment, the selling price minus the cost price of $25,000, will be one of three values $E_1 = \$40,000$, $E_2 = \$10,000$, $E_3 = \$25,000$, and the investment returns will be influenced by the state of the economy according to $e_1 =$ bull market, $e_2 =$ fair market, and $e_3 =$ depression.

The joint probabilities of the return and the state of the economy $P(E_i|e_j)$ are given as follows:

		State of Economy		
		e_1	e_2	e_3
Return	r_1	0.12	0.06	0.02
	r_2	0.10	0.30	0.10
	r_3	0.04	0.12	0.14

For a price of $1,000 an economic consultant may be employed to predict the state of the economy. The conditional probabilities of the economist's estimate of the state of the economy given the true state, $P(\hat{e}_i|e_j)$, are

		True Economic State		
		e_1	e_2	e_3
Estimated	$\hat{e}_1$	0.7	0.1	0.1
Economic	$\hat{e}_2$	0.2	0.8	0.2
State	$\hat{e}_3$	0.1	0.1	0.7

Find the optimal decision regarding hiring the economist and buying the stock. Also find the expected value of perfect information and the expected value of the economist's information.

5.10 A quantity called the risk aversion function is defined as the second derivative of the utility function, or $r(x) = d^2U(x)/dx^2$, and indicates how large a risk premium a decision maker will pay to eliminate uncertainty in a given situation. For cases in which the risk aversion function $r(x)$ is a constant R, we may find the utility function as

$$U(x) = a - (\text{sgn } R)\beta \exp(-Rx), \qquad R \neq 0$$
$$U(x) = a + \beta x, \qquad R = 0$$

Discuss possible uses for a risk aversion factor such as this. How could you conduct a lottery to assess $U(x)$? Contrast and compare your results with those in Table 5.1.

5.11 Use of the exponential is suggested as an approximation to many utility functions. Figure 5.30, on page 446, illustrates a suggested exponential utility function. How could you assess R for this utility function? Please compare and contrast this result with results in Problem 5.10 and Table 5.1.

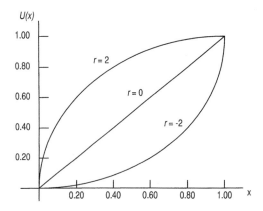

Figure 5.30 Illustration for $U(x) = [1 - \exp(-rx)]/[1 - \exp(-r)]$.

5.12 Suppose you have your choice of two lotteries. The first lottery costs $1.00 to play, the probability of winning is 0.5, and the payoff if you win is $2.10. There is no payoff if you lose. You may play this lottery 10^8 times. The second lottery may be played only once. It costs 10^8 to play this lottery. The probability of winning is 0.5, and the payoff is 2.1×10^8 if you win, but nothing if you lose. Use the theory of normative decision analysis to thoroughly explore the varied aspects of these two lotteries. Which would you choose to play?

5.13 Suppose that Problem 5.12 is changed such that the probability of losing the second lottery is given by 10^{-6}. The payoff for winning the second lottery is given precisely by the expression $[5(10^{12} - 10^8)]/(10^6 - 1)$. How are the results of the previous problem changed? Are there any difficulties in assessment of utilities in this problem or the previous one that are not present in many decision analysis problems? Both this problem and the previous one have interesting relations with Problem 5.14 to which you are now invited to examine.

5.14 A simplistic view of the coal versus nuclear fuel for electric power plant controversy views the probability of environmental damage from coal as much, much higher than that from nuclear energy. However, the likely destruction from a nuclear accident causing environmental damage would be far greater than that from coal. How can decision analysis be used to help resolve this controversy? Please relate your discussion to your readings concerning risk and hazard and Problems 5.12 and 5.13.

5.15 There are firms in the oil exploration industry that drill for oil with little scientific knowledge concerning the seismic characteristics of a given area. Instead, these firms, known as wildcatters, base their drilling decisions on what other firms have discovered while drilling in similar areas. We will consider a simplified version of the wildcatter problem and add complexity and a degree of practicality as we analyze the problem.

(a) Initially suppose that a wildcatter is considering the decision to drill in a section where oil has been found in 20% of the past efforts of other firms. The cost of drilling will initially be assumed to be $100,000. If oil is discovered, the gross return over the life of the producing well is $400,000. Assume that the wildcatter is an expected monetary value player. Should the wildcatter drill? What is the expected net return from drilling?

(b) Suppose now that our wildcatter has a perfect seismic device that will predict perfectly the existence or nonexistence of oil beneath the surface of the earth. What is the expected return using the perfect seismic predictor? What is the expected value of this perfect information?

(c) We will now consider a more realistic case in which we have an imperfect seismic predictor. We will assume that performance records of this device have been kept. The output of the imperfect seismic detector is a set of two lights. A green light is turned on when the seismic indicator predicts that oil is present, and a red light is turned on when it predicts that no oil is present. A summary of the performance record of the seismic device is given by the following conditional probabilities

If oil	If no oil	Then
0.75	0.40	Green light on
0.25	0.60	Red light is on

What is the expected net return using this seismic device? What is the expected value of this seismic device?

(d) If the imperfect seismic predictor costs a certain rental fee to use, what is the maximum fee that the wildcatter would be willing to pay in order to use the device?

5.16 Suppose that the price of crude oil increases such that the gross return over the life of the well in Problem 5.15 is increased from $400,000 to $600,000. Repeat the calculations in Problem 5.15, and contrast the results obtained here with those in the previous problem.

5.17 Reexamine Problem 5.15 for the case where the prior probability of locating oil (without tests) is set equal to P. How sensitive are the results of Problem 5.15 to the value of P?

5.18 Suppose that the wildcatter of Problem 5.15 is risk-averse. How will this affect the answers in Problem 5.15? What will be the wildcatter's decisions when the gross return from a successful well is changed as in Problem 5.16?

5.19 What is your certainty equivalent for the following lotteries:

$$L_1 = (\$1000, \ 0.50; \ \$0, \ 0.50)$$

$$L_2 = (\$2000, \ 0.25; \ \$0, \ 0.75)$$

$$L_3 = (\$5000, \ 0.10; \ \$0, \ 0.90)$$

$$L_4 = (\$667, \ 0.75; \ \$0, \ 0.25)$$

$$L_5 = (\$556, \ 0.90; \ \$0, \ 0.10)$$

On the basis of this, please plot your utility function for money.

5.20 You are in charge of ordering shoes for resale in the store that you manage. You can order these in unit quantities of 50 boxes of shoe pairs. If you order 50 boxes, the cost per box is $10. If you order 100 boxes, the cost is $9 per box. If you order 150 boxes, the costs are $8 per box. If you order 200 or more boxes, the cost if $7 per box. The selling price per box is $14. Any unsold shoes at the end of the period in question will have to be sold at your discount outlet at a price of $6 per box. Suppose that the probability for various demands for shoes is $P(0) = 0$, $P(50) = 0.1$, $P(100) = 0.3$, $P(150) = 0.2$, $P(200) = 0.2$, $P(250) = 0.1$, and $P(300) = 0.1$. What is the order quantity that will max*i*mize the expected profit (neglecting any merchandising costs)? What is the value of perfect information for this problem?

5.21 Suppose four voters have the preference structure $A \mathcal{P} C \mathcal{P} B$, three have $B \mathcal{P} C \mathcal{P} A$, and the last two voters have preferences $C \mathcal{P} B \mathcal{P} A$. In a plurality election, where each voter votes only for their top-ranked candidate, the outcome is $A \mathcal{P} B \mathcal{P} C$ with a vote score of 4:3:2 for candidates $A:B:C$. In many ways, this is a very strange outcome in that the binary preference order structure for these candidates is reversed because these voters will choose pairs in the reversed $C \mathcal{P} B$, $C \mathcal{P} A$, $B \mathcal{P} A$ order with respective 6:3, 5:4, and 5:4 scores. Thus, rather than being top-ranked, the voters' collective judgment may be that candidate A is their inferior choice. Please discuss this result in some detail using the various voting procedures discussed here.

5.22 Please prepare a case study report on the use of MAUT to aid a person in selection of a job.

5.23 Prepare a discussion of group decision analysis — that is, decisions that

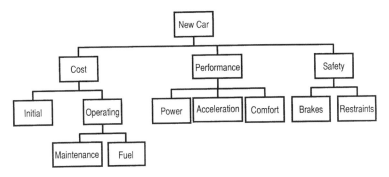

Figure 5.31 Attribute tree for buying a car.

impact a group of people. Discuss the thought that group decision analysis is just a special case of MAUT in which each individual, or special interest group, is represented by one attribute.

5.24 Figure 5.31 represents one possible attribute tree for buying a car. Please write a case study concerning the use of MAUT and related approaches to support the decision of an individual, or of a group such as a family or an organization seeking a fleet of automobiles, regarding an appropriate purchase.

5.25 Under suitable independence assumptions [86], a utility function may be expressed in either the additive form

$$U(x_1, x_2, x_3, \ldots, x_n) = \sum_{i=1}^{n} k_i U_i(x_i)$$

or the multiplicative form

$$1 + KU(x_1, x_2, x_3, \ldots, x_n) = \prod_{j=1}^{n} [1 + Kk_i U(x_i)]$$

Show that the multiplicative form reduces to the additive form whenever $\sum_{i=1}^{n} k_i = 1$, in which case $K = 0$. If $\sum_{i=1}^{n} k_i \neq 1$, then $K \neq 0$, and K can be evaluated by utility assessment procedures at the extreme points of $U(x_i)$. Prepare a study of this multiplicative utility function for the case where there are two attributes and where there are three attributes. In particular, consider the case where the attributes do not allow compensatory tradeoffs, such as in natural resource depletion, and contrast the results with those that might result from the linear additive case.

REFERENCES

[1]   Sage, A. P., *Systems Engineering*, Wiley, New York, 1992.

[2]   Klein, G., *Sources of Power: How People Make Decisions*, MIT Press, Cambridge, MA, 1998.

[3]   Hammond, J. S., Keeney, R. L., and Raiffa, H., *Smart Choices: A Practical Guide to Making Better Decisions*, Harvard Business School Press, Boston, 1999.

[4]   Janis, I. L., and Mann, L., *Decision Making: A Psychological Analysis of Conflict, Choice, and Commitment*, Free Press, New York, 1977.

[5]   Janis, I. J., *Groupthink: Psychological Studies of Policy Decisions and Fiascoes*, Houghton Mifflin, Boston, 1982.

[6]   Janis, I., J., *Crucial Decisions: Leadership in Policy Making and Crisis Management*, Free Press, New York, 1989.

[7]   Dutta, P. K., *Strategies and Games: Theory and Practice*, MIT Press, Cambridge, MA, 1999.

[8]   Fishburn, P. C., *Nonlinear Preference and Utility Theory*, Johns Hopkins University Press, Baltimore, 1988.

[9]   Fishburn, P. C., Generalization of Expected Utility Theories: A Survey of Recent Proposals, *Annals of Operations Research*, Vol. 19, No. 1, 1989, pp. 3–28.

[10] Brown, R., Kahr, A., and Peterson, C., *Decision Analysis for the Manager*, Holt, Rinehart, and Winston, New York, 1974.

[11] Howard, R. A., Foundations of Decision Analysis, *IEEE Transactions on Systems Science and Cybernetics*, Vol. SSC-4, No. 3, 1968, pp. 211–219.

[12] Raiffa, H., *Decision Analysis*, Addison-Wesley, Reading, MA, 1968.

[13] Schlaifer, R., *Analysis of Decision Under Uncertainty*, McGraw-Hill, New York, 1969.

[14] Fishburn, P. C., Foundations of Decision Analysis: Along the Way, *Management Science*, Vol. 15, No. 4, April 1989, pp. 387–405.

[15] Goldstein, W. M., and Hogarth, R. M. (Eds.), *Research on Judgment and Decision Making*, Cambridge University Press, New York, 1997.

[16] Wheeler, D. D., and Janis, I. L., *A Practical Guide for Making Decisions*, Free Press, New York, 1980.

[17] Beyth-Marom, R., and Dekel, S., *An Elementary Approach to Thinking Under Uncertainty*, Lawrence Erlbaum Associates, Hillsdale, NJ, 1985.

[18] Norman, D. A., *The Psychology of Everyday Things*, Basic Books, New York, 1988.

[19] Behn, R. D. and Vaupel, J. W., *Quick Analysis for Busy Decision Makers*, Basic Books, New York, 1982.

[20] Russo, J. E., and Schoemaker, P. H., *Decision Traps: The Ten Barriers to Brilliant Decision Making and How to Overcome Them*, Simon and Schuster, New York, 1989.

[21] Gilovich, T., *How We Know What Isn't So: The Fallibility of Human Reason in Everyday Life*, Free Press, New York, 1991.

[22] Keeney, R. L., *Value Focused Thinking: A Path to Creative Decisionmaking*, Harvard University Press, Cambridge, MA, 1992.

[23] Sage, A. P., *Decision Support Systems Engineering,* Wiley, New York, 1991.

[24] Adelman, L., *Evaluating Decision Support and Expert Systems,* Wiley, New York, 1992.

[25] Hadley, G., *Introduction to Probability and Statistical Decision Theory,* Holden-Day, San Francisco, 1967.

[26] Sage, A. P., and White, C. C., ARIADNE: A Knowledge Based Interactive System for Planning and Decision Support, *IEEE Transactions on Systems, Man, and Cybernetics,* Vol. 14, No. 1, January/February 1984, pp. 35–47. Also in Sage, A. P. (Ed.) *System Design for Human Interaction* IEEE Press, 1987, pp. 415–428.

[27] von Neumann, J., and Morgenstern, O., *Theory of Games and Economic Behavior,* Princeton University Press, Princeton, NJ, 1944.

[28] Savage, L. J., *The Foundations of Statistics,* Wiley, New York, 1954.

[29] North, D. W., A Tutorial Introduction to Decision Theory, *IEEE Transactions on Systems Science and Cybernetics,* Vol. 4, No. 3, Sept. 1968, pp. 200–210.

[30] Holloway, C. A., *Decision Making Under Uncertainty: Models and Choices,* Prentice-Hall, Englewood Cliffs, NJ, 1979.

[31] Kirkwood, C. W., *Strategic Decision Making: Multiobjective Decision Analysis with Spreadsheets,* Duxbury Press, Belmont, CA, 1997.

[32] Luce, R. D., and Raiffa, H., *Games and Decisions: Introduction and Critical Survey,* Wiley, New York, 1957.

[33] Dyer, J. S., and Sarin, R. K., Relative Risk Aversion, *Management Science,* Vol. 28, 1982, pp. 875–886.

[34] Huber, G. P., Multi-attribute Utility Models: A Review of Field and Field-like Studies, *Management Science,* Vol. 20, June 1974, pp. 1393–1402.

[35] Johnson, E. M. and Huber, G. P., The Technology of Utility Assessment, *IEEE Transactions on Systems, Man, and Cybernetics,* Vol. SMC-7, No. 5, May 1977, pp. 311–325.

[36] Farquhar, P. H., Utility Assessment Methods, *Management Science,* Vol. 30, 1984, pp. 1283–1300.

[37] Clemen, R. T., *Making Hard Decisions: An Introduction to Decision Analysis,* 2nd edition, Duxbury Press, Belmont, CA, 1996.

[38] Hammond, K. R., McClelland, and Mumpower, J., *Human Judgment and Decision Making: Theories, Methods, and Procedures,* Praeger, New York, 1980.

[39] Hammond, K. R., Rohrbaugh, J., Mumpower J., and Adelman, L., Social Judgment Theory: Applications in Policy Formation, in *Human Judgment and Decision Processes: Applications in Problems Setting,* Kaplan, M. F., and Schwartz, S. (Eds.), Academic Press, New York, 1977.

[40] Hammond, K. R., Mumpower, J. L., and Cook, R. L., Linking Environmental Models with Models of Human Judgment: A Symmetrical Decision Aid, *IEEE Transactions on Systems, Man and Cybernetics,* Vol. 7, 1977, pp. 358–367.

[41] Mumpower, J. L., Veirs, V., and Hammond, K. R., Scientific Information, Social Values, and Policy Formation: The Application of Simulation Models and Judgment Analysis to the Denver Regional Air Pollution Problem, *IEEE Transactions on Systems, Man, and Cybernetics,* Vol. 9, 1979, pp. 464–476.

[42] Anderson, D. F. and Rohrbaugh, J., Some Conceptual and Technical Problems in Integrating Models of Judgment with Simulation Models, *IEEE Transactions on Systems, Man, and Cybernetics*, Vol. 22, No. 1, January 1992.

[43] Brehmer, B., and Joyce, C. R. B. (Eds.), *Human Judgment: The STJ View*, Elsevier, Amsterdam, 1988.

[44] von Winterfeldt, D., and Edwards, W., *Decision Analysis and Behavioral Research*, Cambridge University Press, New York, 1986.

[45] Merkhofer, M. W., Quantifying Judgmental Uncertainty: Methodological Experiences and Insights, *IEEE Transactions on Systems, Man, and Cybernetics*, Vol. SMC 17, No. 5, September 1987, pp. 741–752.

[46] Wallsten, T. S. and Budescu, D. V., Encoding Subjective Probabilities: A Psychological and Psychometric Review, *Management Science*, Vol. 29, No. 2, February 1983, pp. 151–173.

[47] Borcherding, K., Eppel, T., and von Winterfeldt, D., Comparison of Weighting Judgments in Multiattribute Utility Measurements, *Management Science*, Vol. 37, No. 2, December 1991, pp. 1603–1619.

[48] Keeney, R. L., von Winterfeldt, D., and Eppel, T., Eliciting Public Values for Complex Policy Decisions, *Management Science*, Vol. 36, No. 9, September 1990, pp. 1011–1030.

[49] Harvey, C. M., Structured Prescriptive Models of Risk Attitudes, *Management Science*, Vol. 36, No. 12, December 1990, pp. 1479–1501.

[50] Harvey, C. M., Model of Tradeoffs in a Hierarchical Structure of Objectives, *Management Science*, Vol. 37, No. 8, August 1991, pp. 1030–1042.

[51] Sampson, A. R., and Smith, R. L., Assessing Risks through the Determination of Rare Event Probabilities, *Operations Research*, Vol. 30, No. 5, September 1982, pp. 839–866.

[52] Krantz, D. H., Luce, R. D., Suppes, P., and Tversky, A., *Foundations of Measurement, Volume I: Representational Theory of Measurement*, Academic Press, Orlando, FL, 1971.

[53] Suppes, P., Krantz, D. H., Luce, R. D., and Tversky, A., *Foundations of Measurement, Volume II: Geometric, Threshold, and Probabilistic Representations*, Academic Press, Orlando, FL, 1989.

[54] Luce, R. D., Krantz, D. H., Suppes, P., and Tversky, A., *Foundations of Measurement, Volume III: Representation, Axiomatization, and Invariance*, Academic Press, Orlando, FL, 1990.

[55] Fishburn, P. C., Utility Theory, *Management Science*, Vol. 14, No. 5, January 1968, pp. 335–378.

[56] Keeney, R., and Raiffa, H., *Decisions with Multiple Objectives*, Wiley, New York, 1976.

[57] Keeney, R. and Raiffa, H., *Decisions with Multiple Objectives: Preferences and Value Tradeoffs*, Cambridge University Press, New York, 1993.

[58] Edwards, W., How to Use Multiattribute Utility Measurement for Social Decision Assessment, *IEEE Transactions on Systems, Man, and Cybernetics*, Vol. SMC-7, May 1977, pp. 326–340.

[59] Miller, J. R., A Systematic Procedure for Assessing the Worth of Complex Alternatives, MITRE Corporation Report AD 662001, New Bedford, MA, 1967.

[60] Farris, D. R., and Sage, A. P., On Decision Making and Worth Assessment, *International Journal of System Sciences*, Vol. 6, No. 12, December 1975, pp. 1135–1178.

[61] Sage, A. P., *Methodology for Large Scale Systems*, McGraw-Hill., New York, 1977.

[62] Brams, S. J., and Fishburn, P. C., Approval Voting, *American Political Science Review*, Vol. 72, 1978, pp. 831–847.

[63] Brams, S. J., and Fishburn, P. C., *Approval Voting*, Birkhauser, Boston, 1983.

[64] Saari, D. G., *The Geometry of Voting*, Springer-Verlag, New York, 1994.

[65] Saari, D. G., *Basic Geometry of Voting*, Springer-Verlag, New York, 1995.

[66] Brams, S. J., and Fishburn, P. C., Alternative Voting Systems, in Maisel, L. S. (Ed.), *Political Parties and Elections in the United States: An Encyclopedia*, Garland Press, New York, 1991, pp. 23–31.

[67] Arrow, K. J., *Social Choice and Individual Values*, 2nd Edition, Yale University Press, New Haven, CT, 1963.

[68] Brams, S. J. Approval Voting and the Good Society, *Political Economy of the Good Society Newsletter*, Vol. 3, No.1, Winter 1993, pp. 10–14.

[69] Keeney, R. L., A Group Preference Axiomatization with Cardinal Utility, *Management Science*, Vol. 23, No. 2, October 1976, pp. 140–145.

[70] Matheson, D. and Mathieson, J., *The Smart Organization: Creating Value through Strategic R&D*, Harvard Business School Press, Boston, 1998.

[71] Goodman, L. A., and Markowitz, H., Social Welfare Functions Based on Individual Rankings, *Am. J. Sociology*, Vol. 58, No. 3, November 1952, pp. 257–262.

[72] Black, D., *The Theory of Committees Elections*, Cambridge University Press, New York, 1968.

[73] Harsanyi, J.C., *Essays on Ethics, Social Behavior, and Scientific Explanation*, D. Reidel, Boston, 1976.

[74] Harsanyi, J. C., *Rational Behavior and Bargaining Equilibrium in Games and Social Situations*, Cambridge University Press, New York, 1977.

[75] Harsanyi, J. C., Bayesian Decision Theory, Rule Utilitarianism, and Arrow's Impossibility Theorem, *Theory and Decision*, Vol. 11, 1979, pp. 289–317.

[76] Rawls, J., *A Theory of Justice*, Harvard University Press, Cambridge, MA, 1971.

[77] Keeney, R. L., and Kirkwood, C. W., Group Decision Making Using Cardinal Social Welfare Functions, *Management Science*, Vol. 22, No. 4, December 1975, pp. 430–437.

[78] Kirkwood, C. W. (Ed.), Decision Analysis Special Issue, *Operations Research*, Vol. 28, No. 1, January 1981, pp. 1–252.

[79] Bonner, J., *Introduction to the Theory of Social Choice*, Johns Hopkins University Press, Baltimore, 1986.

[80] Sen, A. K., *Choice, Welfare, and Measurement*, Harvard University Press, Cambridge, MA, 1997.

[81] Saaty, T. L., *The Analytical Hierarchy Process: Planning, Priority Setting, Resource Allocation*, McGraw-Hill, New York, 1980.

[82]   Saaty, T. L., and Kearns, K. P., *Analytical Planning: The Organization of Systems,* Pergamon Press, Oxford, UK, 1985.

[83]   Dyer, J. S., Remarks on the Analytical Hierarchy Process, *Management Science,* Vol. 30, 1990, pp. 259–258. (Also see Dyer, J. S., A Clarification of Remarks on the Analytical Hierarchy Process, *Management Science,* Vol. 30, 1990, pp. 274–275.

[84]   Harker, P. T., and Vargas, L. G., Reply to Remarks on the Analytical Hierarchy Process, *Management Science,* Vol. 30, 1990, pp. 269–273.

[85]   Winkler, R. L., Decision Modeling and Rational Choice: AHP and Utility Theory, *Management Science,* Vol. 30, 1990, pp. 247–248.

[86]   Keeney, R. L., Multiplicative Utility Functions, *Operations Research,* Vol. 23, No. 6, 1974, pp. 22–34.

CHAPTER 6

Systems Engineering and Systems Engineering Management

In this concluding chapter we will discuss some aspects of the management of systems engineering efforts, especially systems engineering processes, or *systems management* [1, 2]. It is not unusual that this subject is called "Engineering Management" or "Project Management" or "Systems Engineering Management." We have no particular objection to the use of these terms, and they are often used in texts on this subject. However, we prefer the use of the term systems engineering management [3, 4], or systems management, to denote the fact that we wish to employ systems engineering principles throughout the management of the process, or product line, efforts that result in a reliable, trustworthy, responsive, and high quality system.

There are many ways in which systems engineering management efforts can be structured. In terms of people and their roles, it is possible to define such functions as program management, project management, chief systems engineer, project controller, and so forth. There is little commonality of these terms and functions across organizations; thus we will not describe specific roles for these people here. Instead, we will concentrate on the overall effort to be accomplished by the systems engineering management team. These responsibilities will generally include the program and project management responsibilities of planning, scheduling, costing, team building and leadership, and interactions with customers and suppliers.

6.1   INTRODUCTION

There are many definitions that could be provided for the term "management" and the related term "systems management" [1,2]. For our purposes, appropriate definitions are as follows:

- *Management* consists of all of the activities undertaken to enable an organization to cope effectively and efficiently within its environment. This will generally involve planning, organizing, staffing, directing, coordinating, reporting, and budgeting activities in order to achieve identified objectives.
- *Systems management* is the organized and integrated set of procedures, practices, technologies, and processes that will contribute to efficient and effective accomplishment of systems engineering objectives relative to management of the fielding of large systems. Systems management efforts are designed to lead to achievement of overall plans or objectives of a systems engineering organization for the realization of trustworthy systems engineering processes for the fielding of large systems.

The word "organization" appears several times in these definitions, as does the word "plan." It is important to note that a given organization, enterprise, or business will, or at least should, have a plan, very likely a set of plans, to achieve the overall objectives of the organization. Many of these will relate to the way in which the organization provides products and services to its customers or clients. Such a plan generally aims to satisfy both general and specific needs. The general needs are those of the organizational units within the enterprise. Out of these high-level plans and policies will evolve a framework that will enable identification of plans and subsequent activities to fulfill the needs of a specific client for a systems engineering product or service. Each of these concerns is important. The first relates to the way in which the enterprise organizes itself, and the second relates to the way it serves customers.

There are a variety of definitions of an organization. Some very classic ones, which are relevant to our systems engineering and systems management discussions, are as follows:

- A system of consciously coordinated activities of two or more people [5]
- Social units deliberately constructed to seek specific goals [6]
- Collectives that have been established on a relatively continuous basis in an environment, with relatively fixed boundaries, a normative order, authority ranks, communication systems, and an incentive system designed to enable participants to engage in activities in general pursuit of a common set of goals [7]
- A set of individuals, with bounded rationality, who are engaged in the decision making process [8]

The term, *bounded rationality*, generally refers to the practice of suboptimization or limited optimization of objectives. It occurs because of limited time, limited information, or limited human ability to optimize.

Organizations can be viewed from a closed-system perspective. In this view, an organization is an instrument designed to enable pursuit of well-defined and well-specified objectives. A closed-system perspective organization will be concerned primarily with four objectives:

1. Efficiency
2. Effectiveness
3. Flexibility or adaptability to external environmental influences
4. Job satisfaction

Four organizational means or activities follow from this [9]:

5. Complexity and specialization
6. Centralization or hierarchy of authority
7. Formalization or standardization of jobs
8. Stratification of employment levels

In this closed-system view, everything is functional and tuned such that all resource inputs are optimized and the associated responses fit into a well-defined master plan.

March and Simon [10, 11] discuss the inherent shortcomings associated with this closed-system model of humans performing machine-like tasks. Not only is the human-as-machine view believed to be inappropriate, but there are pitfalls associated with viewing environmental influences as "noise," as must necessarily be done in the closed-system perspective. March and Simon's broadened view of an organization is known as the open-systems view. In the open-systems view of an organization, concern is not only with objectives but also with appropriate responses to a number of internal and external influences.

Many other authors have expanded upon these views from a variety of perspectives [12–14]. Most management studies show that, in practice, plans and decisions are the result of interpretation of standard operating procedures. Improvements are obtained by careful identification of existing standard operating procedures and associated organizational structures. The resulting organizational process model, originally due to Cyert and March [15], functions by relying on standard operating procedures that constitute the memory or intelligence bank of the organization. Only if the standard operating procedures fail will the organization attempt to develop new standard operating procedures.

Organizational learning results when members of the organization react to changes in the internal or external environment of the organization by detection and correction of errors [16]. An error is a feature of knowledge that makes action ineffective. Ideally the detection, diagnosis, and correction of error produces learning. A major claim by systems designers is that errors in using systems, as well as in designing systems, are due not simply to probabilistic random events that might be removed through improved system operator or system designer training, or through better system designs [17, 18]. Instead, it is argued that errors are due to two generally more important sources.

1. Errors represent systematic interference and incongruities among models, rules, and procedures.
2. Errors represent some dysfunctionality of the effects of adaptive learning mechanisms.

A very important feature and need in organizations is that of *organizational learning*. Mistakes will occur. It is hoped that individuals and organizations learn from mistakes such that things are done better next time. It is difficult to imagine improvement without learning. Without some forms of learning there is no reason to believe that the same mistakes that were made before will not be made again. So, we all need to learn—continuously and throughout a lifetime. "We" is a generic term here and refers to us, me, you, and organizations.

Learning involves the use of observations of the relationships between activities and outcomes, often obtained in an experiential manner, to improve behavior through the incorporation of appropriate changes in processes and product. Thus learning represents acquired wisdom in the form of skill-based knowledge, rule-based knowledge, or formal-reasoning-based knowledge [19, 20]. Thus, it may involve know-how, in the form of skills or rules, or know-why, in the form of formal-reasoning-based knowledge.

Learning involves the following:

• Situation assessment
• Detection of a problem
• Synthesis of a potential solution to the problem
• Implementation of the solution
• Evaluation of the outcome

and the resulting discovery that eventuates from this. This is a formal description of the learning process. It is also the problem-solving process and involves the basic steps of systems engineering.

Peter Senge [21–23] has devoted major attention to the study and development of what are called *learning organizations*. According to Senge, *learning organizations are "organizations where people continually expand their capacity to create the results they truly desire, where new and expansive patterns of thinking are nurtured, where collective aspiration is set free, and where people are continually learning how to learn together."* Five component technologies, or disciplines, are suggested as now converging to enable this learning. These are as follows:

1. Systems thinking
2. Personal mastery through proficiency and commitment to lifelong learning
3. Shared mental models of the organization markets, and competitors
4. Shared vision for the future of the organization
5. Team learning

Systems thinking is denoted as the *fifth discipline* and is the catalyst and "cornerstone" of the learning organization that enables success through the other four dimensions. Lack of organizational capacity on one of these disciplines is called a learning disability. One of the major disabilities is associated with implicit mental models that result in people having deeply rooted mental models without being aware of the cause–effect consequences that result from use of these models. Another is the tendency of people to envision themselves in terms of their position in an organization rather than in terms of their aptitudes and abilities. This often results in people becoming dislocated when organizational changes are necessary and this leads to disconcertment.

Each of the five learning disciplines can exist at three levels. These are termed as follows:

- *Principles,* the guiding ideas and insights that suggest practices
- *Practices,* the existing theories of action in practice
- *Essences,* the wholistic and future-oriented understandings associated with each particular discipline

These correspond very closely with the principles, practices, and perspectives we have used to describe approaches to knowledge and systems engineering.

Based primarily on works in system dynamics, on an approach for the study and modeling of systems of large scale and scope, and on efforts by Argyris and others, 11 laws of the fifth discipline are stated. We restate these here.

1. Contemporary and future problems often come about because of what were presumed to be past solutions.
2. For every action, there is a reaction.
3. Short-term improvements often lead to long-term difficulties.
4. The easy solution may be no solution at all.
5. The solution may be worse than the problem.

6. Quick solutions, especially at the level of symptoms, often lead to more problems than existed initially. Thus, quick solutions may be counterproductive solutions.

7. Cause and effect are not necessarily related closely, either in time or in space. Sometimes actions implemented here and now will have impacts far away at a much later time.

8. The actions that will produce the most effective results are not necessarily obvious at first glance.

9. Low cost and high effectiveness do not have to be subject to compensatory tradeoffs over all time.

10. The entirety of an issue is often more than the simple aggregation of the components of the issue.

11. The entire system, comprised of the organization and its environment, must be considered together.

Neglect of these laws can lead to any number of problems. Most of these are relatively evident from Senge's description and our interpretation of the 11 laws of the fifth discipline. For example, failure to understand law 11 leads to the fundamental attribution error in which we credit ourselves for success and blame others for our failures.

On the basis of these laws, several leadership facets are suggested. Leaders become

- Designers
- Stewards
- Teachers

These are especially important for learning organizations. Each of these leadership characteristics enables everyone in the organization to improve on their understanding and use of the five important dimensions of organizational learning. This is said to result in creative tension throughout the organization. Planning is one of the major activities of the learning organization, and it is through planning that much learning occurs. Often plans will need to be changed to reflect emerging conditions.

One of the fundamental notions in studies of human errors is that there is an intimate association between human intent and human error. Realistic efforts to discuss human error and to design systems that can cope with human error possibilities, therefore, will consider the different types of human intentions and associated errors. In a very insightful work, Reason [24] indicates the importance of knowing the following:

1. Whether human actions are directed by conscious intent
2. Whether human actions proceed as planned
3. Whether human actions achieve the desired result

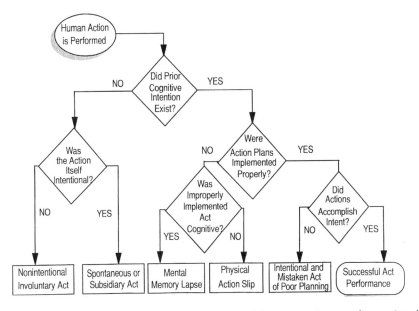

Figure 6.1 Interpretation of Reason's taxonomy of human actions and associated types of human error.

Five types of actions result from this observation, as indicated in Figure 6.1. Successful system designs will create systems that encourage human intentional action and successful act performance. Chapter 9 of reference 25 is very concerned with human error — or, more generally, cognitive ergonomics — and issues in systems engineering and contains a much more extensive discussion and set of references than is possible here.

We have attempted to focus our discussion thus far at a rather conceptual level. We have presented a description and interpretation of some recent studies in behavioral and organizational theory that have direct relevance to systems management. The primary classical organizing principles supporting organizational behavior include the following:

1. Division of labor and task assignment
2. Identifying standard operating principles
3. Top-down flow of decisions
4. Formal and informal channels of communication in all directions
5. The multiple uses of information
6. Organizational learning

We must be conscious of these principles if we are to produce systems management and technical direction perspectives that are grounded in the realities of human desires and capabilities for growth and self-actualization. Also, we must be aware of error possibilities, on the part of both the system architects and developers, and errors on the part of system users. These concerns are very realistic and influence the design of information and decision support systems [26,27], software systems [28], and many other areas within systems engineering that have strong cognitive, human interaction, and human factors concerns [1, 2, 20, 29, 30]. We will return to a discussion of some of these issues later in this chapter.

Much of our discussion to this point may seem oriented more toward management philosophy than toward management practice. Doubtlessly, this is correct. We maintain strongly, however, that successful systems management practice will embody these principles, prescriptions, perspectives and philosophies. Our management philosophy supports the pragmatic management of large systems engineering projects. This philosophy should also be incorporated into the systems that we develop, such that they are suitable for human interaction.

There have been many attempts to classify management functions. Among these is the function-type taxonomy of Anthony [31]. He describes three planning and control functions.

1. *Strategic Planning Function.* Strategic planning is the process of choosing the highest-level policies and objectives, and associated resource allocations and strategies for achieving these. According to Anthony, strategic planning is unsystematic in that the need for strategic decisions may arise at any time and that the threats and opportunities that lead to strategic decisions are not discovered systematically or at uniform intervals.

2. *Management Control Function.* Management control is the process through which managers influence other organizational members in order to help them achieve organizational strategies. Management control decisions are those decisions made for the purpose of ensuring effectiveness in the acquisition and use of resources to achieve strategic plans.

3. *Task Control Function.* The process of task control has as its major objectives the efficient and effective performance of specific tasks. In an earlier work, Anthony described task control in terms of two related functions. Operational control functions were accomplished for the purpose of ensuring effectiveness in the performance of operations. Operational performance functions were associated with the day-to-day decisions made while performing operations.

These three planning and control processes relate to one another as indicated in Figure 6.2. While there is often considerable variation among the many task control systems that may be found in a given organization, nearly all such systems may include interaction between one task manager and a team of nonmanagers, or perhaps with an automated system. Task control functions

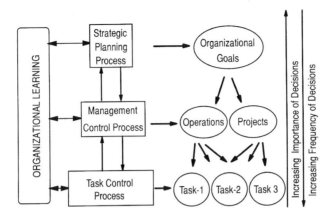

Figure 6.2 Flow of organizational information and associated planning and control decision flow.

are generally well-structured. Management control involves the interaction of managers, generally in resolving unstructured issues in a manner that supports achievement of the strategic plan of the organization. Management control concentrates on the activities that occur within various responsibility centers of the organization. It would appear that the system management function described here, and elsewhere in the systems engineering literature, is fundamentally similar to that of Anthony's management control.

Other discussions of organizational management will concentrate on *activities* of a management team, as contrasted with the *decisions* that they make. It is not unusual to find seven identified management functions or tasks in the classic enterprise management literature:

1. *Planning*, which comprises identification of alternative courses of action that will achieve organizational goals
2. *Organizing*, which involves structuring of tasks that will lead to the achievement of organizational plans, and the granting of authority and responsibility to obtain these
3. *Staffing*, which comprises the selection and training of people to fit various roles in the organization
4. *Directing*, which refers to the creation of an environment and an atmosphere that will motivate and assist people to accomplish assigned tasks
5. *Coordinating*, which involves the integration and synchronization of performance, including the needed measurements and corrective actions that lead to goal achievement
6. *Reporting*, which ensures proper information flow in the organization
7. *Budgeting*, which ensures appropriate distribution of economic resources needed for goal achievement

This POSDCORB theory of management is a very common one and is described in almost all classical management texts.

It is quite clear that these functions are not independent of one another. Because details of these functions are provided in essentially any introductory management guide, we will not pursue these in any detail here. It is important to note that, collectively, these are the tasks of general enterprise management. They apply to systems management in its planning activities, which involve anticipation of potential difficulties and the identification of approaches for detection of problems, diagnosis of causes, and determination of promising corrective actions. They apply also to systems management and its control activities, which involve controls exercised in specific situations in order to improve efficiency and effectiveness of task controls in achieving objectives.

Planning is a prominent word in much of the foregoing. We can identify three basic types of, or levels for, plans:

- Organizational plans
- Program plans
- Project plans

One of the major differences in these types of plans is the duration of the plans. Organizational plans are normally strategic in nature and can be expected to persist over a relatively long time. Program plans are intended to achieve specific results. For large programs, it is generally desirable to disaggregate the program plans and controls into a number of smaller projects. There is, again, no standard set of terminology. The software engineering literature, for example, uses projects where we have used programs here. It is a reasonable speculation that this is done to avoid confusion with program(ming) lines of code.

Three fundamental activities are involved in systems management. These are precisely the steps of the problem-solving or systems engineering process that we have emphasized in this introduction to systems engineering:

1. Formulation of issues
2. Analysis of alternatives
3. Interpretation of the impacts of the alternatives

as we have previously indicated. These steps are encountered at several levels, or phases, of organizational management activities. They begin at the strategic level and result in the preparation of strategic plans, which are then ultimately converted into tactical and operational plans and implemented as management controls or task controls through an effective planning process.

At this highly aggregated level, it may be difficult to envision specific systems management activities. While there are many more finely grained steps into which the aforementioned three steps may be partitioned, three general and nine specific levels of activities appear especially important steps in systems management.

1. Issue formulation
   1.1. Environmental monitoring
   1.2. Environmental understanding
   1.3. Identification of information needs
   1.4. Identification of alternative potential courses of action
2. Issue analysis
   2.1. Identification of the impacts of alternatives
   2.2. Fine-tuning the alternatives for effectiveness
3. Issue interpretation
   3.1. Evaluation of each alternative
   3.2. Selection of a "best" alternative
   3.3. Implementing the selected alternative

The managers of specific systems engineering functional efforts, as well as managers in general, perform each of the activities that we have just identified. These activities are appropriate at each phase in a systems management effort, as well as in such special cases as development of crisis management plans.

There will also generally exist the need for contingency or crisis management plans [32, 33], which will be implemented if the initially intended plans prove unworkable. A *management crisis* exists whenever there is an extensive and consequential difference between the results that an organization hopes to obtain from implementation of a strategic plan and what it actually does obtain. This difference may have a variety of causes, a common element in all is that the organization has, somehow, misjudged either (a) the environment in which it is operating or (b) the impacts of its chosen courses of action on the environment. A *crisis* may occur because of failure to identify a potentially challenging new opportunity, or it may occur as a result of an existing situation that is threatening to the health or survival of the organization. The preferred solution in either case is crisis avoidance. An acceptable, but somewhat less preferred, solution is extrication from crisis situation. The key to each of these "solutions" to a crisis is effective management of the organizational environment. Some activities may be performed at an intuitive level, based on experiential familiarity with particular task requirements. Some should be performed in a formal analytical manner because they are initially unstructured and unfamiliar.

In each of these activities, information is of critical importance. Information is needed about the external environment to facilitate understanding of that environment. The first step in issue formulation is identification of a set of information needs relative to systems management objectives. In parallel with this, preliminary identification of potential alternative courses of action is made, and these potential courses of action further act to frame the information needs for proper analysis and evaluation of these alternatives. All of these activities are accomplished as part of issue formulation. The issue analysis and issue interpretation steps are equally rich in terms of their need for information.

Systems management and management control are vitally concerned with the processing (broadly defined to include acquisition, representation, transmission, and use) of information in the organization. Generally, information is now recognized as a vital strategic resource and will be treated as such here. A simple three-step reasoning process leads to this conclusion.

1. Organizational success depends upon management quality.
2. Management quality depends upon decision quality.
3. Decision quality depends upon information quality and context, such that information becomes actionable knowledge.

It is, of course, necessary that information be interpreted as knowledge in order to lead to effective and actionable knowledge. This suggests a major role for context and experiential familiarity with tasks and the environments into which they are embedded such that information can be processed effectively as information. Earlier we noted that one of the major tasks of management planning and control is that of minimizing the ambiguity of the information that results from the organization's interaction with its external environment. This is accomplished in order to do the following:

1. To better enable the organization to understand its environment
2. To detect or identify problems in need of resolution
3. To diagnose the causes of these problems
4. To identify alternative courses of action or policies to correct or resolve problems
5. To analyze and evaluate the potential efficacy of these policies
6. To interpret these in accordance with the organizational culture and value system
7. To select an appropriate priority order for problem resolution
8. To select appropriate policies for implementation
9. To augment existing knowledge with the new knowledge obtained in this implementation such that organizational learning occurs

This task of minimizing information ambiguity is primarily that of systems management or management control. It is done subject to the constraints imposed by the strategic plan of the organization. In this way, the information presented to those responsible for task control is unequivocal. This suggests that the task control function receives information inputs primarily from those at the management control level. It suggests that there are planning and control activities at each of the three functional levels in an organization. The nature of these planning and control activities are different across these levels, however, as is the information and knowledge that flows into them.

Let us examine some specific implications of systems management for the production of trustworthy systems of hardware along with software, and associated interfaces that facilitate successful human interaction. Regardless of the hierarchical level at which planning is considered, a plan is a statement of what ought to be, together with a set of actions or controls that are designed to cause this to occur. Of course, the interpretation of "ought" may vary considerably as a function of the level at which planning is accomplished. Also, there may be a number of uncertainties that may act to prevent a normally useful set of activities from achieving the objectives that they should achieve.

In general, planning involves the following:

1. Identification of goals to be realized, some of which may already be fulfilled to some extent, at the particular level of planning under consideration

2. Identification of current position relative to the goals such that it becomes possible to specify a set of needs that, when fulfilled, will lead to goal realization

3. Identification of past, present, and future environments such that it becomes possible to understand effects of the constraints on, and alterables of, realistic courses of action

4. Identification of suitable alternative courses of action designed to lead to need fulfillment and goal attainment.

These planning elements are associated with an organization's internal environment, including its culture and standard operating policies, and the external environment. We also need to identify measures or metrics such that we can determine success in need satisfaction, goal attainment, and activity accomplishment. These should be linked together such that we identify and understand relationships among the elements of planning.

Implied in the identification of strategic planning options are the following:

1. An external environment analysis to identify the present context in which the issue being considered is embedded and to forecast possible future situations

2. An internal environment analysis at the level of the organization where planning is being conducted in order to determine available resources and to identify the organizational culture

We can disaggregate this still further. We can speak, for example, of a general environment, or management control environment, as those elements that affect all organizational activities within a specific domain: cultural, demographic, technological, and so forth. Also, we can speak of a task environment as those elements specifically affected by, and affecting, the particular organization and alternative course of action in question.

At this point in a planning effort, we have scoped out the issue considerably and identified a number of possible courses of action. Up to this point, we have accomplished formulation of the issue. The major planning ingredients needed for a complete and useful plan are (1) realistic objectives and (2) identification of a course of action together with suitable and observable activities measures. To achieve these, we need to analyze the options that have been generated to determine their impacts on needs. In dealing with a large and complex issue, a variety of systems analysis tools may need to be used. After this analysis, we need to obtain an interpretation, reflecting the value system of the clients.

Management control efforts are carried out with respect to both (a) programs that ultimately result in the delivery of operational systems and (b) the operational and task-related activities of the systems engineering organization itself. The interpretation activities of management control will include evaluation of these impacts and selection of an alternative course of action, with respect either to program deliverables or to adaptation of operational efforts within the corporation to improve performance.

In any large and complex effort, it will be necessary to break a program down into several project plans. A successful project plan must identify and detail the following:

1. What is to be done?
2. Who will do it?
3. With what resources?
4. In what time period?

The course of action element, what is to be done, must meet the needs of the client and must possess sufficient quality and functionality. Sufficient is a typically very subjective term that depends on the clients needs, priorities, and available resources to meet these.

There are many causes of *systems management failures*:

1. Difficulty of defining work in sufficient detail for the level of skills available
2. Problems with organizing and building the project team
3. Project staff is reassigned prior to project completion
4. Failure of clients to review or understand requirements and specifications
5. No firm agreement on program plan or project plans by management
6. Insufficiently defined project team organization
7. No adequate set of standards
8. No operational level quality assurance or configuration management plans

9. No clear role or responsibilities defined for project personnel
10. Project perceived as not important to individuals or organization
11. No risk management, or crisis management, provisions
12. Inability to measure true project performance
13. Poor communications between management and organization members
14. Poor communications with customer, or client, or sponsor
15. Difficulty in working across functional lines within the organization
16. Improper relations between program and project performance and reward systems
17. Poor program and project leadership
18. Lack of attention to early warning signals and feedback
19. Poor ability to manage conflict
20. Difficulties in assessing costs, benefits, and risks
21. Insensitivity to organizational cultures
22. Insufficient formal program and project guidelines at the level of procedures
23. Apathy or indifference by program or project teams or management
24. Little involvement of project personnel during program planning
25. Rush into project initiation before adequate definition of key tasks
26. Poor understanding of interorganizational interfaces
27. Poor understanding of intraorganizational interfaces
28. Weak assistance and help from upper management
29. Project leader not involved with team
30. Credibility problems with task leaders
31. No mutual trust among team leaders
32. Too much unresolved conflict
33. Unrealistic schedules and budgets
34. Power struggles at various levels in the organization
35. Lack of appropriate attention to strategic quality assurance and quality management
36. Too much reliance on established procedures, which turn out to be inappropriate for the task at hand
37. Lack of appropriate attention to knowledge management efforts and the resulting failure to transfer data to information to knowledge
38. Lack of concern for human dimensions of organizational performance
39. Implementation of reactive policies and with little or no concern for interactive and proactive approaches to everything
40. Concentration of attention on symptoms and their removal, and not upon institutional, infrastructure, and value-related issues
41. Failure to consider the emergent nature of organizations and issues.

Most of these will cause systems engineering projects to fail to finish their scheduled activities on time and within costs. Also, this list is incomplete! You should easily be able to add to it.

There is often disagreement within various groups concerning which of these factors are most important, and which is likely to be responsible for the failure of systems engineering projects. For example, top-level managers will generally indicate that *front-end planning*, including identification of system-level requirements, is very important. Program and project engineers will usually consider this to be less important. On the other hand, program and project engineers will likely perceive *technical complexities* of a program as very important and will wish to devote significant effort to understanding these. Program management will often consider this relatively less important than such factors as identification of requirements and establishment of requirements specifications. Both groups will generally identify customer changes in specifications during program completion as being a major factor in time slippage and cost overruns. On the other hand, clients will doubtlessly not perceive their requirements as changing. They will doubtlessly perceive that these were very poorly identified initially, and perhaps poorly translated to specifications which were then implemented in an error-prone manner. We believe that these differences in perceptions strongly support the use of prototyping techniques and the use of support systems for issue exploration and judgment, so as to enable full understanding of tasks to be undertaken as early as possible in the systems engineering lifecycle. Most importantly, they suggest the incorporation of risk management procedures throughout the life cycle, along with operational and strategic level quality assurance and management of both process and product.

6.2 SYSTEMS ENGINEERING ORGANIZATIONAL STRUCTURES

In general, there are three types of systems engineering organizations. The first and most often found is the *functional organization*. In the functional organization, or functional line organization, one particular group is asked to perform the entire set of activities associated with developing a systems engineering product or service. This does not mean that one person or group is necessarily asked to perform all phases of a systems development effort. Rather, it means that a team of people within a fixed structure are asked to do this. Each manager of a "section" in the organization is given a set of requirements to be met. Each manager is able to exercise more or less complete authority over the activities going on within that particular "section."

A representative functional organization structure is depicted in Figure 6.3. We note that this is a typical hierarchical structure in which management authority for a specific systems engineering program is vested top-level management. The various line supervisors are given tasks, and report to the top management office.

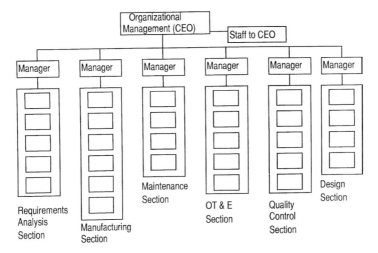

Figure 6.3   Functional line (hierarchical) systems engineering organization.

There are several advantages to this type of management structure:

1. The organization is already in existence prior to the start of a given systems engineering development program, and this enables quick start-up and phase-down of programs.
2. The recruiting, training, and retention of professional people is potentially easier, because these people generally remain with one, often relatively small, unit throughout much of their career. For this reason, functionally structured organizations are often very people-oriented.
3. Standards, metrics, operating procedures, and management authority are already established and have typically become a part of the organizational culture of the systems engineering unit in question.

Of course, there are a number of readily identified limitations also.

1. No one person has complete management responsibility or authority for the program, except at a very high management level. That person will often be in charge of several development programs
2. Interface problems within the organization are often difficult to identify and solve.
3. Often, specific programs are very difficult to monitor and control because there is no single management champion for the effort being undertaken.
4. Functions may tend to perpetuate themselves long after there is any real need for them.

The *program organization* is one possible remedy for this. In a program organization, or project organizations within a given program, one person is vested with overall responsibility to ensure that one specific program is accomplished in a timely and trustworthy manner. In a program management organization, people are usually excised from an existing functional organization and wedded together under the management of the program, or project, manager.

Figure 6.4 illustrates some of the features of this superposition of program authority on a line organization. Figure 6.5 illustrates the resulting program organization, which may and generally will include a number of projects within a given program that, taken together, comprise the program. There are several advantages to such a program or project structure:

1. There is one person in a central position of responsibility and authority for the program, the program manager.

2. That person, or a designee, will generally have authority over all system-level interfaces, especially those that cut across the phases of the development effort and projects within those phases.

3. Decisions can generally be made very quickly at the program level because of the new centralized organization, shown in Figure 6.5, which puts the program at a central management level in the organization.

The limitations to program and project organizations are related to their strengths.

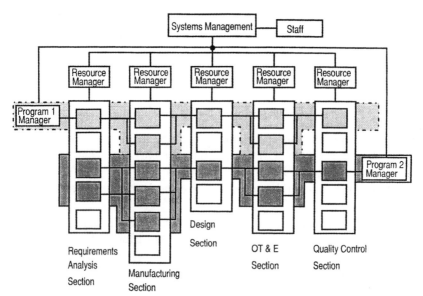

Figure 6.4 Systems engineering matrix project management (obtaining staff from resource managers).

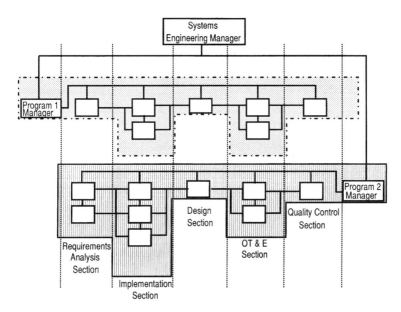

Figure 6.5 Project systems management (after obtaining staff from resource managers).

1. The project organization must be formed from the existing line organization.

2. Recruiting, training, and retention of people to work in a program office is more difficult than in a functional organization, because the resulting programs and the projects within a program are product-oriented rather than people-oriented.

3. The benefits of economy of scale cannot be achieved for other than very large programs, because there will often be only one or two people in a given technical specialty area associated with a program or project.

4. Programs tend to perpetuate themselves, just as a given functional line will tend to perpetuate itself in a strictly functional organization.

5. Often, it will be necessary to develop standards, metrics, techniques, and procedures for each program undertaken. Often, these will not be the same standards across all programs. This will make cost–benefit and other comparisons of efforts across programs very difficult.

The matrix program, (and project) organization has been proposed as a way to, ideally, combine the strengths and minimize the weaknesses of the functional organization and the program organization. Figure 6.6 illustrates how people in a *matrix organization* are managed both by functional line-type supervisors, often called resource managers, and by program or project

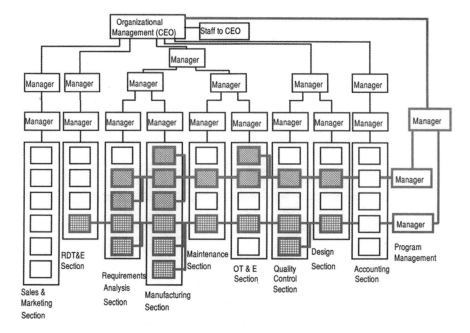

Figure 6.6   Third-generation management: matrix management introduces horizontal management of programs.

managers. In this management structure, any given person in the functional line may be working on more than one program or project at a given time.

In this sort of management structure, the program and project managers have responsibility for short-term supervision of the people working for them on various programs. The resource manager would be responsible for longer-term management of these same people. As with the other two management types, there are advantages and limitations to this form of management structure. Some of the advantages are as follows.

1. There is an improved central position of responsibility and authority over the program being undertaken, as contrasted with that in line, or functional, management.

2. The interfaces between the various specialty functions can be controlled more easily than in line or functional management programs.

3. It is usually easier to start and terminate a program than in program management organizations.

4. Standard operating policies, technical standards, metrics, and procedures are generally already established, unlike in program organization, because these are the responsibility of the typically longer-term functional line management.

5. Professional staffing, recruiting, education, and training are easier, and retention of the best staff members is higher than is typically the case in program management organizations.

6. It is potentially possible to obtain more efficient and more effective use of people, because the structure is more flexible than is possible in either functional or program management.

Potential problems with matrix management are as follows:

1. Responsibility and authority for human resources is shared between the line resource manager and one or more program or project managers.

2. It is sometimes too easy to move people from one project to another, especially compared with the project management organization. This may lead to personnel instability and other morale problems.

3. Because of its complexity, greater organizational understanding and cooperation are required than in either program or functional organizations.

4. There is often greater internal competition for resources than in either the systems program management or functional organization.

There have been a number of proposed extensions to these basic management structures. In an inspiring work, Savage [34, 35] illustrates five recognitional stages, denoted as "days," in the life of many contemporary organizations. These findings may be described as follows:

1. The organization is organized into a set of hierarchically related personnel and applications. These carry such titles as R&D, engineering, manufacturing, sales, service, and accounting. They report in a traditional functional line structure, and the various functional units do not interact.

2. In order to cope with the need for interaction, various application groups are set up. This creates a necessary linkage, or network, between one functional unit in the organization and the others that are needed for a particular application, such as product development or product marketing. Because the people in the various functional units cannot communicate well, or even at all, with one another because they speak different languages than those in other functional units, a "translator" or "expediter" is needed. Because there are many applications in a given organization, the network linkages become numerous, as does the need for expediters and translators.

3. The difficulties of working in parallel across functional units become apparent, and ways are sought to cope with the resulting complexity. Someone suggests having customer expectations as the thematic drivers of considerations that relate to such nonfunctional efforts as process, quality, market, and service.

4. Concerns arise with respect to how the various cross-functional teams are to be managed. Organizational vision is suggested as the monitor and controller of the cross-functional teams through the resultant strategic plans, organizational mission statements and objectives, and realistic management controls. Knowledge is recognized as a valuable resource in this regard in terms of various "knows." This knowledge is responsive to the same sort of questions used in benchmarking, except that it relates to a common knowledge base and capability for describing the various elements needed for each of the applications for which a cross-functional team is responsible. Thus, a knowledge base is needed in terms of

- Know why
- Know what
- Know which
- Know who
- Know where
- Know when
- Know how

This represents the organizational knowledge base. It is what the various cross-functional teams bring to bear on various applications, such as product development. It suggests a role for the original departments as "centers of excellence" or repositories for critical core capabilities, or "virtual resources," but not as actual working line units.

5. In the last stage of development, the potential fragmentation of the organization due to the cross-functional teams is dealt with in terms of strategies for accountability, focus, and coordination. This leads to strategies for integrating the organization through human networking in such a way as to build a continual learning capability. These networks are not just informal networks of humans communicating with one another. This capacity is augmented by networks of information processing systems that enable interrelating various knowledge patterns for enhanced capability and competitiveness.

The strategies for human networking and enterprise integration result from the reality that the traditional resources for production—land, labor, and capital—are now augmented by an information and knowledge resource.

It is primarily this that has led to the major need to replace the traditional steep hierarchical structure found in most organizations by cross-functional teams and human networking so as to enable people empowerment and enterprise integration for enhanced responsiveness and competitiveness. The need is *not* to computerize steep hierarchies. Figures 6.3 through 6.6 indicate the evolution from second- to third-generation management. First generation management involves only land and labor capital is added in second generation management. In second-generation management, steep hierarchies are introduced and we show just a simplified representation of one in Figure 6.3.

Matrix management, or third- generation management, accomplishes the change in organizational structure to enable horizontal communications through adding additional management complexity. In a sense, this is accomplished in fourth-generation management without the additional management complexity through networking the organization. This involves major use of information technology and the need for integrating information technology products and services into organizational environments.

In fourth-generation management, we have extensive networking across the steep hierarchy brought about by second- and third-generation management. Networking is suggested as a fourth-generation management remedy for the dilemmas brought about by steep hierarchies and vertical management. In fourth-generation management, horizontal and vertical communication linkages are established for what was initially the second-generation model of management structural organization. This is not really enterprise integration in that the organization is not truly integrated, except in a narrow technological sense by the wires and software that one person uses in interfacing with another. It is really just computerizing a steep hierarchy, as in Figure 6.7 with perhaps 10 to 20 layers of management. Savage identifies six issues that generally emerge from fourth-generation management:

1. Ownership of information issues, as information becomes "turf" in a steep functional hierarchy

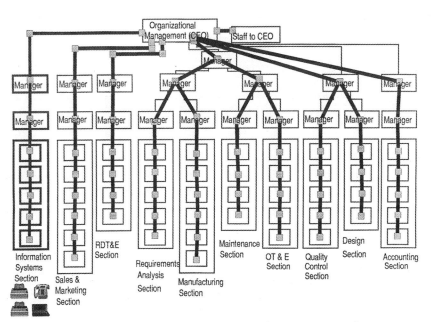

Figure 6.7 Fourth-generation management: One information network structure imposed on steep hierarchy.

2. Managed and massaged information system issues, as various functional units present selective information that best supports their unit

3. Hidden assumptions embedded in various software representations of levels of abstractions and associated information presentations

4. Inconsistent terms and definitions across different applications, which are due to lack of organizational standardization of information architectures and dictionaries

5. Accountability and social value of information

6. Organizational information politics.

Some of these are not inherent issues in information technology and networking, but are aided and abetted by retention of the steep hierarchy form of management. Second-, third-, and fourth-generation management each involve steep hierarchies. Third-generation management attempts to improve on horizontal communication needs through a matrix management structure, but this superimposes additional management layers and brings about the difficulty of one worker reporting to more than one supervisor. There are communication difficulties as well. Fourth-generation management attempts the needed horizontal communication through technology but does not fully ameliorate the other difficulties we have just cited.

Savage presents a number of illustrations in an attempt to show that the computerization of steep hierarchies, or fourth-generation management, really will not work. He presents five needs, in the terms of a set of interrelated conceptual principles, that form the nexus of early fifth-generation management and which will enable the desired transformation.

1. *Peer-to-peer networking* is a major need. This involves three major ingredients: technologies, information, and people. Peer-to-peer networking enables communication from any individual in the organization to any other individual without the necessity of having to go through the conventional steep hierarchical structure. It allows people to work together in a cross-functional manner. Far from eliminating hierarchies, this results in a redefinition of the role and function of the hierarchy and a resulting hierarchy that is much flatter than before networking. While there are major hardware and software difficulties in bringing this about, the human and organizational issues are larger and more complex.

2. *Work as dialogue* is another important need. This involves listening, visioning, remembering, and using knowledge relative to both process and product.

3. The *human time and timing* need is concerned with developing an understanding of past, present, and future patterns such that it becomes possible to see and anticipate future patterns on the basis of experiences and knowledge.

4. An *integrative process* across people, technologies, and the organization to allow for continuous change and teamwork in the organization.

5. *Virtual task-focusing teams* are the final major need and result only from satisfaction of the first four needs.

On the basis of these principles, he suggests ten pragmatic organizational considerations for enabling fifth-generation management

1. Develop a technical networking infrastructure that is flexible and adaptable to organizational needs and continual change.

2. Develop a data integration strategy.

3. Develop functional centers of excellence.

4. Develop and expand the organizational knowledge base.

5. Develop organizational learning, unlearning, and relearning capabilities that are continuously updated and rejuvenated.

6. Develop visioning capacities so that the context for judgments and decisions is visible to all through knowledge of strategic plans and organizational objectives, mission statements, and values.

7. Develop behavior norms, a sense of values, a reward structure, and measurements that support task-focusing teams.

8. Develop the organizational ability to identify, support, and manage multiple functional-task teams.

9. Develop the organizational capacity and capability to support the teamwork of teams.

10. Develop virtual task-focusing teams that are formed of suppliers, customers, and appropriate people from within the organization.

The first two of these are primarily technology-based needs, although there is much need for human interaction with the technologies to be implemented. The last needs have very much to do with people. Those in the middle represent organizational needs. There is not sharp cleavage between technology, organization, and people needs, and these are the major ingredients in any mature systems engineering effort.

We have just begun to scratch the surface of a very important subject, the engineering or reengineering of organizations for enhanced responsiveness. This is an important contemporary subject and there are many new and exciting results being obtained. Among the many currently suggested results are integrated product and process development teams, cross-functional teams, total quality management, benchmarking, and other efforts that are generally associated with reengineering of organizational processes and at the level of systems management. Needless to say, this is an important advanced area for contemporary study and research.

6.3 PRAGMATICS OF SYSTEMS MANAGEMENT

The function of systems control or systems management is needed regardless of the program and project structure that is actually implemented. There are many *responsibilities of top-level systems management*. These include the following:

1. Coordination of issue identification with the client, along with translation of user needs into system specifications
2. Identification of the resources required for trustworthy development of an operational system of hardware and software, as well as appropriate interfaces with human operators of systems and with existing systems
3. Definition and coordination of software and hardware identification, design, integration, and implementation
4. Interfaces with top-level management of both the client and system developer

There will be a number of *front-end problems* facing the typical systems engineering program management team. Generally the most difficult of these involve the following:

- Human communications and interactions
- System requirements that change over time
- Perceptions of systems requirement that change over time when the requirements did not change
- Sociopolitical problems
- The lack of truly useful automated planning and management tools

A *program plan* requires written documentation including sufficient details to indicate that the plan is thoroughly developed. As a minimum, this must include evidence that the following are present:

1. There exists an understanding of the problem.
2. There exists an understanding of the proposed solution.
3. The program is feasible from all perspectives.
4. Each of the associated projects benefits the program.
5. The program and project risks are tolerable.
6. There exists an understanding of project integration needs.
7. The overall effort is cost-effective.
8. The overall effort is conducted from quality assurance and management perspective at both strategic and operational levels in order to ensure processes and products that are trustworthy.

One of the major needs in systems engineering program (and project) management is management planning and control, along with associated monitoring. It is necessary to monitor progress of the program and projects associated with the program, along with their associated controls, in order to know how well the effort, at any specified instant of time, is proceeding according to the schedule and budget that have been set for it. We will attempt change, if and as needed, to ensure that the actual program schedule and performance is as close as possible, or needed, to that planned. These are general systems management needs. They are independent of the structure of the systems engineering organization.

Figure 6.8 illustrates some of the necessary feedback and iteration in the monitoring and controlling of a program that follows from startup of a systems engineering life cycle for fielding or acquisition of a product or service. This assumes that systems planning has been accomplished and that, as a result of the planning effort, an appropriate program and project structure has been obtained. An appropriate sequence of steps is as follows.

1. Monitor progress of the program and projects.

2. Compare actual progress with that contained in the plan.

3. Monitor the quality of the process and the evolving systems engineering product.

4. Revise the plan as needed.

5. Define and utilize operational metrics for evaluation of evolving product quality in terms of the following:

   5.1. Audits—formal examination, generally by an external team, of project management and development

   5.2. Reviews—formal examination, generally by internal project management, of plan and project documents people, and products

   5.3. Inspections—formal examination, generally by an internal peer group, of deliverable parts of the projects

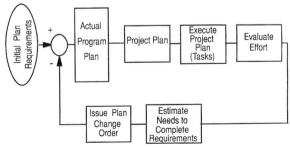

Figure 6.8 Monitoring and control as a feedback and iterative process to aid in systems management.

6. Establish procedures for the following:
   6.1. Understanding of required functions
   6.2. Identification of critical milestones
   6.3. Identification of task responsibility
   6.4. Crisis management

If monitoring and controlling are not performed effectively, it is very likely that both individual project and overall program progress will suffer, and this will not be noticed until it is too late to effectively and efficiently deal with this slippage. We will require an appropriate strategic level approach to process quality assurance and management in order to do this effectively.

Another important need is *organizing and scheduling* the technical details to be performed in a program, along with its constituent projects. The first step in this is to identify the key tasks to be accomplished in terms of the following:

1. The system itself that is under development
2. Documentation for the system
3. Hardware and software tests, reviews, and evaluations
4. Interface and systems integration considerations
5. Configuration management functions
6. System installation and training requirements.

The principal task here is to identify schedules for people and resources such that there will be no unpleasant surprises as the systems engineering effort progresses. Essential in this is the need to communicate these key tasks and requirements to all concerned.

It may appear by now that the critical success factors affecting a systems acquisition or fielding effort, as well as its management and technical direction, need to be identified and resolved at the very initial phases of the systems engineering life cycle. This is, indeed, correct. Effective systems management not only demands the generation of plans but also requires communicating them. There are many reasons for preparation of written plans, which should be accomplished at both the program and project levels. Generally, a good way to do this is through the *systems engineering management plan* and a *configuration management plan*, which may be a part of the SEMP. Table 6.1 presents the generic components of a typical systems engineering management plan.

Specific goals for a systems engineering management plan, and the written documentation concerning it, include sufficient details to indicate that the plan is satisfactory in the sense of producing a cost effective and trustworthy system. As a minimum, this must include evidence that the following are present.

TABLE 6.1   Table of Contents of Typical Systems Engineering Management Plan

Title Page
Preface
Table of Contents
List of Figures
List of Tables
1. Executive summary
2. Introduction
   2.1. Program Objectives
   2.2. Program Description
   2.3. Program Scope
   2.4. Program Overview
   2.5. Program Organization and Responsibilities
   2.6. Program Deliverables
   2.7. The Systems Engineering Management Plan Components
   2.8. Program Milestones
3. Program Management Approach
   3.1. Systems Management Assumptions, Needs, Constraints
   3.2. Systems Management Objectives
   3.3. Program Risk Management
   3.4. Program Staffing
   3.5. Program Monitoring and Control
   3.6. Program Integration Approach
4. Program Technology
   4.1. Technical Description of System Development Projects
   4.2. Project Methods, Tools, and Techniques
   4.3. Project Procedures and Support Functions
5. Program Quality Assurance
   5.1. Quality Assurance Plan
   5.2. Maintenance Plan
   5.3. Documentation Plan
   5.4. Operational Deployment Plan
   5.5. Configuration Management Plan
      5.5.1. CM Objectives and Overview
      5.5.2. CM Organization, Charter, Members, Duties
      5.5.3. CM Methods
         5.5.3.1. Baselines
         5.5.3.2. Configuration Identification
         5.5.3.3. Configuration Control
         5.5.3.4. Configuration Auditing
         5.5.3.5. Configuration Status Accounting
      5.5.4. CM Implementation Plan
         5.5.4.1. Procedures
         5.5.4.2. Personnel
         5.5.4.3. Budget
         5.4.4.4. Implementation milestones
   5.6. System Security Plan

Continued

TABLE 6.1 Table of Contents of Typical Systems Engineering Management Plan (*Continued*)

6. Budget and Resource Management Plan
   6.1. Work Breakdown Structure
   6.2. Cost Breakdown Structure
   6.3. Resource Requirements
   6.4. Budget and Resource Allocation
   6.5. Program Performance Schedule
Additional Contents
Reference Materials
Definitions and Acronyms
Index
Appendex
Appendices of Supporting Material

1. There exists an understanding of the problem.

2. There exists an understanding of the proposed solution.

3. The system acquisition program is feasible from all perspectives: including costs, effectiveness, and timeliness

4. The projects within the program each benefit the program

5. The system acquisition risks are tolerable and explicable

6. There exists an understanding of system integration and maintenance needs

7. The fielded system is a quality system

If the needed quality assurance cannot be provided, system development should probably not proceed.

One of the major needs in systems management is monitoring and control of programs and projects. It is necessary to monitor progress of the program and projects associated with the program in order to know if the effort, at any specified instant of time, is proceeding according to schedule. If not, either because of a change in the requirements or for other reasons, we can take steps to get back on schedule. Thus, systems management and change management are each inseparable parts of a systems engineering management plan. Figure 6.8 illustrated the necessary feedback and iteration in monitoring and controlling.

Staffing is a very important need for systems engineering success at all levels and phases of effort. Organizations are no better than the people who belong to them. This is as true of systems engineering organizations as it is of any other type. One fundamental and often overlooked notion in productive staffing is what we choose to call the *staffing quality principle*. In its simplest form, this states that six people who can each jump one foot does not equal one person who can jump six feet. Few would disagree with this notion as stated. Yet, it is often very hard to apply this principle to programs where a few really good people are really needed for program success, and a larger number with lesser talents would not be especially useful.

6.4 SYSTEMS ENGINEERING METHODS FOR SYSTEMS ENGINEERING MANAGEMENT

There are a relatively large number of systems engineering tools and methods that are appropriate for systems management efforts. In this section we will briefly describe a few of these. Our emphasis will be on those approaches that are most suitable for identification and communication of requirements, risk, and quality management efforts.

6.4.1 Network-Based Systems Planning and Management Methods

Network-based systems planning and management methods were developed in the late 1950s in response to the need for a systematic tool for planning and management of large-scale projects. For the most part, these methods enable project planning, scheduling, and controlling. Project completion schedules generally incorporate the order of activities, time constraints, and resource availability constraints. The most widely known of these approaches are the Program Evaluation and Review Technique (PERT) and the Critical Path Method (CPM). Details concerning these approaches are discussed in a rather large number of texts including some initial treatments of the subject [36–38] and some more modern works [39–43]. We will describe salient features of these approaches here. These, and both other methods that have been developed since, are based on three assumptions.

1. A large project can be disaggregated into a number of separate activities, also called tasks, or jobs.
2. There is a particular sequence in which the tasks must be accomplished.
3. Time for completion, or duration can be estimated for each activity.

Figure 6.9 indicates a typical listing of activities, precedence or sequence relations, and durations for a hypothetical project involving concurrent development of hardware and software. The three assumptions are fulfilled for this hypothetical project. One of the very first efforts in use of network management procedures is to establish these precedence relations.

In a network management model, the identified precedence relations between various activities are displayed graphically in the form of a network. Often, this is called arrow diagramming. An arrow diagram displays precedence relations among the activities that comprise the network. The nodes in the network may represent various events, such as completion of an activity.

Activity	Duration	Predecessor	Successor
a. Requirements	3	Begin	b
b. Specifications	3	a	c,h
c. Hardware Design	2	b	d
d. Breadboard Construction	3	c	e
e. Test Breadboard	2	d	f
f. Design Prototype Hardware	2	e	g
g. Test Prototype Hardware	3	f	k
h. Design Software	4	b	h
i. Code Software	5	h	j
j. Test Software	5	i	k
k. Integrate Sware & Hware	3	g, j	l
l. Operational Evaluation	3	k	m
m. Deploy System	3	l	n
n. Document System	2	m	o
o. Train System Users	3	n	p
p. Maintain System	6	o	End

Activity Duration and Precedence for System Fielding

Figure 6.9   Phased activities for system fielding effort.

The activities themselves would then be illustrated by the branches between nodes. Sometimes, it is necessary to insert dummy activities, or dummy branches, in order to avoid possible ambiguities concerning precedence. This is known as the activity on arrow representation.

An alternate approach to the activity-on-arrow convention is known as the activity on node convention and is generally preferred. In it, the nodes represent activities and the arrows represent events or the time points that follow activity completion. Figure 6.10 illustrates simple sequences of activities and events for these two conventions. While the diagrams of Figure 6.10(a) and 6.10(b) are basically the same, the activity-on-arrow convention will require the insertion of a dummy node in order that no more than a single arrow be needed to represent a single activity. A dummy node is also needed in the representation of Figure 6.10(d) in order to properly represent precedence orderings. In Figure 6.10(c), completion of activities *a* and *b* is sufficient to enable the start of activity *d*, while completion of activities *a*, *b*, and *c* is needed to enable the initiation of activity *e*. This is the activity-on-node representation and does not require the insertion of a dummy node. Such a dummy node is needed in the equivalent activity on arrow representation. Dummy nodes are not needed in the activity-on-node representation, but may be used for convenience or clarity in display of a complicated sequence of activities. We will use the activity-on-node representation in our efforts to follow.

We use nodes to represent activities or jobs. Directed lines connecting two nodes are used to represent the time events that occur upon completion of an activity. Several definitions are of interest for the activity on node representation. Figures 6.10 and 6.11 illustrate some of the following definitions and

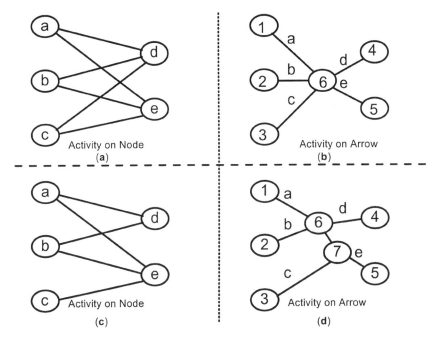

Figure 6.10 Comparison of "activity on node" and "activity on arrow" representations.

relations:

1. An *activity* represents some defined sequence of jobs or tasks that is necessary to complete a project. For convenience, we will generally use lowercase letters to represent activities.

2. A *predecessor activity* is an activity that must immediately precede an activity being considered. In Figure 6.11(a), activity *a* is a predecessor to *b*, and activity *b* is a predecessor to *d*. While activity *a* surely must come before activity *d* in this figure, activity *a* is not called a predecessor of activity *d*. In directed graph terminology, we are asking for a description of a minimum-edge adjacency matrix among the activities. Sometimes this may be difficult to initially establish. A reachability matrix may be established first, and the minimum-edge adjacency matrix may be obtained from this.

3. A *successor activity* is an activity that immediately follows the activity that is being considered. Activities *b* and *c* are successor activities to activity *a* in Figure 6.11(a), and activity *d* is a successor to activity *b*.

4. An *antecedent activity* is any activity that must precede another activity under consideration. Activities a and b are antecedent activities of *d* in

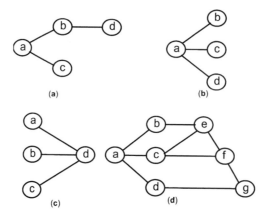

Figure 6.11   Fundamental concepts in "activity on node" representation.

Figure 6.11(a). Activity c is not an antecedent of activity d, and activity a has no antecedents.

5. A *descendent activity* is one that must follow an activity being considered. In Figure 6.11(a), activities b, c, and d are all descendants of activity a.

6. A *burst point*, or burst, is a node with two or more successor activities. Figure 6.11(b) shows a burst point at node a.

7. A *merge point*, or merge node, is a node where two or more activities are predecessors to a single activity. Figure 6.11(c) illustrates a merge node.

We may use these node and arrow symbols, together with the information contained in the activity analysis, to enumerate paths for the network. In general, approaches to construction of a network planning model involve three efforts:

1. Activity analysis

2. Arrow diagramming

3. Path enumeration

which comprise the network planning phase. Following this, a scheduling phase of effort is initiated. In this, the particular sequence of activities that takes the longest time to complete is identified. This is called the critical path. It is important to note that "critical" is used here in a very special and restricted sense. It refers to time only. Figure 6.12 illustrates the diagram that results from use of the activity analysis indicated in Figure 6.9 and the arrow diagramming conventions described by Figure 6.11. We will call this a PERT (program evaluation and review technique) diagram.

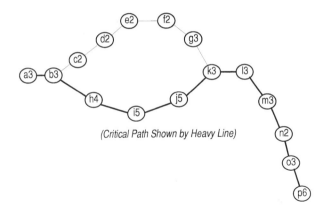

(Critical Path Shown by Heavy Line)

Figure 6.12 Activity on node representation of the system life-cycle schedule.

The time length of the critical path determines the project duration. The critical path is indicated by the thicker branches in Figure 6.12. Any change in an activity on the critical path will affect project duration. Thus, this method focuses attention on "critical" activities. An approach such as this can also be used to estimate project time to completion and can be used to schedule jobs under limited money and human resource availability, as well as to investigate time–cost tradeoffs. This approach is very useful for operational level management and task control of a project. These network planning methods have been used in the preparation and management of large projects in many application areas, many of which are described in the references provided in this section.

In the aerospace and defense industries, network management approaches are usually called by the name PERT. In civil engineering systems areas, such as transportation and construction, the term CPM is often used. Some related names are *precedence diagramming method* (PDM) and *line of balance* (LOB). *Precedence networking* (PN) allows other precedence relationships than "Job B cannot start before job A has been completed," such as "Job B can only start *n* time units after job A has started," and "job B cannot finish until *n* time units after job A has finished." Other methods allow for the inclusion of decision points, such as decision CPM, or network probabilistic elements, such as the *graphic evaluation and review technique* (GERT).

CPM and PERT may be viewed as variations of one another. CPM analysis typically includes forward-pass and backward-pass computations. The forward-pass computation uses a given starting point and determines the earliest completion time for the project. This is obtained by determining the shortest time to arrive at each node and then summing these. The backward pass is the reverse. It starts with a given completion date and then determines the last allowable starting time. The computation starts with time required to reach the final node from its predecessor and then works backward, node by node. A potential use for the forward and backward pass is in determining slack time associated with the noncritical paths. This provides a basis for possible tradeoffs among time and effort.

There are some extensions to this basic description of CPM, such as (a) allowing for crash schedules to complete the project by a specified time and (b) separating *total slack* time into specific sequences of slack time.

PERT differs from CPM primarily in that it permits varying estimates of time required to complete each activity. Thus, for example, if optimistic, pessimistic, and likely activity completion times are given (t_0, t_p, t_l), we compute the expected time and variance for each activity from the relations $t_e = (t_o + 4t_l + t_p)/6$ and $\sigma^2 = [(t_p - t_o)/6]^2$. The earliest overall expected time for completion of the project, T_e, is then the sum of all the incremental expected times, t_e, for each activity along the critical path. In those cases where several activities lead to the same event, the largest expected sum of activity completion values is used. It is now possible to calculate a forward and backward pass, as with CPM. The slack time is the difference between the longest period permitted for activity completion and the expected time, or $T_S = T_L - T_E$. These values can be computed, as well as the probability of meeting any completion time requirement.

All of these computations are based on some assumptions that are subject to a number of questions. First, it is difficult to provide the three estimates of project completion: optimistic, pessimistic, and likely. There is no truly appropriate way to determine the appropriate statistical distribution for activity duration, even though the classical PERT approach uses, or imposes, a special statistical distribution called a beta distribution. That the longest time path is the most critical one may even be subject to question. What do we do with a longest path of 100 months and a variance of 1 month, as contrasted with a second longest path of 50 months and a variance of 100 months? Despite these potential concerns, the PERT approach is very useful and very often used. As with use of any tool, we should be aware of the associated risks and limitations.

The typical products of this network planning and management approaches are as follows:

1. Forecasts of project duration
2. Forecasts of project resource requirements
3. Identification of activities that are critical to project duration and, as a result, those activities that deserve the greatest systems management focus
4. An activity schedule that will not violate time, resource, or other constraints
5. A framework for cost-accounting
6. A basis for monitoring and control of progress in a large project
7. Estimates of the minimum additional costs of change in overall project specifications
8. Increased insight into various critical milestones, as well as identifica-

tion of possible inconsistencies or conflicts that require additional configuration management efforts

9. Improved communication among those responsible for the project
10. Improved documentation of the project
11. Provision of a basis for configuration management and management control

Thus, network-based methods are particularly useful for the operational and task level controls that follow from systems management. They can be particularly useful for configuration and project management efforts.

The outputs of this approach include graphic displays and network diagrams that show sequential relations among activities. This graphic display of the critical path normally identifies the earliest and latest possible starting times for each activity. The use of a network for planning and management of a project normally accomplishes the following.

1. *Formulation, Including Identification of Goals and the Substance of the Project to be Undertaken.* It is essential that a clear definition of the beginning and end of the project, along with its objectives, be stated.

2. *Identification of the Specific Activities to be Included in the Project.* As one task in configuration management, the project should be disaggregated into packages of tasks that require similar resources. These tasks should be performed by the same group of people, and should succeed each other directly, and only on completion of a single precedent package of activities.

3. *Identification of Precedence Relations.* A list is made of the immediate predecessors for each activity. The set of precedence relations is often called the project logic, which reflects the fact that precedence requirements are almost always based on logical requirements.

4. *Construction of the Network Diagram.* In activity-on-node networks, each activity is represented by a node, and the start of the next activity is denoted by an arrow leading to the successor activity. In activity-on-arrows networks, each activity is represented by an arrow connecting two nodes that represent events. Generally, the events denote the completion of an activity. In activity-on-arrows networks, the event at the tail of the arrow must have occurred before the activity can start, while the event at its head can only occur after completion of the activity. Arrows generally point from the left to the right, by convention, thereby indicating one representation of the direction in which time proceeds. Diagrams may be drawn manually, but computer software packages have been developed to aid in handling large projects. Typical software packages include Super Project, ARTEMIS PROJECT, Harvard Total Project Manager, and Microsoft Project. Many integrated operations research software packages, such as STORM, contain a project management module. Generally,

the software is acceptably easy to use and quickly reveals such inconsistencies in approach as improper precedence relations.

5. *Identification of Time Requirements for Each Activity.* Project personnel or others familiar with the project provide estimates of the time required for completion of each activity. These may include only the most likely, or normal, or the most cost-effective time estimates. In the case of great uncertainty, it may be desirable to provide three estimates: a pessimistic; an optimistic, and a most likely estimate. Often, the estimates are indicated in graphical representations.

6. *Identification of the Critical Path(s) and Critical Activities.* The critical path is that sequence of activities in the project that takes the longest time to complete. The activities on the critical path are easily determined. We begin this computation at the beginning of the network, where for convenience we set time = 0. An earliest possible start time is computed for each succeeding activity. This leads to determination of an earliest possible completion time for the project, and this time is equal to the length of the critical path. Then, a latest possible start time is assigned to each of the activities, working back from the earliest possible completion time. The latest possible start time is the latest time the activity can be started without delaying the completion of the entire project. The critical activities are those activities for which the earliest and the latest possible start times are equal. The difference between earliest and latest start times for the other, noncritical activities indicates the amount of float or slack associated with them. This slack represents the time span by which they may be shifted forward or backward without affecting project duration. There may be more than one critical path in a given network. Activities on the critical path deserve most management attention from this perspective of project completion time management. This is the case because delays in their progress will directly affect the project completion time.

7. *Use of the Project Network and Associated Critical Path as a Basis for Planning.* It is possible to make several uses of the just-computed critical path.

   a. *Estimation of uncertainty in forecasted project duration.* Uncertainty estimates for each of the critical path activities indicates where statistical methods should be employed to help estimate overall project duration, and its standard deviation. When changes in the length of the critical path occur, other activities not on the initial critical path may become critical.

   b. *Scheduling of jobs within an activity.* This can be done directly from the network, given some criterion to schedule jobs with float or slack time. The often requested completion time is *as soon as possible* (ASAP). Float occurs whenever jobs on some paths through the network may be delayed without affecting the time to completion on the critical path. The result can be represented graphically as a Gantt or bar chart, as we will discuss in our next subsection.

c. *Activity and job scheduling under resource constraints.* Activities, and the jobs that comprise them, that can be executed at the same time will often require the similar resources in labor and equipment. The float in activities and jobs can be used to reschedule them such as to minimize some of the peak demands for these resources. Various scheduling problems can be solved. We can, for example, find the schedule that does not exceed specific peak resource requirements while keeping the total project duration at some minimum value. We can find the schedule that equalizes resource requirements throughout the entire project duration. Mathematical programming algorithms have been developed to solve problems such as these. Most of these problems can and have been formulated and solved using linear, integer, and mixed-integer programming approaches.

d. *Time–cost minimization.* In general, the time duration of activities can be shortened, for a price. Obviously, shortening of tasks only along the critical path can reduce overall project duration and potentially justify the increased costs. If the same quality level is maintained, it would be rare that a project completion schedule could be compressed without increasing completion costs. When such critical jobs are compressed, others may become critical and themselves require shortening. The network can be used as a basis to compute the additional costs of reducing total project time. If costs for each job increase linearly with compression, the schedule giving minimum additional costs for a specified overall compression can be found using linear programming methods. The results can be used for planning or bidding purposes, or to help compute cost–benefit trade-offs.

e. *Cost accounting and budgeting.* The distinction of separate activities provides a basis for cost accounting through such efforts as the cost breakdown structure (CBS) and work breakdown structure (WBS) efforts that we will soon discuss. A network-based management schedule is a good starting point for planning project costs.

8. *Use of the Network as a Basis for Project Management and Control.* Actual and planned project progress, with respect to both schedule and expenditures, can be easily compared. Causes of schedule delays and/or cost overruns can be diagnosed directly. As long as its logic remains unchanged, an existing project network can also be used to reschedule activities and event completion times to reflect changing circumstances. This will enable the project manager to identify changes in criticality of activities and jobs, to resolve emerging resource competitions, and to find out which activities need to be modified such as to meet impending deadlines.

There are a number of conditions in a project which would normally call for a network approach to management. The most obvious is large size, in which case it would be very helpful to disaggregate the project into a set of predictable, well-defined independent activities with certain precedence relationships. Network management approaches provide a systematic approach that focuses on the overall project configuration effort, while taking relevant and separate project activities into account. This approach provides a unifying framework for planning, scheduling, budgeting, and controlling a project. Network planning methods can also provide a very useful way to examine feasibility of alternative systems engineering proposals, particularly with regard to their costs and schedules.

There are a number of useful byproducts that result from use of network planning models. The structure of the project and the interdependencies of activities are generally clarified, as are the critical activities. There are some caveats, however. The validity of time and cost projections depends on the validity of estimates used for the individual tasks. These estimates may be very uncertain. Estimates provided by those responsible for execution of tasks may be severely biased. It may be difficult to clearly distinguish and assign precedence to all activities before start of a project. The network planning method is usually not appropriate when iterations or cycles occur in a project. In this sense, the presence of cycles in a project management network is beneficial. It allows us to detect severe problems that will exist if the project is fielded as configured. It may not always be possible to start or stop activities independently, and activity times may not be independent of each other. Even if they are independent, it is generally very difficult, in not impossible, to identify all activities prior to the start of a project. Finally, network planning methods do not replace human management. As with other systems engineering methods and tools, they can only provide a tool to support human skills.

6.4.2 Bar Charts

A bar chart, sometimes called a Gantt chart [44] after one of the pioneers in the very early days of "scientific management," is a simple visual-chart-like aid for project planning and management. It was first developed by Henry L. Gantt in the beginning of the twentieth century. Gantt (1861–1919) was a disciple of Frederick Taylor and created a class of charts in which progress is plotted against time. This effort apparently had a major impact on ammunition production and delivery efforts during World War I. The charts are used to display a schedule of activities as they evolve over time, as well as to compare actual and planned progress in those activities. Traditionally, a time scale is listed along the horizontal axis, while activities are listed along the vertical axis. Horizontal lines or bars show the times during which each activity is planned to be carried out. The length of such a line is a measure of the time required for the activity, and can also indicate a quantity of work. Occasionally, milestone dates are illustrated on the vertical axis. These are often indicated by a triangle-like symbol.

Gantt charts are most useful for synthesis and analysis of plans in the project planning and development phases of a systems effort. The Gantt chart is a prime component of the planning for action and implementation step. It can also be helpful for analysis of the feasibility and impacts of alternative plans. Although Gantt charts can be designed and used separately, they are most effective when used in close connection with network planning methods. While Gantt charts emphasize time required to complete tasks, network models enable us to analyze project logic and precedence requirements. The two may best be combined into an integrated project development and management approach. This is generally the approach taken today.

Inspection of a Gantt chart easily shows which activities are planned to be carried out at each specific point in time. It is particularly useful for displaying the results of a PERT or CPM analysis. Thus, such a chart is helpful in forecasting workload, scheduling the resources required to complete a project as a function of time, and other similar uses. Possible shifts in the schedule may also be indicated on the chart, and used to reduce peak demands or resolve conflicts over resource allocation issues.

While execution of activities is in progress, the actual portions of work completed at a specific time may also be shown on the chart in some distinctive manner, such as through use of heavy lines. This enables the person viewing the chart to quickly identify items of potential interest, such as those areas where progress differs from that initially planned in the PERT or CPM schedule. Because of their simplicity, use of Gantt charts can considerably enhance communication about and participation in planning large projects. The charts have been used for a wide variety of projects over many decades now.

There are actually two closely related, but somewhat different, types of Gantt charts that are in use. In one type, various departments, crews, pieces of equipment, and so on, are listed along the vertical axis. Horizontal lines or bars show the timing and amounts of work to be performed by each work grouping. In the other type, activities are listed on the vertical axis, and lines or bars show the time during which they are planned to be carried out. Because it is more useful for project planning, we shall discuss the latter type of chart in some detail here.

The following steps will lead to construction of a Gantt chart. Just as with the network management efforts, we may describe these efforts as planning, scheduling, and control efforts.

1. *Planning.* A decision is made as to which alternative project will be carried out. The plan is worked out in detail and a time scale or completion date specified. The project is disaggregated into a set of independent activities or work packages, perhaps in a cost breakdown structure (CBS) and work breakdown structure (WBS). Precedence

relations between the activities are identified, and estimates for time requirements are obtained. Network planning methodology, typically PERT or CPM, is used to compute expected overall project duration, to identify activities critical with respect to project duration, and to derive earliest and latest possible start or completion times for each activity.

2. *Scheduling.* The time scale along the horizontal axis is specified, and the activities are listed along the vertical axis, usually in ascending order of start times from top to bottom. Planned start and completion time for each activity are linked by a horizontal line showing the time during which the particular task is supposed to be active. Possible earlier start or later completion times are also indicated, for example, by broken extensions of the line indicating what is planned. The completed chart shows, at a glance, which tasks will be active at any point in time. When estimates of requirements for each job are available, graphs showing total workload, equipment requirements, expenditures, and other items of importance for project scheduling and management are illustrated as a function of time. Some analytical effort is possible using these charts. For example, peak loads may be reduced by shifting the active periods of tasks over the allowable range. The Gantt chart can be used as a visual aid to directly evaluate the effects of these possible shifts. However, for complicated projects consisting of hundreds of activities, so many combinations of shifts are possible that it becomes infeasible to evaluate them all graphically. Most of the network software described in the previous section have available options that allow display and printing of Gantt charts.

3. *Control.* While execution of the project is underway, the Gantt chart is usually updated regularly to show actual accomplishments to date. Usually, a solid line is drawn parallel to or over the original line to reflect the fraction of the task that has been completed. If, for example, 50% of a specific task is completed, a solid line is drawn that is half as long as the line showing total activity duration. This enables comparison of actual and planned progress and revision of plans if necessary. A glance at the updated chart quickly reveals those activities that are on schedule, that are ahead of schedule, and that lag behind the initially planned schedule. Depending on the particular situation, this may lead to a reallocation of priorities, or even a revision of the schedule for those tasks that are not yet completed. Using this information, preparation of a Gantt chart to document the execution of the project becomes possible. If required for documentation, the realized start and completion times of all activities can be shown in a Gantt chart in the same format in which the planned schedule was shown.

We shall use the same example here that is used in the description of network planning. Inspection of Figures 6.9 and 6.12 easily results in the Gantt

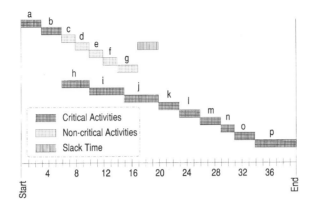

Figure 6.13   Gantt chart corresponding to Figures 6.9 and 6.12 life cycle.

chart illustrated in Figure 6.13. It would be easy to show various critical milestones, both original milestones and those rescheduled as the project evolves, on the diagram. Clearly, the notions described here are very simple ones. And, they are also very useful ones, as are many simple notions.

There have been a number of extensions of these early notions. One of these is based on the *Pareto Principle*, which states that a small number of problems are generally responsible for the majority of the ill effects. In terms of economic systems analysis, it states that a few of the contributors to costs are responsible for most of the costs. In terms of quality, it states that a few of the contributors to quality loss costs are responsible for most of these costs. Thus, a chart that illustrates the most important areas in terms of costs would have potential value in indicating which problem areas should be addressed first. It would certainly be possible to plot a Pareto chart showing the number of occurrences of a problem of a given type versus the type. The classical name given to such a chart is a *histogram*. The biggest potential problem with doing this is that the inferred suggestion visually presented by a Pareto chart is that we should work on the most important elements first, and that the most important elements are those with the largest number of occurrences. A cost of problem-type measure is generally more relevant, or possible some multiple attribute measure, rather than simply number of occurrences. This is called a Pareto chart [45]; a prototypical Pareto chart is shown in Figure 6.14.

The steps involved in constructing a Pareto chart are, conceptually, quite simple. We first identify as many of the relevant influencing elements of interest. This might well be in the form of a matrix of costs. The dimensions of the matrix could be product categories and type of defect, life-cycle phase, and type of error. The entries in the matrix would be the measure of influence, typically cost entries. We would then sum these horizontally or vertically, depending upon the horizontal dimension selected for the Pareto chart. Then we order these elements according to the selected measure of influence, such as cost. A

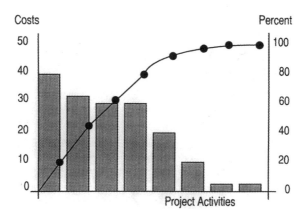

Figure 6.14 Hypothetical Pareto chart.

cumulative distribution of the measure of influence, generally costs, would be illustrated in the form of a line graph and associated with the descending order of the problem elements, such as illustrated in Figure 6.14.

6.4.3 Cost Estimation Methods and Work/Cost Breakdown Structures

There are several approaches that we can use to estimate costs of a project under analysis.

1. Analogy
2. Bottom up
3. Expert opinion
4. Parametric models
5. Top down, or design to cost
6. Price to win

Each is reasonable in particular circumstances. There are also a number of approaches that may be used to estimate effort or cost rates. These include:

1. Direct labor rate
2. Labor overhead rate, or burden,
3. General and administrative rate
4. Inflation rate
5. Profit

These are not mutually exclusive, nor are they collectively exhaustive. There are a number of approaches to cost as pricing strategies. These include:

1. Full cost pricing
2. Investment pricing
3. Promotional pricing

Another approach to costs is determining the cost required to achieve functional worth — that is, to fulfill all functional requirements that have been established for the system. While this is easily stated, it is not so easily measured. A major difficulty is that there are essential and primary functions that a system must fulfill, along with ancillary and secondary functions that, while desirable, are not absolutely necessary for proper system functioning.

After the functional worth of a system has been established in terms of operational effectiveness, perhaps using the methods of decision analysis you studied in the last chapter, it is necessary to estimate the costs of bringing a system to operational readiness. If this cost estimate is to be useful, it must be made before a system has been produced. It is easily possible to think conceptually of three different costs as we have noted in Section 4.5.3:

1. *Could cost* — the lowest reasonable cost estimate to bring all the essential functional features of a system to an operational condition
2. *Should cost* — the most likely cost to bring a system into a condition of operational readiness
3. *Would cost* — the highest cost estimated that might have to be paid for the operational system if significant difficulties ensue

It is interesting to relate "nonfunctional value adding costs" to risks. If the system is well-designed in the first place, all costs of implementation should contribute to the functional worth of the system. Any additional costs would then fall into the category of "risk," — that is, unanticipated expenses. Quite obviously, it is very difficult to estimate each of these costs. The "should" cost estimate is the most likely cost that results from meeting all essential functional requirements in a timely manner. "Could" cost is the cost that would result if no potential risks materialize and all nonfunctional value adding costs are avoided. "Would" cost is the cost that will result if risks of functional operationalization materialize and if non-value-adding costs are not avoided. There is a strong notion of uncertainty in any discussion of costs such as these, and various probabilistic notions need to be used in obtaining useful cost estimates.

As we have often noted, there are three fundamental phases in the systems acquisition life cycle: system definition, system development, and system deployment. These phases may be used as the basis for a work breakdown structure (WBS), or cost breakdown structure (CBS), for depicting cost element structures.

	Requirements Identification
	Specifications Development
	Research and Advanced Development
	Design and Development Plan
System Definition	Prototype Production
	System Test and Evaluation Plan
	Configuration Management
	Operation and Maintenance Plan
	Facilities

Figure 6.15   Work breakdown structure for system definition.

Figures 6.15, 6.16, and 6.17 illustrate some of the many components that comprise a WBS or CBS for the generic three-phase systems engineering life cycle we have used here. The WBS is a family tree that is comprised of the hardware and software, including facilities and services, that results from the systems management efforts associated with engineering of the specific process that will be used to produce a given product. Thus, the WBS provides a structure that is useful for guidance of the team that will implement the system, including associated cost tracking and control. There are a number of related questions which, when answered, provide the basis for reliable cost estimation for the WBS, sometimes called a *contract work breakdown structure* (CWBS) in the DoD literature.

	Production	Manufacturing
		Quality Control
		Detailed Design and Development
		Test and Evaluation
System Development		Concurrent Engineering
	Program Support	Configuration Management
		Project Management
		Quality Assurance
	Nonrecurring	Facilities
	Costs	Tools

Figure 6.16   Work breakdown structure for system development.

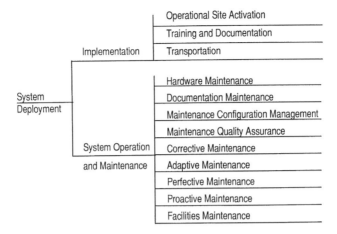

Figure 6.17 Work breakdown structure for system deployment.

There are a number of questions that can be posed, and the responses to these provide valuable input for estimating WBS and costs. Figure 6.18 illustrates a partial work break down structure for a hypothetical effort. Generally, the WBS is displayed in levels, as represented by Figures 6.15 through 6.18. In level 1, the total program scope is identified and provides the basis for release for all of the work on a program. Level 2 identifies the various projects or categories of activity that, taken together, comprise the program. This may include major elements of the system, such as subsystems and program management support efforts. Often budgets for the various elements of a program are prepared at this level. Level 3 identifies the major functions and components of the system that are subordinate to the level 2 items. The number of levels can be extended. In constructing a WBS, it is important to be

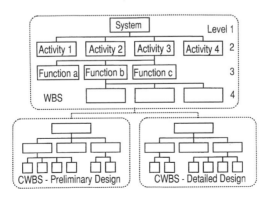

Figure 6.18 Partial work breakdown structure and contract work breakdown structure.

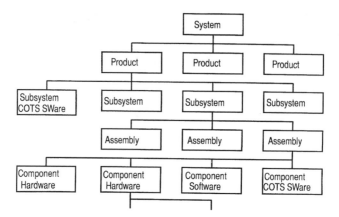

Figure 6.19   The physical architecture system breakdown structure (SBS). [COTS is an acronym for "commercial off-the-shelf" and denotes items that are available before the system is engineered and that are engineered into the system.]

sure that a continuous flow of work-related information is provided in a top-down manner such that (a) all work is represented, (b) a sufficient number of levels are provided to enable the identification of well-defined work packages, and (c) duplication of effort does not occur. If there are an insufficient number of levels, the integration of work effort and management oversight may be very difficult. If there are too many levels, a great deal of time may be expended in performing work review and control at a micro level. The WBS is very similar to, and may be influenced by a systems breakdown structure (SBS), such as shown in Figure 6.19. In a typical SBS, which corresponds to the physical architecture of the system to be engineered, the hierarchical levels of the physical system to be produced are illustrated.

Cost estimation [46] and the methods discussed in Chapter 4, especially Section 4.5, are very relevant to analysis of systems engineering projects and to management of these projects.

6.5   HUMAN AND COGNITIVE FACTORS IN SYSTEMS ENGINEERING AND SYSTEMS MANAGEMENT

The systems engineering life cycle is intended to enable evolution of high-quality, trustworthy systems that have appropriate structure and function support for identified purposeful objectives. Many major efforts called for throughout all phases of a systems engineering life cycle require human interaction with various aspects of the system.

This section continues our discussion of systems engineering, and it concludes this chapter. The focus is on the role of the human in systems, as well as on the information support aids that should be made available to improve development and use of systems. We can only begin to scratch the surface of this subject here. Chapter 9 of reference 27, along with reference 2 and the references therein, provides a much more complete discussion.

Advances in information technology, together with the desire to improve productivity and the human condition, render physiological skills that involve only, or primarily, strength and motor abilities relatively less important than in the past. They significantly increase the importance of cognitive and intellectual skills for the use of systems of all types. The need for humans to monitor and control conditions necessary for effective operation of systems is greater than ever. In many cases today, primarily cognitive efforts are ultimately translated into physical control efforts. Often, systems that accomplish this are called *human*–machine systems.

A second major concern is the design of information technology-based support systems to aid human performance. The first concern, monitoring and human controller effort, has been addressed for a much longer period of time than the second. Such task categories as controlling and problem solving describe typical human activities in human–machine systems. We are particularly concerned here with system design for human interaction, the cognitive tasks involved in problem solving, the physiological tasks involved in controlling complex systems, and the interfaces between humans and systems that are an integral part of this. We will call this area of interest *cognitive ergonomics*.

Some of the roles for a human in a human–machine system include the following:

1. Assessing the situation at hand, in terms of system operation, such as to identify needs for human supervisory control, objectives to be fulfilled, and issues to be resolved
2. Identifying task requirements, such as to enable determination of the issues to be examined further and the issues not to be considered further
3. Identifying alternative courses of actions which may resolve the identified issues
4. Identifying probable impacts of the alternative courses of action on the functioning of the system
5. Interpreting these impacts in terms of the objectives or needs that are to be fulfilled
6. Selecting an alternative for implementation and implement the resulting control
7. Monitoring performance such as to enable determination of how well the human–machine system is performing

Classical human–machine systems engineering, often called *human factors engineering* or *ergonomics*, was focused on training for skill-based behavior and associated physiological concerns. Contemporary efforts necessarily also emphasize the integration of physiological concerns with cognitive concerns. This has been motivated by the great deal of evidence that humans do not react to task requirements in a way capable of being easily stereotyped, but rather in a fashion that is very much a function of the task, the requirements perceived for the task, the environment into which the task is embedded, and the experiential familiarity of the human performing the task with the task, the task requirements, and the environment into which these are embedded — in other words, the *contingency task structure.*

Human tasks in a human–machine system generally involve controlling, which will usually involve some physiological effort, and problem detection and diagnosis, which generally involves cognitive efforts. In addition, monitoring or feedback loops will also exist in order to support learning and more precise control.

Issue, or fault, detection concerns the identification of a potential difficulty that impedes operation of a system. Issue, or fault diagnosis, is concerned with (a) identification of a set of hypotheses concerning the likely cause of a system malfunction and (b) the evaluation and selection of a most likely cause. Issue or fault resolution, correction, or control is concerned with resolving issues or solving problems in actual situations. While detection and diagnosis are primarily cognitive efforts, correction will often also involve physiological efforts. In the classic human–machine system, correction or controlling is accomplished with the objective of returning the overall system to a satisfactory operating state [47–49]. It is not uncommon to distinguish the cognitive effort in controlling from the physiological effort [50]. The efforts involved in detection and diagnosis [51], which are primarily cognitive, may also be called problem formulation or situation assessment. The subsequent efforts at correction may be termed solution execution. Together, the situation assessment and (re)solution execution comprise problem solving. Then, the human–machine systems problem may be regarded as one of problem finding and solution execution.

There are many other tasks in human-machine systems, such as monitoring of performance and communications. These are subtasks, or supporting tasks, for the two primary activities of problem solving and controlling. Communicating falls into two very different categories. The first is the verbal communication among the human members of the group responsible for aspects of system operation. The second involves communication between humans and the technological system that is being controlled. Often this latter form of communication is often called human–computer dialog [52–54], and the design of dialog generation and management systems is an important part of information systems engineering efforts [55].

Even with such slow response systems as ships and industrial process plants, it has become obvious that human operator behavior is not easily explained by well-established control-theoretic methods. Human controller behavior is highly nonlinear and error-prone, even in these simple and slow response cases. Human operator workload in physical task performance has been investigated for many years [56]. Although appropriate definitions and a common understanding concerning the importance of this research area exist, the reliability and validity of most workload measures was initially quite poor. A primary reason for this was that human performance modeling was primarily based on physiological considerations. Initially, there existed a neglect of human cognitive dynamics. Current efforts attempt to alleviate these deficiencies; it is with a very brief summary of these that our efforts in this section are concerned.

There are several approaches to explaining how individuals and organizations acquire, represent, and use information to describe their perceptions of the world around them, as well as issues and situations that are of importance. Human information processing is a vital and crucial ingredient in effective decision making. Information processing theories of problem solving, judgment, and decision making are normally based on the assumption that individuals have the following:

1. An input mechanism for acquisition of information
2. An output mechanism for interpretation and choice making
3. Internal processes for filtering and other analysis efforts associated with information
4. Memories for long- and short-term storage of information

Insights into the nature of cognitive development, including a conceptual model of cognitive activity are contained in the works of Piaget, who founded *genetic epistemology* in the 1930s, and in more recent accounts of this development [57]. According to Piaget, there are four stages of intellectual development in a human being:

- *Sensory motor*
- *Preoperational*
- *Concrete operational*
- *Formal operational*

The last two of these are of particular importance to our efforts here. In the writings of Piaget, intellectual development is seen as a function of four variables: maturation, experience, education, and self-regulation. Self-regulation is a process of mental struggle with discomforting information until identification of a satisfactory mental construct allows intellectual growth or learning. In Piaget's model of intellectual development, both "formal" and concrete operational thinkers can deal logically with empirical data, manipulate symbols, and organize facts toward the solution of well-structured and personally familiar problems. But concrete-only thinkers lack the formal thinker's capacity to reason hypothetically and to consider the effect of different variables or possibilities outside of personal experience. This suggests that concrete operational (only) thinkers may be capable of learning skills through the repeated use of rules to which they are exposed. It suggests that they will be unable to formally reason relative to such things as possible inapplicability of these rules in some situations.

The mature adult decision maker will typically be capable of both formal and concrete operational thought. As we will argue, selection of a formal or concrete cognitive process will depend upon the decision-maker's diagnosis of need with respect to a particular task. That diagnosis will depend upon a decision-maker's maturity, experience, and education with respect to a particular problem. Each of these influence cognitive strain or stress. Ordinarily, a decision maker will prefer a concrete operational thought process and will make use of a formal operational thought process only when concrete operational thought is perceived inappropriate for the task at hand. In general, a concrete operational thought process involves less stress and may involve repetitive and previously learned behavioral patterns. Familiarity and experience, with the issue at hand or with issues perceived to be similar or analogous, play a vital role in concrete operational thought. In novel situations, which are initially unstructured and where new learning is required, formal operational thought is typically more appropriate than concrete operational thought.

It is of interest to describe some salient features of these two processes. In concrete operational thought, people use concepts that (a) are drawn directly from their personal experiences, (b) involve elementary classification and generalization concerning tangible and familiar objects, (c) involve direct cause-and-effect relationships, typically in simple two-variable situations, (d) can be taught or understood by analogy, algorithms, affect, standard operating procedures or recipes, and (e) are "closed" in the sense of not demanding exploration of possibilities outside the known environment of the person and the given data.

In formal operational thought, however, people use concepts that (a) may be imagined, hypothetical, based on alternative scenarios, and/or counterintuitive, (b) may be "open ended," in the sense of requiring speculation about unstated possibilities, (c) may require deductive reasoning using unverified and perhaps flawed hypotheses, (d) may require definition by means of other concepts or abstractions that may have little or no obvious correlation with contemporary reality, and (e) may therefore require the identification and structuring of intermediate concepts not initially specified. These processes blend together, and concrete operational thought may evolve from a learning process that initially involves formal operational thought, or from training. The formal operational thought processes are accomplished through reflective observation, abstract conceptualization, and the testing of the resulting concept implications in new situations. It is in this way that the divergence produced by discomforting new experiences allows the inductive learning of new developments and concepts to be "stored" in memory as part of one's concrete operational experiences.

The formal operational thought process of Piaget appears not fundamentally different from the systems engineering approach that is comprised of the following:

1. Identification of hypotheses, laws, assumptions, or generalizations that may be in the form of axioms or postulates
2. Inductive and deductive reasoning, including the use of both formal and informal, or default, logic or rules of inference
3. Observation, experiment, and confirmation or verification of degrees of support for hypotheses
4. Induction and abduction, using any of a variety of approaches that deal with imperfect information, such as to result in generalizations

These processes have evolved over the history of systems engineering and greatly support information processing, both in systems and organizations [58]. They support the interpretation process of going from model to theory, as well as the formalization process of going from theory to model. They provide the basis for a number of axioms, postulates, inference rules, the confirmation or verification of degrees of factual support to hypotheses, and a number of inductive and deductive reasoning approaches.

There are a number of models that support this assertion. In 1981, Joe Wohl [59] developed a dynamic model for tactical decision processes called SHOR, based upon the following:

1. **S**timulus arrival, suggesting a potential issue associated with some detected decision situation
2. Generation of **H**ypotheses concerning diagnosis of the situation
3. Generation of a number of **O**ptions or response alternatives
4. Some approach to evaluation of the options and selection of a **R**esponse of course of action for implementation

This model was intended to prescribe a portion of the foundation required for a theory of military command and control, including guiding principles for system development in this environment. Wohl is careful to relate this model, which appears very much as a normative model, to the descriptive realities of decision making. A number of decision assessment prescriptions have been based on this model, which is illustrated in Figure 6.20. There have been a number of applications of this model, particularly in command and control environments, and SHOR has led to a number of innovative designs of sensors and data fusion hardware elements. A number of SHOR applications are documented in reference 60, which also contains a reprint of the original Wohl paper.

One particularly useful taxonomy developed by Jens Rasmussen [18, 20, 61] conceptualizes three distinct types of problem-solving behavior, or knowledge

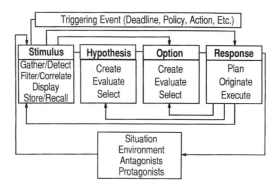

Figure 6.20   Essential features of the SHOR human decision model.

to support reasoning:

- Formal-reasoning-based behavior
- Rule-based behavior
- Skill-based behavior

The choice of which type of reasoning to employ is made by the problem solver on the basis of (a) experiential familiarity with the task at hand and (b) the environment in which this task is embedded. All three cognitive control modes of reasoning can exist at the same time. However, the primary control mechanisms shift toward skill-based as expertise increases. This model was initially devised to describe the judgment and choice processes of process control operators and has since been applied to a large number of other situations. Of importance in a model such as this is the transition from one form of knowledge to another, especially the dynamic learning over time which enables a person to transfer formal rule-based reasoning results to a set of rule-based judgments and then, in turn, to skill-based reasoning.

Conceptual foundations have been developed [62] for human–machine interface design, based primarily on supporting these three cognitive levels. In general, humans use skill-based knowledge, rule-based knowledge, and formal-reasoning-based knowledge in an attempt to keep processing effort at the lowest cognitive level that trustworthy performance of the task, and presumably the human experiential familiarity with the task and environment as well, requires. The ecological interface design construct attempts to minimize the difficulty of controlling a complex system while, at the same time, supporting the entire range of activities that specific users may require.

Ken Hammond [63] has evolved a definitive model of this cognitive continuum that is most appropriate for systems engineering efforts. He indicates that the meaning of analytical cognition in ordinary language is clear; it signifies a step-by-step, conscious, logically defensible process of problem solving. He indicates that the ordinary meaning of intuition signifies a cognitive process that somehow permits the achievement of an answer, solution, or idea without the use of a conscious, logically defensible, step-by-step process. There are numerous normative models for formal analytical cognition to which system engineers may turn. There are a considerably smaller number of fully worked-out models of intuitive cognition. Hammond [64,65] is much concerned with developing cognitive continuum models that enable integration of the facets of intuitive and analytical cognition. Hammond also identifies and describes properties of intuitive and analytical cognition. These follow from three central features of Hammond's theory of dynamic task cognition.

1. Cognition is defined in terms of a cognitive continuum that moves from analysis, at one extreme of the continuum, to intuition, at the other extreme of the continuum.

2. There is a correspondence-accuracy principle which maintains that judgment veridicality is greater when there is an appropriate correspondence between cognitive activity and task properties than where this does not exist.

3. Changes in cognition will be induced by changes in the properties of the contingency task system. The location of the type of cognition on the cognitive continuum will change, and the environmental scanning mechanism will shift from pattern seeking, or symptomatic search, to function–relation seeking, or topographic search.

This model is based on the precept that decision-making tasks are generally *dynamic decision-making tasks* in that they possess the following characteristics.

1. A series of decisions are required.

2. These decisions are interrelated.

3. The decision situation changes, both autonomously and as a consequence of the decision-maker's actions over time.

4. The decisions are made in real time.

Brehmer [66] has also suggested that decision making in dynamic tasks should be seen as a matter of trying to achieve control, instead of only an attempt to resolve discrete choice dilemmas. He proposes a conceptualization of dynamic decision tasks in terms of six basic characteristics: complexity, feedback quality, feedback delay, possibilities for decentralization, rate of change, and prerequisites for control. These six characteristics are indeed present in many empirical studies on dynamic decision making and are, clearly, very important for any study of cognitive ergonomics. This is especially the case when the dynamic nature of appropriate cognitive continuum adaptation is also considered.

The dynamic nature of information and knowledge is very important. Information access and utilization, as well as management of the knowledge resulting from this process, are complicated in a world with high levels of connectivity and a wealth of data, information, and knowledge. Often, the now ubiquitous networks and data warehouses are indicated to be the means to enable us to take advantage of this situation. However, providers of such "solutions" seldom address the basic issue of what information users really need, how this information should be processed and presented, and how it should be subsumed into knowledge that reflects context and experiential awareness of related issues.

The underlying problem is the usually tacit assumption that more information is inherently good to have. What users should do with this information and how value is provided by this usage are seldom clear. The result can be large investments in information technology with negligible improvements of productivity [67]. One of the major needs in this regard is for organizations to develop the capacity to become learning organizations [23, 24, 68, 69] and to support bilateral transformations between tacit and explicit knowledge [70,71]. In effect, this requires knowledge fusion and integration as an alternate approach to knowledge generation which brings together people with different perspectives to resolve an issue and determine a joint response. The result of knowledge fusion efforts may be creative chaos and a rethinking of old presumptions and methods of working. Significant time and effort are often required to enable group members sufficient shared knowledge and to work effectively together and to avoid confrontational behavior. This is why we examined approaches to group dialog as a part of our issue formulation efforts in Chapter 3. Addressing these dilemmas should begin with the recognition that information is only a means to gaining knowledge and that information must be associated with the contingency task structure to become knowledge. This knowledge is the source of desired advantages in the marketplace. Thus, understanding and supporting the transformations from information to knowledge to advantage are central challenges to enhancing information access and utilization in organizations [72, 73].

In many contemporary organizational perspectives, often called the open systems view of an organization, concern is not only with objectives but with appropriate responses to a number of internal and external influences. Weick [74, 75] describes organizational activities of *enactment*, *selection*, and *retention* which assist in the processing of ambiguous information that results from an organization's interactions with ecological changes in the environment. The overall result of this process is the minimization of information equivocality such that the organization is able to

1. Understand its environment
2. Recognize problems
3. Diagnose the causes of problems

4. Identify policies to potentially resolve problems

5. Evaluate efficacy of these policies

6. Select a priority order for problem resolution

These are the primary steps in the systems engineering formulation, analysis, and interpretation efforts we have discussed throughout this text.

The first activity in this open systems description results in the enacted environment of the organization. This enacted environment contains an external part, which represents the activities of the organization in product markets, and an internal part, which is the result of organizing people into a structure to achieve organizational goals. Each of these environments is subject to uncontrollable economic, social, and other influences. Selection activities allow perception framing, editing, and interpretation of the effects of the organization's actions upon the external and internal environments, such as to enable selection of a set of relationships believed of importance. Retention activities allow admission, rejection, modification of the set of selected knowledge in accordance with existing knowledge, and integration of previously acquired organizational knowledge with new knowledge. There is a potentially large number of cycles that may be associated with enactment, selection, and retention. These cycles generally minimize informational equivocality and allow for organizational learning, such that the organization is able to cope with complex and changing environments.

A very important feature of many models of organizations is that of *organizational learning*. Much of the organizational learning that occurs in practice is not necessarily beneficial or appropriate in a descriptive sense. For example, there is much literature which shows that organizations and individuals use improperly simplified and often distorted models of causal and diagnostic inferences, and improperly simplified and distorted models of the contingency structure of the environment and task in which these realities are embedded.

This surely occurs in group and organizational situations, as well as in individual information processing and judgment situations. Individuals often join "groups" to enhance survival possibilities and to enable pursuit of career and other objectives. These coalitions of like-minded people pursue interests that result in emotional and intellectual fulfillment and pleasure. The activities that are perceived to result in need fulfillment become objectives for the group. Group cohesion, conformity, and reinforcing beliefs often lead to what has been called *groupthink* [76]. This is an information acquisition and analysis structure that enables processing only in accordance with the belief structure of the group. The resulting selective perceptions and neglect of potentially disconfirming information preclude change of beliefs.

Organizational learning results when members of the organization react to changes in the internal or external environment of the organization by detection and correction of errors [77]. An error is a feature of knowledge that makes action ineffective; detection and correction of error produces learning. Individuals in an organization are agents of organizational action and organizational learning. In the studies just referred to, Argyris cites two information-related factors that inhibit organizational learning:

1. The degree to which information is distorted such that its value in influencing quality decisions is lessened
2. Lack of receptivity to corrective feedback

Two types of organizational learning are defined and illustrated in Figure 6.21. *Single-loop learning* is learning that does not question the fundamental objectives or actions of an organization. It is essential to the quick action often needed. Members of the organization discover sources of error and identify new strategic activities which might correct the error. The activities are analyzed and evaluated, and one or more is selected for implementation. Environmental control and self protection through control over others, primarily by imposition of power, are typical strategies. The consequences of this approach may include defensive group dynamics and low production of valid information.

This lack of information does not result in disturbances to prevailing values. The resulting inefficiencies in decision making encourage frustration and an increase in secrecy and loyalty demands from decision makers. All of this is mutually self-reinforcing. It results in a stable autocratic state and a self-fulfilling prophecy with respect to the need for organizational control. Thus,

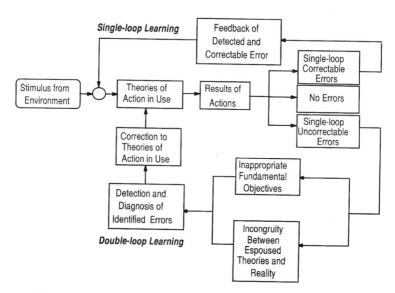

Figure 6.21 Representation of single- and double-loop learning.

while there are many desirable features associated with single-loop learning, there are a number of potentially debilitating aspects as well.

Double-loop learning involves identification of potential changes in organizational goals and of the particular approach to inquiry that allows confrontation with and resolution of conflict, rather than continued pursuit of incompatible objectives leading to intergroup conflict. Not all conflict resolution is the result of double-loop learning, however. Good examples of this are conflicts settled through imposition of power rather than inquiry. Double-loop learning is the result of organizational inquiry that resolves incompatible organizational objectives through the setting of new priorities and objectives. New understandings are developed which result in updated cognitive maps and scripts of organizational behavior. Studies show that poor performance organizations learn primarily on the basis of single-loop learning and rarely engage in double-loop learning. This provides an important motivation for the study of systems engineering methodology and processes, along with efforts to improve the acquisition and procurement of systems through improvement in process, as contrasted with direct effort only at improvement of the product itself.

6.5.1 Rationality Perspectives

In this subsection we describe some rationality perspectives concerning individual and organizational information processing. In order to do this, we will assume some rather stereotypical perspectives and environments in which people acquire, represent, and use information. These perspectives are not, in any sense, mutually exclusive.

The research of Diesing [78] and Steinbrunner [79], and many others, have dealt with *rationality* because it pertains to human information processing and other judgment and choice activities. The various forms of rationality are very helpful in providing at least a partial explanation of why people seek the information they do. Understanding these will help support system design that results in more effective ways to determine information needs. Although a variety of *rationality* forms may be defined, the following appear to be the most common ones.

1. *Economic Rationality.* Maximum goal achievement with respect to technical production of a single product, subject to a production cost constraint, is the typical desired end of very classic microeconomic rationality. Economic rationality extends this concept to a number of products. It seeks to maximize the overall worth, in an economic sense, of a number of investments [80]. This is possible if desired goals are well-defined and measurable, if the techniques employed to attain these goals are not limited in scope or hindered in application, if supply and demand operates in a stable manner, and if the interrelationships of supply and demand are known and available to all. In other words, the requirements

for a "perfectly competitive economy" are satisfied. This is a very useful but incomplete form of rationality.

2. *Technical Rationality.* The activities of an individual are determined in such a way as to maximize the return, or benefit, to the individual from the investment cost of that activity. In a similar way, we can define technical rationality of an organization. All activities within an organization are formed in a manner so as to reach the goals set forth by the organization. Most traditional engineering and organizational analysis has presumed, at least implicitly, technical rationality. Systems are presumed to be designed such as to achieve "optimal attainment" of objectives. The presence of multidimensional and noncommensurate objectives will often prevent this sort of optimization from being easily accomplished. The need for coordination and communication among people in modern decentralized organizations also makes technical rationality very difficult of attainment. There are a number of other reasons as well. Implementation of technical rationality results in what is called the rational actor model. However, this is by no means *the* rational actor model. In the technical rationality model, the decision maker becomes aware of a problem, studies it, carefully weighs alternative means to a solution, and makes a choice or decision based on an objective set of values. In formal rational planning or decision making, the following steps are typically performed.

(a) The decision maker is confronted with an issue that can be meaningfully isolated from other issues.

(b) Objectives, which will result in need satisfaction, are identified.

(c) Possible alternative activities to resolve needs are identified.

(d) The impacts of action alternatives are determined.

(e) The utility of each alternative is evaluated in terms of its impacts upon needs.

(f) The utilities of all alternatives are compared, and the policy or activity with the highest utility is selected for action implementation.

From the perspective of a *Technical Rational Actor Model*, the decision maker becomes aware of a problem, structures the problem space, gathers information, identifies the impacts of alternatives, and implements the best alternative based on a set of values. Because a complete identification of all needs, alterables, objectives, and so on, is not usually possible, one cannot be completely rational in the purest unconstrained sense. This observation, and the more important observation that, in a descriptive sense, humans often do not attempt, due to cognitive limitations, to follow the normative implications of the rational actor model led Simon to develop the satisficing or bounded rationality framework. Often the economic rationality model and the technical rationality model are combined into a single technoeconomic rationality model.

3. *Satisficing or Bounded Rationality*. Decisions are implemented based on a minimum set of requirements to provide a degree of acceptable achievement over the short term. The decision maker does not attempt to extremize an objective function, not even in a substantive way, but rather attempts to achieve some aspiration level. The aspiration level may possibly change due to the difficulty in searching for a solution. It may be lowered in this case, or raised if the goal or aspiration level is too easily achieved. Simon [81, 82] was perhaps the first to make use of the observation, about two decades ago, that unaided decision makers may not be able to make complete substantive use of the economic and technically rational actor model possible. In these situations, the concepts of bounded rationality and satisficing represent much more realistic substantive models of actual decision rules and practices. According to the *satisficing or bounded rationality* model, the decision maker looks for a course of action that is basically good enough to meet a minimum set of requirements. The goal, from an organizational perspective, is "*do not shake the system*" or "*play it safe*" by making decisions primarily on the basis of short-term acceptability rather than seeking a long-term optimum. Simon introduced the concept of satisficing or bounded rationality as an effort to "*... replace the global rationality of economic man with a kind of rational behavior that is compatible with the access to information and computational capabilities that are actually possessed by organisms, including man, in the kinds of environments in which such organisms exist.*" He suggested that decision makers compensate for their limited abilities by constructing a simplified representation of the problem and then behaving rationally within the constraints imposed by this simplified model. We may satisfice by finding either optimum solutions in a simplified world or satisfactory solutions in a more realistic world. As Simon says, "*neither approach dominates the other.*"

4. *Social Rationality*. A potentially simplistic or idealistic look suggests that society functions as a unit seeking betterment for itself. All its energy is directed toward the realization of this goal. The social system is cohesive in that all its activities reinforce achievement of the desired goal. Present decisions are related to those of the past and are projected into the future. While these actions and decisions are usually not efficient and sometimes not even effective, the cohesiveness of society provides continuity for the system. The roles and structure of society are reinforced from previous results, both good and bad, lending credence to the fact that a social system is rather intractable. It maintains a conservative appearance and avoids risk. That it ought to be adaptive to change perhaps can be shown by sudden changes in the morality or consciousness of the members of the society through a violent opposition to the status quo, such as opposition to or blatant disregard of the law by leaders of the society. Social rationality would, for example, not exist independently of some

facets of political and legal rationality, or economic and technical rationality. These are the five rationality perspectives delineated in Diesing's treatise.

5. *Political Rationality.* The decision-making structure is assumed to be influenced by embedded beliefs, values, and interpersonal relationships, the interaction of which define roles under which actions and decisions are based. The three characteristics of this rationality are that all actors remain independent regardless of the pressures to be dependent on one another, the workload is distributed among all members so as to balance and moderate actions of the group, and future decisions are chosen in such a way that the impacts of these decisions will act to bind the group further together and increase participation.

6. *Legal Rationality.* A system exhibiting this form of rationality operates on the basis of rules that are complex, consistent, precise, and detailed. As a result, no ambiguous conflict can occur. It is effective in preventing disputes even though the rules of this system apply differently, to some extent, to each person. The prevention of disputes is accomplished through a "legal" framework which provides a means for settlement of disputes that do result and which sets precedents to guide members of society.

There are other approaches that we could use to characterize rationality. Simon [83] has developed a two-element categorization of rationality as a function of whether an act is rational from an input–output perspective or whether the internal components of the process used are rational. This leads to an identification of *substantive rationality* and *procedural rationality*. These rationality constructs are very similar to the single- and double-loop learning concept we discussed earlier and to approaches to systems engineering at the level of product and at the level of process. They have major implications for the way in which experientially familiar people go about the process of decision making [84].

6.5.2 Human Error and Systems Engineering

Humans are able to do many things. Usually these are done in a quite correct and appropriate manner. Occasionally, errors are committed. More often than not, the errors are small. Committing errors is not necessarily bad, *if* no major harm is done *and* if humans accomplish single- and double-loop learning through error commission in order to avoid future errors. Occasionally, errors can be catastrophic. Generally, we wish to design systems that assist in amelioration of the effects of human error, and ultimately of eliminating most possibilities for human error. Of course, we also wish to support human efforts in such a way that errors, to the extent possible, do not occur.

Human factors engineering, as indicated by Tom Sheridan [85] and Ron Hess [86], is the science and art of interfacing people with the machines they operate and interact with, such as the displays, controls, and computer hardware that send, receive, and process information. These are each illustrations of machine–machine interfaces or human–machine interfaces. Computer software and administrative procedures that determine how people allocate their effort, make decisions, and take actions, is another type of interface. A third relevant interface is between people and the seating, temperature, humidity, lighting, sound, radiation, vibration, and other environmental factors that determine their ability to function. Finally, there is the selection and training of people for their machine interactive tasks. Because human factors engineering includes humans and machines in a common system, it emphasizes system performance over and above performance of either person or machine by itself. It is this integrated performance of people and machines that is of interest here.

Human factors research on the interactions between users and systems is very important. With careful attention to the human factors in information presentation on computers, for example, it is possible to build appropriate systems for use by both experts and novices. The human–machine interface can be significantly improved by screen formats, data entry and display methods, menu structures, and graphical interfaces that reflect the cognitive abilities of humans and also reflect how humans are influenced by the contingency task structure.

There are three primary categories of errors;

- Errors in *detection* that there is a problem
- Errors in *diagnosis* of the cause of problem
- Errors in *correction* of the problem, or errors in execution

We wish to discuss each of these errors, as well as errors in planning for problem solving. These are especially important in situations involving poor information, which implies the need for good information requirements and an appropriate information acquisition system to supply higher-quality information. They are also important where mistakes are costly. This implies the need for very small type III (wrong problem) errors and good decision options and selection processes. Finally, they are important where recovery from missed opportunities is difficult. This implies the need for very small error in opportunity categorization.

The approaches to the study of problem identification and solution fall into the following categories. Error analysis is characterized by analyzing, generally through hindsight, errors that could have produced disorders. Process analysis is generally characterized by an attempt to disaggregate problem solution into finer components that can be more easily subject to detailed scrutiny. This is the formal systems engineering and scientific approach to problem solution. If not handled well, however, this can simply result in the replacement of one set of vague constructs with another. Also, failure to consider contingency task situational contingencies can result in many difficulties. Task analysis is a process to characterize human behavior by an analysis of the interaction of individual goals, or inner psychological environment, with the outer environment. Task analysis considers the outer, or task, environment in explaining individual adaptation to task demands. These approaches are not at all mutually exclusive. Each approach is important, and an integrated approach would appear better than one that considers only one type of analysis. Errors generally arise from one of two important sources of error.

1. Errors represent systematic interference and incongruities among models, rules, and procedures. This apparently could represent a deficiency in single-loop learning.

2. Errors represent some dysfunctionality of the effects of adaptive, or double-loop, learning mechanisms.

From this perspective, trustworthy human–system interactions are achieved through the design of systems that minimize and correct for these difficulties that cannot be eliminated through error recovery or correction approaches.

We have just introduced some elementary notions of human error. Many more are provided in Reason [25]. Figure 6.1, (on page 461) based on the Reason taxonomy of human error, illustrates the variety of action types that are possible. We may act in an unintended manner, either spontaneously or involuntarily and without prior cognitive intent in either case. Our concern here with human actions that are performed without prior cognitive intent is one of providing support aids for situation assessment such that humans do form prior cognitive intents when it is appropriate that this be done.

Thus, our concerns are, mostly, directed at actions for which a prior cognitive intent exists. There are two fundamental types of error in situations where a prior cognitive intent exists. There may exist planning errors, or there may be detection and diagnosis mistakes. These represent failure to identify either suitable objectives or a suitable course of action to obtain it. Alternately, there may exist execution, or controlling, errors. An execution error, which results from actions that do not accomplish their intent, is a lapse when there is an unintended memory error that leads to poor action, but it is a slip when the physical action taken is itself improper and not congruous with the prior cognitive intent.

In other words, actions based on inappropriate intentions are mistakes, and inappropriate actions that are based on proper or improper plans are lapses or slips [87, 88]. There may be errors associated with (a) either detection or diagnosis, or (b) correction or control. Generally, slips and lapses are associated with control, and mistakes are associated with detection and/or diagnosis. Specific error categories, representing potential errors of omission and commission, may be identified for a given application domain. Generally, this would consider inherent human limitations and inherent system limitations. It would also consider specific contributing conditions and events that are particular to the application domain. It would involve the following [89]:

1. Error *identification,* or detection, in which histories of user behavior and system response are correlated with operational procedures and scripts such as to enable us to detect deviations between observed behavior and nominally expected behavior
2. Error *classification,* or diagnosis, in which the causal factors leading to errors and the consequences of various errors are tabulated, structured, organized, and perhaps prioritized
3. Error *remediation,* or correction, in which various monitoring, feedback, and control possibilities are delineated and the most suitable procedures for remediation of errors of various types determined

This can be accomplished from a structural, functional, or purposeful perspective. All three horizons are desirable in order to explore the detailed form of issues and solutions, to conceptualize issues and solutions in terms of input–output relations, and to explore issue and solution needs and requirements. Often, it would also be desirable to be able to allocate these identification, classification, and remediation efforts across humans and machines.

Mistakes, of failures of detection and diagnosis, may be further categorized as due to failures of expertise or lack of expertise. In the first case, a skill-based reasoning approach to situation assessment is applied inadequately. In the second case, lack of skill may force cognitive control at the level of formal-reasoning-based knowledge, and this may turn out to result in a poor solution. Alternately, the lack of sufficient experiential familiarity needed to use skill-based reasoning may be unrecognized. A skill-based or perhaps rule-based knowledge approach may be improperly used, and this may lead to significant error.

In this section we have illustrated a number of models for human cognitive control. Also we considered a number of models of rationality that considered a variety of perspectives relative to judgment and choice situations. Finally, we examined some human error taxonomies. These are very important for systems engineering in general because we recognize that the major purpose of systems engineering is to support human capabilities and ameliorate human shortcomings, on both an individual and an organizational basis. There is an extensive literature on this subject [90], and we have only discussed a portion of that here. For an application of some of these concepts to systems engineering and integration, see reference 91.

6.6 SUMMARY

In this chapter we have presented a number of elementary concepts concerning systems engineering management, including systems management of human related issues. Needless to say, effective systems management is quite essential to the engineering of high-quality, reliable, and trustworthy systems. There is much that could be said concerning the subject of professional practice of systems engineering and management. Some of the pitfalls of analysis and the various craft issues associated with systems analysis are discussed in references 92 and 93. We have also commented upon some of the pitfalls associated with improper use of the systems approach throughout this text, especially in Chapter 1. It is fitting here to conclude our presentation with a recent set of guidelines for effective analysis [94]. The essential ingredients of a good analysis are as follows.

1. A clear, concise, and coherent problem definition
2. A transparent representation of the problem structure and accompanying analytical model
3. Choice of an appropriate problem solving tool
4. Choice of an appropriate computational vehicle
5. Explicit statement of all assumptions
6. Systematic model development and documentation of problem- solving logic
7. Sensitivity analysis of the analysis with assumption changes
8. Understanding analysis in intuitive nontechnical terms, because an analysis may be correct and not be complicated, and it may be complicated and not correct
9. Credible and effective communications of all justified recommendations

While there is much more to systems engineering than analysis, systems engineering does critically depend upon analysis, and it is analysis that we have emphasized in this introduction to systems engineering. These ingredients are very reasonable guidelines for the professional practice of analysis and also provide a sound basis for the management of analysis efforts. This is a vital part of systems engineering and management.

Based upon our efforts in these six chapters and the extensive literature in systems engineering and systems management, we identify some necessary ingredients that must exist in order to develop large systems, solve large and complex problems, or manage large systems:

1. A way to deal successfully with problems involving many considerations and interrelations, including change over time
2. A way to deal successfully with areas in which there are far-reaching and controversial value judgments
3. A way to deal successfully with problems, the solutions to which require knowledge principles, practices, and perspectives from several disciplines

4. A way to deal successfully with problems in which future events are difficult to predict

5. A way to deal successfully with problems in which structural and human institutional and organizational elements are given full consideration.

We believe, strongly, that the systems engineering framework presented in this chapter and throughout this book possesses these necessary characteristics.

Systems engineering is potentially capable of exposing not only technological perspectives associated with large-scale systems, but also needs and value perspectives. Furthermore, it can relate these to knowledge principles and practices such that the result is the successful engineering of high-quality trustworthy systems. We have concentrated on systems engineering management frameworks to enable this in this chapter.

The result of a systems engineering effort is a product or service. The process must necessarily be embedded into the environment that surrounds the systems engineering organization. It must necessarily be concerned with people, organizations, and technology. This introductory text attempts to illustrate this large-scale and large-scope view of systems engineering.

PROBLEMS

6.1 Please reexamine the list of 41 causes of systems management failures presented on page 468. Can you structure this list hierarchically? Can you identify other elements of failure that should be on the list?

6.2 Identify an organization with which you are familiar. Identify the organizational structure. How does this structure compare with the functional line, program management, and matrix management structures described here?

6.3 What sort of systems engineering efforts are best managed with a functional line organization, a project management organization, and a matrix management organization?

6.4 Discuss systems engineering management reasons for disaggregating the systems fielding process into a number of phases.

6.5 Please identify a past systems fielding program that encountered major difficulties. Identify at least three reasons why the program encountered these difficulties. Suggest corrective measures for these difficulties that would remediate the problem and avoid the problem.

6.6 For the data in the following table, please prepare a bar chart and an arrow network, indicate the critical path and its length, and suggest the staffing level profile on the assumption that all tasks begin as early as possible.

Task	Follows Task	Work Day Duration	Staffing Level
a	start	30	4
b	start	25	6
c	start	25	4
d	a	15	6
e	a, b	30	4
f	d, e	20	5

6.7 Identify a work breakdown structure for the definition, development, and deployment phases associated with a contemporary systems engineering fielding effort. Please associate costs with each of the elements in your WBS such that you obtain a cost breakdown structure (CBS) for the project.

6.8 Marvin [95] identifies ten key factors to success, which are very similar to the phases for systems management the we have discussed here. The ten factors, stated in the form of attributes, are as follows.

1. Take time for development by concentrating on the development of new products, rather than doing this as an afterthought.

2. Know the phases of the process in order to ensure successful product development program.

3. Join knowledge with ability, because all of the required talents and abilities should be brought together and incorporated in the group responsible for new product development.

4. Draw on experience because experience will support efficiency and enable developments to move swiftly and surely toward the desired objectives.

5. Develop planning techniques, such as to provide tested systems of organized attack on problems and make the efficient application of experience and ability in problem solving possible.

6. View problems in perspective, such as to be able to see oneself and the product or service under consideration as others see them.

7. Know the competition, because familiarity with the state-of-the-art technology and trade customs provides the background for evaluating the soundness of specific system and product development ventures.

8. Break away from the past, to enable system and product development to be undertaken in an atmosphere of complete independence and with the view that the old ways of doing a job should perhaps be cast aside.

9. Protect product ideas as assets, by protecting basic ideas from being revealed and corporate interests from being disclosed indiscriminately.

10. Provide adequate facilities, because fully maintained and utilized facilities are essential to the success of any product development undertaking.

Marvin identifies ten questions, the answers to which are useful in analyzing a new-product development program

1. Do we have time to do the job?
2. Do we understand all the problems involved?
3. Do we have the ability (technical knowledge and skills) to tackle product development programing?
4. Do we have the experience necessary?
5. Do we know how to plan a successful product development program?
6. Will we be able to put development programs in their proper perspective?
7. Are we familiar with the practices of our competitors?
8. Can we break away from past practices, concepts, and viewpoints?
9. Are proprietary product ideas protected?
10. Do we have the plant and facilities for product development?

The desired answer to all of these questions is, of course, "yes." The more "no"s one has, the lower will be the probability of success. Thus, we see that the suggested life cycle development approach can be converted into a set of metrics to be used for evaluation of proposed new product development strategies. Contrast and compare the implicit life cycle identified here with some that we have obtained in this text. How can the 10 attributes be used to evaluate a proposed technology development? How do these compare with the approaches suggested in this chapter?

6.9 Kuczmarski [96] suggests that the expanded process described here better enables people to cope with the two distinct phases for new-product development that emphasize the preliminary effort of goal or direction setting.

1. Direction setting involves identifying the following:
   1.1. Corporate objectives and strategies
   1.2. New product blueprints

    1.3. New product diagnostic audits;

    1.4. New product strategies

    1.5. Categories of application for new the new product

    1.6. Category analysis and screening

2. New-product development involves the following:

    2.1. Category selection, through analyzing and ranking potentially attractive categories by studying the role of new products in the company and choosing categories that provide the most attractive possibilities for idea generation

    2.2. Idea generation, by generating ideas in selected categories through a variety of problem-solving and creative approaches

    2.3. Concept development, by developing concepts, conducting initial screens and setting priorities by taking ideas that pass the initial screens and developing descriptions of the product

    2.4. Business analysis, by conducting business analysis of selected concepts through formulating a potential market and conducting competitive assessments

    2.5. Screening, or filtering concepts to determine prototype candidates while keeping in mind financial forecasts developed in the business analysis and filtering the remaining concepts through all performance criteria

    2.6. Prototype development, or developing an operating prototype of the product and run product-performance tests

    2.7. Market testing, or determining customer acceptance and running marketing trials in order to determine consumer purchase intent, and testing the product in either a simulated market or actual market trials

    2.8. Plant scale-up and manufacturing testing, or initiating plant scale-up and production to determine roll-out equipment needs and manufacturing the product in large enough quantities to identify bugs and problems and run product-performance tests

    2.9. Commercialization, or developing plans to introduce the new product to the trade and consumers

    2.10. Post-launch checkup, or performance monitoring of the new product

Illustrate a life-cycle systems management model that incorporates these activities.

6.10 We have already noted that systems management is very important to new-product development. Kuczmarski [96] provides an example of a company president's perception of an ideal new-product environment:

    1. Top-level endorsement and high visibility for new products

2. New products tied to long-range corporate objectives and financial plans

3. Agreed-upon new-product charter covering objectives, category areas, screening criteria

4. "Small Company" flexibility and expediency

5. Clear identification of responsibilities

6. High level of communications and interdepartmental cooperation

7. Collegial, teamwork involvement

8. Front-loaded process

9. Portfolio of product improvements, line extensions, and "new" products to balance risk

10. Atmosphere where failures are accepted along with the successes

11. Resources and consistent commitment to do the job, in terms of money and people

Attempt to evaluate the structural models for systems management discussed in this chapter in terms of this environmental characterization.

6.11 In the Directed Research for Product Development Model, Gruenwald [97] presented a systems management life cycle to address new product development objectives. There are seven phases to this life cycle and they may be described as follows:

1. *Phase 1: The Search for Opportunity and the Compilation of Available Data.* Information on the industry, sales data, technology, consumer interests, and the competition are required. The effort is begun by performing an industry analysis that should include sales volume and trends, basic technology, competition, customer definition, and other pertinent factors such as foreign trade, regulatory restrictions, and so on. The next step is to identify opportunities by defining targets, forecasting rough volume and share, and performing a risk-ratio analysis, performing a feasibility study, perform a war-gaming like assessment of the competitive reactions, examining technical hurdles, and consider legal and policy issues. On the basis of this, a decision is made with respect to whether to proceed, that is to go or no go.

2. *Phase 2. Conception.* In this phase, we translate market facts into product concepts and customer positioning communications before making commitments to exhaustive product oriented and product-directed research and development (R&D). There are six steps in this phase.

3. *Phase 3. Modeling (Prototypes).* This involves bringing proposed new products closer to reality in the form of prototype products and prototype communications. The steps in this phase involve developing descriptors and developing prototypes, and a go or no-go decision is made on the basis of the information obtained from using these.

4. *Phase 4. Research and Development.* This phase covers a number of different activities including checking outside scientific resources, pilot plant production, analyzing the factors necessary to scale up from pilot plant production to full-scale commercialization, controlled tests, and feasibility studies. On the basis of the information that results from these steps, a decision is made whether to proceed to the next phase.

5. *Phase 5. Marketing Plan.* This involves the development of a marketing plan and a go or no-go decision to proceed to the next phase is made on the basis of the information obtained at this phase.

6. *Phase 6. Market Testing.*

7. *Phase 7. Major Introduction.* This expands from the market testing phase.

Gruenwald has attempted, through use of this relatively exhaustive lifecycle methodology, to minimize risks and maximize success probability. This is clearly a systems-engineering-based approach to address new products goals in an organized, phase-by-phase iterative method. The important decision considerations that are addressed after a new product has been defined are as follows:

1. Is there a latent demand for the product?
2. Can a product be made that will satisfy the market?
3. Can the company be competitive with the product and within the field?
4. Will the entry be profitable and satisfy the corporate charter, as well as other company objectives?

Contrast and compare this lifecycle with others that have been discussed in Chapter 2 and the literature.

6.12 Please write a report indicating how you might go about determining appropriateness of assignment of various aspects of task performance to humans or to automated machines. How do knowledge representations relate to this?

6.13 A major goal of the systems engineering approach is to assist in formulation, design, development test, and evaluation of systems, based upon knowledge of human roles, capabilities, and tasks performed; such that these systems are designed for human interaction. A framework for system design for human interaction should assist systems engineers in determining the following:

(a) The implications of cognitive limitations on the requirements specifications for systems

(b) Efficient techniques for evaluation of cognitive issues

(c) Approaches for incorporating human interaction concerns into the analysis, design, and evaluation procedures for systems

(d) Methods for structuring and presenting this framework to system developers and system users

6.14 Describe a systems engineering process that incorporates these concerns. Each of the four items in Problem 6.13 serves a very appropriate purpose. It will be appropriate and desirable also to cope with a number of related issues that also support these purposes.

1. There are a large variety of environments into which various fielded system activities are embedded. These include a number of perspectives such as economic, technical, legal, political, and social. This results in a number of "natural" architectures or environments for system use. The garbage-can architecture is one of these that has been recently used for modeling decision assessment environments.

2. The experiential familiarity of the decision-making individual or group with the tasks at hand and the environment into which tasks are embedded will strongly influence the time–stress and resulting cognitive control mode chosen for such information processing tasks as issue detection, issue diagnosis, and issue resolution. Support systems for command and control must be accommodating to these cognitive control mode realities in order to enhance the resulting decision perspectives, rather than to inhibit them.

3. Errors can occur in formulation, analysis, or interpretation (or detection, diagnosis, or correction) efforts. There is need for an error taxonomy that relates both human and human–machine error to environment, cognitive-control mode, and other facets.

4. A computer-based support system to improve human functioning will necessarily have to interface and be interactive with humans and hardware and software in the operational environment. This brings about the needs for (a) system design for human interaction and for (b) open systems architectures that enable the fielded system to be successfully maintained and proactively adapted to new environments. To do this results in operational-level quality assurance.

The systems engineering process must itself be accommodating to these four needs as well as to others. This leads to the need for strategic- or process-level quality assurance and other systems management efforts. Please describe illustrative situations in which each of these apply.

REFERENCES

[1]   Sage, A. P., *Systems Management for Information Technology and Software Engineering*, Wiley, New York, 1995.

[2]   Sage, A. P., and Rouse, W. B. (Eds.), *Handbook of Systems Engineering and Management*, Wiley, New York, 1999.

[3]   Blanchard, B. S., *Systems Engineering Management*, Wiley, New York, 1998.

[4]   Eisner, H., *Essentials of Project and Systems Engineering Management*, Wiley, New York, 1997.

[5]   Bass, B. M. (Ed.), *Stogdill's Handbook of Leadership*, Free Press, New York, 1981.

[6]   Etzioni, A., *Modern Organizations*, Prentice-Hall, Englewood Cliffs, NJ, 1964.

[7]   Hall, R. H., *Organizations, Structure, and Process*, Prentice-Hall, Englewood Cliffs, NJ, 1977.

[8]   Mintzberg, H., *The Nature of Managerial Work*, Harper and Row, New York, 1973.

[9]   Hayre, J., An Axiomatic Theory of Organizations, *Administrative Science Quarterly*, Vol. 10, No. 3, December 1965, pp. 289–320.

[10] March, J. G., and Simon, H. A., *Organizations*, Wiley, New York, 1958.

[11] March, J. G., *A Primer on Decision Making: How Decisions Happen*, Free Press, New York, 1994.

[12] March, J. G. (Ed.), *Decisions and Organizations*, Basil Blackwell, Cambridge, MA, 1988.

[13] Dreyfus, H. L., and Dreyfus, S. E., *What Computers Still Can't Do: A Critique of Artificial Intelligence*, MIT Press, Cambridge, MA, 1992.

[14] Mintzberg, H., Ahlstrand, B., and Lampel, J., *Strategy Safari: A Guided Tour Through the Wilds of Strategic Management*, Free Press, New York, 1998.

[15] Cyert, R. M., and March, J. G., *A Behavioral Theory of the Firm*, Prentice-Hall, Englewood Cliffs, NJ, 1963.

[16] Argyris, C., *Reasoning, Learning and Action: Individual and Organizational*, Jossey-Bass, San Francisco, CA, 1982.

[17] Rasmussen, J., Duncan, K., and Leplat, J. (Eds.), *New Technology and Human Error*, Wiley, Chichester UK, 1987.

[18] Rasmussen, J., and Vicente, K. J., Coping with Human Error Through System Design: Implications for Ecological Interface Design, *International Journal of Man-Machine Studies*, Vol. 31, 1989, pp. 517–534.

[19] Rasmussen, J., *On Information Processing and Human–Machine Interaction: An Approach to Cognitive Engineering*, North-Holland, New York, 1986.

[20] Rasmussen, J., Pejtersen, A., and Goodstein, L. G., *Cognitive Systems Engineering*, Wiley, Inc., New York, 1994.

[21] Senge, P. M., *The Fifth Discipline: The Art and Practice of the Learning Organization*, Doubleday, New York, 1990.

[22] Senge, P. M., The Leaders New Work: Building Learning Organizations, *Sloan Management Review*, Vol. 32, No. 1, Fall 1990, pp. 7–23.

[23] Senge, P., Kleiner, A., Roberts, C., Ross, R., Roth, G., and Smith, B., *The Dance of Change: The Challenge to Sustaining Momentum in Learning Organizations*, Currency Doubleday, New York, 1999.

[24] Reason, J., *Human Error*, Cambridge University Press, Cambridge, UK, 1990.

[25] Sage, A. P., *Systems Engineering*, Wiley, New York, 1992.

[26] Sage, A. P. (Ed.), *Concise Encyclopedia of Information Processing in Systems and Organizations*, Pergamon Press, Oxford, UK, 1990.

[27] Sage, A. P., *Decision Support Systems Engineering*, Wiley, New York, 1991.

[28] Sage, A. P., and Palmer, J. D., *Software Systems Engineering*, Wiley, New York, 1990.

[29] Chapanis, A., *Human Factors in Systems Engineering*, Wiley, New York, 1996.

[30] Andriole, S., and Adelman, L., *Cognitive Systems Engineering for User-Computer Interface Design, Prototyping, and Evaluation*, Lawrence Erlbaum Associates, Hillsdale, NJ, 1995.

[31] Anthony, R. N., The Management Control Function, Harvard Business School Press, Boston, MA, 1988.

[32] Sage, A. P., Information Technology for Crisis Management, *Large Scale Systems*, Vol. 11, No. 3, 1986, pp. 193–205.

[33] Pauchant, T. C., and Mitroff, I. I., *Transforming the Crisis-Prone Organization: Preventing Individual, Organizational, and Environmental Tragedies*, Jossey-Bass, San Francisco, 1992.

[34] Savage, C. M., *Fifth Generation Management: Integrating Enterprises through Human Networking*, Digital Press, Burlington, MA, 1990.

[35] Savage, C. M., *Fifth Generation Management: Co-Creating Through Virtual Enterprising, Dynamic Teaming, and Knowledge Networking*, Butterworth-Heinemann, San Francisco, 1998.

[36] Wiest, J. D., Levy, F. K., *Management Guide to PERT/CPM*, Prentice-Hall, Englewood Cliffs, NJ, 1977.

[37] Archibald, R., and Villoria, R., *Network-Based Management Systems*, (PERT/CPM), Wiley, New York, 1967.

[38] Woodgate, H. S., *Planning by Network*, Brandon/Systems, New York, 1964.

[39] Badiru, A. B., *Project Management in Manufacturing and High Technology Operations*, Wiley, New York, 1988.

[40] Blanchard, B. S., and Fabrycky, W. J., *Systems Engineering and Analysis*, Prentice-Hall, Englewood Cliffs, NJ, 1998.

[41] Kezsbom, D. S., Schilling, D. L., and Edward, K. S., *Dynamic Project Management*, Wiley, New York, 1989.

[42] Michaels, J. V., and Wood, W. P., *Design to Cost*, Wiley, New York, 1989.

[43] Kerzner, H., *Project Management: A Systems Approach to Planning, Scheduling, and Controlling*, 4th edition, Van Nostrand Reinhold, New York, 1992.

[44] Clark, W., *The Gantt Chart*, 3rd edition, Pitman & Sons, London, 1957.

[45] Grant, E. L., and Leavenworth, R. S., *Statistical Quality Control*, McGraw-Hill, New York, 1988.

[46] Lederer, A. L., and Prasad, J., Nine Management Guidelines for Better Cost Estimating, *Communications of the Association for Computing Machinery*, Vol. 35, No. 2, February 1992, p. 51–59.

[47] Pelsma, K. H. (Ed.), Ergonomics Sourcebook: *A Guide to Human Factors Information*, Ergosyst Associates, Lawrence, KS, 1987.

[48] Sheridan, T.B., and Ferrell, W. R., *Man–Machine Systems: Information, Control, and Decision Models of Human Performance*, MIT Press, Cambridge, MA, 1974.

[49]   Sheridan, T.B., and Johannsen, G. (Eds.), *Monitoring Behavior and Supervisory Control*, Plenum Press, New York, 1976.

[50]   Johannsen, G., Rijnsdorp, J. E., and Sage, A. P., Human Interface Concerns in Support Systems Design, *Automatica*, Vol. 19, No. 6, November 1983, pp. 595–603.

[51]   Rasmussen, J., and Rouse, W.B. (Eds.), *Human Detection and Diagnosis of System Failures*, Plenum Press, New York, 1981.

[52]   Card, S. K., Moran, T. P., and Newell, A., *The Psychology of Human–Computer Interaction*, Lawrence Erlbaum Associates, Hillsdale, NJ, 1983.

[53]   Shneiderman, B., *Software Psychology: Human Factors in Computer and Information Systems*, Winthrop Publishers, Cambridge, MA, 1980.

[54]   Shneiderman, B., *Designing the User Interface: Strategies for Effective Human–Computer Interaction*, Addison-Wesley, Reading, MA, 1987.

[55]   Sage, A. P., *Decision Support Systems Engineering*, Wiley, New York, 1991.

[56]   Moray, N. (Ed.), *Mental Workload: Its Theory and Measurement*, Plenum Press, New York, 1979.

[57]   Ginsburg, H., and Opper, S., *Piaget's Theory of Intellectual Development*, Prentice-Hall, Englewood Cliffs, NJ, 1979.

[58]   Sage, A. P. (Ed.), *Concise Encyclopedia of Information Processing in Systems and Organizations*, Pergamon Press, Oxford UK, 1990.

[59]   Wohl, J. G. Force Management Decision Requirements for Air Force Tactical Command and Control," *IEEE Transactions on Systems, Man, and Cybernetics*, Vol. 11, No. 9, September 1981, pp. 618–639.

[60]   Andriole, S. J., and Halpin, S. M. (Eds.), *Information Technology for Command and Control: Methods and Tools for System Development and Evaluation*, IEEE Press, New York, 1991.

[61]   Rasmussen, J., Skills, Rules, Knowledge; Signals, Signs, and Symbols; and Other Distinctions in Human Performance Models, *IEEE Transactions on Systems Man and Cybernetics*, Vol. SMC 13, No. 3, May 1983, pp. 257–266.

[62]   Vicente K. J., and Rasmussen, J., Ecological Interface Design: Theoretical Foundations, *IEEE Transactions on Systems, Man, and Cybernetics*, Vol. 22, No. 4, July 1992, pp. 589–606.

[63]   Hammond, K.R., Intuitive and Analytical Cognition: Information Models, in A. P. Sage (Ed.), *Concise Encyclopedia of Information Processing in Systems and Organizations*, Pergamon Press, Oxford, UK, 1990, pp. 306–312.

[64]   Hammond, K. R., Hamm, R. M., Grassia, J., and Pearson, T., Direct Comparison of the Efficacy of Intuitive and Analytical Cognition in Expert Judgment, *IEEE Transactions on Systems, Man, and Cybernetics*, Vol. 17, No. 5, September 1987, pp. 753–770.

[65]   Hammond, K. R., Judgment and Decision Making in Dynamic Tasks, *Information and Decision Technologies*, Vol. 14, No. 1, 1988, pp. 3–14.

[66]   Brehmer, B., Dynamic Decision Making, in Sage, A. P., *Concise Encyclopedia of Information Processing in Systems and Organizations*, Pergamon Press, 1990, pp. 144–149.

[67]   Harris, D. H. (Ed.), *Organizational Linkages: Understanding the Productivity Paradox*. National Academy Press, Washington, DC, 1994.

[68] Chawla, S., and Renesch, J. (Eds.), *Learning Organizations: Developing Cultures for Tomorrow's Workplace*, Productivity Press, Portland, OR, 1995.

[69] Argyris, C. and Schön, D. A., *Organizational Learning II: Theory, Method, and Practice*, Addison Wesley, Reading, MA, 1996.

[70] Nonaka, I., A Dynamical Theory of Organizational Knowledge Creation, *Organizational Science*, Vol. 5, No. 1, February 1994, pp. 14–37.

[71] Nonaka, I., and Takeuchi, H., *The Knowledge Creating Company*, Oxford, New York, 1995.

[72] Elliott, S. (Ed.), *Using Information Technology to Support Knowledge Management*. American Productivity & Quality Center, Houston, TX, 1997.

[73] Rouse, W. B., Thomas, B. S., and Boff, K. R., Knowledge Maps for Knowledge Mining: Application to R&D/Technology Management. *IEEE Transactions on Systems, Man, and Cybernetics*, Vol. 28, No. 3, 1998, pp. 309–317.

[74] Weick, K. E., *The Social Psychology of Organizing*, Addison-Wesley, New York, 1979.

[75] Weick, K. E., Cosmos versus Chaos: Sense and Nonsense in Electronic Contexts, *Organizational Dynamics*, Vol. 14, No. 3, 1985, pp. 50–64.

[76] Janis, I. L., *Groupthink*, Free Press, New York, 1982.

[77] Argyris, C., and Schön, D. A., *Organizational Learning: A Theory of Action Perspective*, Addison-Wesley, New York, 1978.

[78] Diesing, P., *Reason in Society*, University of Illinois Press, Urbana, IL., 1962.

[79] Steinbruner, J. D., *The Cybernetic Theory of Decision*, Princeton University, Princeton, NJ, 1974.

[80] Sage, A. P., *Economic Systems Analysis: Microeconomics for Systems Engineering, Engineering Management, and Project Selection*, Elsevier North-Holland, New York, 1983.

[81] Simon, H. A., From Substantive to Procedural Rationality, in *Method and Appraisal in Economics*, Latsis, S. J. (Ed.), Cambridge University Press, 1976, pp. 129–148.

[82] Simon, H. A., Rational Decision Making in Business Organization, *American Economic Review*, Vol. 69, No. 4, September 1979, pp. 493–513.

[83] Simon, H. A., Rationality as Process and as Product of Thought, *American Economic Review*, Vol. 68, May 1978, 1–16.

[84] Klein, G., *Sources of Power: How People Make Decisions*, MIT Press, Cambridge, MA, 1998.

[85] Sheridan, T. B., Human Factors Engineering, in Sage, A. P. (Ed.), *Concise Encyclopedia of Information Processing in Systems and Organizations*, Pergamon Press, Oxford, UK, 1990, pp. 209–217.

[86] Hess, R., Human Factors Engineering: Information Processing Concerns, in Sage, A. P. (Ed.), *Concise Encyclopedia of Information Processing in Systems and Organizations*, Pergamon Press, Oxford, UK, 1990, pp. 217–223.

[87] Norman, D. A., Categorization of Action Slips, *Psychological Review*, Vol. 88, No. 1, 1981, pp. 1–15.

[88] Norman, D. A., *The Psychology of Everyday Things*, Basic Books, New York, 1988.

[89]   Rouse, W. B., *Design for Success: A Human Centered Approach to Designing Successful Products and Systems*, Wiley, New York, 1991.

[90]   Salvendy, G. (Ed.), *Handbook of Human Factors and Ergonomics*, Wiley, New York, 1997.

[91]   Nieves, J. M., and Sage, A. P., Human and Organizational Error as a Basis for Process Reengineering: With Applications to Systems Integration Planning and Marketing, *IEEE Transactions on Systems, Man, and Cybernetics*, Volume 28A, No. 6, November 1998, pp. 742–762.

[92]   Majone, G., and Quade, E. S. (Eds.), *Pitfalls of Analysis*, Wiley, Chichester, UK, 1980.

[93]   Miser, H. J., and Quade, E. S. (Eds.), *Handbook of Systems Analysis: Craft Issues and Procedural Choices*, Wiley, New York, 1988.

[94]   Dhebar, A., Managing the Quality of Quantitative Analysis, *Sloan Management Review*, Vol. 34, No. 2, Winter 1993, pp. 69–75.

[95]   Marvin, P., *Product Planning Simplified*, American Management Association, New York, 1972.

[96]   Kuczmarski, T. D., *Managing New Products: Competing Through Excellence*, Prentice-Hall, Upper Saddle River, NJ, 1988.

[97]   Gruenwald, G., *New Product Development, What Really Works*, NTC Business Books, Lincolnwood, IL, 1985.

Index

533